GW00499764

Bacteria–Cytokine Interactions

Interactions

in Health and Disease

Bacteria–Cytokine Interactions

in Health and Disease

Brian Henderson
Stephen Poole
Michael Wilson

PORTLAND PRESS
London and Miami

Published by Portland Press Ltd,
59 Portland Place, London WIN 3AJ, U.K.
Tel: (+44) 171 580 5530; e-mail: edit@portlandpress.co.uk

In North America orders should be sent to Ashgate Publishing Co., Old Post
Road, Brookfield, VT 05036-9704, U.S.A.

ISBN 1 85578 114 X

British Library Cataloguing-in-Publication Data
A catalogue record for this book is available from the British Library

Typeset by Portland Press Ltd.
Printed in Great Britain by Cambridge University Press, Cambridge

*Cover illustration: Enteropathogenic Escherichia coli adhered to enterocytes by means of
the adhesin intimin. Reproduced with permission from
Rosenshine, I. (1996) EMBO J. issue 11, front cover, Oxford University Press*

Contents

3 Bacterial virulence factors

4 Cytokines, homeostasis, networks and disease

7

Bacteria and cytokines: beyond lipopolysaccharide

8

Pharmacological modulation of cytokines

9

Bacteria–cytokine interactions in health and disease: a new synthesis

Preface

The prokaryotic cells of the superkingdom Bacteria have a fossil record which dates back 3 billion years. The other kingdoms of living things, including Eukarya, in which *Homo sapiens* is found, have co-evolved with the ubiquitous bacteria in a process which we can only assume was 'red in tooth and claw'. At the time of writing, only two generations after the introduction of antibiotics, we are approaching the millennium with the certainty that bacteria have won this chemical contest, and we thus look forward to an uncertain future with regard to infections and infectious diseases. This is clearly emphasized by the epidemic of MRSA (methicillin-resistant *Staphylococcus aureus*) infections in hospitals and in the community, the outbreaks of *Escherichia coli* 0157 food poisoning, and by the rapid resurgence of tuberculosis in the first world. In addition to the reappearance of well known bacteria as public health problems, there is the constant recognition of new infectious agents. For example, during the past 20 years or so bacteria such as *Borrelia burgdorferi*, *Helicobacter pylori* and *Legionella pneumophila*, viruses such as human immunodeficiency virus and Ebola virus, and the mysterious proteins known as prions (responsible for diseases such as bovine spongioform encephalopathy), have been recognized as major human pathogens and have also become household words. With the recognition of these diseases, we are presumably seeing the constant process of the co-evolution of the microbial world with that of the multicellular kingdoms. This process has probably been accelerated by the predilection of *Homo sapiens* for migration and rapid transport.

The study of microbial disease has largely been the province of the microbiologist, virologist and pathologist. However, during the past decade a new discipline, which has been termed `cellular microbiology', has started to emerge from a fusion of these fields with modern cellular and molecular biology. Cellular microbiology is concerned with the defining of the interactions which occur between micro-organisms and eukaryotic cells, particularly those of man.

Much of the interaction which occurs between the microbial and the multicellular eukaryotic world is controlled by a plethora of multifarious proteins which act as local hormones and have been termed cytokines. These proteins were discovered by pioneering workers who were attempting to establish how exogenous bacterial pyrogens such as endotoxin induced fever. This led to the discovery of 'endogenous pyrogen' which, after many different guises, was named 'interleukin-1' (IL-1) and became the prototypic cytokine. Since the recognition of IL-1 as both a pyrogen and immune mediator, a vast number of additional cytokines and their receptors have been isolated, sequenced, cloned and expressed. The large number of cytokines and their receptors and the additional mediators that can be induced by cytokines has led to the concept that cells in multicellular organisms are controlled not by single cytokines but by the summation of their responses to cytokine networks. We now recognize cytokines (and the networks they induce) as important homeostatic mediators with a paradoxical role. On the one hand they have vital roles in normal cell development and in our all-important innate and acquired responses to infections. On the other hand they are exquisitely potent mediators of tissue pathology, well recognized by the world's pharmaceutical industry as key therapeutic targets in a growing variety of diseases, including infectious conditions such as septic shock, leprosy and malaria.

Another paradox in the cellular microbiology of man lies in the fact that the average human is composed of 10^{13} cells but contains 10^{14} bacteria. How is it possible to live harmoniously with these vast numbers of bacteria, which contain (as has recently been discovered and will be reviewed in detail) a wide variety of potent cytokine-inducing molecules in addition to that well known cytokine-inducing component, lipopolysaccharide? Recent transgenic knockout mice bring into focus the importance of individual cytokines in controlling the host's response to its normal bacterial populations. This opens up the possibility that the commensal microflora and the cells of the host engage in a continuous cross-talk made up of proteinaceous signals. The host signals are mainly cytokines and the signals produced by the bacteria may be the evolutionary ancestors of these cytokines to which the term `bacteriokine' may be applied.

The aim of this monograph is to bring together the disparate literature on bacterial virulence, bacterial structure and function, the normal microflora, cytokine biology and pathology and pharmacology into a modern synthesis of the interactions that occur between bacteria and host cells, with particular emphasis on the role of cytokine networks. As stated, such interactions may be vitally important in maintaining the stable association that exists between multicellular organisms and their normal microflora.

This monograph should be of interest to microbiologists, bacteriologists, virologists, immunologists, pathologists, cytokine biologists and pharmacologists.

Brian Henderson and Michael Wilson
Eastman Dental Institute, University College London, and
Stephen Poole,
Division of Endocrinology, National Institute for Biological Standards and Control

Abbreviations

ACP	acyl carrier protein
ACTH	adrenocorticotrophic hormone
AIDS	acquired immunodeficiency syndrome
AOAH	acyloxyacyl hydrolase
AP	activator protein
AR	amphiregulin
ARIA	acetylcholine receptor-inducing activity
ATF	activating transcription factor
BCRF-1	BamHIC fragment rightward reading frame
BMP	bone morphogenetic protein
BPI	bactericidal/permeability-increasing protein
BSE	bovine spongiform encephalopathy
C-6	carbon number 6
CAPK	ceramide-activated protein kinase
CD	cluster of differentiation
CDR	complementarity-determining region
CETP	cholesterol ester-transfer protein
CF	cytotoxic factor
CGRP	calcitonin-related gene peptide
CHO	Chinese hamster ovary
CIPD	chronic inflammatory periodontal disease
CMC	critical micellar concentration
CMI	cell-mediated immunity
CMP	cytidine monophosphate
CNTF	ciliary neurotrophic factor
ConA	concanavalin A
COX	cyclo-oxygenase
cpn	chaperonin
CRH	corticotrophin-releasing hormone
Crm	cytokine response modifier
CRP	C-reactive protein
CSAID	cytokine-suppressive anti-inflammatory drug
CSBP	CSAID-binding protein
CSF	colony-stimulating factor
CTL	cytotoxic T-lymphocyte

DAG	2,3-diamino-2,3-dideoxy-D-glucose
DIC	diffuse intravascular coagulation
DPLA	diphosphoryl lipid A
DTH	delayed type hypersensitivity
E-selectin	endothelial selectin (CD62E)
EAP	endotoxin-associated protein
EBV	Epstein–Barr virus
EC	endothelial cell
ECH	E. coli haemolysin
EF	oedema factor
EGF	epidermal growth factor
EGFR	epidermal growth factor receptor
EHV2	equine herpes virus type 2
ELISA	enzyme-linked immunosorbent assay
ENA	epithelial-derived neutrophil chemoattractant
ENP	endotoxin-neutralizing protein
EP	endogenous pyrogen
EPO	erythropoietin
ERK	extracellular-signal-regulated kinase
ET	endothelin
Fc	constant region of immunoglobulin
FGF	fibroblast growth factor
FHA	filamentous haemagglutinin
FMLP	formyl-Met-Leu-Phe
FSH	follicle-stimulating hormone
GAPD	glyceraldehyde-3-phosphate dehydrogenase
Gb_3	globotriaosylceramide
Gb_4	globotetraosylceramide
GAG	glycosaminoglycan
GCF	gingival crevicular fluid
GCP	granulocyte chemotactic protein
G-CSF	granulocyte colony-stimulating factor
GGF	glial growth factor
GH	growth hormone
GlcNAc	N-acetyl-D-glucosamine
GM-CSF	granulocyte-macrophage colony-stimulating factor
GOD	generation of diversity
GPI	glycosyl phosphatidylinositol
Gro	growth-regulated oncogene

HAMA	human anti-mouse antibody
HB-EGF	heparin-binding epidermal growth factor-like factor
HCMV	human cytomegalovirus
HDL	high-density lipoprotein
HGF	hepatocyte growth factor
HIRM	host-immune response modifying
HIV	human immunodeficiency virus
HNP	human neutrophil peptide
HOG	high-osmolarity glycerol protein kinase
hpbMC	human peripheral blood mononuclear cell
HRG	heregulin
HSP	heat-shock protein
HVS	herpesvirus saimiri
IAP	inhibitor of apoptosis
IBD	inflammatory bowel disease
ICAM	intercellular adhesion molecule (CD54)
ICE	pro-IL-1β-converting enzyme
IFN	interferon
IFNR	interferon receptor
IGF	insulin-like growth factor
IgG	immunoglobulin G
IL	interleukin
IL-1ra	interleukin-1 receptor antagonist
IL-2R	IL-2 receptor
Ins(1,4,5)P_3	inositol 1,4,5-trisphosphate
IP-10	IFNγ-inducible protein
IRF-1	interferon regulatory factor-1
IVET	in vitro expression technology
ivi	in vivo-induced gene
JAK	Janus kinase
JNK	c-Jun N-terminal kinase
KDO	2-keto-3-deoxyoctulonic acid
KGF	keratinocyte growth factor
KSHV	Kaposi's sarcoma-associated herpesvirus
LAF	lymphocyte-activating factor
LAL	Limulus amoebocyte lysate assay
LALF	Limulus anti-lipopolysaccharide factor
LAM	lipoarabinomannan

LAP	lipid A-associated protein
LBP	lipopolysaccharide-binding protein
lc	long-chain segment
LDL	low-density lipoprotein
LEM	leucocyte endogenous mediator
LESTR	leukocyte-derived 7-transmembrane domain receptor
LF	lethal factor
LFA	lymphocyte function-associated antigen
LIF	leukaemia inhibitory factor
LLO	listeriolysin O
LPS	lipopolysaccharide
LT	leukotriene
LT	lymphotoxin
LTA	lipoteichoic acid
MØ	macrophage
mAb	monoclonal antibody
MAF	macrophage-activating factor
MALT	mucosal-associated lymphoid tissue
MAP	mitogen-activated protein
MAPK	mitogen-activated protein kinase
MAPKAPK	MAPK-activated protein kinase
MBP	mannose-binding protein
MCAF	monocyte chemotactic and activating factor
MCF	monocyte cytotoxicity-inducing factor
M-CSF	macrophage colony-stimulating factor
MCP	monocyte chemotactic protein
MDO	membrane-derived oligosaccharide
MDP	N-acetyl-muramyl-L-alanyl-D-isoglutamine
MEK	MAPK/ERK kinase
MEKK	MAPK/ERK kinase kinase
MHC	major histocompatibility complex
MIF	migration-inhibitory factor
MIG	monocyte induced by γ-interferon
MIP	macrophage inflammatory protein
MIS	Mullerian-inhibiting substance
MKK	MAPK kinase
αMM	α-methyl-D-mannoside
MMP	matrix metalloproteinase
MNC	mononuclear cell
MPLA	monophosphoryl lipid A
MRSA	methicillin-resistant *Staphylococcus aureus*

α-MSH   α-melanocyte-stimulating hormone

NAD   nicotinamide adenine dinucleotide
NAP   neutrophil activating protein
NCAM   nerve cell adhesion molecule
NDF   new differentiation factor
NF   nuclear factor
NF-κB   nuclear factor κB
NF-IL-6   nuclear factor interleukin-6 (IL-6 transcription factor)
NGF   nerve growth factor
NO   nitric oxide
NSAID   non-steroidal anti-inflammatory drug

OBD   oligonucleotide-based drug
OMP   outer-membrane protein
OSM   oncostatin M

PA   protective antigen
PA   phosphatidic acid
PAF   platelet-activating factor
PAI   plasminogen-activator-inhibitor
PB   polymyxin B
PBP   platelet basic protein
PCR   polymerase chain reaction
PDE   phosphodiesterase
PD-ECGF   platelet-derived endothelial cell growth factor
PDGF   platelet-derived growth factor
PF-4   platelet factor 4
PGE_2 prostaglandin E_2
PHA   phytohaemagglutinin
PKA   protein kinase A
PKC   protein kinase C
PLA   phospholipase A
PLC   phospholipase C
PLTP   phospholipid-transfer protein
PMA   phorbol 12-myristate 13-acetate
PMN   polymorphonuclear neutrophil
PP   peptidoglycan-polysaccharide
PPAR   peroxisome proliferator-activated receptor
PRL   prolactin
PSGL   P-selectin glycoprotein ligand
PT   pertussis toxin

PtdIns(4,5)P_2	phosphatidylinositol 4,5-bisphosphate
PTK	protein tyrosine kinase

RA	rheumatoid arthritis
RANTES	regulated on activation, normal T expressed and secreted
RSLA	*Rhodobacter sphaeroides* lipid A
RT-PCR	reverse transcriptase-polymerase chain reaction
RTK	receptor tyrosine kinase
RTX	repeats-in-toxin

SAA	serum amyloid A
SAM	surface-associated material
SAP	stress-activated protein
SAPK	stress-activated protein kinase
sc	short-chain segment
SCDGF	Schwaan cell-derived growth factor
SCF	stem cell factor
SCIP	staphylococcal cytokine-inhibiting protein
SDS/PAGE	sodium dodecyl sulphate/polyacrylamide gel
SEA	staphylococcal enterotoxin A
SEB	staphylococcal enterotoxin B
SEE	staphylococcal enterotoxin E
SEK	stress-activated protein kinase/ERK kinase
SFV	Shope fibroma virus
SH	Src-homology
sIgA	secretory immunoglobulin A
SLT	Shiga-like toxin
SMase	sphingomyelinase
STATs	signal transducers and activators of transcription

TβR-I	TGFβ type I receptor
TAP	tracheal antimicrobial peptide
TB	tuberculosis
TCR	T-cell receptor
TCT	tracheal cytotoxin
TGF	transforming growth factor
Th	T helper subset (0, 1 or 2)
TIMP	tissue inhibitor of metalloproteinase
Tk	tyrosine kinase
TNBS	2,4,6-trinitrobenzene sulphonic acid
TNF-R55	55 kDa TNF receptor

TNF	tumour necrosis factor
TNF-R75	75 kDa TNF receptor
TNF-α	tumour necrois factor-α
TNFR	tumour necrosis factor receptor
tPA	tissue plasminogen activator
TRADD	TNF receptor-associated death domain
TRAF	TNF receptor-associated factor
TSH	thyroid-stimulating hormone
TSS	toxic shock syndrome
TSST	toxic shock syndrome toxin
Tyk	tyrosine kinase
UDP-Glc	uridine diphosphate glucose
UDP-GlcN	UDP-glucosamine
UDP-GlcNaC	UDP-N-acetyl-glucosamine
VCAM-1	vascular cell adhesion molecule-1 (CD106)
VEC	vascular endothelial cell
VEGF	vascular endothelial growth factor
VLDL	very-low-density lipoprotein
VVGF	vaccinia virus-derived growth factor
Yop	*Yersinia* outer protein

Introduction

"we live in evolutionary competition with microbes, bacteria and viruses —
there is no certainty that we will be the winners"

Joshua Lederberg

1.1   Background

In the middle of the 14th Century the Crimean port of Caffa (now known as
Feodosiya) was under siege by Janiberg, Khan of the Kipchak Tartars. Since
the 1330s the plague had been spreading from central Asia, eventually reaching
the Crimea around 1346 and killing the besieging Tartars. As more and more
of his troops died, Janiberg was forced to withdraw, leaving their corpses
behind. In what may have been the first recorded incident of germ warfare, the
departing Khan used his siege catapults to hurl the infected corpses of his
troops over the walls of Caffa, thus sealing the fate of the city [1].

This story may be apocryphal, but it encapsulates the fear that still exists
in the popular imagination about infectious diseases. The recent furore in the
press over *Streptococcus pyogenes*, the so-called flesh-eating bacterium, is a case
in point. It is therefore paradoxical that in the 1990s, after the longest and most
pronounced retrenchment of infectious diseases in recorded history, we should
find ourselves (at least metaphorically) back with Janiberg the Khan. Between
the end of the Second World War and the late 1970s the developed nations of
the world had seen a striking increase in their living standards and quality of
life. The introduction and continued development of antibiotics in the 1950s,
60s and 70s contributed dramatically to the quality of life of Western nations
and, in particular, resulted in the loss of fear of bacterial infectious diseases [2].
Indeed, Mims has likened the populations of the developed nations to specific
pathogen-free animals, i.e. those in which pathogenic micro-organisms have
been eliminated [3]. However, by the late 1980s popular unease about
infectious diseases was re-appearing. This started in the early 1980s with the
appearance of the human immunodeficiency virus (HIV) and the consequences

of infection — i.e. acquired immunodeficiency syndrome (AIDS) — and
quickly rose to a crescendo of panic which has, surprisingly, abated as AIDS
moved into the third world and became pandemic. It is estimated that by the
Millennium there will be tens of millions of cases of AIDS worldwide.

While HIV and AIDS have gripped the popular imagination, a large
number of other infectious agents have appeared on the scene since the Second
World War (Table 1.1). These include conditions which are now well known,
such as Legionnaire's disease, Ebola fever and toxic shock syndrome, and also
a whole range of diseases which have not entered the public imagination
including Pontiac fever and Hantavirus pulmonary syndrome. In the U.K. the

Table 1.1 Infectious agents and diseases recognized since 1950

Agent	Disease
Hantavirus	Korean haemorrhagic fever
Flavivirus	Dengue haemorrhagic fever
Arenavirus	Argentine haemorrhagic fever
Kyasanur virus	Kyasanur forest disease
Chikungunya virus	Chikungunya
O'Nyong-Nyong virus	O'Nyong-Nyong fever
Haemorrhagic virus	Oropouche
Encephalitis virus	LaCross encephalitis
Marburg virus	Marburg disease
Legionella micdadei	Pontiac fever
Arenavirus	Lassa fever
Toxoplasmosis gondii	Human toxoplasmosis
Borrelia burgdorferi	Lyme disease
Ebola virus	Ebola disease
Campylobacter jejuni	Enteritis
Clostridium difficile	Pseudomembranous colitis
Cryptosporidium parvum	Diarrhoea
Staphylococcus aureus	Toxic shock syndrome
Gardnerella vaginalis	Bacterial vaginosis
Helicobacter pylori	Gastritis (cancer?)
Legionella pneumophila	Legionnaires' disease
Mobiluncus spp.	Bacterial vaginosis
Escherichia coli 0157:H7	Food poisoning
Hepatitis C virus	Hepatitis
Human herpesvirus 6	Exanthum subitum
Human immunodeficiency virus	AIDS
Parvovirus	Fifth disease
Rotavirus	Infantile diarrhoea
'Small round' virus	Gastro-enteritis
Bovine spongiform encephalitis (BSE)	BSE prion

whole population (including the authors of this book) is living with the possibility that there may be an epidemic of spongiform encephalopathy caused by
one of the most fascinating, and least understood, infectious agents — the
prion [in this case the bovine spongiform encephalopathy (BSE) prion]. We
hope that we finish this book before the symptoms of the new variant
Creutzfeldt–Jakob disease take their toll.

Even well known infectious organisms can appear in new guises. The
common gut bacterium *Escherichia coli* can, by incorporating the genes for
various so-called virulence factors (see Chapter 3), become pathogenic, and a
number of such strains causing a wide range of syndromes are known. In 1993,
in Seattle, a large number of children were admitted to hospital with acute
kidney failure following a bout of painful bloody diarrhoea. A number of
those admitted died. Similar cases were found in other parts of the U.S.A. and
all were traced to the eating of hamburgers from a single fast-food outlet. The
organism responsible turned out to be a strain of *E. coli* designated 0157:H7.
The symptoms caused by this strain, mainly due to the bacterium causing
kidney failure, are only now being recognized as occurring worldwide. The
most recent episode occurred in Scotland during the writing of this book, and
resulted in 16 deaths. The pathology is due to this bacterial strain acquiring a
particular toxin, called a Shiga-like toxin (SLT), which among other actions
induces the production of cytokines. The mechanism of action of this toxin
will be described in more detail in Chapter 8.

Another fascinating organism is *Helicobacter pylori*. Up until 1983 the
stomach was not thought to harbour bacteria. *H. pylori* was subsequently
discovered to colonize the stomach of at least 25% of the adult population and
now appears to be an essential factor in the pathogenesis of peptic ulceration,
an inflammatory condition of the gastric mucosa [4]. Moreover, there is now
the possibility that *H. pylori* is a risk factor in gastric cancer [5], thus widening
the pathological base of diseases caused by bacteria.

Is the appearance of these new diseases a cause for concern? The Nobel
laureate microbiologist Joshua Lederberg summed up this situation when he
wrote "we live in evolutionary competition with microbes, bacteria and viruses
— there is no certainty that we will be the winners". Micro-organisms first
appeared on this planet some 3.5 billion years ago and have presumably been
in evolutionary competition with one another ever since then. The large
number of secondary metabolites isolated from soil bacteria, and shown to
have potent pharmacological actions (such organisms are also the source of
many of our antibiotics), gives a measure of this evolutionary struggle for
survival. The appearance of multicellular organisms some one to two billion
years ago gave prokaryotic cells additional habitats in which to live and
compete. The current appearance of new diseases presumably reflects this
ongoing Darwinian evolutionary competition for habitats and resources. Many
of the new infectious diseases reported since the 1950s are due to the exposure
of the particular micro-organism to a host population with which it has not

previously been in contact. A good historical example of this is the lethal effects of diseases such as measles in African and South American populations not previously exposed to this virus. A modern equivalent is Lyme disease, a debilitating condition caused by a spirochaete, *Borrelia burgdorferi*, which is borne by ticks which live on deer. The symptoms of Lyme disease had been reported for decades in Europe, but cases were rare. It was only in the mid-1970s that an outbreak was reported in Old Lyme, Connecticut, and the incidence of this disease has continued to grow with the condition now ranging from New England to California. It is thought to result from greater contact between man and deer in the U.S.A. Of course, evolutionary competition works both ways. Diseases such as bubonic plague (which was epidemic) are now rare, and, apart from sporadic outbreaks (such as in India in 1994), appear to exist only in wild animal populations. Interestingly, certain infectious diseases are the result of current treatments. A good example is pseudomembranous colitis which was a rare disease before the 1970s. In this condition, which can be lethal, the administration of particular antibiotics (e.g. clindamycin) inhibits the growth of the major colonic bacteria, thereby allowing the overgrowth of the Gram-positive bacterium *Clostridium difficile*. This, in turn, produces toxins that damage the mucosa of the colon causing release of large amounts of cellular debris and so forming the pseudomembrane [6].

Evolutionary change is the driving force in biological diversity. The rate of evolutionary change is generally slow but it can respond rapidly in the right circumstances. Antibiotics were generally introduced into clinical practice in the 1950s and there has been an acceleration of their usage since that time. Antibiotics have also been used widely in agricultural practice, where they have applied significant evolutionary pressure on bacteria, although it has been surprising to learn just how quickly the bacterial genome has responded. Antimicrobial chemotherapy can be traced back to the introduction of the sulphonamides in 1935 and penicillin in 1941. In 1938 virtually all strains of *Neisseria gonorrhoeae* were susceptible to sulphonamides. However, 10 years later less than 20% of clinical isolates retained susceptibility. A similar story unfolds with penicillin. When this, most famous of all drugs, was introduced into the clinic, less than 1% of strains of *Staphylococcus aureus* were resistant. Yet by 1946, 14% of hospital strains were resistant and a year later the figure was 38%. Today the figure is greater than 90% and there is an epidemic of methicillin-resistant *Staph. aureus* (MRSA) in Europe and North America. Indeed, at the time of writing, many of the antibiotics introduced into clinical practice during the past two decades have engendered resistant strains, and diseases such as tuberculosis, which had been assumed to have been defeated, are returning to the developed world with a vengeance [7]. The development of novel antibiotics has slowed down worldwide largely due to the realization that, ultimately, bacteria will 'defeat' any new compound. Instead, increasing thought is being given to more radical approaches to the treatment of bacterial infections, such as targeting virulence factors (e.g. proteases) or selectively blocking damaging aspects of the host's response to infections.

1.2   Bacterial–host interactions in health and disease

The development of antibiotics had one major deleterious effect with regard to research into infectious diseases — it put the brake on the study of the interactions which occur between bacteria and the host tissues. Such studies were also discouraged by the attitude which arose in the 1950s and was maintained up until the 1980s that bacteria were simple cellular systems lacking significant cell–cell interactions. It is now becoming clear that bacteria–bacteria interactions and bacteria–eukaryotic cell interactions are involved in the induction and maintenance of pathology in bacterial infections. Evidence for the latter has been clearly defined in two recent reports in the magazine *Science* [8,9]. In the first report it was shown that pathogenic strains of *Yersinia*, which harbour a 70 kb plasmid encoding a group of highly regulated secreted virulence proteins, known as Yops, only up-regulate the synthesis and secretion of these proteins when the bacterium makes contact with host eukaryotic cells [8]. The second report illustrated that the binding of uropathogenic strains of *E. coli* via P-pili up-regulated genes in the bacterium required for the altered environmental conditions (in this case lowered environmental iron levels), which occur when bacteria bind to eukaryotic cells. The bacteria induce the transcription of proteins involved in the synthesis of low-molecular-mass, iron-binding, siderophores and of receptors for these compounds, thus ensuring a normal supply of this vital metal [9]. In addition, as will be discussed throughout this book, the authors propose that such cell–cell interactions between our resident normal microflora and our own eukaryotic cells are vital for the maintenance of a healthy bodily state.

As it is obvious that there are increasing numbers of bacteria which are becoming resistant to antibiotics, a knowledge of how bacteria cause pathology at the cellular and molecular level could be used to develop effective symptomatic therapeutics. Most of our knowledge about bacterial virulence mechanisms has focused on the exotoxins which many bacteria produce. Example of such exotoxins include cholera toxin, anthrax toxin and the Shiga-like toxins, and these proteins have proved to be extremely powerful probes with which to study aspects of the molecular mechanisms used by eukaryotic cells in, for example, cell signalling or intracellular transport. Indeed, this study of the interaction between bacteria and mammalian cells is now termed 'cellular microbiology' [10] and is, in reality, the topic of this book. This newly emerging scientific discipline illustrates the close interactions which occur between the prokaryotic and eukaryotic worlds. An interesting example of this close relationship is the enzyme lysozyme. This enzyme was originally discovered in tears by Alexander Fleming, and was shown to be able to degrade bacterial peptidoglycan. Thus it is accepted that this enzyme has antibacterial actions. However, it has recently been found that a number of bacteria contain a conserved family of genes that encode proteins which are related to lysozyme and have enzymic activity, and that these enzymes contribute to the virulence of certain bacteria [11].

Studies of the host defence systems of innate and acquired immunity have revealed that all multicellular organisms produce a vast range of proteins which have been categorized as 'cytokines'. These proteins are now recognized as: (i) vital local hormones, involved in homeostatic control of cells, tissues and of the organism as a whole; (ii) key mediators of the host defence systems against infections; and (iii) the major mediators of tissue pathology in bacterial infections and inflammatory disease generally. Cytokines were first discovered in the 1950s by investigators studying the fever caused by bacterial extracts. They were rediscovered in the 1960s by immunologists prising apart the cellular basis of acquired immunity. The past two decades have seen an explosion of interest in cytokines. This interest has been fuelled largely by the belief that cytokines are the therapeutic targets in diseases ranging from asthma to the zoonoses. However, as is obvious from the last few sentences, cytokines have paradoxical actions, able to both protect the host from and cause the symptoms of infection. For example, in the U.S.A., around 300 000 individuals die each year from the consequences of infection in a condition known as septic shock. This occurs when the bacteria causing infection release specific constituents which induce the rapid production of particular cytokines, such as interleukin (IL)-1, IL-6 and tumour necrosis factor (TNF)α. These cytokines, in turn, cause activation of the vasculature and inhibit the action of the heart, thus leading to organ failure, circulatory collapse and death. Yet it is these same cytokines that are believed to protect us, day-to-day, from the large number of infectious bacteria which exist in our environment. A good example of this is the recent finding that the family of low-molecular-mass cytokines known as chemokines are important in controlling HIV-1 infection [12,13].

The discussion thus far has centred on disease and the role of bacteria. Of course we live in a world which is literally groaning under the weight of its bacterial populations. The majority of these bacteria cause us no concern, and, indeed, only about 400 micro-organisms (including viruses, bacteria and other parasites) are able to induce infectious diseases in man [3]. Compare this with the fact that the average human body is host to around 1000 different bacterial species present on the skin and on all the epithelial surfaces of the mouth, respiratory tract, gastrointestinal tract and urogenital tract. Counting the bacteria which populate the average human body gives rise to a rather surprising statistic. The number of eukaryotic cells in the average human body is of the order of 10^{13}. However, the number of bacteria populating that body is in the region of 10^{14}. Thus, at any one time there are 10 bacteria for every human cell in the human body [14]. Of course, as these bacteria are present only on epithelial surfaces the local ratio of bacteria to epithelial cells will be very much larger than 10:1. This knowledge of the numbers of bacteria populating our bodies leads to a very interesting paradox: why do we not respond to these bacteria? After all, from the point of view of their complement of inflammation-inducing components, there is nothing special about this so-called normal or commensal microflora. The bacterial population which constitutes the normal microflora is composed of Gram-negative and Gram-positive species, all of which contain pro-inflammatory substances which should have effects on the cells with which they are in contact. Yet, this patently does not occur. The authors have suggested the following revolutionary hypothesis: failure of the commensal bacteria to cause inflammation is due to the fact that these bacteria produce proteins which act like host cytokines and function to damp down any inappropriate host tissue immune or inflammatory responses. We have termed these proteins 'bacteriokines', and suggest that they may be evolutionary ancestors of the cytokines which organize the homeostatic interactions in eukaryotic multicellular organisms.

This book will therefore focus on the interactions which occur between bacteria (both infectious and the normal microflora) and the host, and the roles that cytokines play in these cell–cell interactions. The next two chapters will concentrate on the bacteria and their potential for virulence. In Chapter 4 the biology of cytokines will be reviewed. Chapters 5 and 6 will consider the chemistry, biochemistry, molecular biology and physiology of lipopolysaccharide. In Chapter 7 the recent literature showing that bacteria produce a very wide range of cytokine-modulating molecules will be reviewed, while the therapeutic potential of modulating cytokines will be considered in Chapter 8. In the final chapter a synthesis of the new information about bacteria–host–cytokine interactions is attempted in order to suggest novel ways of looking at the biology of host–microbial interactions in health and disease.

References

1. Ziegler, P. (1970) The Black Death, Penguin Books, London
2. Franklin, T.J. and Snow, G.A. (1991) Biochemistry of Antimicrobial Action, Chapman and Hall, London
3. Mims, C., Dimmock, N., Nash, A. and Stephen, J. (1995) Mims' Pathogenesis of Infectious Diseases, 4th edn., Academic Press, London
4. Blaser, M.J. (1997) Ecology of *Helicobacter pylori* in the human stomach. J. Clin. Invest. **100**, 567–570
5. Parsonnet, J., Friedman, G., Vadndersteen, D.P., Chang, Y., Vogelmam, J.H., Orentreich, N. and Sibley, R.K. (1991) *Helicobacter pylori* infections and the risk of gastric carcinoma. N. Engl. J. Med. **325**, 1127–1131
6. Knoop, F., Owens, M. and Crocker, I. (1993) *Clostridium difficile*: clinical disease and diagnosis. Clin. Microbiol. Rev. **6**, 251–265
7. Greenwood, D. (1995) Antimicrobial Chemotherapy, 3rd edn., Oxford University Press, Oxford
8. Pettersson, J., Nordfelth, R., Dubinina, E., Bergman, T., Gustafsson, M., Magnusson, K.E. and Wolf-Watz, H. (1996) Modulation of virulence factor expression by pathogen target cell contact. Science **273**, 1231–1233
9. Zhang, J.P. and Normark, S. (1996) Induction of gene expression in *Escherichia coli* after pilus-mediated adhesion. Science **273**, 1234–1236

10. Cossart, P., Boquet, P., Normark, S. and Rappuoli, R. (1996) Cellular microbiology emerging. Science **271**, 315–316

11. Mushegian, A.R., Fullner, K. J., Koonin, E.V. and Nester, E.W. (1996) A family of lysozyme-like virulence factors in bacterial pathogens of plants and animals. Proc. Natl. Acad. Sci. U.S.A. **93**, 7321–7326

12. Cocchi, F., DeVico, A.L., Garzino-Demo, A., Arya, S.K., Gallo, R.C. and Lusso, P. (1995) Identification of RANTES, MIP-1α and MIP-1β as the major HIV-suppressive factors produced by CD8^{+} T cells. Science **270**, 1811–1815

13. Bates, P. (1996) Chemokine receptors and HIV-1: an attractive pair. Cell **86**, 1–3

14. Tannock, G.W. (1995) Normal Microflora, Chapman and Hall, London

2

Bacteria and infectious diseases

"The infection raged and the people were now frightened and terrified to the last degree so that they gave themselves up and abandoned themselves to their despair".
A Journal of the Plague Year
Daniel Defoe, 1722

2.1   Introduction

Bacteria, the smallest organisms capable of independent existence, have a diameter of approximately 1.0 μm. Apart from their overall shape [spherical (coccus), rod-like (bacillus) or spiral] and their cellular arrangement (pairs, clusters, chains, etc.), little can be discerned of their structure by light microscopy. Examination of the internal structure of these prokaryotic micro-organisms by electron microscopy generally reveals only a DNA-containing region and ribosomes. In some species, however, additional features can be seen — storage granules and invaginations of the cytoplasmic membrane known as mesosomes. Membrane-bound organelles and a nuclear membrane are notably absent from all bacteria. In contrast, the cell envelope of bacteria is far more complex than that of eukaryotic cells, consisting of a cytoplasmic membrane and a rigid cell wall which, in some species (Gram-negative bacteria), is multi-layered. Many, if not most, bacteria also have an additional layer (capsule) external to their cell wall and may have one or more of a variety of surface appendages. The only exceptions to this generalized description of a bacterial cell are the mycoplasmas which do not have a cell wall. In the case of bacteria adapted to living in, or on, another organism, the cell envelope forms the interface between the bacterium and its host. This is the structure which the host may, or may not, recognize as being 'not-self' and, if it is recognized as being 'foreign', the host may respond by attempting to eliminate the intruder. On the other hand, this is the structure which will be used by the bacterium to adhere to the host, to avoid its defences and possibly to damage it in some way. We believe that cytokines are key mediators of this interplay between bacteria and their host, and this concept is discussed in greater detail in subsequent chapters. Investigations of the ability of bacteria to influence cytokine synthesis in host cells have involved, almost exclusively, bacterial-cell-surface components or, to a lesser extent, secretory products of bacteria. Further discussion of bacterial ultrastructure, therefore, will be confined to the cell envelope and associated surface components.

2.2   Bacterial ultrastructure

2.2.1   The bacterial cell envelope

2.2.1.1 Cytoplasmic membrane

The innermost layer of the envelope in all bacteria is the cytoplasmic membrane and this has the classic unit membrane structure, although it differs from the cytoplasmic membrane of eukaryotes in that it does not contain sterols. As in the cells of higher organisms, it functions mainly as an osmotic barrier and in the transport of nutrients and waste products. However, in bacteria it is also involved in ATP production (as it contains the components of the electron-transport chain) and in the synthesis and export of cell wall components. There are remarkably few studies relating how components of the cytoplasmic membrane, when liberated by lysis, for example, interact with host cells, and none concerning their ability to affect cytokine synthesis by such cells. However, studies of this type have been carried out on the cell membranes of *Mycoplasma* spp. These cell-wall-less bacteria are bounded by a cytoplasmic membrane, the composition of which differs from that of the cytoplasmic membrane of other prokaryotes in that it contains cholesterol. A number of studies (see Chapter 7) have shown that membrane lipoproteins of mycoplasmas are able to stimulate the release of IL-1β, TNFα and IL-6 from monocytes [1,2].

2.2.1.2 Cell wall

Unlike the homeostatically controlled environment of cells of higher organisms, free-living bacteria are subject to the vagaries of their external environment. Protection against such fluctuations is provided by the cell wall which prevents osmotic lysis and also functions as a molecular sieve [3]. On the basis of a simple staining procedure developed more than 100 years ago, bacteria can be divided into two major groups, Gram-positive and Gram-negative, and this reflects major differences in the structure of the cell wall of these two groups. Virtually every component of the bacterial cell wall has some role to play in the interaction between the bacterium and its host, and many cell wall components have been shown to affect the synthesis and/or release of cytokines from host cells.

2.2.1.2.1 Gram-positive cell wall

Electron microscopic examination of the walls of Gram-positive bacteria shows a simple structure consisting of a single layer with a thickness of between 20 and 50 nm. Chemical and immunological analysis, however, has revealed greater complexity (Figure 2.1a). The main structural component of the cell wall of Gram-positive bacteria is a heteropolymer (known as peptido-

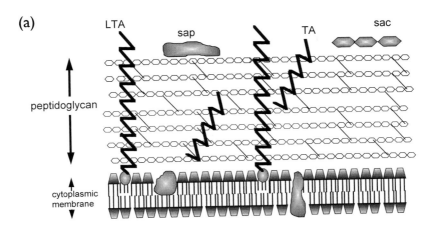

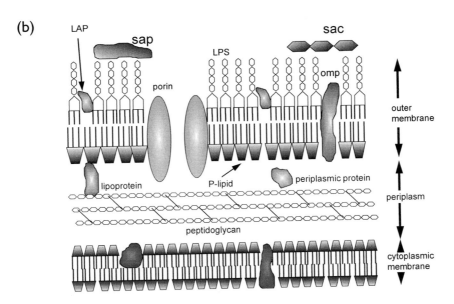

Figure 2.1 Diagrammatic representation of a cross-section through the wall of a typical Gram-positive (a) and Gram-negative (b) bacterium

Abbreviations used: LAP, lipid A-associated protein; LPS, lipopolysaccharide; LTA, lipoteichoic acid; omp, outer-membrane protein; P-lipid, phospholipid; sac, surface-associated carbohydrate; sap, surface-associated protein; TA, teichoic acid.

glycan) which consists of chains of alternating N-acetyl glucosamine and N-acetyl muramic acid residues, which are cross-linked by short peptide chains. The composition of these peptides is species-dependent and is unusual in Nature in that D-amino acids are present. Cross-linking occurs in all three planes, resulting in one macromolecule (between 20 and 40 layers thick) enclosing the whole bacterial cell. Peptidoglycan is the main component of the Gram-positive cell wall (comprising at least 40% of its mass), is responsible for its mechanical strength and acts as a permeability barrier [4,5]. Peptidoglycans from several species, as well as their degradation products, are able to stimulate the release of a number of cytokines from monocytes and other host cells [6]. This is reviewed in Chapter 7.

The other major components of the wall are anionic polymers such as teichoic and lipoteichoic acids [7]. Teichoic acids consist of chains of glycerol, ribitol, mannitol or sugars linked by phosphodiester bonds, and are attached to muramic acid residues in the peptidoglycan. D-Alanine or L-lysine are common substituents of the chains, which usually contain approximately 40 residues. Lipoteichoic acids (LTAs) consist of chains of glycerol phosphate, with D-alanine and sugar substituents, attached to a glycolipid (or diglyceride) in the cytoplasmic membrane. Both teichoic acids and LTAs are antigenic and often constitute the major somatic antigens of Gram-positive bacteria. LTAs are amphiphilic molecules with a number of biological activities which include the ability to stimulate (albeit at relatively high concentrations) the release of cytokines from monocytes [8]. In some ways, LTAs can be regarded as the Gram-positive equivalent of lipopolysaccharide (LPS), although the latter has a much greater potency and range of biological activities (reviewed in detail in Chapters 6 and 7).

Although disregarded until comparatively recently, proteins covalently linked to the peptidoglycan are now recognized as being important cell wall components, particularly in the case of protein A in *Staphylococcus aureus* and the M-proteins of streptococci. In some streptococci they can account for 16% of the dry mass of the cell wall. Many of these proteins function as adhesins, are involved in evading host defence systems, are important mediators of host tissue destruction and are known to induce cytokine release from host cells [9].

2.2.1.2.2   Gram-negative cell wall

In Gram-negative bacteria, the cell wall is a more complex multi-layered structure consisting of a thin peptidoglycan layer (possibly only one sheet of peptidoglycan) and a lipid bilayer forming an outer membrane external to this [10]. The outer membrane is linked to the peptidoglycan by a lipoprotein which spans a gelatinous region (containing proteins, enzymes and highly hydrated peptidoglycan with a low degree of cross-linking), known as the periplasm (Fig. 2.1b).

The innermost leaflet of the outer membrane consists principally of phospholipids together with some proteins, while the outer leaflet is composed of

LPS molecules of which there are approximately 3.5×10^6 per cell in *Escherichia coli* [11]. LPS is an amphiphilic molecule consisting of three regions: (i) a glycolipid (lipid A) embedded in the outer layer of the membrane, which is responsible for most of the biological activities characteristic of LPS; (ii) a 'core' oligosaccharide containing a characteristic sugar acid, 2-keto-3-deoxyoctulonic acid (KDO) and a heptose; and (iii) an antigenic polysaccharide (O-antigenic side chain) composed of a chain of repeating oligosaccharide units, which projects away from the cell interior and can render the bacterial surface extremely hydrophilic. LPS is the main somatic antigen of Gram-negative bacteria and its structure is very much strain-dependent. LPS can provoke a wide range of responses in whole organisms and many of these are attributable to its ability (at extremely low concentrations, e.g. pg/ml) to induce the release of a variety of cytokines from a range of host cells [12,13]. The structure of LPS and its biological actions are discussed in detail in Chapters 5 and 6. Approximately 50% of the dry mass of the outer membrane is protein and more than 20 immunochemically distinct proteins (termed outer-membrane proteins; OMPs) have been identified in *E. coli*. Some of these proteins (porins) form trimers, which span the outer membrane and contain a central pore with a diameter of about 1 nm. These porins (e.g. OmpC and OmpF of *E. coli*) are permeable to molecules with molecular masses up to approximately 500–600 Da [14]. They are also potent inducers of cytokine release from monocytes and lymphocytes [15], as described in Chapter 7. Some methods of extracting LPS from bacteria result in an LPS–OMP complex (to which the term 'endotoxin' should be applied). The protein(s) associated with the LPS is known as lipid A-associated protein (LAP) or endotoxin-associated protein (EAP) and is known to have biological activities distinct from those of LPS [16]. As detailed in Chapter 7, LAPs are able to stimulate cytokine release from a number of host cells including fibroblasts and monocytes [17] (Chapter 7).

2.2.1.2.3   Cell wall of *Mycobacterium* species

In the genus *Mycobacterium* the peptidoglycan layer is covered by lipid-rich layers so that up to 60% of the dry weight of the cell wall may consist of lipids, rendering it extremely hydrophobic [18]. A variety of lipids, glycolipids and lipoproteins have been isolated from mycobacterial cell walls and several of these contain mycolic acids, which are unique to the mycobacteria, nocardiae and corynebacteria. One of the major constituents of the cell wall is lipoarabinomannan which, as well as playing an important role in enabling survival within the host, is also a potent inducer of cytokine release from monocytes and macrophages [19] (Chapter 7).

2.2.2   Cell-surface appendages

Some, but not all, bacteria (both Gram-positive and Gram-negative) have one
or more of a number of appendages anchored in the cell wall and/or cytoplasmic membrane. These include flagella, fimbriae, conjugative pili and fibrils.

2.2.2.1   Flagella

Flagella are organelles responsible for bacterial motility [20]. They consist of a
globular protein (flagellin), which aggregates in a helical arrangement to form a
long, hollow filament (usually many times the length of the bacterial cell). This
filament terminates in a complex 'basal body' embedded in the cell wall and
cytoplasmic membrane. It has been known for many years that the flagellins
are highly antigenic (they constitute the H-antigens of motile bacteria), but
little is known about their cytokine-inducing ability. To date there appear to
be only two reports concerning this: the first [21] describes how flagella of
Salmonella typhimurium can increase the level of mRNA for granulocyte-
macrophage colony-stimulating factor (GM-CSF) in mouse peritoneal
macrophages; and the second [22] shows that flagellin from *Pseudomonas
aeruginosa* stimulates the release of IL-8 from a human epithelial cell line.

2.2.2.2   Fimbriae

Fimbriae (also termed pili) are rod-shaped structures originating in the cytoplasmic membrane and are composed of a hydrophobic protein, pilin. They are
shorter and thinner than flagella and their main function is to enable adhesion
of the organism to host cells or to other bacteria. A bacterium may be able to
elaborate a number of fimbriae, each with a specific adhesin at its tip, to enable
adhesion to a particular host receptor [23]. They are found mainly on Gram-
negative bacteria, although they have also been detected on streptococci and
actinomycetes. The nature of the receptor for some of the adhesins of various
organisms is shown in Table 2.1.

A number of *in vitro* and *in vivo* studies (see Chapter 7) have shown that
fimbriae are able to induce the release of a variety of cytokines from epithelial
cells, monocytes, macrophages, fibroblasts and lymphocytes [6,24].

Table 2.1 Receptors for bacterial adhesins

Organism	Receptor for adhesin
Actinomyces naeslundii	Galactosides
Bordetella pertussis	Galactose
Pseudomonas aeruginosa	L-Fucose
Vibrio cholerae	L-Fucose
Escherichia coli	Oligomannosides

2.2.2.3   Conjugative pili

Pili are similar in structure to fimbriae (although much thicker and longer), but are involved in the transfer of DNA during bacterial conjugation.

2.2.2.4   Fibrils

Fibrils are shorter and much thinner than fimbriae or pili and may cover the whole cell surface [25]. They are found on the surfaces of oral streptococci, *Bacteroides* spp. and other bacteria. Their function is unknown.

2.2.3   Cell-surface-associated components

Most bacteria, especially when first isolated from their natural habitat, have additional layers of material on their surface, which are generally only loosely associated with the cell wall. Examples of such components include capsules, slime layers and S-layers. While these are generally regarded as being composed mainly of carbohydrates, our own studies have revealed the presence of large numbers of proteins loosely associated (i.e. easily extracted with saline) with the cell surfaces of a number of oral, and other, bacteria. As described in Chapter 7, some of these proteins are remarkably potent modulators of cytokine synthesis by host cells.

2.2.3.1   Capsules

The capsule of a bacterium is a gelatinous, highly hydrated matrix composed of polysaccharide(s) and/or protein(s). It is loosely associated with the cell wall and is often shed into the environment [26,27]. Although generally regarded as being amorphous, there is evidence to suggest a highly ordered secondary structure in the capsule of *E. coli*. There is also evidence implying that capsular material may be linked covalently to the cytoplasmic membrane by means of phospholipid substituents. Capsules act as molecular sieves and adhesins, and also protect the organism against phagocytosis. Capsular materials are highly antigenic (constituting the K antigens of bacteria) and are able to induce cytokine release from host cells [28,29] (Chapter 7).

2.2.3.2   Slime layers

Slime layers are similar to capsules except that they have a higher water content and are even more loosely attached to the cell surface than capsules [30].

2.2.3.3   S-layers

The surfaces of many species of bacteria are completely covered in a crystalline array of protein, or glycoprotein, self-assembling units [31]. Such layers (so-called 'S-layers') have protective and adhesive functions similar to capsules. To date, there appear to be no reports of their cytokine-inducing abilities.

2.3   The normal bacterial flora of man

The normal microflora comprises those microbes, the majority being bacteria, which are found associated with body surfaces exposed to the external environment, i.e. the skin, oral cavity, respiratory tract, gastrointestinal tract, vagina and urinary tract (Figure 2.2). The number of bacteria inhabiting these regions is enormous and, indeed, outnumbers mammalian cells by a factor of ten. On a simple numerical basis, therefore, we are in fact more microbe than man. The natural reaction to this surprising, and little known, fact is to want to know more about our uninvited partners in a relationship that can be guaranteed to last a lifetime. The next questions to ask concern the nature of the relationship. Is it a matter of caring and sharing between prokaryotes and eukaryotes? What will happen if one of these partners gets out of control and comes to dominate the relationship? How is a healthy balance maintained? How do the partners communicate? We believe that in answering the latter we can begin to understand something about the other questions raised and we have hypothesized that cytokines play a central role in dictating the nature of the prokaryotic–eukaryotic partnership [32] (see Chapters 4, 7 and 9 for a fuller explanation).

The term 'normal microflora' is rather misleading as it tends to imply, incorrectly, that it constitutes a definable, static entity. In fact, this is far from

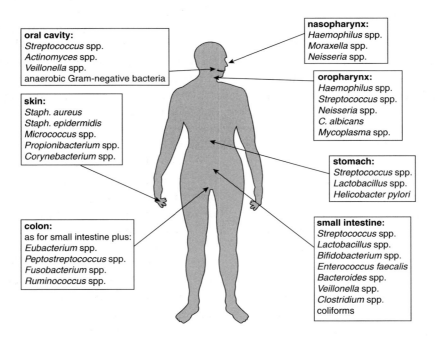

Figure 2.2 The normal microflora of man

being the case. The types of bacteria found associated with a healthy individual vary enormously from site to site within that individual. Hence, it is essential to speak in terms of the normal microflora of the oral cavity, of the skin, etc. Even within a particular anatomical region, for example the oral cavity, there are tremendous differences between the types of bacteria found at particular locations such as the teeth, the tongue, etc. Such variations are the result of environmental selection factors (physical, chemical and biological) operating at each of these sites. The particular combination of physical (e.g. temperature, nature of the colonizing surface available, osmotic pressure), chemical (e.g. types of nutrients available, composition of the gaseous phase, pH) and biological (e.g. host secretions, presence of other bacteria) factors at the site results in a unique environment which limits the types of bacteria that can colonize that particular location. Attempts to define what constitutes the normal microflora of an anatomical region become even more difficult when variations between individuals are taken into account, as it is well established that diet, hygiene practices, gender, age, etc. can all markedly affect the types of bacteria to be found at a particular site in an individual. Two additional factors make the task of defining 'normality' extremely difficult. The first of these is that, throughout the history of the discipline of bacteriology, the efforts of most bacteriologists have been directed at studying disease-inducing bacteria rather than the 'harmless' species that live with us, so that relatively little is known about our habitual prokaryotic partners. Secondly, the studies that have been carried out have revealed the enormous complexity of the microbial communities leading to problems of isolation and identification. With these limitations in mind, this chapter will attempt to summarize what is known of the nature of the bacteria with which we, usually, live in harmony. For each anatomical site, a brief description of the major environmental selection factors operating there will be given in order to provide some insight into why particular organisms are to be found there.

2.3.1 Skin

The skin provides a large, readily accessible area for colonization by bacteria and can, of course, come into contact with a huge number and variety of microbes. It might, therefore, be thought to harbour a large bacterial population. However, the skin is in fact rather sparsely populated in comparison with other body surfaces for reasons discussed below. Nevertheless, because of its large surface area (approximately 2 m^2) the total number of skin bacteria is considerable, amounting to approximately 10^{12} per person. So what is the nature of the environment offered to bacteria attempting to colonize this region? First, it must be recognized that the skin exhibits variation at different anatomical sites. Hence, the moisture content and temperature vary enormously as a result of the distribution of hair and sebaceous glands and the gross anatomy of what is, in fact, our largest organ. Enclosed, hairy regions with many sebaceous glands, such as the armpits, offer a warmer, moister environment than, for example, the back. The former regions support a more dense population (several million per cm^2) than the latter (several hundred per cm^2). From the point of view of maintaining a permanent population on the skin, one of the major problems confronting bacteria is that the outermost layers are continually being shed, carrying with them adherent organisms. The principal sources of nutrients for bacteria are sweat and sebum. However, metabolism of the lipids in sebum can lead to the accumulation of fatty acids which are toxic to many bacterial species and also results in a low pH at which many bacteria cannot survive. Evaporation of sweat can cause the accumulation of high concentrations of sodium chloride on the skin surface which, again, some species cannot tolerate. Bacteria attempting to colonize pores, hair follicles and sweat glands will be confronted by additional antibacterial factors including the peptidoglycan-degrading enzyme lysozyme, antibodies produced by skin-associated lymphoid tissue, and, possibly, antibiotic peptides. Bacteria on the skin have ready access to oxygen so that most of the indigenous species are either obligate aerobes such as micrococci (Gram-positive cocci) or facultative anaerobes including staphylococci (Gram-positive cocci) and corynebacteria (Gram-positive rods). However, within the ducts of sebaceous glands, conditions are more anaerobic, thus permitting the survival of organisms such as propionibacteria (Gram-positive rods).

In order to establish themselves as permanent members of the skin microflora, bacteria would, ideally, possess the following characteristics: aerobic or facultatively anaerobic respiration; the ability to adhere to keratinized epithelial cells; the ability to utilize lipids as a carbon and energy source; and tolerance of high salt concentrations. Generally speaking, the skin microflora is dominated by bacteria with these characteristics, the predominant organisms being species belonging to the following genera: *Staphylococcus, Micrococcus, Corynebacterium* and *Propionibacterium*.

As is the case with all anatomical regions, several members of the normal microflora are able to induce disease under certain circumstances [33]. The most important of these are *Staph. aureus* (boils, wound infections, food poisoning), *Staph. epidermidis* (infective endocarditis) and *Propionibacterium acnes* (acne). These diseases will be considered in greater detail later.

2.3.2 Oral cavity

The oral cavity contains such a varied set of habitats, each with its own characteristic microflora, that to speak of a 'normal oral microflora' is meaningless without specifying which particular oral habitat. The main habitats are the buccal mucosa, the tongue, the gingival crevice and the teeth. Each of these has a number of microhabitats, each offering a different environment in terms of oxygen content, redox potential, pH and access to nutrients. One of the unique features of the oral cavity is that it provides the only non-shedding surfaces in the body — the surfaces of the teeth — thus enabling bacteria to accumulate to form dense aggregates known as biofilms, i.e. dental plaque.

Because of the constant flow of saliva into the oral cavity, swallowing, tongue movements and chewing, bacteria are subjected to powerful forces tending to remove them so that the ability to adhere to oral surfaces (or to already adherent bacteria) is an essential pre-requisite for an organism to become established as a member of the normal microflora. While ostensibly an aerobic environment, the vast majority of organisms in the oral cavity are facultative or obligate anaerobes, with the relative proportions of each varying with the nature of the microhabitat. The gingival crevice, for example, because of its low redox potential, restricted access to oxygen and plentiful supply of nutrients from a serum-like exudate (gingival crevicular fluid), harbours high proportions of Gram-negative obligate anaerobes such as *Porphyromonas* spp., *Fusobacterium* spp., *Veillonella* spp. and spirochaetes. In contrast, the microflora of a freshly colonized tooth surface is dominated by aerobic and facultatively anaerobic bacteria such as *Neisseria* spp. and streptococci. With regard to epithelial surfaces, the dorsum of the tongue supports approximately 100 bacteria per cell and these are mainly streptococci, *Actinomyces* spp. and *Veillonella* spp. In contrast, the cheek epithelium has a more sparse population (less than 25 bacteria per cell) comprising mainly streptococci.

2.3.3   Respiratory tract

Apart from the bronchi and bronchioles, which are normally sterile, the other regions of the respiratory tract, i.e. anterior nares, nasopharynx and oropharynx, have a characteristic microflora.

The dominant feature of the respiratory tract affecting the establishment of a normal microflora is an efficient particle- and microbe-excluding system consisting of: (i) hairs in the anterior nares for removing large particles; and (ii) a ciliated-epithelium coated in mucin, in which bacteria are trapped, carried to the back of the pharynx by ciliary action and swallowed. These systems serve to reduce microbial colonization of the respiratory tract and maintain the lower regions almost microbe-free — a considerable achievement when it has been estimated that the average individual inhales approximately 10 000 bacteria per day. In order to establish themselves as part of the normal microflora, any bacteria resisting expulsion by the above mechanisms must be able to adhere to the epithelium lining the respiratory tract and overcome antibacterial factors present in mucin (see Chapter 3). These include lysozyme, lactoferrin (a protein which binds iron, thus limiting bacterial growth), secretory IgA (which prevents attachment of bacteria to epithelial cells) and superoxide radicals generated by lactoperoxidase. Bacteria able to attach to the epithelium which (as in the case of skin) is continually being shed, and survive the host defences listed above, must then be able to grow and reproduce in this aerobic environment in which the only source of nutrients are those secreted by host cells. The microflora is dominated by aerobic and facultatively anaerobic species.

2.3.3.1 Anterior nares

The epithelium lining this region is similar to skin and so it is not surprising that the normal microflora is comprised of organisms normally found on the skin, i.e. *Staph. epidermidis* and corynebacteria. *Staph. aureus* may also be present in as many as 85% of the population.

2.3.3.2 Nasopharynx

The nasopharynx has a more complex microflora than that found in the anterior nares [34]. In addition to staphylococci and corynebacteria, a number of Gram-negative species are present, including *Haemophilus influenzae, Moraxella catarrhalis* and *Neisseria* spp. The most frequently encountered strains of *H. influenzae* are those that do not possess a capsule, and these strains can cause diseases such as otitis media and sinusitis. However, capsulated strains of the organism may also be present, and these are responsible for more serious infections such as meningitis, acute epiglotitis and pneumonia. While most of the *Neisseria* spp. found in the nasopharynx are unable to cause disease in healthy individuals, as many as 5% of the population harbour *N. meningitidis* which can cause a life-threatening meningitis.

2.3.3.3 Oropharynx

The microflora of the oropharynx is very different from that of the two regions already described and exhibits an even greater diversity than that of the nasopharynx [35]. The reasons for this are not clear, but a contributing factor may be the availability of a wider range of nutrients from food ingested by the host. The most distinguishing feature of the microflora is the dominance of members of the genus *Streptococcus*, a group of facultatively anaerobic, Gram-positive, chain-forming cocci. Traditionally this genus is sub-divided on the basis of effect on blood-containing agar into three groups: non-haemolytic, β-haemolytic (complete destruction of red blood cells) and α-haemolytic (partial destruction of red blood cells accompanied by a green coloration). α-Haemolytic and non-haemolytic species are numerically dominant and these can give rise to endocarditis when they gain access to the bloodstream of individuals with damaged heart valves. As many as 10% of the population harbour *Strep. pyogenes*, a β-haemolytic organism. This is the most frequent bacterial cause of pharyngitis which, in some cases, can be followed by rheumatic fever or acute glomerulonephritis. It is also the causative agent of impetigo, cellulitis, lymphangitis and erysipelas. Up to 70% of the population may have the α-haemolytic *Strep. pneumoniae* in their oropharynx and this is responsible for a wide range of infections including pneumonia, meningitis, otitis media and sinusitis. Other bacteria constituting the normal microflora of the oropharynx are *Mycoplasma* spp.

2.3.3.4 Lower respiratory tract

Because of the efficiency of the ciliated epithelium in expelling microbes, the lower respiratory tract is usually sterile. Any organisms which do reach the alveoli are usually destroyed by phagocytic macrophages.

2.3.4   Gastro-intestinal tract

The gastro-intestinal tract has a large surface area (approximately 200 m^2) and contains most of the bacteria (approximately 10^{14}) inhabiting man [36]. It consists of several distinct regions with a number of key environmental selection factors in common, which differ markedly from those affecting the anatomical regions discussed so far. First, as there is little ingress of air, these ecosystems are predominantly anaerobic and have a low redox potential. Secondly, bacteria are not dependent on host secretions for nutrients as an enormous variety of nutrients is available from food ingested by the host. Thirdly, the tract consists of a number of fluid-filled cavities so that the ability of an organism to adhere to a surface within these ecosystems is not an essential pre-requisite for it becoming established as a member of the normal microflora. One school of thought holds that only attached cells should be regarded as true members of the microflora, the others being considered as transients spending between 12 and 18 h (the normal intestinal transit time) inside the host. Finally, extremes of pH, at least in the case of the stomach, are encountered.

In addition to the normal antibacterial mechanisms associated with mucosal surfaces, the gastro-intestinal tract also produces proteolytic enzymes and bile salts which exert an antibacterial effect.

2.3.4.1 Stomach

Because of its low pH and the secretion of proteolytic enzymes, the microbial content of the stomach is sparse (approximately 10^3 bacteria/ml) and consists mainly of members of the aciduric genera *Streptococcus* and *Lactobacillus*. Whether they should be considered as constituting a normal microflora is debatable and many would regard them simply as transients. More controversial is the situation regarding *Helicobacter pylori*, a Gram-negative, spiral-shaped, motile bacterium which attaches itself to the mucosal wall or embeds itself in the mucus associated with the wall. This organism causes gastritis and peptic and duodenal ulcers, but is also found in the stomach of healthy individuals.

2.3.4.2 Small intestine

The duodenum and jejunum also have a sparse microflora (approximately 10^5 bacteria/ml), although it is more complex than that of the stomach. In addition to those organisms found in stomach contents, members of the genera *Bacteroides* (Gram-negative anaerobic rods) and *Bifidobacterium* (Gram-positive anaerobic rods) are also present.

The ileum, the next region of the small intestine, has a more substantial (up to 10^9 bacteria/ml) and diverse bacterial flora consisting of lactobacilli, bifidobacteria, *Enterococcus faecalis*, *Bacteroides* spp., *Veillonella* spp., *Clostridium* spp. (Gram-positive anaerobic rods) and Gram-negative facultative rods such as *E. coli*.

2.3.4.3   Colon

Large numbers of bacteria are attached to the mucosal surface of the colon and are present in the lumen, where they constitute approximately 55% of the solids. Nearly 500 bacterial species have been isolated from the colon, although only approximately 40 are regularly encountered with one species, *Bacteroides vulgatus*, regularly comprising 10% of the microflora. Obligate anaerobes comprise greater than 90% of the microflora with five genera accounting for most of these: *Bacteroides*, *Eubacterium*, *Bifidobacterium*, *Peptostreptococcus* (Gram-positive cocci) and *Fusobacterium* (Gram-negative rods). Other numerically important genera include *Ruminococcus* (Gram-positive anaerobic cocci), *Streptococcus*, *Clostridium* and *Enterococcus*. Species belonging to the following genera are regularly isolated, although they comprise smaller proportions of the microflora: *Escherichia*, *Enterobacter* (Gram-negative facultative rods), *Proteus* (Gram-negative facultative rods), *Lactobacillus* and *Veillonella*.

Common inhabitants of the colon known to cause infections include *Clostridium perfringens* (gangrene), *Bacteroides* spp. (peritonitis, intra-abdominal abscesses), *Clostridium difficile* (pseudomembranous colitis) and *E. coli* (gastroenteritis, urinary tract infections, neonatal meningitis).

2.3.5   Urogenital tract

2.3.5.1   Urinary tract

The urinary tract consists of the urethra and the bladder and, apart from the distal portion of the urethra, does not have a normal microflora as it is regularly flushed by sterile urine. Bacteria from the skin, anus and vagina often colonize the distal region of the urethra so that staphylococci, lactobacilli and *E. coli* are often found there.

2.3.5.2   Vagina

Prior to puberty and after the menopause, vaginal secretions are alkaline and the microflora is dominated by staphylococci and streptococci. In contrast, between puberty and the menopause, the secretions are acidic (pH 5). This is the result of the fermentation of glycogen (by host cells or bacteria), which accumulates in epithelial cells due to the action of oestrogens. The low pH encourages colonization by aciduric species such as lactobacilli, which constitute the dominant members of the vaginal microflora. This is a good example of the way in which host factors (in this case oestrogens) can have a major effect on the composition of the normal microflora. A variety of other bacteria have also been isolated from vaginal secretions, including staphylococci, corynebacteria, streptococci, *Peptostreptococcus* spp., *Peptococcus* spp., *Eubacterium* spp., *Propionibacterium* spp. and anaerobic Gram-negative bacilli [37].

2.4   Bacteria and disease

2.4.1   Types of host–bacteria perturbations

In comparison to the prokaryotes, we are relative newcomers to this planet and it is a credit to the prokaryotic component of *Homo sapiens* that it has accommodated the eukaryotic newcomers so well that an equilibrium acceptable to both of these kingdoms has evolved. In general, the relationship between the prokaryotes and eukaryotes comprising *H. sapiens* is a harmonious one and constitutes an example of mutualism, i.e. where both partners benefit and neither suffers. Hence, the host provides a warm, moist nutrient-rich environment for the bacteria which in turn provide protection against colonization by pathogenic species and also synthesize a range of vitamins (niacin, thiamine, riboflavin, vitamin K) which may be utilized by the host. However, for a variety of reasons, this relationship may turn into one in which the prokaryotes gain at the expense of the eukaryotes. This constitutes what is often termed an 'endogenous' or 'opportunistic' infection and will be examined in more detail later in this section. Furthermore, there are certain bacteria with which we do not have such a cosy relationship and when these organisms, which do not usually constitute part of our normal microflora, are encountered, the result is a parasitic relationship in which the prokaryotes gain considerably at the expense of *H. sapiens*. These constitute what most readers will recognize as the classic (exogenous) infectious diseases.

2.4.2   Diseases caused by members of the normal microflora

First, a few words about terminology. Should diseases caused by organisms which are normally present on man be classed as infections? The dictionary definition of the term 'infection' is "the state produced by the establishment of an infective agent in or on a suitable host". In the case of diseases caused by organisms already established on the host as part of the normal microflora, the term infection would seem inappropriate. However, as the terms opportunistic, or endogenous, infections are in general use to describe diseases caused by members of the normal microflora, these terms will be adopted in this book.

In the previous sections we have described the types of organisms associated with particular anatomical sites and have pointed out that many of these organisms are capable of initiating an infectious process but, in general, do not do so. In considering such diseases, therefore, the first question to ask is certainly: what upsets this balanced state of affairs, thereby precipitating the disease? The changes resulting in a switch from a mutualistic association to a disease-inducing symbiotic one have, in some cases, been identified although many remain obscure. Some of the most important means by which opportunistic infections arise are described in the following sections.

2.4.2.1   Breaching the epithelial barrier

The enormous numbers of bacteria that inhabit our body surfaces are separated from the underlying sterile tissues by an epithelial barrier which, at some sites, consists of no more than a single layer of cells. When this barrier is compromised in any way there are invariably, as pointed out above, a number of organisms with the ability to take advantage of the situation. Examples include infections of wounds and burns with *Staph. aureus*, peritonitis due to *Bacteroides fragilis* following rupture of the appendix or bowel surgery, septicaemia caused by a wide range of organisms following puncture wounds, and gangrene due to contamination of wounds with *Cl. perfringens*.

   Toothbrushing may also traumatize the junction between the teeth and associated soft tissues resulting in the access of bacteria to underlying tissues and a transient bacteraemia. For most healthy members of the population this, apparently, does not result in any infectious process. However, in individuals with congenital cardiac abnormalities or prosthetic heart valves, or those who have suffered attacks of rheumatic fever, the organisms may colonize the cardiac tissue or prosthesis resulting in endocarditis.

2.4.2.2   The presence of a foreign body which interferes with normal host functions

Prosthetic heart valves and joints and catheters provide surfaces on which bacteria can accumulate and form biofilms. Biofilms are less susceptible to phagocytosis and to the bactericidal effects of serum and blood because of the presence of large amounts of extracellular material [38,39]. Foreign bodies obstructing tracts (e.g. the urinary and respiratory tracts) not only act as a nidus for the accumulation of bacteria, but also interfere with blood and lymphatic flow in neighbouring tissues, hence rendering the host less able to deal with the accumulated organisms [40]. They can also interfere with mechanisms designed to remove bacteria from these sites, e.g. the flushing action of urine and the mucociliary escalator in the respiratory tract. The type of organisms associated with such infections will depend on the anatomical site involved, but will invariably include species such as *Staph. epidermidis, Staph. aureus, Pseudomonas* spp. and *Candida* spp. [41].

2.4.2.3   The transfer of bacteria to sites where they do not constitute part of the normal microflora

Females between the ages of 20 and 40 years are particularly susceptible to urinary tract infections by members of the normal microflora of the colon, such as *E. coli, Proteus* spp. and *Klebsiella* spp., with *E. coli* being the most frequent causative organism. Infection is facilitated by the close proximity of the colon to the urethra. The causative organisms colonize the peri-urethral

area and then ascend the urethra to the bladder. Certain strains of *E. coli* pre-
dominate as the causative agents in a particular geographical area, suggesting
that the organism can be transmitted from person to person (e.g. via food) and
then become established in the gastro-intestinal tract.

Aspiration pneumonia results from the aspiration of food, liquid or gastric
contents into the upper respiratory tract. The disease is usually polymicrobial
and a wide range of organisms is encountered (being dependent on the source
of the aspirate), including anaerobes, Gram-negative bacilli such as *Klebsiella*
spp. and *Pseudomonas* spp. and staphylococci [42,43].

2.4.2.4 Suppression of the immune system by drugs or radiation during cancer therapy

Cancer therapy involves the use of radiation or drugs which are designed to
kill rapidly growing cells. Unfortunately, this includes polymorphonuclear
neutrophils (PMNs), resulting in the patient becoming neutropaenic. As
PMNs constitute one of the major host defences against infections, these
treatments leave the patient prone to infection by a wide variety of micro-
organisms which normally do not cause problems in individuals with an intact
defence system. Gram-negative bacilli (e.g. *E. coli*, *Klebsiella* spp. and
Pseudomonas spp.), *Staph. aureus* and *Candida albicans* are the most frequent
causes of infections in such patients [44].

2.4.2.5 Impairment of host defences following infection by an exogenous pathogen

Infection with HIV, the causative agent of AIDS, seriously impairs the
immune response and so allows infection by members of the normal
microflora such as *C. albicans*, *Pneumocystis carinii*, *Strep. pneumoniae*, *Staph.
aureus*, *Corynebacterium* spp., etc. [45,46].

One of the great scourges of mankind, influenza, is caused by orthomyxo-
viruses which destroy cells lining the upper and lower respiratory tracts, hence
reducing the ability of the epithelium to exclude bacteria. Furthermore, the
viruses also inhibit phagocytosis of bacteria by alveolar macrophages, thereby
permitting the survival of members of the normal microflora such as *Staph.
aureus*, *Strep. pneumoniae* and *H. influenzae* [47]. These organisms can then
cause an often fatal pneumonia.

2.4.2.6 Elimination of constituents of the normal microflora by antibiotics

The normal microflora at most anatomical sites, particularly the gastrointestinal tract and the oral cavity, is very complex and the relative numbers of
the various constituents of the microflora are controlled by competition for
adhesion sites and nutrients, interdependence due to food webs, the
production of antimicrobial factors, etc. This delicate balance can be easily
upset by broad-spectrum antimicrobial agents such as the tetracyclines,
resulting in overgrowth by resistant organisms such as *C. albicans* and *Cl.*

difficile which can cause candidosis and pseudomembranous colitis respectively [48].

2.4.2.7 Unknown precipitating factor

The most common diseases of mankind are the inflammatory periodontal diseases which include gingivitis (when only the gingivae, i.e. gums, are involved) and periodontitis (when tooth-supporting structures such as the alveolar bone are involved). These diseases affect everyone at one time or another. From a bacteriological perspective these are fascinating diseases, not only because they are so prevalent but also because they are caused by members of the normal oral microflora and are the result of biofilm (dental plaque) formation on the only naturally occurring, non-shedding surfaces of the body, i.e. the teeth. So what enables this normally benign oral microflora to induce such profound tissue destruction? The answer to this can be best understood by taking an ecological perspective as outlined in the ecological plaque hypothesis of Marsh [49]. Bacteria able to adhere to the saliva-covered tooth surface form a pioneer community (consisting mainly of *Streptococcus* spp., *Neisseria* spp., *Haemophilus* spp. and *Actinomyces* spp.), which alters the physical, chemical and biological characteristics of this habitat as a result of its metabolic activities. These changes in the nature of the habitat then provide the opportunity for colonization by different organisms ('secondary colonizers') more suited to this altered environment, so that a secondary community becomes established which comprises some organisms from the pioneer community and the new secondary colonizers. This process by which organisms alter their environment and provide a habitat suitable for different organisms is termed 'microbial succession'. Microbial succession ultimately results in a community whose composition can remain stable (termed a 'climax community') due to a variety of interactions (both positive and negative) involving food webs, competition for nutrients, bacteriocin production, co-aggregation and physicochemical factors such as pH, redox potential, etc. In the climax community of organisms within the subgingival region (i.e. the gap between the tooth and gum), there may well be low numbers of organisms (periodontopathogenic species) capable of initiating tissue destruction. Some environmental perturbation may then occur which could favour the proliferation of one (or more) of these periodontopathogens (e.g. *Porphyromonas gingivalis*, *Prevotella intermedia* and *Actinobacillus actinomycetemcomitans*) to disease-inducing levels. One possible event leading to such a scenario would be an increased flow of nutrient-rich, antibody-containing gingival crevicular fluid (GCF) due to an inflammatory response to plaque accumulation. The GCF may supply a key nutrient, the absence of which may have been limiting the growth of one of the periodontopathogens, or may provide antibodies against a non-periodontopathogenic species whose elimination may serve to allow periodontopathogens to proliferate.

2.4.3   Exogenous infections

Bacteria that are capable of inducing disease in individuals with intact specific and non-specific defence systems are generally referred to as being pathogens. Many of the most dangerous of these organisms are not usually part of the normal human microflora, although man is the natural host, e.g. *Bordetella pertussis*, *Shigella* spp., *Mycobacterium tuberculosis* and *Vibrio cholerae*. Others are primarily inhabitants of other animals (e.g. *Mycobacterium bovis*, *Bacillus anthracis*, *Yersinia pestis*) or are found in the soil (e.g. *Cl. tetani*) or in water (*Pseudomonas* spp.). There are, however, some exceptions to this generalization as a number of organisms can colonize certain individuals (termed 'carriers') without causing disease, and these can then be transmitted to susceptible individuals, e.g. *Salmonella typhi* or *Corynebacterium diphtheriae*. Quite often patients recovering from an infection will harbour the organisms for some time and so can act as a source of infection for other individuals. The classic example of this is Typhoid Mary, the first recognized carrier of typhoid fever, who was responsible for nine separate outbreaks of the disease (over at least two decades), which resulted in 54 cases and four deaths.

One complication associated with designating a particular bacterial species as being a pathogen is that within a species different strains can be recognized. These strains, usually differentiated on the basis of phenotypic characteristics, can exhibit considerable differences in virulence. As the phenotype of an organism can be profoundly affected by environmental factors, there is nowadays more interest in determining genetic variability within a species. Such analyses have established the existence of large numbers of clonal types in some pathogenic species. For example it has been shown that in North America 104 clonal types of *H. influenzae* serotype b exist. Some clones of a particular organism may constitute part of the normal human microflora, whereas others are highly virulent and colonization by the latter often results in disease. The classic example of the wide variation in virulence displayed by strains of a particular species is *H. influenzae*, which exists as six serotypes distinguishable on the basis of the antigenicity of their capsular polysaccharides. Whereas most individuals harbour *H. influenzae* serotypes a, c, d, e and f in the upper respiratory tract without causing any disease, the highly virulent serotype b is rarely found in healthy individuals. Furthermore, genetic analysis of serotype b strains has shown that only six of the 104 clonal types are responsible for 81% of disease outbreaks. At the other extreme is the enteric pathogen *Shigella sonnei* in which only one clonal type has been identified.

Table 2.2 lists some important pathogenic bacteria and their natural reservoirs.

2.5   Disease classification

A remarkable feature of most infections is the predictability of the course of the disease. This gives rise to a characteristic set of features in the patient which

Table 2.2 Pathogens and their natural reservoirs

Organism	Disease	Natural reservoir
Clostridium tetani	Tetanus	Soil
Legionella pneumophila	Pneumonia	Water
Borrelia burgdorferi	Lyme disease	Ticks
Bacillus anthracis	Anthrax	Other animals
Yersinia pestis	Plague	Rats
Vibrio cholerae	Cholera	Water
Neisseria meningitidis	Meningitis	Man
Salmonella typhi	Typhoid fever	Man
Burkholderia pseudomallei	Melioidosis	Soil, water
Treponema pallidum	Syphilis	Man
Mycobacterium tuberculosis	Tuberculosis	Man
Haemophilus influenzae	Meningitis, upper respiratory tract infections	Man
Corynebacterium diphtheriae	Diphtheria	Man

is of enormous use to a clinician trying to diagnose the infection. Hence, a particular organism will usually infect a certain tissue or organ, will either be limited to that tissue or spread throughout the body and will produce a characteristic pattern of tissue destruction. This predictable behaviour is due to the possession by any particular organism of a limited set of virulence factors which dictate where the organism can initiate the infection, whether it can invade and what damage it can do when it has established itself. A useful classification of bacterial infections, which emphasizes the underlying virulence mechanisms, is one based on their site of action, whether or not invasion of underlying tissues accompanied by dissemination is involved, and whether the organisms release exotoxins (Figure 2.3). Examples of each of the major types of infection, based on this classification system, are given below.

2.5.1 Infections which are not usually accompanied by tissue invasion

Many bacterial infections are confined to external or internal body surfaces (underlying tissue may be invaded to a limited extent) without the organism being disseminated throughout the body. Examples include diphtheria, pharyngitis, cholera, whooping cough and gastroenteritis. Although all of these surface conditions manifest as inflammation, the bacterial factors responsible for eliciting this response have, in many cases, not yet been identified.

2.5.1.1 Infections confined to mucosal surfaces
2.5.1.1.1 Infections involving exotoxin production
2.5.1.1.1.1 *Diphtheria*
Diphtheria is a disease that originates in the pharynx, following the inhalation of a bacteria-laden aerosol from an infected individual, and is invariably

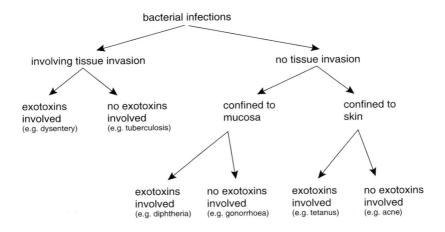

Figure 2.3 Classification of bacterial infections

confined to the upper respiratory tract [50]. It is caused by *Coryn. diphtheriae*, a Gram-positive, aerobic non-spore-forming rod, which secretes a toxin that kills host cells, induces heart failure and adversely affects the nervous system. On recovering from the infection, a patient will often harbour the organism in the throat or nose for several weeks and so spread this bacterium.

2.5.1.1.1.2   Cholera
The causative agent of cholera, *V. cholerae*, is a Gram-negative, facultatively anaerobic, motile curved rod, which is transmitted in faeces-contaminated water or food. It colonizes the mucosa of the small intestine and there it secretes an enterotoxin which stimulates the secretion of chloride, bicarbonate and water from epithelial cells resulting in a profuse, watery diarrhoea [51].

2.5.1.1.1.3   Whooping cough
Whooping cough is caused by *Bord. pertussis*, a Gram-negative, aerobic, non-motile rod, which is spread in aerosols created by infected individuals. It adheres to the cilia of the respiratory epithelia and secretes several toxins which kill epithelial cells, cause dysfunction of the central nervous system and induce hypoglycaemia [52].

2.5.1.1.1.4   Urinary tract infections
Urinary tract infections are more common in females and involve the colonization of the urethra by organisms such as *E. coli* and *Proteus* spp., which are normal inhabitants of the colon [53]. In order to resist the flushing action of urine, one of the prime requirements of a urinary pathogen is an ability to adhere strongly to mucosal cells, and in *E. coli* this is mediated by pili. Many

uropathogenic strains of *E. coli* also produce an exotoxin which can kill host cells and stimulate an inflammatory reaction.

2.5.1.1.1.5   Peptic ulcers
It is now recognized that the main cause of gastric and duodenal ulceration is infection with *H. pylori* [54]. This Gram-negative, curved motile organism has been isolated from as many as 75% of patients with gastric ulcers and 95% with duodenal ulcers. It is able to penetrate the protective mucus layer and attach to mucus-secreting cells, where it secretes a urease which, because of the buffering power of the resulting ammonia liberated, enables it to survive at the low pH of the stomach, but also damages epithelial cells. It also secretes a vacuolating cytotoxin which plays a role in ulcer development.

2.5.1.1.2   Infections involving no exotoxin production
2.5.1.1.2.1   Mycoplasma pneumonia
Mycoplasma pneumoniae, an aerobic, pleomorphic bacterium without a cell wall, is one of the most common causes of pneumonia in children and adolescents. The organism adheres to epithelial cells of the upper respiratory tract, where it initially inhibits ciliary action, so enabling colonization of the lower regions of the tract, and then kills the cells.

2.5.1.1.2.2   Gonorrhoea
Neisseria gonorrhoeae, the causative organism of gonorrhoea, is a Gram-negative, aerobic, non-motile coccus which does not survive for long outside its host, so that transmission of the infection is invariably by direct person-to-person contact [55]. The bacteria adhere to columnar epithelial cells of the cervix and urethra by means of pili and induce a strong inflammatory response giving rise to a characteristic purulent discharge. This is presumably due to the induction of cytokines, although the bacterial constituents responsible have not yet been defined. Following firm attachment, bacteria are taken up by the epithelial cells and transported to the basement membrane. However, in only approximately 1% of all cases does the infection become systemic, resulting in arthritis and, less commonly, endocarditis.

2.5.1.2   Infections confined to the skin
2.5.1.2.1   Infections involving exotoxin production
2.5.1.2.1.1   Tetanus
Tetanus is caused by *Cl. tetani*, a Gram-positive, anaerobic, spore-forming, motile rod found in soil and faeces. The disease results from the contamination of minor puncture wounds, providing that conditions there are suitable for spore germination and toxin production [56]. The bacteria remain at the site of the wound, but the toxin binds to nerve terminal membranes, is internalized and then migrates to the central nervous system. Its principal effect is to block

the normal inhibition of spinal motor neurons, which results in characteristic muscular spasms.

2.5.1.2.1.2   *Cutaneous abscesses*
Cutaneous abscesses are invariably caused by *Staph. aureus*, a member of the normal skin microflora [57]. This organism produces a range of toxins and extracellular enzymes of which the most likely to be involved in abscess formation are: (i) the Panton–Valentine leukocidin and δ-haemolysin, which kill polymorphonuclear leucocytes; (ii) coagulase, which walls off the lesion and protects bacteria from phagocytes; and (iii) lipases, which release inflammatory fatty acids. The ability of *Staph. aureus* toxins to induce cytokine synthesis is discussed in Chapter 7.

2.5.1.2.1.3   *Impetigo*
Impetigo is a superficial skin infection, due to either *Staph. aureus* or *Strep. pyogenes*, which affects the epidermis just below the outer layers and results in the formation of blisters [58].

2.5.1.2.2   Infections involving no exotoxin production
2.5.1.2.2.1   *Acne*
Acne is caused by *P. acnes*, a Gram-positive, facultatively anaerobic, non-spore-forming, non-motile rod, which is a member of the normal skin microflora. These bacteria inhabit sebaceous glands and can hydrolyse the lipids present in sebum to produce fatty acids which induce the inflammation characteristic of the condition [59].

2.5.1.2.2.2   *Erythrasma*
Corynebacterium minutissimum infects the stratum corneum of moist areas of the skin, such as the toe webs and axillae, producing red scaly patches — a condition known as erythrasma [60].

2.5.2   Infections accompanied by tissue invasion
2.5.2.1   Infections involving exotoxin production
2.5.2.1.1   Dysentery
Bacterial dysentery can be caused by any of the four species of the genus *Shigella*, which are Gram-negative, facultatively anaerobic, non-motile rods. These bacteria are essentially human pathogens and are usually transmitted via contaminated food or water although, as the infectious dose is so low (a few hundred organisms), it can also be transmitted by person-to-person contact [61]. Following adhesion to mucosal cells of the colon (mediated by outer-membrane proteins), the bacteria are ingested and then spread to adjacent cells, killing these as they go. All four *Shigella* species secrete toxins, but *Sh. dysenteriae* also releases an intracellular Shiga toxin on lysis, which is enterotoxic,

neurotoxic and cytotoxic and is thought to be responsible for acute kidney failure which is often fatal.

2.5.2.1.2   Gangrene
Gangrene is an aggressive, often lethal, infection of muscle caused mainly by *Cl. perfringens*, a member of the normal intestinal microflora, following contamination of deep wounds or after abdominal surgery [62]. The bacteria germinate, multiply and spread into adjacent groups of muscle, releasing a number of potent toxins which results in extensive tissue destruction.

2.5.2.1.3   Pneumococcal meningitis
Strep. pneumoniae is found only in man and is present in the nasopharynx of up to one-third of the population. Little is known concerning the means by which the organism invades the epithelium and enters the bloodstream to reach the meninges. However, this may involve a toxin (pneumolysin) released when the organism lyses and which can kill host cells. The presence of several potent cytokine-inducing cell-wall components may account for the organism's ability to damage endothelial cells and so gain entry to the cerebrospinal fluid. However, it should be noted that pneumolysin is one of the most potent inducers of monocyte cytokine synthesis yet described.

2.5.2.2   Infections involving no exotoxin production
2.5.2.2.1   Tuberculosis
Tuberculosis originates in the lungs following inhalation of the causative organism, *M. tuberculosis*, a Gram-positive, aerobic, non-motile, non-spore-forming rod [63]. The organism is able to survive and multiply in alveolar macrophages, possibly because of the waxy layer on its cell wall, which then fuse to form giant cells. Eventually a walled-off lesion, known as a granuloma, is formed consisting of bacteria, macrophages and dead epithelial cells. This usually limits the primary infection, as most of the mycobacteria cannot tolerate the low oxygen content and pH at the centre of the granuloma. In some cases the organism spreads to infect the bones, joints, kidneys and meninges. Some organisms also survive in a dormant form and, when re-activated, can give rise to post-primary tuberculosis resulting in severe damage to the lungs and possibly to other regions of the respiratory tract, the intestinal tract, bladder and kidneys.

2.5.2.2.2   Meningitis due to *Haemophilus influenzae*
H. influenzae is a Gram-negative, facultatively anaerobic, non-motile, non-spore-forming rod which forms part of the normal microflora of the upper respiratory tract. Most of these normal residents are strains of low virulence, but colonization by serotype b strains can result in a number of serious infections, including meningitis [64]. Colonization of the nasopharynx may lead to a nasopharyngitis which can result in a bacteraemia. However, little is

known concerning how the organism then invades the meninges to cause meningitis.

2.5.2.2.3   Typhoid fever

This results from the ingestion of *Sal. typhi* in contaminated food or water [65]. The organisms invade the outermost cells (M cells) of the Peyer's patches and lymphoid follicles, multiply in the submucosal layers and then enter the bloodstream. They then multiply in the spleen and liver and re-enter the bloodstream and exit through the gut epithelium which becomes severely ulcerated. Most of the symptoms of typhoid fever are attributable to cytokine induction by LPS and other cell-wall components.

2.5.2.2.4   Brucellosis

Brucellosis is a zoonosis caused by *Brucella* spp. (mainly *Brucella abortus*) which normally infect domesticated animals [66]. The infection is transmitted via direct contact with infected milk, vaginal secretions or carcases. It is a chronic disease characterized by recurrent episodes of fever, sweating, aches and pains, and hence is also known as undulant fever. The bacteria can penetrate mucous membranes or skin abrasions and are then disseminated widely to become intracellular parasites of cells of the mononuclear-phagocyte system.

2.6   Concluding remarks

This chapter has provided a brief description of the composition of the normal bacterial flora of man and has given examples of the diseases caused by some of its constituents as well as those caused by exogenous species. It remains now to address the question of how bacteria can induce an infectious process against the determined efforts of the host to prevent this occurring. This ability is attributable to the possession by the organism of one or more features known as virulence factors (or determinants) and these are the subject of the next chapter.

References

1.   Herbelin, A., Ruuth, E., Delorme, D., Michel-Herbelin, C. and Praz, F. (1994) *Mycoplasma arginini* TUH-14 membrane lipoproteins induce production of interleukin-1, interleukin-6 and tumour necrosis factor alpha by human monocytes. Infect. Immun. **62**, 4690–4694

2.   Rawadi, G. and Roman-Roman, S. (1996) Mycoplasma membrane lipoproteins induced pro-inflammatory cytokines by a mechanism distinct from that of lipopolysaccharide. Infect. Immun. **64**, 637–643

3.   Poxton, I.R. (1993) Prokaryote envelope diversity. J. Appl. Bacteriol. **74**, 1S–11S

4.   Doyle, R.J. and Marquis, R.E. (1994) Elastic, flexible peptidoglycan and bacterial cell wall properties. Trends Microbiol. **2**, 57–60

5.   Dijkstra, A.J. and Wolfgang K. (1996) Peptidoglycan as a barrier to transenvelope transport. J. Bacteriol. **178**, 5555–5562

6.   Henderson, B., Poole, S. and Wilson, M. (1996) Bacterial modulins: a novel class of virulence factors which cause host tissue pathology by inducing cytokine synthesis. Microbiol. Rev. **60**, 316–341

7.   Baddiley, J. (1989) Bacterial cell walls and membranes. Discovery of the teichoic acids. Bioessays **10**, 207–210

8.   Heumann, D., Barras, C., Severin, A., Glauser, M.P. and Tomasz, A. (1994) Gram positive cell walls stimulate synthesis of tumour necrosis factor alpha and interleukin-6 by human monocytes. Infect. Immun. **62**, 2715–2721

9.   Tufano, M.A., Cipollaro de L'Ero, G., Ianniello, R., Galdiero, M. and Galdiero, F. (1991) Protein A and other surface components of *Staphylococcus aureus* stimulate production of IL-1α, IL-4, IL-6, TNF and IFN-γ. Eur. Cytokine Network **2**, 361–366

10. Nikaido, H. (1988) Structure and functions of the cell envelope of gram-negative bacteria. Rev. Infect. Dis. **10**, 279S–281S

11. de Maagd, R.A. and Lugtenberg, B.J. (1987) Outer membranes of gram-negative bacteria. Biochem. Soc. Trans. **15**, 54S–62S

12. Wilson, M. (1995) Biological activities of lipopolysaccharides from oral bacteria and their relevance to the pathogenesis of chronic periodontitis. Sci. Prog. (Oxford) **78**, 19–34

13. Manthey, C.L. and Vogel, S.N. (1994) Interactions of lipopolysaccharide with macrophages. Immunol. Ser. **60**, 63–81

14. Nikaido, H. (1992) Porins and specific channels of bacterial outer membranes. Mol. Microbiol. **6**, 435–442

15. Tufano, M.A., Rossano, F., Catalanotti, P., Ligouri, C., Capasso, C., Ceccarelli, T. and Marinelli, P. (1994) Immunobiological activities of *Helicobacter pylori* porins. Infect. Immun. **62**, 1392–1399

16. Hitchcock, P.J. and Morrison, D.C. (1984) The protein component of bacterial endotoxins. In Handbook of Endotoxin, vol. 1. Chemistry of Endotoxin (Rietschel, E.T., ed.), pp. 339–374, Elsevier/North-holland Publishing Co., Amsterdam

17. Reddi, K., Poole, S., Nair, S., Meghji, S., Henderson, B. and Wilson, M. (1995) Lipid A-associated proteins from periodontopathogenic bacteria induce interleukin-6 production by human gingival fibroblasts and monocytes. FEMS Immunol. Med. Microbiol. **11**, 137–144

18. Shinnick, T.M., King, C.H. and Quinn, F.D. (1995) Molecular biology, virulence, and pathogenicity of mycobacteria. Am. J. Med. Sci. **309**, 92–98

19. Barnes, P.F., Chatterjee, D., Abrams, J.S., Lu, S., Wang, E., Yamamura, M., Brennan, P.J. and Modlin, R.L. (1992) Cytokine production induced by *Mycobacterium tuberculosis* lipoarabinomannan: relationship to chemical structure. J. Immunol. **149**, 541–547

20. Jones, C.J. and Aizawa, S-I. (1991) Bacterial flagellum and flagellar motor: structure, assembly and function. Adv. Microbial Physiol. **32**, 110–172

21. Yamamoto, Y., Klein, T.W. and Friedman, H. (1996) Induction of cytokine granulocyte-macrophage colony-stimulating factor and chemokine macrophage inflammatory protein 2 mRNAs in macrophages by *Legionella pneumophila* or *Salmonella typhimurium* attachment requires different ligand-receptor systems. Infect. Immun. **64**, 3062–3068

22. DiMango, E., Zar, H.J., Bryan, R. and Prince, A. (1995) Diverse *Pseudomonas aeruginosa* gene products stimulate respiratory epithelial cells to produce interleukin-8. J. Clin. Invest. **96**, 2204–2210

23. Paranchych, W. and Frost, L.S. (1988) The physiology and biochemistry of pili. Adv. Microbial Physiol. **29**, 53–114

24. Kreft, B., Bohnet, S., Carstensen, O., Hacker, J. and Marre, R. (1993) Differential expression of interleukin-6, intracellular adhesion molecule I, and major histocompatibility complex class II molecules in renal carcinoma cells stimulated with S. fimbriae of uropathogenic *Escherichia coli*. Infect. Immun. **61**, 3060–3063

25. Handley, P.S. (1991) Negative Staining. In Microbial Cell Surface Analysis: Structural and Physicochemical Methods (Mozes, N., Handley, P.S., Busscher, H.J. and Rouxhet, P.G., eds.), pp. 63–86, VCH Publishers, New York

26. Costerton, J.W., Irvin, R.T. and Cheng, K.J. (1981) The bacterial glycocalyx in nature and disease. Annu. Rev. Microbiol. **35**, 299–324

27. Weiner, R., Langille, S. and Quintero, E. (1995) Structure, function and immunology of bacterial exopolysaccharides. J. Indust. Microbiol. **15**, 339–346

28. Otterlei, M., Sundan, A., Skjak-Braek, G., Ryan, L., Smidsrod, O. and Espevik, T. (1993) Similar mechanisms of action of defined polysaccharides and lipopolysaccharides: characterization of binding and tumor necrosis factor alpha induction. Infect. Immun. **61**, 1917–1935

29. Soell, M., Diab, M., Haan-Archipoff, G., Beretz, A., Herbelin, C., Poutrel B. and Klein, J. (1995) Capsular polysaccharide types 5 and 8 of *Staphylococcus aureus* bind specifically to human epithelial (KB) cells, endothelial cells, and monocytes and induce release of cytokines. Infect. Immun. **63**, 1380–1386

30. Hussain, M., Wilcox, M.H. and White, P.J. (1993) The slime of coagulase-negative staphylococci: biochemistry and relation to adherence. FEMS Microbiol. Rev. **10**, 191–207

31. Sleytr, U.B., Messner, P. and Sara, M. (1993) Crystalline bacterial cell surface layers. Mol. Microbiol. **10**, 911–916

32. Henderson, B. and Wilson, M. (1996) *Homo bacteriens* and a network of surprises. J. Med. Microbiol. **45**, 1–2

33. Yagupsky, P. (1993) Bacteriologic aspects of skin and soft tissue infections. Pediatric Annals **22**, 217–224

34. Aniansson, G., Alm, B., Andersson, B., Larsson, P., Nylen, O., Peterson, H., Rigner, P., Svanborg, M. and Svanborg, C. (1992) Nasopharyngeal colonisation during the first year of life. J. Infect. Dis. **165**, S38–S42

35. Stoll, P., Huber, H., Pelz, K. and Weingart, D. (1993) Antimicrobial effects of the tetrachlorodecaoxygen-anion complex on oropharyngeal bacterial flora: an in vitro study. Chemotherapy **39**, 40–47

36. Berg, R.D. (1996) The indigenous gastrointestinal microflora. Trends Microbiol. **4**, 430–435

37. Domingue, P.A., Sadhu, K., Costerton, J.W., Bartlett, K. and Chow, A.W. (1991) The human vagina: normal flora considered as an *in situ* tissue-associated, adherent biofilm. Genitourinary Med. **67**, 226–231

38. Jensen, E.T., Kharazmi, A., Lam, K. and Costerton, J.W. (1990) Human polymorphonuclear leukocyte response to *Pseudomonas aeruginosa* biofilms. Infect. Immun. **58**, 2383–2385

39. Anwar, H., Strap, J.L. and Costerton, J.W. (1992) Susceptibility of biofilm cells of *Pseudomonas aeruginosa* to bactericidal actions of whole blood and serum. FEMS Microbiol. Lett. **92**, 235–242

40. Jansen, B. and Peters, G. (1993) Foreign body associated infections. J. Antimicrob. Chemother. **32**, 69–75

41. Pittet, D., Hulliger, S. and Auckenthaler, R. (1995) Intravascular device-related infections in critically ill patients. J. Chemother. **7**, 55–66

42. Vincent, M.T. and Goldman, B.S. (1994) Anaerobic lung infections. Am. Family Phys. **49**, 1815–1820

43. Mier, L., Dreyfuss, D., Darchy, B., Lanore, J.J., Djedaini, K., Weber, P., Brun, P. and Coste, F. (1993) Is penicillin G an adequate initial treatment for aspiration pneumonia? A prospective evaluation using a protected specimen brush and quantitative cultures. Intens. Care Med. **19**, 279–284

44. Cohen, J. (1996) Infection in the immunocompromised host. In Oxford Textbook of Medicine, 3rd edn., vol. 1 (Weatherall, D.J., Ledingham, J.G.G. and Warrell, D.A., eds.), pp. 1027–1035, Oxford University Press, Oxford

45. Gradon, J.D., Timpone, J.G. and Schnittman, S.M. (1992) Emergence of unusual opportunistic pathogens in AIDS: a review. Clin. Infect. Dis. **15**, 134–157

46. Berger, B.J., Hussain, F. and Roistacher, K. (1994) Bacterial infections in HIV-infected patients. Infect. Dis. Clin. N. Am. **8**, 449–465

47. Takala, A.K., Meurman, O., Kleemola, M., Kela, E., Ronnberg, P.R., Eskola, J. and Makela, P.H. (1993) Preceding respiratory infection predisposing for primary and secondary invasive *Haemophilus influenzae* type b disease. Pediatr. Infect. Dis. J. **12**, 189–195

48. Fekety, R. (1995) Antibiotic-associated diarrhoea and colitis. Cur. Opin. Infect. Dis. **8**, 391–397

49. Marsh, P.D. (1994) Microbial ecology of dental plaque and its significance in health and disease. Adv. Dent. Res. **8**, 263–271

50. Galazka, A.M., Robertson, S.E. and Oblapenko, G.P. (1995) Resurgence of diphtheria. Eur. J. Epidemiol. **11**, 95–105

51. Kaper, J.B., Morris, J.G. and Levine, M.M. (1995) Cholera. Clin. Microbiol. Rev. **8**, 48–86

52. Mathis, R.D., Shoaf, B. and Weiner, T.I. (1993) Pertussis: the return of a bad penny. Pediatr. Emerg. Care **9**, 218–220

53. Ang, B.S. (1995) Urinary tract infections. Singapore Med. J. **36**, 314–317

54. Peura, D.A. (1996) *Helicobacter pylori* and ulcerogenesis. Am. J. Med. **100**, 19S–25S

55. Judson, F.N. (1990) Gonorrhea. Med. Clin. N. Am. **74**, 1353–1366

56. Roos, K.L. (1991) Tetanus. Seminars Neurol. **11**, 206–214

57. Williams, R.E. and MacKie, R.M. (1993) The staphylococci. Importance of their control in the management of skin disease. Dermatol. Clinics **11**, 201–206

58. Darmstadt, G.L. and Lane, A.T. (1994) Impetigo: an overview. Ped. Dermatol. **11**, 293–303

59. Leung, A.K. and Robson, W.L. (1991) Acne J. Royal Soc. Health **111**, 57–60

60. Golledge, C.L. and Phillips, G. (1991) *Corynebacterium minutissimum* infection. J. Infect. **23**, 73–76

61. Gilman, R.H. (1984) Bacillary dysentery. Comp. Ther. **10**, 14–19

62. Canoso, J.J. and Barza, M. (1993) Soft tissue infections. Rheum. Dis. Clinics N. Am. **19**, 293–309

63. Phelan, J.A., Jimenez, V. and Tompkins, D.C. (1996) Tuberculosis. Dent. Clin. N. Am. **40**, 327–341

64. Kornelisse, R.F., de Groot, R. and Neijens, H.J. (1995) Bacterial meningitis: mechanisms of disease and therapy. Eur. J. Pediatr. **154**, 85–96

65. Cvjetanovic, B. (1986) Typhoid and its control. J. Diarr. Dis. Res. **4**, 139–143

66. Young, E.J. (1995) An overview of human brucellosis. Clin. Infect. Dis. **21**, 283–289

3

Bacterial virulence factors

"...the calamity was spread by infection, that is to say, by some certain steams, or fumes, which the physicians call effluvia, by the breath, or by the sweat, or by the stench of the sores of the sick persons, or some other way, perhaps, beyond even the reach of the physicians themselves, which effluvia affected the sound, who come within certain distances of the sick, immediately penetrating the vital parts of the said sound persons, putting their blood into an immediate ferment, and agitating their spirits to that degree which it was found they were agitated; and so those newly infected persons communicated it in the same manner to others."

A Journal of the Plague Year
Daniel Defoe, 1722

In any infectious disease, several stages can be recognized from the moment the host comes into contact with the bacterium until the adverse consequences for the host become apparent. First, the organism adheres to the host and this is followed, in some cases, by invasion of the underlying tissues. The organism then grows and multiplies and, at the same time, protects itself from the host defence strategies that will have been activated. Many of these will be cytokine-driven. This is followed by some host tissue destruction, due to the effect of the bacterium itself, or as a consequence of collateral damage by host defence mechanisms. Those attributes of a bacterium which enable it to accomplish these stages are termed virulence factors and these have traditionally been grouped into the following classes: adhesins (responsible for adhesion of the organism to host tissues); invasins (responsible for tissue invasion); impedins (enabling the organism to overcome host defences); and aggressins (inducing damage to host tissues). Recently, we have proposed that a further class of virulence factor should be recognized: namely, modulins. These are bacterial components or products which affect cytokine synthesis (see Chapters 6 and 7 for further details). The development of molecular genetic approaches during the past two decades has enabled considerable progress in the identification and characterization of bacterial virulence factors [1], and their biology will be addressed in this chapter.

3.1   Adhesion

The first stage in a bacterial infection involves the organism adhering to some body surface. This may be one of a variety of host cells, the surface of a tooth or the surface of an already adherent microbe. Bacteria unable to adhere will be removed by the natural cleansing mechanisms operating at all body surfaces. Subsequently, bacteria may detach themselves from the colonized surface and move (or be transported) to another location within the host, or be disseminated from the host to infect a new individual. The means by which bacteria adhere to, or detach themselves from, a surface are therefore of funda-mental importance in any understanding of the sequence of events leading to the development of an infectious disease. Furthermore, knowledge of these mechanisms could also lead to the development of new anti-infective strategies based on the ability to interfere with bacterial adhesion.

As it is generally held that the surfaces of all prokaryotic and eukaryotic cells, as well as naturally occurring substrata, are negatively charged, bacteria must first of all overcome the resultant strong repulsive forces when attempting to adhere to a surface. Regardless of the nature of the substratum, bacterial adherence is dependent on the ability of a limited range of physico-chemical forces to counteract these repulsive forces, including ionic interactions, hydrogen bonding, hydrophobic forces and metal ion-mediated complexes [2]. Whichever of these are employed, adherence of the bacterium to the substratum is mediated by the interaction between a molecule (an adhesin) on the bacterial surface and a complementary molecule (a receptor) on the substratum. There can be no doubt that these interactions are highly specific given the remarkable ability of bacteria to: (i) colonize cells of hosts susceptible to a particular infectious agent, but not to similar types of cells in animals unsusceptible to the infection [3]; and (ii) exhibit tissue tropism, i.e. adhere to certain tissues within a particular host [4]. Such species and tissue tropisms, however, cannot be explained merely in terms of the existence of a particular adhesin on an organism and a corresponding receptor on the eukaryotic cell. Other factors, particularly environmental considerations, must be involved.

3.1.1   Bacterial adhesins
3.1.1.1 Lectins
One of the most extensively studied type of bacterial adhesion mechanism involves the interaction between a bacterial lectin (a protein capable of binding specifically to carbohydrate-containing molecules) and its corresponding receptor on a substratum [5]. Lectins may be found on the end of pili, in capsules or attached to the bacterial cell wall, and have been studied mainly in

Table 3.1 Examples of bacteria that contain lectin-binding proteins and the identity of the corresponding carbohydrate receptor in the host tissue

Organism	Receptor	Reference
Salmonella typhimurium	Galβ(1→3)GalNAc	[6]
Escherichia coli	N-Acetyl-D-glucosamine	[7]
Fusobacterium nucleatum	Galactose	[8]
Pseudomonas aeruginosa	GalNAcβ(1→4)Gal	[9]
Pseudomonas aeruginosa	Gal-β(1→4)GlcNAc	[10]
Neisseria gonorrhoeae	GalNAcβ(1→4)Gal	[11]
Helicobacter pylori	NeuNAcα(2→3)Galβ	[12]
Escherichia coli	Galβ(1→4)Glcβ	[13]
Klebsiella pneumoniae	N-Acetylneuraminic acid and N-acetyl-D-glucosamine	[14]
Pseudomonas aeruginosa	Thiogalactosides	[15]

Gram-negative bacteria. Table 3.1 lists some bacteria in which lectins have been identified, together with the identity of the carbohydrate to which the lectin binds.

Lectin–carbohydrate interactions have been shown to be involved in the adhesion of bacteria to intestinal epithelial cells [16], pharyngeal epithelial cells [17], buccal epithelial cells [18], erythrocytes [19], urinary tract epithelial cells [20], other bacteria [21] and teeth [22].

3.1.1.2 Lipoteichoic acid
This cell-wall component of Gram-positive bacteria has been shown to be an important adhesin for a number of species, including *Streptococcus pyogenes* [23], *Staphylococcus aureus* [24], *Staphylococcus epidermidis* [25] and viridans streptococci [26]. In each of these organisms, the receptor for the LTA has been shown to be fibronectin, a glycoprotein produced by a number of host cells including epithelial cells [27,28].

3.1.1.3 Proteins
Staph. aureus has been shown to produce a 210 kDa surface protein which mediates adhesion of the organism to fibronectin [29]. The organism also binds to a number of other host proteins including fibrinogen, mediated by a 59 kDa protein [30], and laminin, mediated by a 57 kDa protein [31]. *Streptococcus pneumoniae* adheres to glycoproteins of the nasopharyngeal epithelium and this is thought to be mediated by a 37 kDa protein (psaA) [32]. However, it is known that antibodies reacting with carbohydrates comprising the capsule of the organism can interfere with adhesion. It may be that binding of antibodies to the capsule results in it 'swelling' so that the psaA no longer protrudes from the capsule, thus masking the organism's adhesin [33].

3.1.1.4 Hydrophobins

The hydrophobicity of the bacterial cell surface is thought to be an important determinant of adhesion to surfaces, although the identity of the molecules responsible for the hydrophobic character of the surfaces (hydrophobins) have, generally, not been identified [34]. The involvement of hydrophobic interactions in adhesion has been studied most extensively in oral bacteria, mainly from the point of view of the adhesion of streptococci to tooth surfaces [35]. In general, the more hydrophobic an organism the greater its ability to adhere to saliva-coated hydroxyapatite. Adhesion of *Porphyromonas gingivalis* to collagen is also thought to be mediated by hydrophobic interactions [36]. There is also evidence suggesting that adhesion of staphylococci to solid surfaces, such as those used in prosthetic devices, is mediated by hydrophobic interactions [37].

3.1.1.5 Carbohydrates

Alginate, an exopolysaccharide of *Pseudomonas aeruginosa*, appears to play a role in adhesion of the organism to tracheal cells and mucins as it binds to both buccal and tracheal cells and to bronchotracheal mucin. Furthermore, antibodies to the alginate inhibit binding of the organism to tracheal cells [38,39].

Streptococcus oralis, a major constituent of dental plaque, adheres to the tooth surface by means of a polysaccharide composed of hexasaccharide-repeating units containing glycerol linked through a phosphodiester to C-6 of an α-galactopyranosyl residue and joined end-to-end through galactofuran-osyl-β(1→3)-rhamnopyranosyl linkages [40].

Many infections due to *Staph. epidermidis* are associated with prosthetic devices and adhesion to such materials has been shown to be mediated by surface polysaccharides [41,42]. Biofilm formation on such devices has also been shown to involve the production of an intercellular polysaccharide adhesin [43].

3.1.1.6 Lipopolysaccharide

Several studies have implicated LPS as being important in the adhesion of *Campylobacter jejuni* to epithelial cells. Strains which are highly negatively charged and only weakly hydrophobic bind in greater number to a human intestinal cell line than strains that are more hydrophobic [44]. A whole-cell lysate of the organism partially inhibited binding to Hep-2 cells and this inhibition was unaffected by heating the lysate to 100°C for 30 min or by treating it with proteases. However, treatment of the lysate with periodate abolished the inhibitory activity [45]. Adhesion of the organism to intestinal epithelial cells was partially inhibited by fucose and mannose and completely inhibited by LPS from the organism. The binding of *C. jejuni* LPS to the cells was inhibited by fucose and by treating the LPS with periodate. Other organisms in which LPS is thought to function as an adhesin include

Table 3.2 Bacterial adhesins and the cells/tissues to which they bind

Organism	Target cell/tissue	Bacterial adhesin/structure responsible for adhesion
Bacteroides fragilis	Epithelial cells	Fimbriae
Campylobacter jejuni	M cells	OMP
Pseudomonas aeruginosa	Tracheal cells	Alginate
	Erythrocytes	Hydrophobins
	Buccal epithelial cells	Fimbriae
Campylobacter jejuni	Intestinal epithelial cells	LPS, OMPs
Streptococcus mitis	Oral epithelial cells	Surface protein
Streptococcus salivarius	Oral epithelial cells, hydroxyapatite	Fibrillar protein
Streptococcus pyogenes	Epithelial cells	LTA
Lactbacillus fermentum	Intestinal epithelial cells	LTA
Neisseria gonorrhoeae	Urethral epithelium	Pili, OMPs
Vibrio cholerae	Intestinal epithelium	Haemagglutinin, LPS
Salmonella typhimurium	Intestinal epithelium	Fimbriae
Actinomyces viscosus	Hydroxyapatite	Fimbriae
Haemophilus influenzae	Mucins	Fimbriae
Chlamydiae	Conjunctival epithelium	Glycosaminoglycan
Mycoplasma pneumoniae	Respiratory epithelium	Surface protein
Neisseria meningitidis	Nasopharyngeal epithelium	Pili
Pseudomonas aeruginosa	Laminin	OMPs
Escherichia coli	Intestinal epithelium	Pili

Helicobacter pylori, Pseudomonas aeruginosa, Salmonella typhi, Shigella flexneri and *Escherichia coli* [46].

In Table 3.2 representative examples are given of the bacterial adhesin responsible for attachment of organisms to a particular substratum.

3.1.2 Adhesion mechanisms

Although there is considerable evidence to show that bacterial adhesion to host cells involves specific interactions between an adhesin and a receptor, much less is known concerning the actual mechanism involved during the adhesion process. One complicating factor is that many organisms contain a number of adhesins with different receptor specificities, suggesting that adhesion may be a sequential process involving more than one adhesin. For example, at least 17 different adhesins have been identified in *E. coli* [47]. Alternatively, the organism may express only one of these adhesins under the environmental conditions existing when initial contact with the target cell takes place. Subsequent release of the bacterium into underlying tissues may trigger the expression of an adhesin with a different specificity, enabling the targeting of a different type of host cell. Unfortunately, little is known concerning the factors regulating the expression of adhesins by bacteria.

As most infectious diseases involve adhesion to mucosal surfaces, the adhesion of a non-invasive pathogen, *Bordetella pertussis*, will be described in detail to illustrate the adhesive processes involved in a typical infection, with particular emphasis on the multiple adhesins displayed by this organism and their regulation.

Bord. pertussis produces a number of adhesins that can mediate binding to ciliated and non-ciliated cells and to macrophages: a haemagglutinin, pertussis toxin, pertactin and two types of pili [48]. The most extensively studied of these are: (i) the haemagglutinin which forms 12 nm thick filaments (filamentous haemagglutinin; FHA), can exist as a microcapsule and is secreted in a soluble form; and (ii) pertussis toxin (PT), which is mainly secreted into the growth medium, although small amounts do remain associated with the cell wall [49,50]. FHA is a 220 kDa protein with domains able to mediate attachment to a variety of host-cell receptors including galactose-*N*-acetylglucosamine (present in lactosylceramide and non-sialylated glycolipids in the membranes of ciliated cells), heparin, sulphated polysaccharides (enabling binding to erythrocytes) and leucocyte integrin CR3 (enabling binding to macrophages) [51,52].

PT is a 105 kDa hexamer with two of the subunits (S2 and S3) being involved in adhesion. S2 (22 kDa) binds to lactosylceramide which is present in ciliated epithelial cells, while S3 (22 kDa) binds to α2,6-linked sialic acid found in phagocytes [53,54].

The importance of FHA in adhesion has been demonstrated in a number of ways. First, anti-FHA antibodies inhibited adhesion of the organisms to cells in tissue culture [55]. Secondly, mutants unable to produce FHA could not adhere to ciliated human tracheal cells [56] but could do so in the presence of exogenous FHA [57]. Mutants defective in the production of PT behaved in a similar manner, i.e. their ability to adhere to ciliated epithelial cells was restored by the addition of PT. The importance of both adhesins in achieving colonization has also been demonstrated *in vivo* as mutants lacking both FHA and PT failed to adhere to the respiratory epithelium of rabbits [58]. The adhesin containing an Arg-Gly-Asp sequence (pertactin, a 69 kDa protein) mediates binding to Chinese hamster ovary cells, which is blocked by synthetic peptides containing an Arg-Gly-Asp sequence [59]. It also enables binding of the organism to leucocytes via a protein–protein interaction with the integrin CD11b/CD18 (CR3, αmβ2) [60]. This would enable the adhesion of the organism to alveolar macrophages, thereby facilitating colonization of the lung. In an animal model of whooping cough, mutation of the Arg-Gly-Asp domain of the FHA resulted in a 1000-fold reduction in the number of *Bord. pertussis* colonizing the lungs [61]. *Bord. pertussis*, therefore, has a number of adhesins which could account for the sequence of events leading from initial contact with the host in the upper respiratory tract to final colonization of the lung. One adhesin may permit attachment to mucin and heparin-containing glycosaminoglycans and proteoglycans (via the heparin-binding domain on the FHA), FHA and PT could then mediate binding to ciliated epithelium, and finally the integrin-binding domain on FHA could permit subsequent colonization of the lung by enabling adhesion to alveolar macrophages. Some progress has been made in understanding the regulation of a number of these adhesins. Expression of the genes for FHA (*fha*) and PT (*ptx*) is favoured by low concentrations of Mg^{2+} and nicotinic acid and a temperature of 37°C. Activation of these genes is under the control of a locus consisting of two genes, *bvgA* and *bvgS*. The protein encoded by *bvgS* is a histidine kinase which responds to environmental signals and phosphorylates the *bvgA*-encoded protein. This, in turn, acts as the transcriptional activator of the genes responsible for the synthesis (*ptxS1–5*) and secretion (*ptsA* and *ptsB*) of PT.

3.1.3   Post-adhesion events
Apart from enabling an organism to maintain itself on host surfaces, adhesion has other consequences, both for the bacterium and its host. A considerable number of studies have shown that adhesion of a bacterium to an inanimate substratum, resulting in the formation of a biofilm, has profound effects on the organism in terms of its structure, growth, metabolism and susceptibility to antimicrobial agents [62,63]. However, epithelial surfaces generally cannot support dense biofilms of this type (a more likely consequence is the formation of a monofilm) and less is known about the effects on bacteria of adhesion to host cells. Nevertheless, evidence suggests that adhesion-induced changes in bacteria do occur. For example, Finlay et al. [64] have reported that a number of new proteins were induced in *Salmonella typhimurium* following its adhesion to an epithelial cell line. It has also been shown that *Neisseria gonorrhoeae* attached to HeLa cells grow at a faster rate than unattached cells [65]. This is an important, but neglected, area of research.

Adhesion of bacteria to host cells is a prelude to a whole series of events leading to the development of disease pathology. But does adhesion *per se* have any direct pathological consequences for the host in terms of cytokine production? Adhesion of *Sal. typhimurium* to human intestinal epithelial cells has been shown to stimulate the release of the potent chemoattractant, IL-8, and to induce the transepithelial migration of human PMNs. Such migration was found not to be due to the classic formyl-peptide-induced directed migration pathway, and neutralization of the IL-8 did not block PMN transmigration. This led to the suggestion that a novel transcellular chemotactic factor is induced by interaction of bacteria with epithelial cells [66].

Yamamoto et al. [67] have also reported that binding of *Legionella pneumophila* to the surface of murine macrophages (treated with cytochalasin D to prevent phagocytosis) resulted in increased levels of mRNA for IL-1α, IL-1β, IL-6, TNFα and GM-CSF, but not for interferon (IFN)β. Interestingly, at least two ligand–receptor systems appear to be involved in cytokine induction, one being α-methyl-D-mannoside (αMM)-dependent, and the other αMM-independent. Hence, induction of macrophage mRNAs for IL-1β, IL-6 and GM-CSF was inhibited by αMM, while the levels of mRNA for the chemokines macrophage inflammatory protein (MIP)-1β and MIP-2 were unaffected [68]. That induction of mRNA for different cytokines was dependent on activation of different ligand–receptor systems was supported by studies of the signalling pathways involved. Induction of mRNA for GM-CSF, but not for MIP-2, was found to involve calmodulin and myosin light-chain kinase. Stimulation by oral viridans streptococci of the release of IL-8 from epithelial cells and IL-6 and IL-8 from endothelial cells is also thought to be mediated by interactions between protein (a cell-surface antigen designated I/IIf) and carbohydrate (a rhamnose–glucose polymer) adhesins and host-cell receptors [69]. The protein-induced cytokine release from endothelial cells was inhibited by N-acetylneuraminic acid and fucose, suggesting the involvement of an adhesin–lectin interaction between the bacteria and host cells. Intravesicular inoculation of mice with E. coli strains which have P fimbriae results in the release of IL-6 in urine [70]. Cytokine release was dependent on the presence of the Pap G adhesin on the fimbriae (which is specific for Galα1,4Gal residues), as no IL-6 was released by strains bearing S-fimbriae or type 1 fimbriae. Glucosphingolipids, rather than glycoproteins, were shown to be the receptors for the adhesin, as inhibition of the synthesis of these compounds resulted in a reduction in bacterial adherence to the cells and a reduction in IL-6 release [71]. Deliberate colonization of the human urinary tract with E. coli has been shown to lead to an increase in urinary IL-6 levels (but not levels of circulating IL-6) within 1 h, and patients with natural episodes of bacteriuria caused by E. coli were found to secrete IL-6 into their urine [72].

Although there have been relatively few reports of cytokine release due to the adhesion process, the ability of adhesins (or components mediating adhesion) to induce cytokine release is well recognized. Hence, as will be described in Chapters 6 and 7, LPS, LTA, OMPs, surface polysaccharides, fimbriae and fibrils have all been shown to stimulate cytokine release from a variety of host cells.

The cytokines released following adhesion have been shown to have a marked effect on the expression of host-cell receptors for bacterial adhesins. For example, pneumocytes and vascular endothelial cells express two types of receptor for Strep. pneumoniae, but when stimulated by TNF or IL-1, receptors with a different specificity are expressed [52]. In order to remain adherent, therefore, bacteria must be able to produce a new adhesin. Cytokines produced at mucosal surfaces can also trigger the release of antimicrobial peptides [73], while cytokine overproduction may have pathological consequences for the host.

3.2   Invasion

Following adhesion to a host cell, the next step in the pathogenesis of many bacterial infections involves invasion of that cell. Although invasion is usually associated with entry into epithelial cells, entry into phagocytes will also be considered in this section when such entry is accompanied by survival of the organism rather than its destruction. Invasion involves host-cell cytoskeletal rearrangements which are induced following adhesion of the organism. Two general mechanisms appear to operate which differ depending on whether or not adhesion to the host cells is mediated by integrins. Examples of organisms utilizing the former mechanism include *Yersinia* spp., *Bordetella* spp. and *Leg. pneumophila*, while the latter is exemplified by *Salmonella* spp., *Listeria monocytogenes* and *Shigella* spp.

3.2.1   Invasion initiated by integrin binding

Yersinia spp. have an outer-membrane protein, invasin, which binds to β1 integrins on host cells and is thought to be responsible for invasion of these cells [74,75]. Invasion appears to be dependent on there being large numbers of integrin receptors as well as a high-affinity interaction between the adhesin and its receptor. Hence, invasion is less efficient when the bacteria are coated with low-affinity ligands for the α5β1 integrin and more efficient when the integrin is overexpressed [76]. Another adhesin of this organism, YadA, has also been shown to mediate invasion of epithelial cells via integrins [77]. Exactly how binding to integrins initiates bacterial uptake is not known. Conformational changes in the receptor for the adhesin may initiate transmembrane signalling events responsible for bacterial entry. In support of this, it has been reported that tyrosine kinase inhibitors inhibit invasin-mediated uptake by epithelial cells [78].

3.2.2   Invasion not mediated by integrin binding

Considerably more is known about invasion of epithelial cells by bacteria that do not bind to integrins on the host cell. In the case of many enteric pathogens, entry is via the brush border, rather than through tight junctions, by a process resembling phagocytosis. This involves cytoskeletal rearrangements of columnar epithelial cells, resulting in the formation of extrusions (membrane ruffling) which engulf and internalize the bacteria [79]. Actin, α-actinin, talin, tubulin, tropomyosin and ezrin localize near the adhesion site of the bacterium shortly after adhesion of the bacteria and cytochalasin blocks bacterial uptake showing the need for functional actin filaments. Several studies have shed light on the means by which *Sal. typhimurium* induces such cytoskeletal rearrangements [80,81]. An early stage in the process is activation of the receptor for epidermal growth factor (EGFR) and this initiates the cascade of reactions that occurs when the receptor is activated by its usual ligand (see Chapter 4). This results in the phosphorylation of phospholipase A_2, which stimulates the release of arachidonic acid from membrane phospholipids. The arachidonic acid is converted into leukotrienes, which open Ca^{2+} channels in the membrane resulting in an increase in the level of Ca^{2+} ions. This, in turn, activates actin-depolymerizing proteins, which could result in depolymerization of actin filaments of the microvilli, hence releasing monomers to be used in membrane ruffling. Phospholipase A_2 also stimulates the release from the host-cell membrane of profilin, which is also involved in actin reorganization.

Once the bacteria have been internalized they are able to multiply within the vesicle and the latter can fuse to form larger vesicles containing many cells. Invasion by *Sal. typhimurium* appears to be controlled by a number of genetic loci and it has been estimated that as much as 50 kb of DNA is required for entry into epithelial cells [82,83]. One of these loci, the *inv* locus, consists of 14 contiguous genes (*invA–invO*, but not *invB*), and mutations in these genes result in an inability of the bacterium to invade cells but not to attach to them. Expression of genes in this locus is regulated by DNA superhelicity, which is affected by oxygen tension and osmolarity [84,85].

Invasion of epithelial cells by *N. gonorrhoeae* is mediated by an outer-membrane protein (opacity factor; Opa) which induces a phagocytic-like uptake of the bacteria [86]. Following adherence of the bacteria (via Opa) to the epithelial cell, F-actin accumulates at the site of adhesion. While invasion was unaffected by monodansylcadaverine (an inhibitor of clathrin-mediated endocytosis) or taxol (a microtubule-stabilizing agent), it was completely blocked by cytochalasin D.

Although considerable progress has been made in determining the mechanisms involved in the invasion of epithelial cells by bacteria, far less is known about the means by which bacteria invade other tissues. For example, despite the enormous interest in the pathogenesis of bacterial meningitis, little is known about how bacteria cross the blood–brain barrier. However, it has been shown recently [87] that invasion of brain microvascular endothelial cells by *E. coli* is mediated by a 35 kDa outer-membrane protein (OmpA). Although the invasion mechanism has not yet been established, it has been shown that the OmpA binds to GlcNAcβ1-4GlcNAc epitopes on the

Table 3.3 Bacterial invasins and their target cells

Organism	Target cell	Invasin
Bordetella pertussis	Macrophages	Filamentous haemagglutinin
Shigella spp.	M cells of Peyer's patches	Invasion plasmid antigen (Ipa)B, IpaC
Listeria monocytogenes	M cells or intestinal crypt cells	Internalin
Legionella pneumophila	Macrophages	Macrophage infectivity potentiator
Yersinia spp.	Intestinal epithelial cells	Invasin
Yersinia pseudotuberculosis	Epithelial cells	YadA

endothelial cell surface and that invasion can be blocked by wheat germ agglutinin and chitin derivatives [88].

Table 3.3 lists the invasins which have been identified in a number of important pathogens.

3.2.3   Invasion and cytokine synthesis

A number of studies have shown that invasion of host cells by bacteria is accompanied by cytokine synthesis. For example, invasion of fibroblasts by *Sh. flexneri* induced the production of IFNβ, while isogenic variants, which were not invasive, did not. This suggests that bacterial uptake may require interferon synthesis [89]. Similarly Eckmann et al. [90] concluded, from their studies of the invasion of epithelial cells by *Salmonella* spp. and *L. monocytogenes,* that the process of cell entry, and not simply the presence of the bacteria, was the signal inducing invaded cells to synthesize and secrete IL-8. As epithelial cells are the first site of entry of bacteria, the activation of IL-8 synthesis would act as an early warning system at a time when bacterial products are unavailable to stimulate circulating leucocytes. Invasion of human colon epithelial cells by a number of invasive strains of bacteria is accompanied by secretion of the same four pro-inflammatory cytokines [IL-8, monocyte chemotactic protein (MCP)-1, GM-CSF and TNFα], whereas non-invasive strains did not affect the levels of these cytokines [91]. The cells did not express mRNAs for IL-2, IL-4, IL-5, IL-6, IL-12p40 or IFNγ in response to these bacteria. and IL-1 and IL-10 levels were not increased. As IL-8, MCP-1, GM-CSF and TNFα are all involved, directly or indirectly, in chemotaxis and activation of inflammatory cells, the up-regulation of these cytokines in epithelial cells would serve to activate the mucosal inflammatory response to deal with the invading organisms.

It has also been reported that the invasin protein of *Yersinia pestis* interacts with β$_1$ integrins and acts as a co-stimulatory signal for the proliferation of CD4 lymphocytes, a process accompanied by the synthesis of TNFα and IFNγ [92]. Endothelial cells also respond to bacterial invasion by releasing cytokines [93]. Internalization of *Staph. aureus* by human umbilical vein endothelial cells resulted in the synthesis of mRNA for IL-6 after 3 h and then IL-1β mRNA after 12 h. Treatment of the endothelial cells with cytochalasin D, an inhibitor of endocytosis but not of bacterial adhesion, inhibited IL-6 and IL-1β gene expression, demonstrating that cytokine induction was a consequence of bacterial invasion rather than adhesion.

3.3   Growth *in vivo*

Once a bacterium has established itself on or in its host, its persistence there will depend on its subsequent ability to grow and multiply. The nutrients available to the organism and the physicochemical conditions prevailing will depend on the particular site of colonization or invasion, and a wide range of such environments can be envisaged, e.g. the bladder, the stomach, the surface of a tooth, interstitial spaces, the bloodstream, the interior of a macrophage, a sebaceous gland, etc. Each of these will offer a distinct set of environmental conditions to the organism under which it will have to grow and reproduce. Although innumerable studies have determined the nutritional and environmental factors governing growth of bacteria *in vitro,* there are surprisingly few equivalent studies *in vivo* [94]. *In vitro,* bacteria are invariably cultured under conditions which would rarely be encountered *in vivo.* Hence, they have excess nutrients, optimal physico-chemical conditions (pH, osmolarity, redox potential), and competing organisms are absent. All of these are known to influence bacterial growth and the expression of virulence factors. Although it should be possible to determine the nutrient availability and the conditions prevailing at a particular site, and to reproduce these in the laboratory, studies of this type are few and far between. We remain, therefore, remarkably ignorant of the nature of the nutrients actually used by bacteria when growing *in vivo,* and how these nutrients are obtained from host cells or secretions. Nevertheless, the tissue tropism displayed by bacteria has been shown to be due, in part, to the presence of particular secretory products of the host cells which are essential growth requirements for, or stimulate the growth of, certain bacteria. For example, in bovine infectious abortion, the causative organism, *Brucella abortus,* localizes in the placenta and the chorion due to the presence of erythritol which stimulates the growth of the organism. Temperature is also believed to restrict infections such as leprosy to cooler areas of the body (e.g. skin, nasal mucosa, testicles), as the optimum temperature for growth of *Mycobacterium leprae,* the causative organism, is less than 37°C. More recently, the environment inside a vacuole within an epithelial cell has been determined using *lacZ* transcriptional fusions made to genes regulated by Fe^{2+}, Mg^{2+} and pH in *Sal. typhimurium* [95]. The vacuoles were found to be slightly acidic and to contain low levels of Fe^{2+} and Mg^{2+}.

One of the few areas in this field which has been extensively studied concerns the effect of iron on bacterial growth and virulence and the competition between bacteria and host cells for this essential nutrient [96,97]. Iron is required by most bacteria for the activation of ribonucleotide reductase, aconitase and a number of enzymes involved in electron transfer and oxygen metabolism. It is also involved in the regulation of the synthesis of many bacterial toxins [98]. An important means by which the host tries to control the growth of pathogenic bacteria is by limiting the availability of free iron through the formation of iron complexes with proteins such as transferrin (in plasma, cerebrospinal fluid, perspiration), lactoferrin (in tears, saliva, bronchial mucus, seminal fluid, milk) and ferritin. As well as these constitutive mechanisms for the withholding of iron, a number of other systems come into operation during an infectious process, including: (i) the suppression of assimilation of dietary iron; (ii) increased synthesis of ferritin; (iii) binding of iron by apolactoferrin from neutrophils; and (iv) release of haptoglobin and haemopexin to bind haemoglobin and haemin [99]. The first two of the above processes are activated by IL-1, IL-6 and TNFα. Bacteria, of course, have evolved means of circumventing these defence mechanisms and these are discussed later.

3.4   Overcoming host defence mechanisms

As well as having to obtain nutrients for growth and reproduction, survival of an adherent/invasive organism also depends on its ability to overcome an extensive range of host defence mechanisms. These can be grouped broadly into two categories: non-specific (or innate) and specific (or acquired). The former include antibacterial substances, phagocytes and complement, while the latter include antibodies and cytotoxic T-cells.

3.4.1   Non-specific defence mechanisms
3.4.1.1   Mechanisms involving skin and mucosal surfaces
3.4.1.1.1   Normal microflora

One of the most important defence systems against exogenous pathogens is not due directly to the host itself, but is mediated by the normal microflora colonizing the host. Hence, the normal inhabitants of skin and mucosal surfaces tend to prevent colonization by exogenous pathogens through the occupation of adhesion sites, the utilization of available nutrients and the production of inhibitory substances [100,101]. While there are many studies showing that elimination, or interference with, the normal microflora increases susceptibility to infection with exogenous pathogens, the mechanisms involved are poorly understood [102,103]. Nevertheless, there is considerable interest in using members of the normal microflora to increase host resistance to disease, particularly to intestinal infections [104].

3.4.1.1.2   Host-derived antibacterial substances

Secretions by various cells of the skin and mucosa contain a number of antibacterial molecules, including fatty acids (skin), HCl (stomach), lysozyme (tears, saliva, mucin, sweat), pepsin (gut), lactoperoxidase (saliva, mucin) and bile salts (gut). It has also been shown that epithelial cells from a variety of sites in mammals secrete antibacterial peptides with a broad spectrum of activity. These include cryptidins (in the intestine), lingual antimicrobial peptide (in the tongue) and tracheal antimicrobial peptide [105,106] (see Chapter 9 for further detail). Pathogenic bacteria have evolved to withstand many of these toxic chemicals, although the means by which they do so have not generally been determined. Gram-negative bacteria, because of their outer membrane, are generally less susceptible to these noxious chemicals than Gram-positive organisms.

The host also produces a number of iron-binding proteins (e.g. lactoferrin, transferrin) which, while not antibacterial *per se*, bind iron, an essential micronutrient for bacteria, thereby limiting bacterial growth [107]. Bacteria have, of course, developed means of overcoming this defence mechanism. For example, *Por. gingivalis*, one of the causative agents of periodontitis, can bind and degrade lactoferrin [108]. Some pathogenic bacteria can acquire iron from the lactoferrin–Fe complex (and also from transferrin, present in serum and milk) by binding and extracting Fe from the complex. *N. gonorrhoeae*, for example, has receptors for both transferrin and lactoferrin, although the mechanism by which the iron is removed from these proteins and internalized by the organism remains unknown [109]. *Ps. aeruginosa* produces a siderophore, pyoverdin, which enables the organism to grow in the presence of apotransferrin and which, *in vitro*, can remove iron from ferritransferrin [110]. Other pathogens such as *L. monocytogenes, Haemophilus influenzae, Neisseria meningitidis* and *Bord. pertussis* can also acquire iron from host iron-binding proteins [99].

3.4.1.1.3 Mechanical expulsion of bacteria

The mucociliary escalator is an important means of expelling bacteria from the respiratory tract. *Ps. aeruginosa*, a respiratory pathogen in patients with cystic fibrosis and bronchiectasis, produces pyocyanin and L-hydroxyphenazine which are able to slow the movement of cilia, thereby impeding the clearance of the organism from the respiratory tract [111]. Pneumolysin from *Strep. pneumoniae* also exhibits a similar action.

3.4.1.2 Mechanisms involving underlying tissues

3.4.1.2.1 Complement

The complement system consists of a group of serum proteins (designated C1–C9) which, when activated, results in a reaction cascade providing molecules capable of directly killing Gram-negative bacteria, opsonizing bacteria for engulfment by phagocytes and attracting phagocytes [112,113]. Activation can take place by three distinct mechanisms: (i) attachment of mannose-binding protein to the bacterial surface (lectin pathway); (ii) binding of one of the complement proteins (C3) to the bacterial surface (alternative pathway); or (iii) binding of antibody to a bacterial antigen (classic pathway). The first two of these pathways constitute part of the non-specific host defence system. Bacteria have evolved a number of strategies to protect themselves from the adverse effects of complement activation, including: inhibition of complement activation by masking the bacterial components responsible; interruption of the cascade by binding essential intermediates; degradation of complement-bound components; and blocking the formation of the membrane-attack complex responsible for bacterial lysis. These strategies are described in a number of recent reviews [114,115].

Table 3.4 Bacterial products or components capable of killing phagocytic cells

Organism	Toxin	Target cell(s)	Mode of action
Streptococcus pyogenes	Streptolysin O	PMNs, macrophages	Lysosomes discharge into cytoplasm
	Streptolysin S	PMNs, macrophages	Lysosomes discharge into cytoplasm
Actinobacillus actinomycetemcomitans	Leukotoxin	PMNs, macrophages	Forms pores in membrane
Staphylococcus aureus	Panton–Valentine leukocidin	PMNs, macrophages	Induces a K^+-specific channel in membrane
Clostridium perfringens	α-Toxin	PMNs	Lecithinase, destroys membrane
Bacillus anthracis	Anthrax toxin	PMNs, macrophages	?
Shigella spp.	?	Macrophages	Apoptosis
Yersinia spp.	YopE	PMNs	Destruction of cytoskeleton
Clostridium difficile	Toxin A	PMNs	?
Legionella pneumophila	?	Macrophages	?
Chlamydia spp.	?	Macrophages	Lysosomes discharge into cytoplasm
Campylobacter rectus	Leukotoxin	PMNs	?

3.4.1.2.2   Phagocytes
Bacteria have adopted a variety of ways of dealing with PMNs and macrophages: (i) killing the phagocyte; (ii) inhibiting chemotaxis; (iii) resisting phagocytosis; (iv) resisting killing; (v) inhibiting lysosomal fusion; and (vi) colonization of phagocytes [116].

3.4.1.2.2.1   Killing of phagocytes
Many successful pathogens secrete toxins able to kill PMNs and/or macrophages and examples of these are listed in Table 3.4.

3.4.1.2.2.2   Inhibition of chemotaxis
Phagocytic cells are attracted to certain molecules secreted by invading bacteria and also to chemotactic factors such as C5a produced following complement activation. A number of bacterial secretory products able to interfere with this process have been identified. Streptolysins and the θ toxin of *Clostridium perfringens* inhibit PMN chemotaxis, while toxins from *Staph. aureus* inhibit PMN locomotion. Stationary-phase cells of *H. influenzae* have been shown to secrete N-acetyl-D-glucosamine-containing glycopeptides which are able to inhibit neutrophil migration towards the chemotactic factors N-formyl-L-methionyl-L-leucyl-L-phenylalanine and leukotriene B4 [117].

Two toxins produced by *Bord. pertussis*, PT and invasive adenylate cyclase, inhibit the migration of monocytes. YopH, a tyrosine phosphatase and YpkA, a serine/threonine kinase, are proteins secreted by *Yersinia* spp. which interfere with phagocyte signal transduction machinery and so can inhibit chemotaxis in phagocytes. The periodontopathogenic bacterium, *Actinobacillus actinomycetemcomitans*, secretes a low-molecular-mass compound which inhibits PMN chemotaxis towards the bacterial chemotactic peptide formyl-Met-Leu-Phe (FMLP). This bacterial component does not interfere with phagocytosis and is not cytotoxic, but functions by inhibiting binding of FMLP to PMN receptors [118].

3.4.1.2.2.3   *Resisting phagocytosis*
A variety of bacteria produce surface components which enable them to evade being phagocytosed. For example, many possess carbohydrate or proteinaceous capsules external to their cell walls which protect them from phagocytosis unless they are first opsonized . Examples are given in Table 3.5.

Another means of preventing phagocytosis is exemplified by *Strep. pyogenes*, the surface of which has proteinaceous fibrils composed of M protein which binds factor H in preference to factor B. This results in the displacement of factor B from the bound C3b, rendering the latter susceptible to inactivation by factor I, a serine protease. Degradation of C3b, the opsonizing component of complement, means that the bacterium is more difficult to phagocytose [119]. It has also been shown that certain strains of this organism resist opsonization and phagocytosis by binding fibrinogen to their surfaces [120].

3.4.1.2.2.4   *Resisting killing*
While some bacteria may be unable to prevent phagocytosis, they have evolved mechanisms to resist killing by the antibacterial compounds produced by the phagocyte. It has been shown that as many as 200 genes may be involved in

Table 3.5 Bacterial components responsible for protection against phagocytosis

Organism	Anti-phagocytic component
Streptococcus pyogenes	Hyaluronic acid
Yersinia pestis	Fra1, a capsular protein
Salmonella typhi	N-Acetylglucosamine uronic acid polymer
Pseudomonas aeruginosa	Alginate
Streptococcus pneumoniae	Complex polymer of sugar alcohols, amino sugars, sugars and choline
Haemophilus influenzae type b	Polymer of ribose, ribitol and phosphate
Bacillus anthracis	Poly-D-glutamic acid
Neisseria meningitidis	Poly-N-acetylneuraminic acid

enabling *Sal. typhimurium* to survive in phagocytes and that uptake by macrophages involves the up-regulation of more than 40 proteins and the down-regulation of approximately 100 proteins [121]. However, the functions of most of these proteins remain to be determined. Two important characteristics of *Sal. typhimurium*, which contribute to its ability to survive in phagocytes, are its resistance to reactive oxygen species and to the antibacterial peptides known as defensins. The former is attributable to the induction of catalase and superoxide dismutase by the organism when subjected to oxygen stress. Nearly all aerobic bacteria possess these enzymes to protect themselves against reactive oxygen species and many also have systems for repairing damage caused to DNA and membranes by these species [122].

The ability of *Mycobacterium tuberculosis* to survive and grow in PMNs and macrophages is a key virulence factor of the organism, although how it manages to accomplish this is uncertain. One contributing factor may be the ability of the organism to prevent acidification of the phagosome, possibly by producing ammonia [123]. Following fusion with the lysosome, the less acidic pH may interfere with activation of lysosomal enzymes. It has also been shown that cell-wall glycolipids of mycobacteria can act as scavengers of oxygen radicals and so may protect the organism in the phagolysosome [124]. This organism also has cell-wall components, sulphatides and lipoarabinoman-nan, which block the priming of macrophages by IFNγ, LPS and IL-1β, thereby preventing the enhancement of macrophage superoxide production and phagocytosis [125,126].

Brucella abortus induces a reduced respiratory burst and degranulation in PMNs and this has been attributed to its LPS [127]. A weak oxidative burst has also been observed during phagocytosis of *Erysipelothrix rhusiopathiae* by macrophages and this is thought to contribute to its ability to survive in these cells [128]. A protein from *A. actinomycetemcomitans* has been shown to inhibit the protein kinase C-dependent production of H_2O_2 by PMNs stimulated with phorbol myristate [129]. The organism also appears to be resistant to a mixture of the three defensins [human neutrophil peptide (HNP)-1, -2 and -3] found in human neutrophils, as well as to the individual peptides [130]. *Sal. typhimurium* also displays resistance to defensins and this is under the control of a two-component regulatory system encoded by phoPQ, although the genes responsible for resistance have not been identified [131]. *Burkholderia pseudomallei*, the cause of melioidosis, has been shown to be not only resistant to human defensin HNP-1, but also to grow in its presence [132], while *Brucella* spp. are resistant to lysosome extracts and a range of cationic peptides, including defensin-NP-2, bactenecins 5 and 7, CAP18 peptide, cecropin P1 and lactoferrin [133]. Although little is known concerning the mechanism of bacterial resistance to non-oxidative killing, the work of Visser et al. [134] has shown that, in the case of *Yersinia enterocolitica*, resistance to antimicrobial peptides from human PMNs is attributable to the expression of an outer-membrane protein, YadA. This forms a polymeric fibrillar matrix covering the whole cell [134]. Other ways in which bacteria resist killing by host-derived peptides have been reviewed recently [135].

3.4.1.2.2.5   Inhibiting lysosomal fusion

Once bacteria have been engulfed by a phagocyte, the resulting phagosome then fuses with a lysosome containing a range of antibacterial compounds. Interference with this fusion process will increase the chances of the organism surviving in the phagocytic cell. Organisms able to inhibit fusion of the phagosome and lysosome include *Bord. pertussis*, *N. gonorrhoeae*, *Chlamydia* spp., *M. tuberculosis* and *Sal. typhimurium* [116,136,137].

3.4.1.2.2.6   Colonization of phagocytes

A completely different strategy adopted by some bacteria for dealing with phagocytes is to actually use these cells as a habitat, either in the short or long term [138]. Because of their much greater life span, this strategy usually involves macrophages rather than PMNs. For example, *Bord. pertussis* uses two of its adhesins to induce its uptake by macrophages using a process involving mimicry of host selectins and integrins. The PT adhesin binds to selectin receptors, on the macrophage surface resulting in activation of CD11b/CD18 receptors which bind to the filamentous haemagglutinin of *Bord. pertussis* inducing internalization of the bacterium [50,61]. Internalization is therefore achieved without the organisms being confronted by an oxidative burst. Other bacteria capable of growing in macrophages include *Leg. pneumophila*, *L. monocytogenes*, *M. tuberculosis*, *M. leprae* and *Brucella* spp.

3.4.2   Specific defence mechanisms

3.4.2.1   Mechanisms involving skin and mucosal surfaces

All mucosal surfaces have lymphoid tissue (mucosal-associated lymphoid tissue; MALT) which constitutes a specific defence system, as they produce secretory immunoglobulin A (sIgA). It has been estimated that approximately 75% of the total immunoglobulin produced by humans is IgA. Secretory IgA exerts a protective role against infecting bacteria as it can help to trap bacteria in mucin, prevent bacterial adhesion to epithelial cells and neutralize bacterial exotoxins [139]. Inevitably, some bacteria have evolved strategies for coping with this defence system, including capsule formation (to resist antibody and complement deposition), antigenic variation (to avoid host immune surveillance), the production of IgA proteases and the shedding of bound antibodies. In this section only the latter two mechanisms will be considered; the others have either been described previously or will be discussed later, as they involve mechanisms also designed to deal with organisms which invade underlying tissues.

Several bacterial species produce IgA proteases including *Strep. pneumoniae*, *N. gonorrhoeae*, *H. influenzae*, *Vibrio cholerae*, some urogenital pathogens and oral streptococci [140–142]. By cleaving human IgA in the hinge region, the antigen-binding domain is separated from the effector region of the molecule. The production of these proteases enables the bacteria to colonize mucosal surfaces in the presence of sIgA antibodies. Another interesting strategy for circumventing IgA-mediated inhibition of colonization has been developed by *Streptococcus mutans*. Secretory IgA binds to the immuno-dominant antigen (P1) on the surface of this organism and inhibits its adhesion to salivary-coated hydroxyapatite. However, it has been shown that the bacteria can release the bound antibody in the form of antigen–antibody complexes, enabling the organism to adhere to the hydroxyapatite [143].

3.4.2.2 Mechanisms involving underlying tissues

The effector molecules/cells of the specific host defence systems with which an invasive bacterium is confronted are antibodies (mainly IgG and IgM), activated macrophages (i.e. macrophages with increased phagocytic and digestive abilities following activation by IFNγ) and cytotoxic T cells which are responsible for killing bacteria-infected host cells. The principal means by which bacteria circumvent these defences include: (i) antigenic variation; (ii) antibody degradation; (iii) evasion of phagocytosis (discussed above); and (iv) immunosuppression.

3.4.2.2.1 Antigenic variation

An effective way of confusing and evading the humoral immune response is to continually produce new versions of an immunogenic surface molecule [144]. This strategy of antigenic variation is exhibited by a number of pathogens including *N. gonorrhoeae*, *E. coli*, *Borrelia hermsii* and *Mycoplasma* spp. For the first two organisms, the antigens involved are located on the pili [145,146]. Adhesion of *N. gonorrhoeae* to host cells is mediated by pili, the main subunit of which is a protein, PilE. It has been estimated that the organism can produce as many as 10^6 different antigenic variants of this protein, enabling it to continually circumvent the protection afforded by anti-(adhesive sIgA) antibodies [147]. Antigenic variation in an immunodominant lipoprotein is displayed by one of the causative agents of relapsing fever, *Borr. hermsii* [148].

3.4.2.2.2 Immunosuppression

An important means by which a number of bacteria can suppress the immune response is by the production of superantigens. Superantigens are molecules that activate all T-cells expressing particular β-chain segments of the variable region (Vβ) of the α/β T-cell receptor (TCR) for antigen, and so are potent T-cell mitogens. They also affect B-cells resulting in polyclonal antibody production, and stimulate the release of cytokines (including IL-1, IL-2, IL-6, TNFα, TNFβ and IFNγ) from a number of cell types [149]. The resulting effects can include a diversion or suppression of the immune response, direct toxicity and the induction of immune tolerance [150]. The immunosuppression is due to a combination of the effects of cytokine release and the deletion of specific (dependent on the particular superantigen) Vβ TCR-bearing T-cells. A wide range of bacteria are known to produce superantigens, including *Staph. aureus*, *Strep. pyogenes*, *Ps. aeruginosa*, *M. tuberculosis*, *Cl. perfringens*, *Yersinia* spp. and *Mycoplasma* spp. [151]. Cytokine induction by superantigens is discussed in detail in Chapter 7.

M. leprae can suppress either cell-mediated or humoral immunity and this gives rise to the two different forms of the disease — lepromatous and tuberculoid leprosy respectively. In the former, there is an effective humoral antibody response, but a defective cell-mediated immune response to antigens of the organism, and many bacteria are present. In the disease lesions, cytokines typically produced by Th$_2$ cells (IL-4, IL-5, IL-10) dominate. In the tuberculoid form of the disease, there is no defect in cell-mediated immunity, but only low levels of antibody are detectable. Few viable bacilli are found in the tissues, and the cytokine pattern (IL-2, IFNγ and TNFβ) is typical of that due to Th$_1$ cells. The bacterial factors responsible for inducing these different responses remain to be determined.

The cell wall of *M. tuberculosis* contains lipoarabinomannan, which can suppress several aspects of cell-mediated immunity, including the antigen responsiveness of human lymphocytes and antigen-induced proliferation of human CD4$^+$ T-cell clones [152,153].

A heat-resistant, intracellular 14 kDa protein from *A. actinomycetemcomitans* has been reported to have immunosuppressive properties [154]. Hence, it can suppress the production of IL-2, IFNγ, IL-4 and IL-5 from CD4$^+$ T-cells and can inhibit the proliferation of mouse splenic T-cells stimulated with concanavalin A.

The exotoxin A of *Ps. aeruginosa* is immunosuppressive in that it inhibits lymphocyte proliferative responses and lymphokine production and this may enable the organism to persist in the lung [155].

Leg. pneumophila produces a 40 kDa zinc metalloproteinase which inactivates IL-2 and the CD4 receptor on T-cells, both of which are involved in T-cell activation [156]. Lysates from two pathogenic strains of *E. coli* have been reported to contain a protein or proteins able to inhibit mitogen-stimulated expression of IL-2, IL-4, IL-5 and IFNγ by lymphocytes in a dose-dependent manner, although the expression of cytokines produced predominantly by monocytes was unaffected [157].

A 90–100 kDa protein consisting of two subunits (44 kDa and 48 kDa) has been isolated from lysates of *Fusobacterium nucleatum*. This protein, as well as its 44 kDa component, was able to inhibit human T-cell responses to mitogens and antigens [158]. Analysis of the protein's mode of action has revealed that the T-cells were activated, in that they expressed CD69, CD25 and CD71, as well as secreting IL-2, but were unable to exit the G1 phase of the cell cycle. If this were to occur *in vivo* it would result in a state of local, or possibly systemic, immunosuppression [159].

M. tuberculosis is killed by activated macrophages and activation is dependent on IFNγ production by T-cells. Lipoarabinomannan suppresses T-cell proliferation and blocks transcriptional activation of IFN-inducible genes in macrophages without affecting their viability. It may, therefore, prevent IFN-induced activation of macrophages [160].

3.5 Cell and tissue damage

The many ways in which bacteria can damage the host can be classified into two broad groups: (i) direct damage due to the production of toxins and hydrolytic enzymes; and (ii) indirect (self-inflicted) damage as a consequence of the release from host cells of inflammatory mediators and enzymes (from lysed cells) and the production of antigen–antibody complexes.

3.5.1 Bacterial products that damage the host directly

3.5.1.1 Exotoxins

Bacteria produce a wide range of toxins which vary with respect to their structure, their target cell/tissue, their effects on the cell/tissue and their mode of action, all of which have formed the basis of classification schemes. As exotoxins have always been considered to be important virulence factors, a considerable amount of data has been accumulated on their structure and mode of action, but this is beyond the scope of this chapter and so only a brief outline of the distinguishing features of the major groups of exotoxins will be provided. Apart from their activities as toxins, many of these proteins are able to stimulate the release of cytokines from host cells (see Chapter 7 for further discussion).

3.5.1.1.1 RTX toxins

These toxins are so-called because they contain a tandem duplication of nine amino acids (RTX; repeats-in-toxin). They constitute the largest family of exotoxins and are produced by numerous genera of Gram-negative bacteria (Table 3.6). The distinguishing characteristics of these exotoxins [161] are as

Table 3.6 Examples of bacteria that produce RTX toxins

Organism	Activity of toxin
Escherichia coli	Haemolysin
Actinobacillus actinomycetemcomitans	Leukotoxin
Bordetella pertussis	Haemolysin
Proteus vulgaris	Haemolysin
Pasteurella haemolytica	Leukotoxin
Serratia marcescens	Metalloprotease
Pseudomonas aeruginosa	Alkaline protease
Pseudomonas fluorescens	Lipase
Serratia marcescens	Lipase

follows: (i) they consist of an unknown number of a single polypeptide (RTX A), each containing a tandem duplication of nine amino acids which are located N-proximal to an export-targeting structure; (ii) secretion of the protein across the cytoplasmic membrane is not accompanied by cleavage of an N-terminal leader peptide; (iii) a structure within the C-terminus of the exported protein targets the protein for export; (iv) a specific secretory apparatus exists, consisting of an inner-membrane protein (RTX B protein), a transmembrane channel (RTX D protein) and an outer-membrane TolC-like protein; and (v) the RTX A protein does not accumulate in the periplasm during export. Despite these common features, members of this family of toxins display a range of activities and include exotoxins that are cytolytic, metallo-dependent proteases and lipases [162].

3.5.1.1.2 A-B toxins

The characteristic feature of these toxins, which constitute one of the largest families of bacterial exotoxins, is that they contain a host cell receptor-binding domain (B) which determines their host cell specificity, and another region which mediates their biological activity (A). This group includes the ADP-ribosylating toxins, which transfer ADP-ribose from NAD to a target protein on the host cell, thereby either inactivating it or interfering with its usual function [163]. Most of these toxins exhibit one of three basic structures [164]: (i) a single protein with the A and B domains covalently linked (e.g. diphtheria toxin); (ii) a multi-protein complex with the A and B domains non-covalently associated (e.g. PT); or (iii) two separate, non-associated proteins, one of which has the A domain while the other contains the B domain (e.g. C2 toxin of *Cl. botulinum*). Examples of A-B toxins are given in Table 3.7.

Table 3.7 Examples of the A-B family of toxins

Toxin	Bacterium	Target cell/tissue	Physiological effect
Cholera toxin	*Vibrio cholerae*	Intestinal cells	Diarrhoea
Shiga toxin	*Shigella dysenteriae*	Intestinal cells	Diarrhoea
		Endothelial cells	Kidney failure
Pertussis toxin	*Bordetella perutussis*	Ciliated respiratory epithelial cells	Increased mucus production
Tetanus toxin	*Clostridium tetani*	Neurons	Paralysis
Shiga-like toxin	Enterohaemorrhagic *Escherichia coli*	Intestinal cells Endothelial cells	Diarrhoea Kidney failure
Diphtheria toxin	*Corynebacterium diphtheriae*	Several	Heart damage, nerve paralysis
Heat-labile enterotoxin	*Escherichia coli*	Intestinal cells	Diarrhoea

Table 3.8 Pore-forming toxins of Gram-positive bacteria

Toxin	Organism	Target cells
Listeriolysin O	*Listeria monocytogenes*	Erythrocytes, macrophages
Perfrinolysin O	*Clostridium perfringens*	Erythrocytes, PMNs and many others
α-Toxin	*Staphylococcus aureus*	PMNs, monocytes and others
δ-Toxin	*Staphylococcus aureus*	Erythrocytes, leucocytes, others
α-Toxin	*Clostridium septicum*	Erythrocytes, others?
Streptolysin O	*Streptococcus pyogenes*	Erythrocytes, PMNs and others

3.5.1.1.3   Pore-forming toxins of Gram-positive bacteria

This group includes exotoxins with a diverse range of effects on a wide range of cell types (Table 3.8). They are generally produced as soluble proteins, but then aggregate to form large complexes on target cell membranes, resulting in pore formation and cytolysis [165].

One pore-forming toxin worthy of a more detailed description is the pneumolysin of *Strep. pneumoniae*. This 53 kDa intracellular protein is an example of an endotoxin, but not in the generally accepted sense of the term, i.e. a protein–LPS complex [166]. It is a thiol-activated toxin which interacts with target cell membranes, possibly via cholesterol, and then oligomerizes to form transmembrane pores, resulting in cell lysis. It displays a range of *in vitro* and *in vivo* activities (including cytotoxicity to endothelial cells, epithelial cells, monocytes and PMNs), which may play a role in the pathology of diseases caused by this organism [167]. It is cytotoxic to alveolar epithelial cells and so may be responsible for lung damage characteristic of pneumonia, as well as facilitating entry of the organism into the bloodstream [168]. It also activates the classic complement pathway in the absence of specific antibodies [169]. The cytotoxic and complement-activating activities are determined by different regions of the molecule, with an 11-amino-acid domain (residues 427–437) being responsible for the former activity. Using mutants of *Strep. pneumoniae* in a mouse intraperitoneal challenge model, it has been shown that the contribution made by pneumolysin to the virulence of the organism is attributable mainly to its cytotoxic properties [170]. Pneumolysin is also an extraordinarily potent inducer of IL-1β and TNFα in human monocytes with concentration as low as 3 pg/ml and 10 pg/ml being able to induce the synthesis of IL-1β and TNFα respectively in a human monocytic cell line [171].

3.5.1.2   Extracellular enzymes

Many bacteria secrete hydrolytic enzymes which, *in vitro*, are capable of damaging tissues by degrading extracellular matrix components.

Table 3.9 Examples of protease production by bacteria

Organism	Protease	Substrate	Possible role *in vivo*
Pseudomonas aeruginosa	Las A (serine protease) and Las B (zinc metalloproteinase)	Elastin	Degradation of elastin in lung
Yersinia pestis	Pla protease	?	Dissemination of infection
Vibrio vulnificus	Metalloproteinase	Elastin, casein, collagen	?
Clostridium perfringens	Collagenase	Collagen	Breaks down muscle
Legionella pneumophila	Zinc metalloproteinase	Elastin	Damages lung tissue
Streptococcus pyogenes	Cysteine protease	Fibrin, IL-1β precursor	Systemic shock?

3.5.1.2.1   Proteases

By degrading host proteins, bacterial proteases undoubtedly act as virulence factors from the point of view of enabling growth *in vivo*. Inevitably this activity will have an adverse effect on host tissue integrity, but there is little definitive evidence to support the role of any particular protease as a key virulence factor contributing to host damage in any infectious disease of man [172]. However, proteases may contribute to bacterial virulence by interfering with host defence systems, for example by degrading complement components and immunoglobulins [173].

The ability to secrete proteases is widely distributed among bacteria and is evident in species from the following genera: *Streptococcus, Staphylococcus, Clostridium, Pseudomonas, Escherichia, Vibrio, Salmonella, Porphyromonas* and *Yersinia* [174]. Collectively, and in some cases individually, these organisms produce proteases capable of degrading a broad range of substrates, as listed in Table 3.9.

As well as providing a means of obtaining nutrients *in vivo* [175] and, possibly, inducing tissue damage directly, bacterial proteases may also contribute to tissue damage in an indirect manner (this is discussed later). Proteases may have an additional role to play during the initial colonization of the organism. For example, one of the consequences of the degradation of host matrix proteins by the proteases of *Por. gingivalis* is the exposure of cryptic ligands which enable enhanced binding of the organism via its fimbriae [176].

There is increasing evidence that bacterial proteases can also interfere with the cytokine networks responsible for both the maintenance of health and for mounting an effective response to challenge by infecting bacteria. Both the alkaline protease and elastase of *Ps. aeruginosa*, for example, are able to inactivate human recombinant IFNγ and human recombinant TNFα, but not IL-1α or IL-1β [177]. If this were to occur *in vivo* it would have a significant effect on host immune and inflammatory responses, resulting in immunosuppression. We have also shown that supernatants from *Por. gingivalis* cultures are able to hydrolyse IL-1β, IL-6 and IL-1 receptor antagonists *in vitro* and that this was accompanied, in the case of IL-1β, by a loss of biological activity [178]. Inhibition of cytokine production may represent another means by which proteases can function as a virulence factor. The *pla* protease of *Y. pestis*, for example, has been shown to play a key role in the dissemination of the organism in mice [172]. When injected subcutaneously into mice the resulting lesion is remarkably devoid of PMNs, while *pla* mutants produce lesions containing many inflammatory cells. One possible explanation for this is that the *pla* protease inhibits synthesis of the chemotactic cytokine IL-8. Another intriguing means by which bacterial proteases can manipulate the host cytokine network is by molecular mimicry. *Strep. pyogenes* secretes a cysteine protease (streptococcal pyrogenic toxin B), which mimics the activity of pro-IL-1β-converting enzyme (ICE) in that it can cleave biologically inactive human IL-1β precursor to produce IL-1β [179]. The action of this enzyme on IL-1β precursor (either secreted or released on cell necrosis) could result in an overproduction of IL-1β, which is consistent with the characteristic features of some streptococcal diseases, i.e. hypotension, shock and multiorgan failure.

One area that has received little attention is the possibility that the end products of protein degradation may have cytokine-inducing activities. Recently Engel et al. [180] have reported that an 80 kDa protease from *Por. gingivalis* can generate Fc fragments from human IgG1 which, in turn, are able to induce the release of IL-6, IL-8 and TNFα from human peripheral blood monocytes. We have also recently shown that a self-generated breakdown product of one of the proteases of *Por. gingivalis* (RI protease) stimulates IL-6 release from human monocytes.

3.5.1.2.2 Other hydrolytic enzymes

In addition to proteins such as collagen, the pericellular and extracellular matrices of animal tissues contain proteoglycans and hyaluronic acid (a gly-

Table 3.10 Examples of bacterial extracellular matrix-degrading enzymes

Enzyme	Substrate	Organism
Neuraminidase	Neuraminic acid	*Vibrio cholerae, Streptococcus pneumoniae*
Hyaluronate lyase	Hyaluronic acid	*Clostridium perfringens, Streptococcus pneumoniae, Streptococcus pyogenes, Staphylococcus aureus, Propionibacterium acnes, Peptostreptococcus* spp., *Streptococcus disgalactiae, Treponema denticola*
Chondroitin lyase	Chondroitin sulphate	*Flavobacterium heparinum, Proteus vulgaris, Bacteroides thetaiotamicron, Pseudomonas* spp., *Proteus mirabilis, Streptococcus intermedius, Treponema denticola*
Keratanase	Keratan sulphate	*Pseudomonas* spp., *Citrobacter freundii*
Heparin lyase	Heparin	*Flavobacterium heparinum*

cosaminoglycan) as major components [181]. Proteoglycans are composed of glycosaminoglycan chains attached to a protein core and may be classified on the basis of the nature of the glycosaminoglycan as galactosaminoglycans (e.g. chondroitin sulphate, dermatan sulphate) and glucosaminoglycans (e.g. heparan sulphate, heparin, keratan sulphate). A number of bacteria are able to degrade these polymers *in vitro* (Table 3.10) and this may contribute to the tissue destruction accompanying infections with these organisms.

3.5.2   Bacterial products that act as inducers of self-inflicted damage
3.5.2.1   Non-immunological products
3.5.2.1.1   Liberation of lysosomal enzymes from phagocytes
The lysosomes of phagocytes contain a number of potent antimicrobial agents (reactive oxygen species, hypochlorite, enzymes, etc.), which are also potentially harmful to host tissues. Killing of phagocytes by bacteria, or frustrated attempts at phagocytosis, can result in the release of lysosomal contents into surrounding tissues with deleterious effects. In some cases, the accumulation of phagocytes, lysed host cells and dead bacteria results in the formation of pus.

3.5.2.1.2   Stimulation of inflammatory mediators
3.5.2.1.2.1   Direct stimulation of cytokines by bacterial components or products
A wide variety of bacterial components and products are able to stimulate the release of cytokines from a number of host cell types [182,183]. These are discussed in detail in Chapter 7.

3.5.2.1.2.2   Liberation of cytokines as a consequence of bacterial activation of host defence systems
In addition to the direct stimulation of cytokine release by bacterial components discussed above, a number of effector cells of the host's defence

Table 3.11 Liberation of cytokines as a consequence of bacterial activation of host defence systems

Host defence component	Cytokine-inducing process	Cytokines released
Mast cell	Binding of IgE	IL-4, TNFα and others
Naive T cell	Activation by antigen-presenting cell	IL-2
Infected macrophage	Interaction with Th₁ cell	IL-10, IL-12
Th₁ cell	Interaction with macrophage-containing bacteria	IFNγ, GM-CSF, TNFα, TNFβ, IL-2, IL-3, MCF, MIF
Th₂ cell	Interaction with antigen-specific B cell	IL-3, IL-4, IL-5, IL-6, IL-10, GM-CSF, TGFβ

Abbreviations used: MCF, monocyte cytotoxicity-inducing factor; MIF, migration-inhibitory factor. For full list, see page xiii.

Table 3.12 Pathological consequences of immune complex reactions

Site of formation or deposition of Ag–Ab complex	Pathological consequences
Extravascular tissue	Inflammation and oedema
Walls of blood vessels	Vascular damage
Walls of blood vessels of skin	Tender, red skin nodules; erythema nodosum
Walls of small arteries	Periarteritis nodosum
Kidney glomeruli	Swelling of basement membrane, loss of albumin and erythrocytes in urine; acute or chronic glomerulonephritis
Lung	Alveolitis
Bloodstream	Disseminated intravascular coagulation
Joints	Swelling and inflammation

system are activated to produce cytokines during the course of an infection. The main cells and processes involved are summarized in Table 3.11.

As described previously, many bacteria produce proteases, and also protease inhibitors, which may have roles in the control of cytokine networks.

3.5.2.2 Immunopathological products

3.5.2.2.1   Immune complex reactions

The formation of antigen–antibody (Ag–Ab) complexes is central to a number of strategies employed by the host to combat bacterial infections, including phagocytosis, prevention of bacterial attachment, neutralization of toxins, activation of complement and the killing of pathogen-infected host cells. However, they also are capable of inducing a considerable amount of damage to host tissues as a consequence of the activation of the complement system, the kinin system and platelet aggregation [184]. Complement activation also results in the release of factors chemotactic for PMNs, which can then degranulate, liberating more inflammatory compounds. The consequences of this depend on where the Ag–Ab complexes are formed as shown in Table 3.12.

Most bacterial infections result in some degree of tissue damage as a result of the formation of immune complexes, and this is particularly so in the case of syphilis and diseases caused by *Strep. pyogenes*.

3.5.2.2.2   Cell-mediated reactions

A cell-mediated immune response accompanies all bacterial infections and is characterized by inflammation, lymphocyte infiltration and the accumulation of macrophages and their activation. However, the magnitude of the response varies depending on the individual and the nature of the causative organism. In the case of diseases such as tuberculosis and tuberculoid leprosy, the pathology is dominated by the cell-mediated immune response to the infecting organisms. This results in the formation of a chronic inflammatory lesion (a granuloma), consisting of a core of live and dead macrophages, together with giant cells (due to macrophage fusion) surrounded by T cells, many of which are CD4-positive [185]. Other diseases resulting in granuloma formation are syphilis, actinomycosis and the chlamydial infection, lymphogranuloma inguinale. Bacterial infections in which cell-mediated immunity is an important cause of tissue destruction are those due to *L. monocytogenes*, *Y. pestis*, *Brucella* spp., *Salmonella* spp., *Leg. pneumophila*, *Rickettsia* spp. and *Chlamydia* spp.

3.5.2.2.3   Cytotoxic reactions

These reactions occur when antibody binds to antigens on host cells, thereby activating complement or inducing cytotoxicity by natural killer cells. The result is death of the antibody-coated host cell, a process termed 'antibody-dependent cell-mediated cytotoxicity'. Some surface components of bacteria have similar epitopes to those displayed on the surface of host cells, so that antibodies produced against the infecting organism can bind to the host cell antigens. For example, the M protein of *Strep. pyogenes* has epitopes similar to those of cardiac myosin and sarcolemma membrane proteins, so that damage to heart tissue (i.e. rheumatic fever) may ensue following infection with this organism [186].

3.6   Regulation of bacterial virulence factors

The previous sections have described how bacteria, in order to successfully initiate and maintain an infectious process, must possess a number of virulence factors. It is important now to address the question of how the production of these virulence factors is regulated. Obviously, it would not be cost-effective, in terms of its energy or materials balance, for a bacterium to synthesize all of these factors at every stage in the disease process. The phenotype displayed by an organism is governed by its genome and by the environment in which it finds itself [187]. Evidence that bacteria produce a particular virulence factor only when it is required, demonstrating environmental control over virulence gene expression, came initially from studies comparing freshly isolated and laboratory grown strains of an organism. For example, freshly isolated strains of *N. gonorrhoeae*, but not laboratory-grown strains, were resistant to complement-mediated killing, and great difficulty was encountered in getting *B. anthracis* to produce an exotoxin in laboratory medium. Since these early studies, considerable progress has been made in elucidating how environmental conditions control the production of virulence factors. That such control is necessary can be appreciated by considering the large number of different environments to which an organism such as *Sal. typhi* will be exposed: water/sewage, the intestinal lumen, epithelial cells, blood, phagocytes, liver and spleen. This list includes only those sites within which the organism actually grows and/or spends an appreciable period of time. It also has to survive in a wider range of environments including the oral cavity and the stomach. In order to survive and grow in each of these environments, the organism must be able to sense the prevailing external conditions and adjust its structure, physiology and behaviour accordingly. Bacteria have evolved sophisticated systems for eliciting adaptive responses to their environment [188]. It would appear that the response of many bacterial species to a variety of environmental stimuli occurs by a similar mechanism, involving a two-component regulatory system. One component of this system (the 'sensor', located in the cytoplasmic membrane) monitors some environmental parameters and signals the second component (the 'response regulator', a cytoplasmic protein), which initiates the response — usually some change in gene expression. Typically, sensors have an N-terminal input domain located in the periplasmic space which, on detecting a stimulus, modulates the signalling activity of its C-terminal transmitter to communicate with its response-regulatory partner. The latter usually has an N-terminal receiver module which detects the incoming signal from the sensor and then alters the activity of its C-terminal output domain, thereby initiating a response. The means by which the sensor detects a particular stimulus is, in some cases, by ligand binding. This probably stimulates a conformational change, resulting in transmission of the signal across the cytoplasmic membrane to the transmitter domain which protrudes into the cytoplasm. Transmission of the signal to the receiver domain occurs by a series of phosphorylation/dephosphorylation reactions, involving phosphorylation of a histidine residue in the transmitter followed by the transfer of this group to an aspartate residue in the receiver, which, in turn, transfers the phosphate group to water. The receiver domain of the response regulator is usually connected to its output domain by flexible linkers. The output domain generally has DNA-binding activity so that phosphorylation of the receiver serves to modulate transcription of one or more target genes [189]. Examples of two-component regulatory systems for the control of virulence gene expression are given in Table 3.13.

A particular environmental signal (e.g. a change in osmolarity, pH or temperature) will often alter the expression of a number of genes encoding virulence factors needed at a particular stage in the infectious process [190,191]. These genes will often be organized in a regulon; i.e. although they are located on different parts of the chromosome they are under the control of the same regulatory protein. For example, *Yersinia* spp. have a temperature-dependent regulon which regulates nearly all the virulence factors of the organism, including adhesins, invasins, enterotoxin and outer-membrane proteins [75]. A group of regulons which are affected by the same environmental stimulus but respond in different ways are said to comprise a stimulon.

Environmental parameters frequently shown to be important in controlling virulence gene expression in many organisms include temperature, pH, osmolarity, the composition of the gaseous phase and the concentration of iron, phosphate and calcium [192].

Table 3.13 Examples of two-component regulatory systems for the control of virulence gene expression

Organism	Signal	Sensor	Response regulator	Response
Bordetella pertussis	Temperature, Mg^{2+}, nicotinic acid	BvgS	BvgA	Synthesis of filamentous haemagluttinin, PT
Shigella flexneri	Osmolarity	EnvZ	OmpR	Porin expression
Pseudomonas aeruginosa	Osmolarity	AlgR2	AlgR1	Alginate synthesis
Klebsiella pneumoniae	N_2	NtrC	NtrA	Urease production
Vibrio cholerae	pH, osmolarity, temperature	ToxS	ToxR	Synthesis of toxin and pili

By subjecting an organism *in vitro* to key environmental parameters operating at each stage in an infectious process and determining the effects these have on virulence gene expression it should, therefore, be possible to identify those virulence factors essential to the organism for adhesion, avoidance of host defence systems, etc. However, as it is virtually impossible to reproduce *in vitro* the conditions prevailing *in vivo*, this strategy has obvious disadvantages. Recently, another approach has been developed which enables the identification of those genes expressed by an organism *in vivo* [193]. This technique, termed *in vitro* expression technology (IVET), involves identifying those genes expressed when the organism is inoculated into an appropriate animal, but which are minimally expressed on normal laboratory media. Such genes will include those required for the infectious process. By inoculating *Sal. typhimurium* into mice, a number of *in vivo*-induced (*ivi*) genes have been identified, including those encoding the two subunits of carbamoyl phosphate synthetase, an enzyme involved in arginine and pyrimidine synthesis [194]. As animal tissues generally have a low pyrimidine content, the induction of these genes probably represents the organism's response to enable it to survive in this new environment. The original IVET system relied on the use of a mutant with a genetically defined nutritional deficiency (e.g. *Sal. typhimurium* defective in purine synthesis), but this has been improved to enable it to be used in a wider range of host/pathogen systems [195]. In this system, clones of the test organism expressing *ivi* genes are identified on the basis of antibiotic resistance, and this has enabled the identification of a gene encoding an enzyme involved in fatty acid oxidation. This could be related to the high fatty acid content of the inflamed environment and could represent the organism's attempt to neutralize the pro-inflammatory activities of these compounds.

Another level of control of virulence gene expression has been recognized following the discovery of quorum sensing in bacteria, i.e the ability of bacteria to monitor their population density [196]. This system functions by virtue of the production of a diffusible compound (an autoinducer), which accumulates in the environment and, at high cell densities, reaches a level sufficient to activate transcription of certain genes. This has been studied extensively in the marine bacterium *V. fischeri*, which secretes the autoinducer *N*-3-(oxohexanoyl)homoserine lactone, and this reaches a sufficient concentration (at a cell density of approximately 10^{10} cfu/ml) to activate the organism's luminescence genes [197]. The autoinducer binds to a protein encoded by the *luxR* gene which is then able to activate the transcription of the operon *luxICDABEG*. *LuxI* encodes the autoinducer synthase while the other genes in the operon code for other components required for luminescence. Cell-density-dependent luminescence is, in effect, mediated by *luxR* and *luxI*. Regulatory systems homologous to the LuxR and LuxI system have been detected in several other species and have been shown to regulate the production of important virulence factors (including elastase and alkaline protease) in *Ps. aeruginosa* [198].

3.7   Concluding remarks

This chapter has outlined the means by which bacteria initiate and perpetuate an infectious process and has been, for obvious reasons, very one-sided in that the machinations of the host during this process have been given little attention. In the next chapter this imbalance will be redressed by considering the response of the host, in terms of the nature and activities of the cytokines it produces when it is confronted by 'troublesome' bacteria.

References

1. Hensel, M. and Holden, D.W. (1996) Molecular genetic approaches for the study of virulence in both pathogenic bacteria and fungi. Microbiology **142**, 1049–1058

2. Ofek, I. and Doyle, R.J. (1994) Principles of bacterial adhesion. In Bacterial Adhesion to Cells and Tissues, pp. 1–15, Chapman & Hall, New York

3. Tuomanen, E.I., Nedelman, J., Hendley, J.O. and Hewlett, E.L. (1983) Species specificity of *Bordetella* adherence to human and animal ciliated respiratory epithelial cells. Infect. Immun. **42**, 692–695

4. Tuomanen, E.I. (1992) Cell adhesion molecules in the development of bacterial infections. In Leukocyte Adhesion. Basic and Clinical Aspects. (Gamberg, C.G., Mandrup-Poulsen, T., Wogensenbach, L. and Hokfelt, B., eds.), pp. 297–306, Elsevier Science Publishers, New York

5. Ofek, I. and Sharon, N. (1990) Adhesins as lectins: specificity and role in infection. Curr. Top. Microbiol. Immunol. **151**, 91–113

6. Giannasca, K.T., Giannasca, P.J. and Neutra, M.R. (1996) Adherence of *Salmonella typhimurium* to Caco-2 cells: identification of a glycoconjugate receptor. Infect. Immun. **64**, 135–145

7. Bertin, Y., Girardeau, J.P., Darfeuille-Michaud, A. and Contrepois, M. (1996) Characterization of 20K fimbria, a new adhesin of septicemic and diarrhea-associated *Escherichia coli* strains, that belongs to a family of adhesins with *N*-acetyl-D-glucosamine recognition. Infect. Immun. **64**, 332–342

8.   Murray, P.A., Kern, D.G. and Winkler, J.R. (1988) Identification of a galactose-binding lectin on
      Fusobacterium nucleatum FN-2. Infect. Immun. **56**, 1314–1319

9.   de Bentzmann, S., Roger, P., Dupuit, F., Bajolet-Laudinat, O., Fuchey, C., Plotkowski, M.C. and
      Puchelle, E. (1996) Asialo GM1 is a receptor for Pseudomonas aeruginosa adherence to regenerat-
      ing respiratory epithelial cells. Infect. Immun. **64**, 1582–1588

10.   Ramphal, R., Carnoy, C., Fievre, S., Michalski, J.-C., Houdret, N., Lamblin, G., Strecker, G. and
      Roussel, P. (1991) Pseudomonas aeruginosa recognises carbohydrate chains containing type I
      (Galβ1-3GlcNAc) or type 2 (Galβ1-4GlcNAc) disaccharide units. Infect. Immun. **59**, 700–704

11.   Stroberg, N., Deal, C., Nyberg, G., Normark, S., So, M. and Karlsson, K-A. (1988) Identification of
      carbohydrate structures that are possible receptors for Neisseria gonorrhoeae. Proc. Natl. Acad.
      Sci. U.S.A. **85**, 4902–4906

12.   Evans, D.G., Evans, D.J., Moulds, J.J. and Graham, D.Y. (1988) N-acetylneuraminyllactose-binding
      fibrillar hemagglutinin of Campylobacter pylori: a putative colonization factor antigen. Infect. Immun.
      56, 2896–2906

13.   Senior, D., Baker, N., Cedergren, B., Falk, P., Larson, G., Lindstedt, R. and Svanborg-Eden, C.
      (1988) Globo-A: a new receptor specificity for attaching Escherichia coli. FEBS Lett. **237**, 123–127

14.   Di Martino, P., Bertin, Y., Girardeau, J.P., Joly, B. and Darfeuille-Michaud, A. (1995) Molecular
      characterization and adhesive properties of CF29K, an adhesin of Klebsiella pneumoniae strains
      involved in nosocomial infections. Infect. Immun. **63**, 4336–4344

15.   Garber, N., Guempel, U., Belz, A., Gilboa-garber, N. and Doyle, R.J. (1992) On the specificity of
      the D-galactose-binding lectin (PA-1) of Pseudomonas aeruginosa and its strong binding to
      hydrophobic derivatives of D-galactose and thiogalactose. Biochim. Biophys. Acta **1116**, 331–333

16.   Firon, N., Ashkenazi, S., Mirelman, D., Ofek, I. and Sharon, N. (1987) Aromatic alpha-glycosides
      of mannose are powerful inhibitors of the adherence of type I fimbriated Escherichia coli to yeast
      and intestinal epithelial cells. Infect. Immun. **55**, 472–476

17.   Andersson, B., Dahmen, J., Frejd, T., Leffler, H., Magnusson, G., Noori, G. and Svanborg-Eden, C.
      (1983) Identification of an active disaccharide unit of a glycoconjugate receptor for pneumococci
      attaching to human pharyngeal epithelial cells. J. Exp. Med. **158**, 559–570

18.   Doig, P., Paranchych, W., Sastry, P.A. and Irvin, R.T. (1989) Human buccal epithelial cell receptors
      of Pseudomonas aeruginosa: identification of glycoproteins with pilus binding activity. Can. J.
      Microbiol. **35**, 1141–1145

19.   Giampapa, C.S., Abraham, S.M., Chiang, T.M. and Beachey, E.H. (1988) Isolation and characteriza-
      tion of a receptor for type I fimbriae of Escherichia coli from guinea pig erythrocytes. J. Biol.
      Chem. **263**, 5362–5367

20.   Leffler, H. and Svanborg-Eden, C. (1980) Chemical identification of a glycosphingolipid receptor
      for Escherichia coli attaching to human urinary tract epithelial cells and agglutinating human ery-
      throcytes. FEMS Microbiol. Lett. **8**, 127–134

21.   Weiss, E.I., London, J., Kolenbrander, P., Kagemeier A.S. and Andersen, R.N. (1987)
      Characterization of lectin-like surface components on Capnocytophaga ochracea 33596 that
      mediate coaggregation with gram-positive oral bacteria. Infect. Immun. **55**, 1198–1202

22.   Demuth, D.R., Golub, E.E. and Malamud, D. (1990) Streptococcal-host interactions; structural and
      functional analysis of a Streptococcus sanguis receptor for a human salivary glycoprotein. J. Biol.
      Chem. **265**, 7120–7126

23.   Nealon, T.J. and Mattingly, S.J. (1984) Role of cellular lipoteichoic acids in mediating adherence of
      serotype III strains of group B streptococci to human embryonic, fetal and adult epithelial cells.
      Infect. Immun. **43**, 523–530

24.   Wyatt, J.E., Poston, S.M. and Noble, W.C. (1990) Adherence of Staphylococcus aureus to cell
      monolayers. J. Appl. Bacteriol. **69**, 834–844

25.   Chugh, T.D., Burns, G.J., Shuhaiber, H.J. and Bahr, G.M. (1990) Adherence of Staphylococcus epi-
      dermidis to fibrin-platelet clots in vitro is mediated by lipoteichoic acid. Infect. Immun. **58**, 315–319

26.   Hogg, S.D. and Manning, J.E. (1988) Inhibition of adhesion of viridans streptococci to fibronectin-
      coated hydroxyapatite beads by lipoteichoic acid. J. Appl. Bacteriol. **65**, 483–489

27. Savoia, D. and Landolfo, S. (1987) Modulation of the adherence of group A streptococci to murine cells. Microbiologia **10**, 281–290

28. Simpson, W.A. and Beachey, E.H. (1983) Adherence of group A streptococci to fibronectin on oral epithelial cells. Infect. Immun. **39**, 275–279

29. Froman, G., Switalski, L.M., Speziale, P. and Hook, M. (1987) Isolation and characterization of a fibronectin receptor from *Staphylococcus aureus*. J. Biol. Chem. **262**, 6564–6571

30. Usui, Y. (1986) Biochemical properties of fibrinogen-binding protein (clumping factor) of the staphylococcal cell surface. Zbl. Bakt. Hyg. **263**, 287–297

31. Mota, G.F.A., Carneiro, C.R.W., Gomes, L. and Lopes, J.D. (1988) Monoclonal antibodies to *Staphylococcus aureus* laminin-binding proteins cross-react with mammalian cells. Infect. Immun. **56**, 1580–1584

32. Sampson, J.S., O'Conner, S.P., Stinson, A.R., Tharpe, J.A. and Russel, H. (1994) Cloning and nucleotide sequence analysis of *psaA*, the *Streptococcus pneumoniae* gene encoding a 37-kilodalton protein homologous to previously reported *Streptococcus* spp. antigens. Infect. Immun. **62**, 319–324

33. Watson, D.A., Musher, D.M. and Verhoet, J. (1995) Pneumococcal virulence factors and host immune responses to them. Eur. J. Clin. Microbiol. Infect. Dis. **14**, 479–490

34. Rosenberg, M. and Doyle, R.J. (1990) Microbial cell surface hydrophobicity: history, measurement, and significance. In Microbial Cell Surface Hydrophobicity (Doyle, R.J. and Rosenberg, M., eds.), pp. 1–37, American Society for Microbiology, Washington

35. Doyle, R.J., Rosenberg, M. and Drake, R. (1990) Hydrophobicity of oral bacteria. In Microbial Cell Surface Hydrophobicity (Doyle, R.J. and Rosenberg, M., eds.), pp. 387–419, American Society for Microbiology, Washington

36. Naito, Y., Tohda, H., Okuda, K. and Takazoe, I. (1993) Adherence and hydrophobicity of invasive and non-invasive strains of *Porphyromonas gingivalis*. Oral Microbiol. Immunol. **8**, 195–202

37. Wadstrom, T. (1991) Molecular aspects on pathogenesis of staphylococcal wound and foreign body infections: bacterial cell surface hydrophobicity, fibronectin, fibrinogen and collagen binding surface proteins determine ability of staphylococci to colonize in damaged tissues and on prosthesis materials. Zbl. Bakt. Suppl. **21**, 37–52

38. Ramphal, R., Guay, C. and Pier, G.B. (1987) *Pseudomonas aeruginosa* adhesins for tracheobronchial mucin. Infect. Immun. **55**, 600–603

39. Baker, N., Hansson, G.C., Leffler, H., Riise, G. and Svanborg-Eden, C. (1990) Glycosphingolipid receptors for *Pseudomonas aeruginosa*. Infect. Immun. **58**, 2361–2366

40. Glushka, J., Cassels, F.J., Carlson, R.W. and van-Halbeek, H. (1992) Complete structure of the adhesin receptor polysaccharide of *Streptococcus oralis* ATCC 55229   (*Streptococcus sanguis* H1). Biochemistry **31**, 10741–10746

41. Tojo, M., Yamashita, N., Goldman, D.A. and Pier, G.B. (1988) Isolation and characterization of a capsular polysaccharide adhesin from *Staphylococcus epidermidis*. J. Infect. Dis. **157**, 713–722

42. Muller, E., Hubner, J., Gutierrez, N., Takeda, S., Goldmann, D.A. and Pier, G.B. (1993) Isolation and characterization of transposon mutants of *Staphylococcus epidermidis* deficient in capsular polysaccharide/adhesin and slime. Infect. Immun. **61**, 551–558

43. Mack, D., Nedelmann, M., Krokotsch, A., Schwarzkopf, A., Heesemann, J. and Laufs, R. (1994) Characterization of transposon mutants of biofilm-producing *Staphylococcus epidermidis* impaired in the accumulative phase of biofilm production: genetic identification of a hexosamine-containing polysaccharide intercellular adhesin. Infect. Immun. **62**, 3244–3253

44. Walan, A. and Kihlstrom, E. (1988) Surface charge and hydrophobicity of *Campylobacter jejuni* strains in relation to adhesion to epithelial HT-29 cells. Acta. Pathol. Microbiol. Immunol. Scand. **96**, 1089–1096

45. McSweegan, E. and Walker, R.I. (1986) Identification and characterisation of two *Campylobacter jejuni* adhesins for cellular and mucous substrates. Infect. Immun. **53**, 141–148

46. Jacques, M. (1996) Role of lipo-oligosaccharides and lipopolysaccharides in bacterial adherence. Trends Microbiol. **4**, 408–410

47. Ofek, I. and Doyle, R.J. (1994) Common themes in bacterial adhesion. In Bacterial Adhesion to Cells and Tissues, pp. 513–561, Chapman & Hall, New York

48. Weiss, A.A. and Hewlett, E.L. (1986) Virulence factors of Bordetella pertussis. Annu. Rev. Microbiol. **40**, 661–686

49. Parker, C.D. and Armstrong, S.K. (1988) Surface proteins of Bordetella pertussis. Rev. Infect. Dis. **10**, S327–S330

50. Sandros, J. and Tuomanen, E. (1993) Attachment factors of Bordetella pertussis: mimicry of eukaryotic recognition molecules. Trends Microbiol. **1**, 192–196

51. Prasad, S.M., Yin, Y., Rozdzinski, E., Tuomanen, E. and Masure, H.R. (1993) Identification of a carbohydrate recognition domain in filamentous haemagglutinin from Bordetella pertussis. Infect. Immun. **61**, 2780–2785

52. Cundell, D.R. and Tuomanen, E. (1995) Attachment and interaction of bacteria at respiratory mucosal surfaces. In Virulence Mechanisms of Bacterial Pathogens (Roth, J.A., Bolin, C.A., Brogden, K.A., Minion, F.C. and Wannemuehler, M.J., eds.), pp. 3–20, American Society for Microbiology, Washington

53. van't Wout, J., Burnette, W.N., Mar, V.L., Rozdzinski, E., Wright, S.D. and Tuomanen, E.I. (1992) Role of carbohydrate recognition domains of pertussis toxin in adherence of Bordetella pertussis to human macrophages. Infect. Immun. **60**, 3303–3308

54. Saukkonen, K., Burnette, W.N., Mar, V., Masure, H.R. and Tuomanen, E. (1992) Pertussis toxin has eukaryotic-like carbohydrate recognition domains. Proc. Natl. Acad. Sci. U.S.A. **89**, 118–122

55. Relman, D.A., Domenighini, M., Tuomanen, E., Rappuoli, R. and Falkow, S. (1989) Filamentous haemagglutinin of Bordetella pertussis: nucleotide sequence and crucial role in adherence. Proc. Natl. Acad. Sci. U.S.A. **86**, 2637–2641

56. Weiss, A.A., Hewlett, E.L., Myers, G.A. and Falkow, S. (1983) Tn5-induced mutations affecting virulence factors of Bordetella pertussis. Infect. Immun. **42**, 33–41

57. Tuomanen, E. and Weiss, A. (1985) Characterization of two adhesins of Bordetella pertussis for human ciliated respiratory epithelial cells. J. Infect. Dis. **152**, 118–125

58. Tuomanen, E., Weiss, A., Rich, R., Zak, F. and Zak, O. (1986). Filamentous haemagglutinin and pertussis toxin promote adherence of Bordetella pertussis to cilia. Dev. Biol. Stand. **61**, 197–204

59. Leininger, E., Roberts, M., Kenimer, J.G., Charles, I.G., Fairweather, N., Novotny, P. and Brennan, M.J. (1991) Pertactin, an Arg-Gly-Asp-containing Bordetella pertussis surface protein that promotes adherence of mammalian cells. Proc. Natl. Acad. Sci. U.S.A. **88**, 345–349

60. Relman, D., Tuomanen, E., Falkow, S., Golenbock, D.T., Saukonnen, K. and Wright, S.D. (1994) Recognition of a bacterial adhesin by an integrin: macrophage CR3 ($\alpha_m\beta_2$, CD11b/CD18) binds filamentous haemagglutinin of Bordetella pertussis. Cell **61**, 1375–1382

61. Saukkonen, K., Cabellos, C., Burroughs, M., Prasad, S. and Tuomanen, E. (1991) Integrin-mediated localization of Bordetella pertussis within macrophages: role in pulmonary colonization. J. Exp. Med. **173**, 1143–1149

62. Costerton, J.W., Lewandowski, Z., Caldwell, D.E., Korber, D.R. and Lappin-Scott, H.M. (1995) Microbial biofilms. Annu. Rev. Microbiol. **49**, 711–745

63. Wilson, M. (1996) Susceptibility of oral bacterial biofilms to antimicrobial agents. J. Med. Microbiol. **44**, 79–87

64. Finlay, B.B., Heffron, F. and Falkow, S. (1989) Epithelial cell surfaces induce salmonella proteins required for bacterial adherence and invasion. Science **243**, 940–943

65. Bessen, D. and Gotschlich, E.C. (1986) Interactions of gonococci with Hela cells: attachment, detachment, replication, penetration and the role of protein II. Infect. Immun. **54**, 154–160

66. McCormick, B.A., Colgan, S.P., Delp-Archer, C., Miller, S.I. and Madara, J.L. (1993) Salmonella typhimurium attachment to human intestinal epithelial monolayers: transcellular signalling to subepithelial neutrophils. J. Cell Biol. **123**, 895–907

67. Yamamoto, Y., Okubo, S., Klein, T.W., Onozaki, K., Saito, T. and Friedman, H. (1994) Binding of Legionella pneumophila to macrophages increases cellular cytokine mRNA. Infect. Immun. **62**, 3947–3956

68. Yamamoto, Y., Klein, T.W. and Friedman, H. (1996) Induction of cytokine granulocyte-macrophage colony-stimulating factor and chemokine macrophage inflammatory protein 2 mRNAs in macrophages by *Legionella pneumophila* or *Salmonella typhimurium* attachment requires different ligand-receptor systems. Infect. Immun. **64**, 3062–3068

69. Vernier, A., Diab, M., Soell, M., Haan-Archipoff, G., Beretz, A., Wachsmann, D. and Klein, J-P. (1996) Cytokine production by human epithelial and endothelial cells following exposure to oral viridans streptococci involves lectin interactions between bacteria and cell surface receptors. Infect. Immun. **64**, 3016–3022

70. Linder, H., Engberg, I., Hoschutsky, H., Mattsby-Baltzer, I. and Svanborg, C. (1991) Adhesion-dependent activation of mucosal interleukin-6 production. Infect. Immun. **59**, 4357–4362

71. Svensson, M., Lindstedt, R., Radin, N.S. and Svanborg, C. (1994) Epithelial glucosphingolipid expression as a determinant of bacterial adherence and cytokine production. Infect. Immun. **62**, 4404–4410

72. Svanborg, C., Agace, W., Hedges, S., Linder, H. and Svensson, M. (1993) Bacterial adherence and epithelial cell cytokine production. Zbl. Bakt. **278**, 359–364

73. Russell, J.P., Diamond, G., Tarver, A.P., Scanlin, T.F. and Bevins, C.L. (1996) Coordinate induction of two antibiotic genes in tracheal epithelial cells exposed to the inflammatory mediators lipopolysaccharide and tumor necrosis factor alpha. Infect. Immun. **64**, 1565–1568

74. Isberg, R.R. and Leong, J.M. (1990) Multiple beta 1 chain integrins are receptors for invasin, a protein that promotes bacterial penetration into mammalian cells. Cell **60**, 861–871

75. Cornelis, G.R. (1992) Yersinias, finely tuned pathogens. Symp. Soc. Gen. Microbiol. **49**, 231–265

76. Tran van Nhieu, G. and Isberg, R.R. (1993) Bacterial internalisation mediated by beta 1 chain integrins is determined by ligand affinity and receptor density. EMBO J. **12**, 1887–1895

77. Bliska, J.B., Copass, M.C. and Falkow, S. (1993) The *Yersinia pseudotuberculosis* adhesin YadA mediates intimate bacterial attachment to and entry into Hep-2 cells. Infect. Immun. **61**, 3914–3921

78. Rosenshine, I., Duronio, V. and Finlay, B.B. (1992) Tyrosine protein kinase inhibitors block invasin-promoted bacterial uptake by epithelial cells. Infect. Immun. **60**, 2211–2217

79. Finlay, B.B., Ruschkowski, S. and Dedhar, S. (1991) Cytoskeletal rearrangements accompanying salmonella entry into epithelial cells. J. Cell Sci. **99**, 283–296

80. Galan, J.E., Pace, J. and Hayman, M.J. (1992) Involvement of the epidermal growth factor receptor in the invasion of cultured mammalian cells by *Salmonella typhimurium*. Nature (London) **357**, 588–589

81. Pace, J., Hayman, M.J. and Galan, J.E. (1993) Signal transduction and invasion of epithelial cells by *Salmonella typhimurium*. Cell **72**, 505–514

82. Galan, J.E. and Curtis III, R. (1989) Cloning and molecular characterization of genes whose products allow *Salmonella typhimurium* to penetrate tissue culture cells. Proc. Natl. Acad. Sci. U.S.A. **86**, 6383–6387

83. Groisman, E.A. and Ochman, H. (1993) Cognate gene clusters govern invasion of host epithelial cells by *Salmonella typhimurium* and *Shigella flexneri*. EMBO J. **12**, 3779–3787

84. Galan, J.E. and Curtis III, R. (1990) Expression of *Salmonella typhimurium* genes required for invasion is regulated by changes in DNA supercoiling. Infect. Immun. **58**, 1879–1885

85. Ernst, R.K., Domboski, D.M. and Merrick, J.M. (1990) Anaerobiosis, type I fimbriae and growth phase are factors that affect invasion of Hep-2 cells by *Salmonella typhimurium*. Infect. Immun. **58**, 2014–2016

86. Grassme, H.U., Ireland, R.M. and van Putten, J.P. (1996) Gonococcal opacity protein promotes bacterial entry-associated rearrangements of the epithelial cell actin cytoskeleton. Infect. Immun. **64**, 1621–1630

87. Prasadarao, N.V., Wass, C.A., Weiser, J.N., Stins, M.F., Huang, S.H. and Kim, K.S. (1996) Outer membrane protein A of *Escherichia coli* contributes to invasion of brain microvascular endothelial cells. Infect. Immun. **64**, 146–153

88. Prasadarao, N.V., Wass, C.A. and Kim, K.S. (1996) Endothelial cell GlcNAcβI-4GlcNAc epitopes for outer membrane protein A enhance traversal of *Escherichia coli* across the blood-brain barrier. Infect. Immun. **64**, 154–160

89. Hess, C.B., Niesel, D.W., Cho, Y.J. and Klimpel, G.R. (1987) Bacterial invasion of fibroblasts induces interferon production. J. Immunol. **138**, 3949–3953

90. Eckmann, L., Kagnoff, M.F. and Fierer, J. (1993) Epithelial cells secrete the chemokine interleukin-8 in response to bacterial entry. Infect. Immun. **61**, 4569–4574

91. Jung, H.C., Eckmann, L., Yang, S.K., Panja, A., Fierer, J., Morzyckawroblewska, E. and Kagnoff, M.F. (1995) A distinct array of proinflammatory cytokines is expressed in human colon epithelial cells in response to bacterial invasion. J. Clin. Invest. **95**, 55–65

92. Brett, S.J., Mazurov, A.V., Charles, I.G. and Tite, J.P. (1993) The invasin protein of *Yersinia* spp. provides co-stimulatory activity to human T cells through interaction with β_1 integrins. Eur. J. Immunol. **23**, 1608–1614

93. Yao, L., Bengualid, V., Lowy, F.D., Gibbons, J.J., Hatcher, V.B. and Berman, J.W. (1995) Internalization of *Staphylococcus aureus* by endothelial cells induces cytokine gene expression. Infect. Immun. **63**, 1835–1839

94. Smith, H. (1990) Pathogenicity and the microbe *in vivo*. J. Gen. Microbiol. **136**, 377–393

95. Garcia del Portillo, F., Foster, J.W., Maguire, M.E. and Finlay, B.B. (1992) Characterization of the micro-environment of *Salmonella typhimurium*-containing vacuoles within MDCK epithelial cells. Mol. Microbiol. **6**, 3289–3297

96. Litwin, C.M. and Calderwood, S.B. (1993) Role of iron in regulation of virulence genes. Chem. Microbiol. Rev. **6**, 137–149

97. Wooldridge, K.G. and Williams, P.H. (1993) Iron uptake mechanisms of pathogenic bacteria. FEMS Microbiol. Rev. **12**, 325–348

98. Weinberg, E.D. (1990) Roles of trace metals in transcriptional control of microbial secondary metabolism. Biol. Metals **2**, 191–196

99. Weinberg, E.D. (1995) Acquisition of iron and other nutrients *in vivo*. In Virulence Mechanisms of Bacterial Pathogens (Roth, J.A., Bolin, C.A., Brogden, K.A., Minion, F.C. and Wanemuehler, M.J., eds.), pp. 79–93, American Society for Microbiology, Washington

100. van der Waaji, D. (1992) Mechnisms involved in the development of the intestinal microflora in relation to the host organism: consequences for colonization resistance. Symp. Soc. Gen. Microbiol. **49**, 1–12

101. Salminen, S., Isolauri, E. and Onnela, T. (1995) Gut flora in normal and disordered states. Chemotherapy **41**, 5–15

102. Hentges, D.J., Pongpech, P. and Que, J.U. (1990) How streptomycin treatment compromises colonisation resistance against enteric pathogens in mice. Microb. Ecol. Health Dis. **3**, 105–111

103. Smith, H. (1995) The revival of interest in mechanisms of bacterial pathogenicity. Biol. Rev. **70**, 277–316

104. Fuller, R. (1992) Probiotics. The Scientific Basis. Chapman & Hall, London

105. Schonwetter, B.S., Stolzenberg, E.D. and Zasloff, M.A. (1995) Epithelial antibiotics induced at sites of inflammation. Science **267**, 1645–1648

106. Boman, H.G. (1995) Peptide antibiotics and their role in innate immunity. Annu. Rev. Immunol. **13**, 61–92

107. Weinberg, E.D. (1993) The iron-witholding defense system. ASM News **59**, 559–562

108. de Lillo, A., Teanpaisan, R., Fierro, J.F. and Douglas, C.W.I. (1996) Binding and degradation of lactoferrin by *Porphyromonas gingivalis, Prevotella intermedia* and *Prevotella nigrescens*. FEMS Immunol. Med. Microbiol. **14**, 135–143

109. Cornelissen, C.N., Biswas, G.D., Tsai, J., Parachuri, D.K., Thompson, S.A. and Sparling, P.F. (1992) Gonococcal transferrin-binding protein I is required for transferrin utilisation and is homologous to Ton-B-dependent outer membrane receptors. J. Bacteriol. **174**, 5788–5797

110. Meyer, J.M., Neely, A., Stinzi, A., Georges, C. and Holder, I.A. (1996) Pyoverdin is essential for virulence of *Pseudomonas aeruginosa*. Infect. Immun. **64**, 518–523

111. Johnson, J.A. (1995) Pathogenesis of bacterial infections of the respiratory tract. Br. J. Biomed. Sci. **52**, 157–161

112. McAleer, M.A. and Sim, R.B. (1993) The complement system. In Activators and Inhibitors of Complement (Sim, R.B., ed.), pp. 1–15, Kluwer Academic Publishing, Dordrecht, The Netherlands

113. Sim, R.B. and Malhotra, R. (1994) Interactions of carbohydrates and lectins with complement. Biochem. Soc. Trans. **22**, 106–111

114. Taylor, P.W. (1995) Resistance of bacteria to complement. In Virulence Mechanisms of Bacterial Pathogens (Roth, J.A., Bolin, C.A., Brogden, K.A., Minion, F.C. and Wannemuehler, M.J. eds.), pp. 49–64, American Society for Microbiology, Washington

115. Moffitt, M.C. and Frank, M.M. (1994) Complement resistance in microbes. Springer Semin. Immunopathol. **15**, 327–344

116. Mims, C., Dimmock, N., Nash, A. and Stephen, J. (1995) The encounter of the microbe with the phagocytic cell. In Mims' Pathogenesis of Infectious disease, pp. 75–105, Academic Press, London

117. Cundell, D.R., Taylor, G.W., Kanthakumar, K., Wilks, M., Tabaqchali, S., Dorey, E., Devalia, J.L., Roberts, D.E., Davies, R.J. and Wilson, R. (1993) Inhibition of human neutrophil migration in vitro by low molecular mass products of nontypeable *Haemophilus influenzae*. Infect. Immun. **61**, 2419–2424

118. Van Dyke, T.E., Bartholomew, E., Genco, R.J., Slots, J. and Levine, M.J. (1992) Inhibition of neutrophil chemotaxis by soluble bacterial products. J. Periodontol. **53**, 502–508

119. Fischetti, V.A. (1989) Streptococcal M protein: molecular design and biological behaviour. Clin. Microbiol. Rev. **2**, 285–295

120. Dale, J.B., Washburn, R.G., Marques, M.B. and Wessels, M.R. (1996) Hyaluronate capsule and surface M protein in resistance to opsonization of Group A streptococci. Infect. Immun. **64**, 1495–1501

121. Abshire, K.Z. and Neidhardt, F.C. (1993) Analysis of proteins synthesised by *Salmonella typhimurium* during growth within a host macrophage. J. Bacteriol. **175**, 3734–3743

122. Farr, S.B. and Kogoma, T. (1991) Oxidative stress responses in *Escherichia coli* and *Salmonella typhimurium*. Microbiol. Rev. **55**, 561–585

123. Gordon, A.H., D'Arcy-Hart, P. and Young, M.R. (1980) Ammonia inhibits phagosome-lysosome fusion in macrophages. Nature (London) **286**, 79–81

124. Neill, M. and Klebanoff, S.J. (1988) The effect of phenolic glycolipid-1 from *Mycobacterium leprae* on the antimicrobial activity of human macrophages. J. Exp. Med. **167**, 30–42

125. Brozna, J.P., Horan, M., Radenacher, J.M., Pabst, K.M. and Pabst, M.J. (1991) Monocyte responses to sulfatide from *Mycobacterium tuberculosis*: inhibition of priming enhances release of superoxide associated with increased secretion of interleukin-1 and tumour necrosis factor alpha and altered protein phosphorylation. Infect. Immun. **59**, 2542–2548

126. Sibley, L.D., Adams, L.B. and Krahenbuhl, J.L. (1990) Inhibition of interferon-gamma-mediated activation in mouse macrophages treated with lipoarabinomannan. Clin. Exp. Immunol. **80**, 141–148

127. Rasool, O., Freer, E., Moreno, E. and Jarstrand, C. (1992) Effect of *Brucella abortus* lipopolysaccharide on oxidative metabolism and lysozyme release by human neutrophils. Infect. Immun. **60**, 1699–1702

128. Shimoji, Y., Yokomizo, Y. and Mori, Y. (1996) Intracellular survival and replication of *Erysipelothrix rhusiopathiae* within murine macrophages: failure of induction of the oxidative burst of macrophages. Infect. Immun. **64**, 1789–1793

129. Ashkenazi, M., White, R.R. and Dennison, D.K. (1992) Neutrophil modulation by *Actinobacillus actinomycetemcomitans*. II. Phagocytosis and development of respiratory burst. J. Periodont. Res. **27**, 457–465

130. Miyasaki, K.T., Bodeau, A.L., Ganz, T., Selsted, M.E. and Lehrer, R.I. (1990) *In vitro* sensitivity of oral Gram-negative facultative bacteria to the bactericidal activity of human neutrophil defensins. Infect. Immun. **58**, 3934–3940

131. Fields, P.I., Groisman, E.A. and Heffron, F. (1989) A Salmonella locus that controls resistance to microbicidal proteins from phagocytic cells. Science **243**, 1059–1062

132. Jones, A.L., Beveridge, T.J. and Woods, D.E. (1996) Intracellular survival of *Burkholderia pseudomallei*. Infect. Immun. **64**, 782–790

133. Martinez de Tejada, G., Pizarro-Cerda, J., Moreno, E. and Moriyon, I. (1995) The outer membranes of *Brucella* spp. are resistant to bactericidal cationic peptides. Infect. Immun. **63**, 3054–3061

134. Visser, L.G., Hiemstra, P.S., van den Barselaar, M.T., Ballieux, P.A. and van Furth, R. (1996) Role of YadA in resistance to killing of *Yersinia enterocolitica* by antimicrobial polypeptides of human granulocytes. Infect. Immun. **64**, 1653–1658

135. Groisman, E.A. (1994) How bacteria resist killing by host-defence peptides. Trends Microbiol. **2**, 444–449

136. Eissenberg, L.G. and Wyrick, P.B. (1981) Inhibition of phagolysosome fusion is localised to *Chlamydia psittaci*-laden vacuoles. Infect. Immun. **32**, 889–898

137. Steed, L.L., Setareh, M. and Friedman, R.L. (1991) Intracellular survival of virulent *Bordetella pertussis* in human polymorphonuclear leukocytes. J. Leukocyte Biol. **50**, 321–330

138. Kaufmann, S.H.E. and Flesch, I.E.A. (1992) Life within phagocytic cells. Symp. Soc. Gen. Microbiol. **49**, 97–106

139. Kilian, M. and Russell, M.W. (1994) Function of mucosal immunoglobulin. In Handbook of Mucosal Immunology (Ogra, P.L., Mestecky, J., Lamm, M.E., Strober, W., McGhee, J. and Bienenstock, J., eds.), pp. 127–137, Academic Press, San Diego

140. Kilian, M., Reinholdt, J., Lomholt, H., Poulsen, K. and Frandsen, E.V. (1996) Biological significance of IgA1 proteases in bacterial colonization and pathogenesis: critical evaluation of experimental evidence. APMIS **104**, 321–338

141. Cole, M.F., Evans, M., Fitzsimmons, S., Johnson, J., Pearce, C., Sheridan, M.J., Wientzen, R. and Bowden, G. (1994) Pioneer oral streptococci produce immunoglobulin A1 protease. Infect. Immun. **62**, 2165–2168

142. Toma, C., Honma, Y. and Iwanaga, M. (1996) Effect of *Vibrio cholerae* non-O1 protease on lysozyme, lactoferrin and secretory immunoglobulin A. FEMS Microbiol. Lett. **135**, 143–147

143. Lee, S.F. (1995) Active release of bound antibody by *Streptococcus mutans*. Infect. Immun. **63**, 1940–1946

144. Roche, R.J. and Moxon, E.R. (1995) Phenotypic variation of carbohydrate surface antigens and the pathogenesis of *Haemophilus influenzae* infections. Trends Microbiol. **3**, 304–309

145. Robertson, B.D. and Meyer, T.F. (1992) Antigenic variation in bacterial pathogens. Symp. Soc. Gen. Microbiol. **49**, 61–73

146. Smyth, C.J. and Smith, S.G.J. (1992) Bacterial fimbriae, variation and regulatory mechanisms. Symp. Soc. Gen. Microbiol. **49**, 267–298

147. Zhang, Q.Y. (1992) Gene conversion in *Neisseria gonorrhoeae*; evidence for its role in pilus antigenic variation. Proc. Natl. Acad. Sci. U.S.A. **89**, 5366–5370

148. Balfour, A.G. (1990) Antigenic variation of a relapsing fever *Borrelia* species. Annu. Rev. Microbiol. **44**, 155–171

149. Sawitzke, A.D., Knudtson, K.L. and Cole, B.C. (1995) Bacterial superantigens in disease. In Virulence Mechanisms of Bacterial Pathogens (Roth, J.A., Bolin, C.A., Brogden, K.A., Minion, F.C. and Wannemuehler, M.J., eds.), pp. 145–169, American Society for Microbiology, Washington

150. Schlievert, P.M. (1993) Role of superantigens in human disease. J. Infect. Dis. **167**, 997–1002

151. Blackman, M.A. and Woodland, D.L. (1995) *In vivo* effects of superantigens. Life Sci. **57**, 1717–1735

152. Kaplan, G., Gandhi, R.R., Weinstein, D.E., Levis, W.R., Patarroyo, M.E., Brennan, P.J. and Cohn, Z.A. (1987) *Mycobacterium leprae* antigen-induced suppression of T cell proliferation in vitro. J. Immunol. **138**, 3028–3034

153. Moreno, C., Mehlert, A. and Lamb, J. (1988) The inhibitory effects of mycobacterial lipoarabino-mannan and polysaccharides upon polyclonal and monoclonal human T cell proliferation. Clin. Exp. Immunol. **74**, 206–210

154. Kurita-Ochiai, T. and Ochiai, K. (1996) Immunosuppressive factor from *Actinobacillus actino-mycetemcomitans* down-regulates cytokine production. Infect. Immun. **64**, 50–54

155. Staugas, R.E., Harvey, D.P., Ferrente, A., Nandoskar, M. and Allison, A.C. (1992) Induction of tumour necrosis factor (TNF) and interleukin-1 (IL-1) by *Pseudomonas aeruginosa* and exotoxin A-induced suppression of lymphoproliferation and TNF, lymphotoxin, gamma interferon and IL-1 production in human leukocytes. Infect. Immun. **60**, 3162–3168

156. Mintz, C.S., Miller, R.D., Gutgsell, N.S. and Malek, T. (1993) *Legionella pneumophila* protease inacti-vates interleukin-2 and cleaves CD4 on human T cells. Infect. Immun. **61**, 3416–3421

157. Klapproth, J.M., Donnenberg, M.S., Abraham, J.M., Mobley, H.L. and James, S.P. (1995) Products of enteropathogenic *Escherichia coli* inhibit lymphocyte activation and lymphokine production. Infect. Immun. **63**, 2248–2254

158. Demuth, D.R., Savary, R., Golub, E. and Shenker, B.J. (1996) Identification and analysis of fipA, a *Fusobacterium nucleatum* immunosuppressive factor gene. Infect. Immun. **64**, 1335–1341

159. Shenker, B.J. and Datar, S. (1996) Fusobacterium nucleatum inhibits human T-cell activation by arresting cells in the mid-G1 phase of the cell cycle. Infect. Immun. **63**, 4830–4836

160. Chan, J., Fan, X., Hunter, S.W., Brennan, P.J. and Bloom, B.R. (1991) Lipoarrabinomannan, a possible virulence factor involved in persistence of *Mycobacterium tuberculosis* within macrophages. Infect. Immun. **59**, 1755–1761

161. Welch, R.A. (1995) Phylogenetic analysis of the RTX toxin family. In Virulence Mechanisms of Bacterial Pathogens (Roth, J.A., Bolin, C.A., Brogden, K.A., Minion, F.C. and Wannemuehler, M.J., eds.), pp. 195–206, American Society for Microbiology, Washington

162. Welch, R.A., Bauer, M.E., Kent, A.D., Leeds, J.A., Moayeri, M., Regassa, L.B. and Swenson, D.L. (1995) Battling against host phagocytes: the wherefore of the RTX family of toxins? Infect. Ag. Dis. **4**, 254–272

163. Krueger, K.M. and Barbieri, J.T. (1995) The family of ADP-ribosylating exotoxins. Clin. Microbiol. Rev. **8**, 34–47

164. Krueger, K.M. and Barbieri, J.T. (1995) Bacterial ADP-ribosylating exotoxins. In Virulence Mechanisms of Bacterial Pathogens (Roth, J.A., Bolin, C.A., Brogden, K.A., Minion, F.C. and Wannemuehler, M.J., eds.), pp. 231–242, American Society for Microbiology, Washington

165. Tweten, R.K. (1995) Pore-forming toxins of Gram-positive bacteria. In Virulence Mechanisms of Bacterial Pathogens (Roth, J.A., Bolin, C.A., Brogden, K.A., Minion, F.C. and Wannemuehler, M.J., eds.), pp. 207–229, American Society for Microbiology, Washington

166. Paton, J.C., Andrew, P.W., Boulnois, G.J. and Mitchell, T.J. (1993) Molecular analysis of the patho-genicity of *Streptococcus pneumoniae*: the role of pneumococcal proteins. Annu. Rev. Microbiol. **47**, 89–115

167. Watson, D.A., Musher, D.M. and Verhoef, J. (1995) Pneumococcal virulence factors and host immune responses to them. Eur. J. Clin. Microbiol. Infect. Dis. **14**, 479–490

168. Rubins, J.B., Duane, P.G., Charboneau, D. and Janoff, E.N. (1992) Toxicity of pneumolysin to pulmonary endothelial cells in vitro. Infect. Immun. **60**, 1740–1746

169. Mitchell, T.J., Andrew, P.W., Saunders, F.K., Smith, A.N. and Boulnois, G.J. (1991) Complement activation and antibody binding by pneumolysin via a region of the toxin homologous to a human acute phase protein. Mol. Microbiol. **5**, 1883–1888

170. Berry, A.M., Alexander, J.E., Mitchell, T.J., Andrew, P.W., Hansman, D. and Paton, J.C. (1995) Effect of defined point mutations in the pneumolysin gene on the virulence of *Streptococcus pneumoniae*. Infect. Immun. **63**, 1969–1974

171. Houldsworth, S., Andrew, P.W. and Mitchell, T.J. (1994) Pneumolysin stimulates production of tumor necrosis factor alpha and interleukin-1 beta by human mononuclear phagocytes. Infect. Immun. **62**, 1501–1503

172. Goguen, J.D., Hoe, N.P. and Subrahmanyam, Y.V.B.K. (1995) Proteases and bacterial virulence: a view from the trenches. Infect. Agents Dis. **4**, 47–54

173. Hong, Y.Q. and Ghebrehiwet, B. (1992) Effect of *Pseudomonas aeruginosa* elastase and alkaline protease on serum complement and isolated components C1q and C3. Clin. Immunol. Immunopathol. **62**, 133–138

174. Travis, J., Potempa, J. and Maeda, H. (1995) Are bacterial proteinases pathogenic factors? Trends Microbiol. **3**, 405–407

175. Homer, K.A., Kelley, S., Hawkes, J., Beighton, D. and Grootveld, M.C. (1996) Metabolism of gly-coprotein-derived sialic acid and N-acetylglucosamine by *Streptococcus oralis*. Microbiology **142**, 1221–1230

176. Kontani, M., Ono, H., Shibata, H., Okamura, Y., Tanaka, T., Fujiwara, T., Kimura, S. and Hamada, S. (1996) Cysteine protease of *Porphyromonas gingivalis* 381 enhances binding of fimbriae to cultured human fibroblasts and matrix. Infect. Immun. **64**, 756–762

177. Parmely, M., Gale, A., Clabaugh, M., Horvat, R. and Zhou, W-W. (1990) Proteolytic inactivation of cytokines by *Pseudomonas aeruginosa*. Infect. Immun. **58**, 3009–3014

178. Fletcher, J., Reddi, K., Poole, S., Nair, S., Henderson, B., Tabona, P. and Wilson, M. (1997) Interactions between periodontopathogenic bacteria and cytokines. J. Periodont. Res. **32**, 200–205

179. Kapur, V., Majesky, M.W., Li, L.L., Black, R.A. and Musser, J.M. (1993) Cleavage of interleukin 1 beta (IL-1 beta) precursor to produce active IL-1beta by a conserved extracellular cysteine protease from *Streptococcus pyogenes*. Proc. Natl. Acad. Sci. U.S.A. **90**, 7676–7680

180. Engel, D., Grenier, D., Hobbs, M., Morgan, E., Hugli, T.E. and Weigle, W.O. (1994) Inflammatory potential of IgG Fc fragments generated by *Porphyromonas gingivalis* protease. FASEB J. **8**, 223

181. Hardingham, T.E. and Fosang, A.J. (1992) Proteoglycans: many forms and many functions. FASEB J. **6**, 861–870

182. Henderson, B., Poole, S. and Wilson, M. (1996) Bacterial modulins: a novel class of virulence factors which cause host tissue pathology by inducing cytokine synthesis. Microbiol. Rev. **60**, 316–341

183. Wilson, M. and Henderson, B. (1996) Cytokine-inducing components of periodontopathogenic bacteria. J. Periodont. Res. **31**, 393–407

184. Hoiby, N., Doring, G. and Schiotz, P.O. (1986) The role of immune complexes in pathogenesis of bacterial infections. Annu. Rev. Microbiol. **40**, 29–53

185. Dannenberg, A.M. (1989) Immune mechanisms in pathogenesis of pulmonary tuberculosis. Rev. Infect. Dis. **11**, S369–S378

186. Baird, R.W., Bronz, M.S., Kraus, W., Hill, H.R., Veasey, L.G. and Dale, J.B. (1991) Epitopes of Group A streptococcal M protein shared with antigens of articular cartilage and synovium. J. Immunol. **146**, 3132–3137

187. Straley, S.C. and Perry, R.D. (1995) Environmental modulation of gene expression and pathogene-sis in Yersinia. Trends Microbiol. **3**, 310–317

188. Parkinson, J.S. (1993) Signal transduction schemes of bacteria. Cell **73**, 857–871

189. Parkinson, J.S. (1995) Genetic approaches for signalling pathways and proteins. In Two-Component Signal Transduction (Hoch, J.A. and Silhavy, T.J., eds.), pp. 9–24, American Society for Microbiology Press, Washington

190. Dorman, C.J. and Bhriain, N.N. (1992) Global regulation of gene expression during environmental adaptation: implications for bacterial pathogens. Symp. Soc. Gen. Microbiol. **49**, 193–230

191. Mekalanos, J.J. (1992) Environmental signals controlling expression of virulence determinants in bacteria. J. Bacteriol. **174**, 1–7

192. Dziejman, M. and Mekalanos, J.J. (1995) Two-component signal transduction and its role in the expression of bacterial virulence factors. In Two-Component Signal Transduction (Hoch, J.A. and Silhavy, T.J., eds.), pp. 305–317, American Society for Microbiology Press, Washington

193. Mahan, M.J., Slauch, J.M., Hanna, P.C., Camilli, A., Tobias, J.W., Waldor, M.K. and Mekalanos, J.J. (1994) Selection for bacterial genes that are specifically induced in host tissues: the hunt for virulence factors. Infect. Agents Dis. **2**, 263–268

194. Mahan, M.J., Slauch, J.M. and Mekalanos, J.J. (1993) Selection of bacterial virulence genes that are specifically induced in host tissue. Science **259**, 686–688

195. Mahan, M.J., Tobias, J.W., Slauch, J.M., Hanna, P.C., Collier, R.J. and Mekalanos, J.J. (1995) Antibiotic-based selection for bacterial genes that are specifically induced during infection of a host. Proc. Natl. Acad. Sci. U.S.A. **92**, 669–673

196. Fuqua, W.C., Winans, S.C. and Greenberg, E.P. (1994) Quorum sensing in bacteria: the LuxR-LuxI family of cell density-responsive transcriptional regulators. J. Bacteriol. **176**, 269–275

197. Kaplan, H.B. and Greenberg, E.P. (1985) Diffusion of autoinducer is involved in regulation of the *Vibrio fischeri* luminescence system. J. Bacteriol. **163**, 1210–1214

198. Latifi, A., Winson, M.K., Foglino, M., Bycroft, B.W., Stewart, G.S., Lazdunski, A. and Williams, P. (1995) Multiple homologues of LuxR and LuxI control expression of virulence determinants and secondary metabolites through quorum sensing in *Pseudomonas aeruginosa* PAO1. Mol. Microbiol. **17**, 333–343

4

Cytokines, homeostasis, networks and disease

"After such knowledge, what forgiveness? Think now
History has many cunning passages, contrived corridors
And issues, deceives with whispering ambitions,
Guides us by vanities"

T.S. Eliot, Gerontion

4.1   Introduction

The average human body is composed of a staggering number of cells (10^{13}) which are arranged into a variety of what may be termed fixed (e.g. muscle, liver, spleen) and motile (cellular elements of the blood) 'tissues'. It is self-evident (at least in the 1990s) that the activities of this multitude of biochemically, physiologically and morphologically distinct cells have to be integrated to produce a normally functioning organism. It is now clear that cytokines are key molecules, controlling cellular integration in all aspects of the organism from conception to death.

Our current understanding of the integrative nature of biological systems owes much to the pioneering work of the Frenchman, Claude Bernard (1813–1878) of the Collège de France, who created the concept of the maintenance of the 'milieu interieur' and suggested that the system of ductless glands, the endocrine system, performed an integrating function to maintain home-ostasis [1]. However, it was not until 1902 that Bayliss and Starling [2] demonstrated that an acid extract of the duodenum, when injected into dogs, induced a marked flow of pancreatic juice. The substance responsible was named secretin and Starling subsequently coined the term 'hormone' (from the Greek word meaning 'I excite') for such intercellular messengers. From this study, the concept developed of hormones as molecules produced by specific glands and secreted directly into the blood, in which they are conveyed to

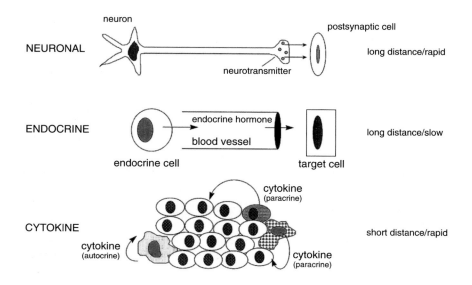

Figure 4.1 Comparison of the three main systems for integrating cell and tissue behaviour in the multicellular organism

Neuronal communication generally works over long distances and is relatively rapid (myelinated fibres can conduct signals at velocities of up to 130 m/s). Endocrine signalling is relatively slower, depending on blood transport and diffusion for transmission of the signal. In contrast, many cytokines act at the local level. Such interactions, if they occur between neighbouring cells, are termed paracrine ('para': nearby) or, if they feedback on to the cytokine-producing cell, are called autocrine ('auto': self).

selected organs or tissues where they exert their effect (Figure 4.1). These ideas were further developed by the American physiologist Walter Cannon, who coined the term homeostasis (from the Greek words: 'homoios', 'the same'; and 'stasis', 'standing'), which literally means 'standing or staying the same' [3]. However, Cannon was aware of the dynamic nature of physiological systems and defined homeostasis as 'a condition that may vary, but which is relatively constant'.

At the end of the Second World War advances in mathematics and the development of the computer had given rise to the concepts of cybernetics, information theory and games theory [4]. Cybernetics, its offshoot General System Theory, and also non-equilibrium thermodynamics [5], have had marked influences on our understanding of the mechanisms used in the homeostatic control of biological systems [4]. It was during this period that the pioneering work was begun on endotoxin-induced fever, an aberration in the control system of temperature homeostasis. This led to the discovery of protein factors which were called endogenous pyrogens, and which we now know as cytokines (and in particular IL-1, TNF, IL-6 and IFNγ). The history of the relationship between the study of endotoxin/lipopolysaccharide (LPS) and the discovery of cytokines is described in more detail in Chapters 5 and 6.

The cytokines are now viewed as key signalling molecules involved in the homeostatic integration of cellular, tissue and organismal activities.

In the 1990s we now know that there are an extremely large number of proteins which fit the loose definition of the term cytokine. Horst Ibelgaufts, in a prodigious feat of erudition, has produced a dictionary of cytokines which, when it was published in 1995, contained 778 pages [6]. Updates of this book are now available on the Internet. More than 100 different cytokine gene products have been identified to date. This compares with only a few dozen protein or peptide hormones. Increasingly, the definitions of cytokines and hormones are being blurred and the overlap between these two groups of (informational) molecules is growing. A good example is the cytokine erythro-poietin (EPO), which is produced primarily by cells in the kidney. This protein circulates in the blood and acts as a stimulator of haematopoiesis. The synthesis of EPO is controlled by the kidney, which senses the blood oxygen tension, essentially a measure of erythrocyte numbers. The synthesis and activity of EPO is controlled by interaction with various cytokines such as IL-1, IL-4 and TNF. Thus, in terms of its presence in the blood and its control by physiological regulatory processes, EPO would appear to merit classification as an endocrine hormone. However, there are two pieces of evidence to support this protein also being classed as a cytokine. First, EPO is produced by a variety of cells in the kidney and liver and there is no specific 'endocrine'-producing cell. Secondly, the EPO receptor is a member of the cytokine receptor superfamily or type I receptor family, which includes receptors for growth hormone and prolactin (both endocrine hormones) and cytokines such as IL-2, IL-3, IL-4, IL-5 and IL-6 [7].

As we approach the Millennium it is becoming evident that for biological systems to function they must have integrated control systems. Such biological control requires information transmission. Endocrine hormones and cytokines, along with the nervous system, are clearly the major systems for integrating cellular behaviour within the multicellular organism (Figure 4.1; Table 4.1). With more than 100 cytokines, and at least an equal number of receptors, the interactions between these proteins are now recognized to be complex, and such interactions (of which we know very little) have been termed cytokine networks. It is likely that the generation of such networks induces stable homeostatic conditions and that perturbation of such networks results in disease. It is not known if these networks have any relationship with neuronal networks in the brain or with artificial neural networks which are increasingly popular with neuroscientists and computer programmers. The authors propose that stable cytokine networks are required to control the response of the host to the vast numbers of bacteria (10^{14}) which constitute the normal microflora of *Homo sapiens*. This idea is developed in more detail in the final chapter of this book.

This chapter will give a brief overview of cytokine biology for those not familiar with this new and rapidly growing area of biology. We are not

Table 4.1 Comparison of endocrine hormones and cytokines

Biological properties	Endocrine hormones	Cytokines
Producing cells	Limited, possibly only one cell type	Many, for some cytokines possibly all cells
Responding cells	Ranges from one target cell (e.g. with TSH) to many (e.g. with steroids or thyroid hormones)	Generally, a range of cell types
Role	Maintenance of homeostasis	Similar, plus control of immune and inflammatory processes plus tissue repair
Redundancy	Low	High
Pleiotropy	Low	High
Present in blood	Yes	Generally no except in pathology (EPO is an exception)
Sites of action	Widespread	Generally local — autocrine and paracrine action
Inducers	Physiological stimuli	As hormones, plus many additional stimuli including infection, injury, etc.

Abbreviation used: TSH, thyroid-stimulating hormone.

attempting to be comprehensive. Readers interested in the details of the structure, genetics, molecular biology and cell biology of cytokines are referred to the following textbooks [6,8–14].

4.2   What are cytokines?

The general term cytokine, while first used by Cohen and co-workers in 1974 [15], has only crept into popular usage during the past decade. Prior to this, the molecules now termed cytokines were known under various classifications — lymphokines, monokines and growth factors being the major types. The terms lymphokines and monokines have largely been dropped because of the findings that many of these so-called molecules were made by cells that were not lymphocytes or monocytes and acted on cells other than those of the lymphoid or myeloid lineages. However, the major determining factor in placing all these molecules under the one generic heading was the realization that the cytokines exhibit a surprising overlap in biological actions which is due to their binding to and modulating the activity of cells. Thus, one can define cytokines as a miscellaneous group of inducible (generally soluble) peptides, proteins or glycoproteins, which have no inherent activity and act as regulators of cell, tissue and organismal function by binding with high affinity to specific cell-surface receptors. This high binding affinity of cytokines (in the nano- to picomolar range) and the generally low receptor density on receptive cells has resulted in a very sensitive cell-control system able to respond to cytokine levels in the nano- to femtomolar range. In the endocrine system, various hormones are stored in readiness for activation of the glandular tissue (e.g. the thyroid hormones). Most cytokines are not stored in cells and their synthesis generally is *de novo* and depends upon the correct signals reaching the producing cell. There are, of course, exceptions to this. Platelets (which are non-nucleated cells) store the cytokines transforming growth factor-β (TGFβ) and platelet-derived growth factor (PDGF). In addition, certain growth factors can be found bound to the connective tissue macromolecules of the extracellular matrix. As described in Chapters 5 and 6, the first cytokines to be recognized were the pro-inflammatory cytokines with pyrogenic activity (particularly IL-1). The actions of such endogenous pyrogens were being actively researched in the 1950s and 1960s. Another group of molecules, now firmly entrenched in our paradigm of host defence response, the interferons, was first described in the late 1950s [16]. However, cytokine research received its greatest boost from the studies of protein factors which could influence the behaviour (movement, proliferation, etc.) of lymphoid and myeloid cell populations. Such studies also demonstrated the activity of the so-called colony-stimulating factors — proteins that were studied using *in vitro* assays, which depended on the formation of colonies of cells for their 'read-out'. These various proteins were termed lymphokines by Dudley Dumonde and colleagues at the Kennedy Institute of Rheumatology in London, U.K. [17]. This branch of immunological studies revealed that acquired immunity involved another group of protein mediators, in addition to antibodies and major histocompatibility complex (MHC) proteins. Studies of the lymphokines led to the introduction of the term 'interleukin' to describe proteins that acted to communicate between leucocytes, and the first two cytokines to which this nomenclature was applied were IL-1 and IL-2 [18]. The 1960s also saw the beginnings of research on growth factors with the discovery of high concentrations of nerve growth factor (NGF) and epidermal growth factor (EGF) in saliva [19]. Thus, it can be seen that the discovery of cytokines has come from a number of biological disciplines: immunology has fostered studies of the ILs, haematologists have investigated the colony-stimulating factors, virologists the interferons, and cell biologists and oncologists the growth factors.

The past two decades have seen an unprecedented upsurge in information about cytokines. This has been the result of: (i) developments in molecular biology, allowing the rapid cloning and expression of novel proteins; (ii) the finding that cytokines are pathological mediators; and (iii) the evolution of the biotechnology/biopharmaceutical industry with its interest in idiopathic disease. This concatenation of events has been responsible for the rapid discovery of the many hundreds of proteins — cytokines and their receptors — which constitute the new branch of biological/physiological/medical science known as cytokine biology. This cornucopia of factors has brought with it a highly confusing terminology, primarily due to the need of the investigator to rapidly name his or her cytokine in order to obtain priority in the publications or patents arena. Thus, cytokines are frequently named in terms of the first biological action attributed to them. However, as cytokines generally have a wide range of actions, this first activity described may only be a minor role. Hence, the name of the protein may fail to describe its major biological function(s). In addition, many cytokines have been discovered simultaneously by different groups and given different names. For example, tumour necrosis factor-α (TNFα) [also known as cachectin, cytotoxic factor, differentiation-inducing factor, etc.] is cytotoxic for a small number of transformed cell lines and human breast cancer transplanted into mice. However, its activity as an anti-tumour agent in man has been disappointing. This is presumably due to the fact that TNFα is an important host-defence pro-inflammatory molecule which has multiple effects on lymphoid, myeloid and mesenchymal cells. TGFβ [also known as cartilage-inducing factor, differentiation-inhibiting factor, epithelial cell growth-inhibiting factor, milk-derived growth factor, etc.] provides an excellent example of the inappropriateness of cytokine nomenclature. Despite the appellation — growth factor — this molecule is a potent inhibitor of the proliferation of a wide variety of cells, including epithelial cells, fibroblasts and lymphocytes. It is no wonder that cytokine biology is found to be a confusing subject.

In the next section, cytokine nomenclature will be considered and the current criteria used to differentiate cytokine families will be reviewed.

Table 4.2 Cytokine families

Family	Examples	Major biological activities
Interleukins	IL-I to IL-18	Mainly lymphoid-lineage growth factors
Cytotoxic cytokines	TNFα, TNFβ, CD40L	Pro-inflammatory molecules with cytotoxic/apoptotic potential
Interferons	α*-, β-, γ-, ω-Interferons	Anti-viral and immunological actions
Colony-stimulating factors	IL-3, M-CSF, G-CSF, GM-CSF	Myeloid growth and differentiation factors
Growth factors	EGF, TGFα, PDGF	Proliferation of various cell types including epithelial and mesenchymal cells
Chemokines	IL-8, MIP-Iα, MCP	Chemotactic proteins for various leucocyte populations

*Many subtypes of α-interferon exist.

Table 4.3 Biological actions common to IL-1 and TNFα

In vivo	*In vitro*
Endogenous pyrogens	Activators of T- and B-lymphocytes
Inducers of acute-phase protein synthesis	Inducers of cyclo-oxygenase II
Inducers of septic shock-like conditions	Inducers of metalloproteinase synthesis and release
Inhibitors of cardiac function	Stimulators of fibroblast proliferation
Inducers of hyperalgesia	Inducers of vascular endothelial cell adhesion molecules (ICAM, E-selectin)
	Modulators of vascular endothelial cell coagulation systems
	Stimulators of cartilage breakdown
	Inhibitors of connective tissue matrix macromolecules
	Stimulators of bone resorption
	Induction of cytokine synthesis including autocrine stimulation of IL-1 and TNF

4.2.1 Cytokine nomenclature

As stated above, a very large number of cytokines are known to exist and their rate of discovery has not abated. The pleiotropy of cytokines, their diverse structures and the finding that known proteins and peptides (e.g. molecular chaperones, antibiotic peptides, etc.) may show functional overlap with cytokines [20] (Chapter 9) make it important to attempt to classify these proteins in order to bring order to this area of biology. At the present time cytokines may be divided into six families (Table 4.2). These subdivisions are based upon the biological actions of the cytokines. However, with the great overlap in activity displayed by certain cytokines (see Table 4.3 for a comparison of the biological actions of IL-1 and TNFα), such divisions are probably artificial and will change with time.

The term interleukin ('inter', between; 'leukin', leucocytes) was coined in 1979 [18] to describe protein factors produced by leucocytes which functioned to modulate the behaviour of other leucocytes. IL-1 and IL-2 were the first cytokines to be given this new nomenclature. There are currently 20 inter-leukins [IL-1 to IL-18, including three members of the IL-1 family, i.e. IL-1α, IL-1β and IL-1ra (IL-1 receptor antagonist)], which have been well character-ized, and two additional proteins have recently been tentatively given the titles IL-17 and IL-18. The cellular sources and functions of these interleukins are listed in Table 4.4. As can be seen, the designation, interleukin, is inappropriate for a number of these proteins. For example, members of the IL-1 family, IL-3, IL-6, IL-8, and IL-11, are made by cells other than leucocytes and some of these cytokines can act on cells other than leucocytes.

The so-called cytotoxic cytokines initially included only TNFα and lym-photoxin (also called TNFβ). In addition to their ability to kill certain cell

Table 4.4 Cellular source, gene location/structure and major function of the interleukins

Interleukin*	Gene location and structure	Cell source	Major functions
IL-1α (17)	2q13–21; seven exons	Many cell types	Multiple functions (see Table 4.3)
IL-1β (17)	2q13–21; seven exons	Many cells types	As IL-1α
IL-1ra (17–22)	2q14–21; four exons	MØ, endothelial cell, keratinocytes	Inhibits biological actions of IL-1 (antagonist)
IL-2 (15)	4q26; four exons	Th$_0$, Th$_1$, memory T-cells	T-cell growth and differentiation
IL-3 (15–17)	5q23–31; five exons	T-cells, mast cells, epithelial cells	PMN/monocyte colony stimulation
IL-4 (20)	5q23-31, 4 exons	Th$_2$ lymphocytes	B-cell proliferation, MØ inhibition
IL-5 (13)	5q23–31; four exons	T-lymphocytes	Eosinophil growth/ differentiation factor
IL-6 (21)	7p21; four exons	Many cell types	Multiple functions (overlaps with IL-1)
IL-7 (25)	8q12-13, 6 exons	Bone marrow stromal cells/ thymic cells	B/T-cell proliferation factor
IL-8 (8)	4q12–21; four exons	MØ, fibroblasts, PMN, etc.	T-cell/monocyte chemokine/ PMN activator
IL-9 (18)	5q31–35; five exons	CD4 T-cells	Mast cell/T-cell growth factor
IL-10 (40 hd)	chr1; five exons	CD4/CD8 T-cells, monocytes	Anti-inflammatory, inhibits MØ, T-cell activity
IL-11 (23)	19q13.3–13.4; five exons	Mesenchymal cells	Haemapoietic growth factor, accessory immune factor
IL-12 (70 htd)	5q31–33, 3p12–q13.2	MØ, B- and T-lymphocytes, etc.	Generates Th$_1$ lymphocytes
IL-13 (10)	5q31; four exons	T-lymphocytes	Stimulates B-cells, inhibits some MØ functions
IL-14 (60)		T- and B-lymphocytes	B-cell growth factor
IL-15 (11)		Peripheral blood mononuclear cells	T-cell growth factor
IL-16 (13)		T-lymphocytes	T-cell chemoattractant

*Parentheses indicate molecular mass (kDa) of biologically active interleukin. Abbreviations used: hd, homodimer; htd, heterodimer; MØ, macrophage.

Table 4.5 Cytotoxic and growth-inhibiting cytokines

Cytokine*	Gene location and structure	Cell source	Major functions
TNFα (17)	6p23–6q12; four exons	MØ, PMN, T-cells	See Table 4.3
TNFβ (20–25)	6p23–6q12; four exons	T-cells, fibroblasts, endothelial cells	Similar to TNFα
Fas ligand† (31)			Induces apoptosis
CD40 ligand† (28)		B-, T-monocytes	Ig class switching, blocks apoptosis
CD30 ligand†			T-cell proliferation; T-cell death
CD27 ligand†		T-cells	T-cell proliferation
NGF (trimer)	1p13	Many	Neurotrophic factor
Oncostatin M (28)	22q12–q12.2; three exons	MØ, T-cells	Anti-proliferative or proliferative depending on cell type
Amphiregulin (9–10)	4q13–21	Placenta, testis, ovary	As above
Mullerian inhibiting substance (145)	19p13.3–13.2; five exons	Sertoli cells, granulosa cells	Regression of Müllerian ducts

*Parentheses indicate molecular mass (kDa) of biologically active cytokine. †Proteins are membrane bound.

lines, these cytokines are also potent pro-inflammatory molecules. Isolation and analysis of the TNF receptor gene has revealed that there is in fact a family of TNF receptors which binds a wide range of ligands including NGF. Of note is the fact that a number of the ligands of the TNF receptor family are themselves membrane-bound proteins (e.g. CD27, CD30 and CD40). As the key role of the TNFs is cell killing or inhibition of cell growth, it might be more logical to designate them as a cell-growth-inhibitor family; this would bring in a number of other cytokines such as amphiregulin and oncostatin M which do not bind to the TNF receptor family (Table 4.5).

The interferons were among the first cytokines to be discovered. These cytokines play a major role in inhibiting the growth and spread of viruses. The mechanism of action of the interferons is shown schematically in Figure 4.2 and the various interferons are classified in Table 4.6. For further information on the role of interferons in viral infections the reader is referred to [21,22].

The colony-stimulating factors (CSFs) are a small group of cytokines involved in the control of the growth and differentiation of granulocytes (PMNs) and monocytes in the bone marrow. These cytokines also act on mature cells and play a role in inflammation. It would be logical to group the CSFs together with other factors involved in haematopoiesis (Table 4.7).

The growth factors now encompass a very large number of proteins, including families such as the TGFβ superfamily [which, in turn, includes the bone morphogenetic proteins (BMPs) and the activins/inhibins] and the

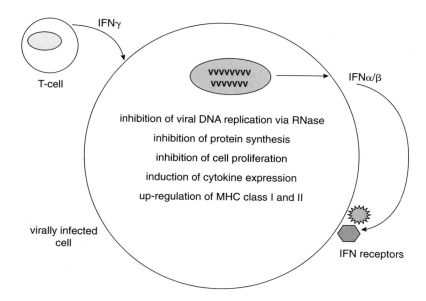

Figure 4.2 The mechanism of action of interferons (IFNs) in virally infected cells
A cell infected by a virus (v) induces the expression of IFNα and IFNβ, which are released by the cell and bind, in an autocrine manner, to the same cell or, in a paracrine manner, to nearby cells. Such cell receptor activation results in cellular changes which inhibit viral replication (by cleaving viral RNA) and cellular proliferation. Activated T-cells can also produce IFNγ which can induce class I and II expression on the virally infected cell, making it a target for immune killing.

Table 4.6 The interferons

Interferon†	Gene location and structure	Cell source	Major functions
IFNα (19–26)*	9p22; no introns	Many	Antiviral, anti-proliferative
IFNβ (20)	9p22; no introns	Fibroblasts, epithelial cells	Regulates non-specific immune responses to virally infected cells
IFNγ (20, 25)	12q24.1; four exons	T-cells	Many immune-regulating functions
IFNω	Similar to IFNα in structure and, potentially, in activity		

*Parentheses indicate molecular mass (kDa) of biologically active interferon. †There are at least 23 different variants of IFNα.

fibroblast growth factor (FGF) family. Again, it must be noted that some of the growth factors are able to inhibit the growth of certain primary cells or cell lines. The reader should also be aware that many of the cytokines already described in the various tables are also growth factors for lymphoid or myeloid populations. A listing of some of the well characterized growth factors is provided in Table 4.8.

Table 4.7 The colony stimulating and haematopoietic factors

Factor*	Gene location and structure	Cell source	Major functions
IL-1 (see Table 4.4)			
IL-3 (see Table 4.4)			
IL-11 (see Table 4.4)			
CD27 ligand (see Table 4.5)			
M-CSF†	1p13–2212; 10 exons	Monocytes, PMN, EC, fibroblasts	Stimulates growth of MØ stem cells; stimulates MØ functions
G-CSF (20)	17q21–q22; five exons	MØ, PMN, EC, fibroblasts	Stimulates growth of granulocyte stem cells, activates mature PMNs
GM-CSF (18–32)	5q22–31; four exons	MØ, T-cells, EC, fibroblasts	Stimulates MØ and granulocyte progenitor cells; has pro-inflammatory actions
EPO (34–37)	7q21–q22; five exons	Kidney EC and interstitial cells	Differentiation factor for erythroid precursor cells
LIF (58)	22q12–q12.2; three exons	MØ, T-cells, fibroblasts	Stimulates cell proliferation, acute-phase protein synthesis, etc.
SCF (36)	12q22–q24; 21 exons	Fibroblasts	Growth factor for primitive lymphoid and myeloid haematopoietic bone marrow progenitor cells
Activin‡		Bone marrow cells	Erythroid growth factor, stimulates endocrine hormone (e.g. FSH) production

*Parentheses indicate molecular mass (kDa) of biologically active factor. †A number of alternatively spliced variants of M-CSF are produced: M-CSFα (256 amino acids); M-CSFβ (554 amino acids); and M-CSFγ (438 amino acids). ‡Three different activins exist, designated A, B and A-B. Abbreviations used: EC, endothelial cell; EPO, erythropoietin; FSH, follicle-stimulating hormone; MØ, macrophage; LIF, leukaemia inhibitory factor; PMN, polymorphonuclear leucocyte; SCF, stem cell factor.

The final subdivision of the cytokines is the large group of peptide chemotactic factors known as chemokines (Table 4.9). The chemokines have molecular masses in the range 8–10 kDa with 20–50% sequence homology at the protein level, and all have four conserved cysteine residues involved in intramolecular disulphide bond formation. This large group of proteins can be

Table 4.8 Growth factors

Growth factor	Molecular mass (kDa)	Comment
TGFβ	25	This superfamily now consists of 30 or more members; the TGFβs (1–3) are inhibitors of cell proliferation
FGF	16–18	Family of nine or more heparin-binding proteins; major action may be as angiogenesis factors
PDGF	30 (heterodimer)	Three forms exist and are found in platelets; multiple actions in addition to acting as growth factor
TGFα	5–20	Growth factor in many tissues for many cell types
EGF	6	Family of EGF-like or EGF domain-containing proteins (betacellulin, amphiregulin, uEGF-1, *Delta, Notch, Lin*-12)
Betacellulin	32	Mitogen for certain epithelial and vascular smooth muscle cells
Amphiregulin	8	Growth factor for human carcinoma cell lines; may be involved in cancer and psoriasis
IGF-1/IGF-2	7	Related to NGF and relaxin; plasma proteins regulated by growth factor. IGF-1 knockout mice die at birth
VEGF	46–48	Mitogen for vascular endothelial cells
KGF	23	Specific mitogen for epithelial cells
HGF	82	Most potent mitogen for hepatocytes; induces endothelial cell proliferation and migration
Heregulin	45	Ligand of *neu* oncogene receptor
CNTF	24	Neurotrophic factor

Abbreviations used: CNTF, ciliary neurotrophic growth factor; HGF, hepatocyte growth factor; KGF, keratinocyte growth factor; VEGF, vascular endothelial cell growth factor; IGF, insulin-like growth factor.

divided into two families based both upon the chromosomal location of the genes and the structure of the protein. The α-chemokine family maps to human chromosome 4q12–21 and the first two cysteine residues are separated by a single amino acid. Because of this, these proteins are called the C-X-C chemokines (C being the single-letter code for cysteine). The β-chemokine family maps to human chromosome 17q11–32. The first two cysteine residues are adjacent in this group of proteins, which are therefore termed the C-C chemokines. The specificity of chemokines for leucocytes is not fully established. IL-8, which is a C-X-C chemokine, is a powerful neutrophil chemoattractant, but will also cause lymphocyte chemotaxis at higher concen-

Table 4.9 Chemokines

Family	Chemokine	Producing cell	Cells chemoattracted
C-X-C	IL-8	MØ, PMN, VEC, etc.	PMN, basophils, eosinophils
	ENA-78	Epithelial cells	PMN
	Gro-α	MØ, PMN, platelet	PMN
	GCP-2	Osteosarcoma cells	PMN
	Platelet factor-4	Platelets, MØ	PMN, monocyte, fibroblasts
	IP-10	MØ, VEC, fibroblasts	PMN, monos, T-cells
	MIG	MØ	PMN
	MIP-2	MØ, Langerhans cells	PMN
C-C	MCP-1	MØ, fibroblasts, VEC	Monos
	MCP-2	Osteosarcoma	Monos
	MCP-3	Osteosarcoma	Monos, eosinophils
	MIP-1α	MØ, mast cell, eosinophil	Monos, CD8 T- and B-cells
	MIP-1β	MØ, mast cell	Monos, CD4 T-cells
	RANTES	Mast cell, platelet, T-cells	Monos, T-cells, eosinophils
	Eotaxin	MØ	Eosinophils
	I-309	T-cells	Monocytes
Lymphotactin [22a]			T-cells

Abbreviations used: ENA, epithelial-derived neutrophil chemoattractant; GCP, granulocyte chemotactic protein; IP, IFNγ-inducible protein; MCP, monocyte chemotactic protein; MØ, macrophage; MIP, macrophage inflammatory protein; PMN, polymorphonuclear leucocyte; RANTES, regulated on activation, normal T expressed and secreted; VEC, vascular endothelial cell; MIG, monocyte induced by IFNγ.

trations. While not a hard and fast rule, C-X-C chemokines tend to attract neutrophils, while C-C chemokines attract monocytes. Eotaxin is a recently described chemokine which is selective for eosinophils. There is a third family of chemokines, first discovered in 1994, with currently one family member called lymphotactin. This protein lacks two of the four characteristic cysteine residues, but in terms of sequence resembles the C-C chemokines. Lymphotactin is a powerful chemoattractant for T-lymphocytes.

The subdivisions of cytokines described were derived using essentially 'historical' criteria and, given our current knowledge about the biological activity, gene and protein structure and receptor-binding of cytokines, other ways of classifying these proteins are now possible. For example, in Figure 4.3 the various cytokines have been subdivided into nine groupings which delineate the major biological actions of the cytokines.

4.2.2   Subdivision of cytokines based on structure

The immense interest in cytokines displayed by the pharmaceutical and bio-pharmaceutical industries has resulted in an outpouring of structural data from X-ray crystallographic and NMR solution studies. The small size of most cytokines has allowed the latter technique to be used to define their structures.

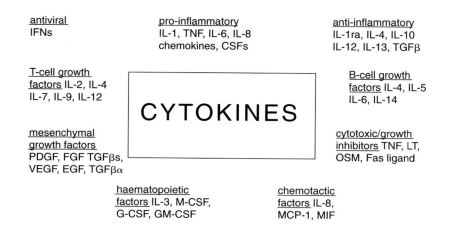

Figure 4.3 Diagram showing how the known cytokines could be subdivided into nine categories, depending on their major biological activities
For a full list of the abbreviations used, see page xiii.

The crystal structure of IL-1β has been attained at a resolution of 0.2 nm [23] and this was followed by determination of its structure using three- and four-dimensional NMR [24]. In contrast, the atomic coordinates of IL-8 were first solved by NMR [25] and these coordinates were then used by crystallographers to solve the crystal structure [26]. The early literature on the structural studies of cytokines is reviewed in style by the Nobel laureate, Max Perutz, in his extremely readable text on protein structure [27]. This is recommended to the reader interested in cytokine structures.

Structurally, cytokines are generally proteins of low molecular mass, ranging from 5 kDa (TGFα) to 145 kDa (Mullerian-inhibiting substance; MIS), consisting of a single polypeptide chain. Cytokines with structures other than a single polypeptide chain include TGFβ, PDGF, IL-12 and MIS. In addition, TNFα and TNFβ form homotrimers in solution. Over the past decade, sufficient cytokine structures have been elucidated to allow the classification of cytokines into four major groups. Members of the first group have a structure composed of four antiparallel α-helical segments. This group has been further subdivided into proteins with long-chain (lc) or short-chain (sc) segments. Cytokines in the second group contain long β-sheet structures and can be subdivided further into those with: (i) β-sheet-rich structures known as cystine knots; (ii) proteins with the conformation known as β-jellyrolls; and (iii) proteins with three sets of 4-β strands which have a Y-shaped or three-leaf (trefoil) structure. The third group has short-chain α/β structures and includes: (i) members of the EGF family with at least two antiparallel β-strands connected to the intervening loops by three disulphide bonds (so-called S–S-rich β-meander structure); (ii) insulin-related cytokines which contain a conserved set of three disulphide bonds linking three short α-helices;

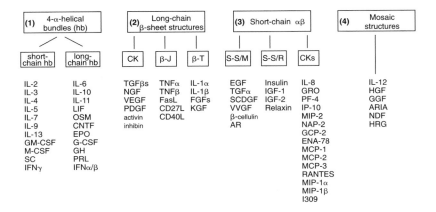

Figure 4.4 Subdivision of the cytokines into four structural classes
The main classes are: (1) cytokines with four α-helical bundles [either short-chain helical bundle (hb) or long-chain hb]; (2) long-chain β-sheet structures [further subdivided into cytokines with cystine knots (CK), β-jellyroll (β-J) or β-trefoil (β-T) structures]; (3) short-chain α/β structures [further subdivided into S-S-rich β-meander (S-S/M); S-S-rich α/β (S-S/R) and chemokine (CK) structures]; and (4) cytokines with mosaic structures. For a full list of the abbreviations used, see page xiii.

and (iii) the C-X-C and C-C chemokines. Members of the fourth group — the mosaic structures — contain a range of domain structures such as immunoglobulin (Ig)-like domains and Kringle domains (found in proteins such as plasminogen). Examples of the four main structural classes of cytokines are provided in Figure 4.4.

Table 4.10 Classification of cytokines on the basis of receptor binding

Cytokine receptor class	Cytokines binding to this receptor class
Class I (including haemopoietin receptors)	IL-2 (to the β-chain), IL-3, IL-4, IL-5, IL-6, IL-7, IL-8, IL-9, IL-11, IL-12, EPO, G-CSF, GM-CSF, LIF, CNTF, prolactin, growth hormone
Class II	Interferons, IL-10
Receptor tyrosine kinase	Various growth factors (EGF, FGF, IGF, KGF, PDGF, VEGF)
Protein serine/threonine kinase receptors	TGFβ superfamily (30 plus proteins)
TNF receptor family	TNFα, TNFβ, Fas ligand, CD27 ligand, CD30 ligand, CD40 ligand
G-protein-coupled receptors	Chemokines (approximately 30 proteins)
Ig super family	IL-1α, IL-1β, IL-1ra

IGF, insulin-like growth factor; KGF, keratinocyte growth factor; VEGF, vascular endothelial growth factor; LIF, leukaemia inhibitory factor; CNTF, ciliary neurotrophic factor.

4.2.3   Subdivision of cytokines based upon receptor binding

With a few exceptions, for example, platelet-derived endothelial cell growth factor (PD-ECGF), which is homologous to the *Escherichia coli* enzyme thymidine phosphorylase [28], cytokines have no inherent biological activity and must bind to a specific cell receptor in order to express such activity. A large number of cytokine receptors have now been identified and cloned, and this has allowed their subdivision into various classes. Thus, it is also possible to classify cytokines on the basis of the receptors to which they bind (Table 4.10).

The classification of cytokines in terms of their biological activities, physical structures or receptor binding allows a certain degree of order to be attained, but it is clear that the three types of classification do not result in as good an overlap as one might have expected. This confirms that we still have a long way to go towards elucidating the biology of this fascinating group of hormone-like molecules.

4.3   Biological actions of cytokines

The previous section introduced the reader to the very large number of cytokines that are now known to exist, and attempted to make order of this menagerie of molecules. In this section the biological actions of cytokines, and the cell populations producing these molecules, will be discussed.

The biological studies of cytokines were initiated by workers interested in infection and immunity and the emphasis in cytokine biology has retained this focus. However, it is important to realize that cytokines are increasingly being found to play key roles in normal physiology and physiological development. For example, it is now established that cytokines are found at all stages of development. The so-called 'pro-inflammatory' cytokines, IL-1, TNF, IL-6 and IL-3, have been found in unfertilized oocytes [29]. Cytokines are also involved in the development of invertebrates, giving clues to the evolutionary history of these proteins. An interesting example is the role of TGFβ in the development of the nematode, *Caenorhabditis elegans*, much beloved of developmental biologists. This little worm undergoes a developmental change under conditions of food scarcity. The gene controlling the development of this so-called dauer-stage has recently been cloned and shown to be a homologue of TGFβ [30].

In addition to the activities of the cytokines, which depend upon binding to specific receptors, the cytokine receptors themselves can have biological activities while resident on the cell membrane. There are a growing number of examples of cytokine receptors acting as portals for the entry of viruses. For example, HIV enters cells by binding to a chemokine receptor [31] and it has also recently been reported that herpes simplex virus-1 enters cells by binding to a novel member of the TNF receptor family [32]. The biological roles of soluble cytokine receptors will be described later in this chapter.

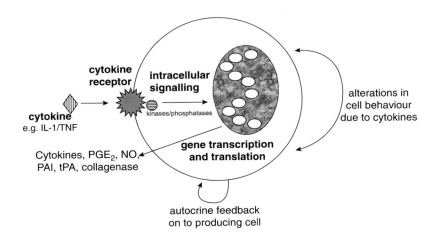

Figure 4.5 Cytokines have no inherent biological activity (such as is possessed by enzymes) and only express their 'activity' when they bind to their respective specific receptors

Binding of a cytokine to its receptor induces a wide variety of intracellular signalling pathways (depending on the receptor), which ultimately result in the transcription of selected genes and changes in cell behaviour (e.g. proliferation, chemotaxis, apoptosis). The changes in cell behaviour are shown schematically in Figure 4.6. Certain of the transcribed gene products (e.g. collagenase and tPA) may be released from the cytokine-activated cells, or may result in the production of low-molecular-mass mediators, such as PGE_2 or NO.

4.3.1 Interactions of cytokines with cells — biological consequences

Like endocrine hormones, cytokines only have biological meaning when they bind to their cell-surface receptor and activate specific genes. The consequences of the binding of a cytokine to its specific receptor is the induction of selective intracellular signalling which results in the switching on (or off) of particular genes and the production and release of the products of such genes. These can either be proteins (other cytokines, proteases, etc.) or lower-molecular-mass mediators produced as a result of the induction of enzymes such as cyclo-oxygenase II (which produces prostaglandins) or nitric oxide (NO) synthase whose actions are self-evident. In Figure 4.5 the binding of a cytokine to its cellular receptor is seen to result in the production of various molecules which could produce pathology [prostaglandin E_2 (PGE_2), NO, tissue plasminogen activator (tPA), plasminogen-activator-inhibitor (PAI), collagenase]. Proteinases such as collagenase and tPA could directly induce tissue damage. The production of NO and PGE_2 would have a range of actions particularly on the vasculature. The most fascinating action of cytokines is their capacity to induce their own synthesis as well as that of other cytokines. It is this ability of a single cytokine to induce cells to synthesize and release a wide variety of other cytokines that makes cytokine biology such a confusing and fascinating area of study. These patterns of cytokine production, and the

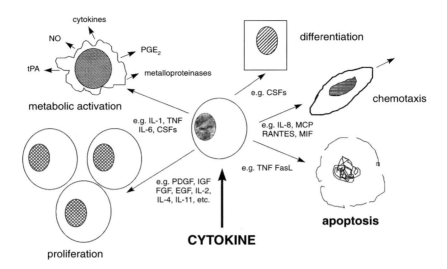

Figure 4.6 The various changes that can occur in cells exposed to cytokines
For a full list of the abbreviations used, see page xiii.

production of secondary mediators induced by cytokines, leads to complicated networks of interactions which will be discussed in more detail at the end of this chapter.

In addition to metabolically activating them, cytokines can modify the behaviour of cells in a wide variety of ways (Figure 4.6). Growth mediators (interleukins, CSFs, growth factors) can either stimulate cells into cell cycle or synergize with other factors to allow entry into and completion of the cell cycle. The CSFs and other cytokines (interleukins) are involved in controlling the differentiation of myeloid and lymphoid cell lineages in the bone marrow. It is now known that cell death is essential for organismal stability and that failure to control homeostatic programmed cell death could lead to various tissue pathologies including cancers. A number of cytokines and cell-bound proteins binding to the TNF receptor family are now recognized to act to induce apoptosis, which is simply another aspect of the cell cycle. Another major group of cytokines — the chemokines — are involved in the recruitment of the correct populations of leucocytes to sites of injury, inflammation and infection.

4.3.1.1 Actions of IL-1 on cells
It is perhaps difficult to grasp the enormous influence and range of effects that cytokines can have on cells. To emphasize this, the biological actions of the prototypic pro-inflammatory cytokine, IL-1, will briefly be described. As will become clear in later chapters, IL-1 has effects on probably every cell type in the body. It is normal to talk of IL-1 as if it were one protein but, in fact, it is a

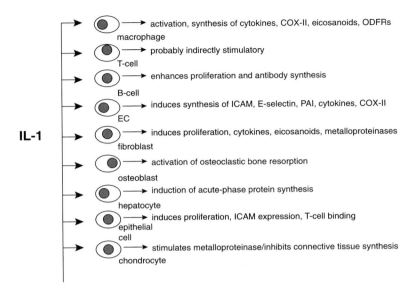

IL-1

activation, synthesis of cytokines, COX-II, eicosanoids, ODFRs
macrophage

probably indirectly stimulatory
T-cell

enhances proliferation and antibody synthesis
B-cell

induces synthesis of ICAM, E-selectin, PAI, cytokines, COX-II
EC

induces proliferation, cytokines, eicosanoids, metalloproteinases
fibroblast

activation of osteoclastic bone resorption
osteoblast

induction of acute-phase protein synthesis
hepatocyte

induces proliferation, ICAM expression, T-cell binding
epithelial cell

stimulates metalloproteinase/inhibits connective tissue synthesis
chondrocyte

Figure 4.7 Some of the main effects of IL-1 on mammalian cells
Abbreviations used: COX-II, cyclo-oxygenase II; ICAM, intercellular adhesion molecule; ODFR, oxygen-derived free radical; PAI, plasminogen-activator-inhibitor; EC, endothelial cell.

family of proteins with, in its widest context, nine molecules. The proteins IL-1α and IL-1β are produced as propeptides of molecular mass 31–33 kDa. The IL-1α propeptide precursor is biologically active, but the IL-1β propeptide is inactive. Activation of IL-1β is by a specific cysteine proteinase called pro-IL-1β-converting enzyme (ICE) [33], which appears also to have a role in the production of IL-1α [34]. The proteolytically cleaved products of IL-1 are both biologically active. The third member of the IL-1 gene family is IL-1ra, which is synthesized either with or without a leader sequence, depending on the cell, resulting in it being either a secreted product or an intracellular molecule. These two forms depend upon IL-1ra having alternative first exons. There are two receptors for the IL-1 molecules. The type I receptor is an 80 kDa signalling receptor, while the type II (60 kDa) receptor does not signal and has been termed a decoy receptor. Adding up all the proteins involved in controlling IL-1 function we find that there are nine molecules in this extended 'family' [35–37]. The importance of IL-1 is illustrated by considering the evolution of this complex system that involves ICE, IL-1ra and the type II receptor. Indeed, this IL-1 family appears to be unrelated to other cytokines and may be a signalling molecule that evolved early on in evolutionary terms. In Chapter 9, the finding that *E. coli* has receptors for this cytokine is described. Whether IL-1 started life as a bacterial signalling molecule is an intriguing, although unlikely, possibility.

The range of actions of IL-1 is shown clearly by the disparate names that have been coined to describe this molecule (see Chapter 6, Figure 6.1). The biological activities demonstrated by the mature IL-1 molecules are shown schematically in Figure 4.7 and will only be briefly described here. Greater detail can be found elsewhere [8,35]. IL-1 is a potent stimulator of macrophage functions and is responsible for up-regulation of a variety of genes, including those for cytokines, eicosanoids, adhesion receptors, etc.

In vitro studies have revealed IL-1 to be a potent stimulator of eicosanoid synthesis by a range of cells, most notably fibroblasts, monocytes and vascular endothelial cells. The capacity of IL-1, and other pro-inflammatory cytokines such as TNF, to increase cellular eicosanoid synthesis is due to the up-regulation of the transcription of the gene for one of the two forms of cyclo-oxygenase (COX) called COX-II. This inducible inflammatory COX has been hailed as the new therapeutic lode-stone of the pharmaceutical industry, in the belief that selective inhibitors of the enzyme will prevent inflammation without paying the cost in side-effects. It has been found that PGE_2, and possibly other eicosanoids, can inhibit the synthesis of IL-1 by up-regulating intracellular cyclic AMP levels. This inhibitory activity is also found with certain bacterial toxins which can block pro-inflammatory cytokine synthesis (reviewed in Chapter 7).

IL-1 was recognized early as a lymphocyte-activating factor (LAF) and for many years was implicated as a major stimulator of T- and B-lymphocytes. There is certainly evidence for IL-1 acting to promote B-cell function. However, studies using IL-1 receptor-neutralizing antibodies or IL-1ra have failed to show any effect on antigen-driven T-lymphocyte activation. Furthermore IL-1 knockout mice do not appear to have immunological defects. Thus, it is not established that IL-1 is particularly important in acquired immunity, although it probably has indirect effects on T-cells. Consequently, IL-1 is probably best seen as a vital regulator of innate immune responses.

IL-1 has pronounced effects on cultured vascular endothelial cells (VECs), altering the adhesiveness of these cells to leucocytes by promoting the synthesis of a number of leucocyte-adhesive molecules including the intercel-lular adhesion molecules (ICAMs), vascular cell adhesion molecules (VCAMs) and selectins. The coagulation system of VECs is also affected by IL-1, with the balance of activity tilted in favour of a procoagulant state. This involves suppression of the expression of thrombomodulin, alterations in plasminogen activator, up-regulation of tissue factor, increased production of eicosanoids, particularly prostacyclin, and the induction of the inducible form of nitric oxide synthase which is responsible for the synthesis of NO. The induction of NO synthase normally requires the interaction between various cytokines (IL-1, TNF, IFNγ) and possibly LPS. IL-1 also causes changes in proteoglycan metabolism by cultured VECs. These molecules are important in regulating vascular permeability and interactions with platelets. IL-1 causes an increased production of soluble glycosaminoglycans (GAGs), but decreased production of cell-associated GAGs. IL-1 induces VECs to synthesize a number of cytokines, including IL-1, G-CSF, GM-CSF and various chemokines, and also stimulates VECs to synthesize the vasoactive peptide endothelin (ET)-1 [38], illustrating yet another interaction between peptide hormones and cytokines. IL-1 also increases the expression of the low-density lipoprotein (LDL) receptor on VECs.

IL-1 has dramatic effects on cells of the connective tissues, including fibroblasts, osteoblasts and chondrocytes. The induction of eicosanoid synthesis has already been mentioned. Other effects of IL-1 include stimulation of proliferation (a process requiring the synthesis and action of PDGF), induction of pro-inflammatory cytokines, induction of the matrixin family members (collagenase, stromelysins and gelatinases) and inhibition of the natural inhibitors of these enzymes — the tissue inhibitors of metalloproteinases (TIMPs). IL-1 also acts to inhibit the synthesis of connective tissue macromolecules by chondrocytes.

IL-1 acts both on the liver and the brain to induce what is known as the acute-phase response. This is a stereotyped response to infection (and injury) involving the synthesis of a variety of liver proteins, which act as opsonins, protease inhibitors and free radical-scavengers, and the stimulation of cells of the innate and acquired immune response. Part of the acute-phase response is an increase in the rate of muscle proteolysis.

As would be expected of a molecule which acts as an endogenous pyrogen, IL-1 has profound effects on the brain. *In vitro*, the effects of IL-1 include: altered activity of hypothalamic neurons, augmented γ-aminobutyric acid (GABA) receptor function in cortical neurons and inhibition of long-term potentiation in hippocampal slices [39].

These various *in vitro* findings are mimicked by the injection of IL-1. Local injection of IL-1 induces leucocytic infiltration, the appearance of low-molecular-mass mediators of inflammation, the induction of tissue breakdown and the inhibition of connective tissue matrix synthesis. Systemic injection of IL-1 induces the acute-phase response and fever. One fascinating finding is that pro-inflammatory cytokines such as IL-1 are extremely potent hyperalgesic agents which could contribute to the pain associated with inflammation. This hyperalgesia does not appear to be dependent on the synthesis of prostanoids [40,41].

4.3.1.2 Cell priming by cytokines

As will be described later in this chapter, the cytokines form networks of interactions. One of the cornerstones of such networks is the phenomenon known as cell priming. This is the ability of cells to respond to a particular agonist with enhanced sensitivity or activity if they have had prior exposure to the priming factor, such as LPS or a particular cytokine. For example, exposure of neutrophils to GM-CSF or TNFα primes them to become much more responsive to chemotactic factors. The mechanism of action of these cytokines appears to be related to their ability to stimulate the synthesis of

Table 4.11 Cytokines produced by different cell types

Cell type	Cytokines produced
Leucocytes	
Monocytes/Macrophages	IL-1α, IL-1β, IL-1ra, IL-3, IL-5, IL-6, IL-8, IL-10, IL-12, IL-15?, TNFα, LIF, OSM, G-CSF, GM-CSF, M-CSF, TGFα, TGFβ, VEGF, HB-EGF, bFGF, EPO, MCAF, Groα, MIP-1α, MIP-1β, NGF, activin
Neutrophils	IL-1α, IL-1β, IL-1ra, IL-3, IL-6, IL-8, IFNα, G-CSF, M-CSF, GM-CSF, Groα, defensins
Eosinophils	IL-1α, IL-3, IL-5, IL-6, TNF, GM-CSF, TGFα, TGFβ, MIP-1α
Mast cells	IL-1α, IL-1β, IL-2, IL-3, IL-4, IL-5, IL-6, IL-7, IL-8, IL-10, IL-11, IL-13, TNFα, IFNγ, GM-CSF, LIF, TGFβ, MIG, IP10, MCP-1, MIP-1α, MIP-1β, RANTES, MCP-3
Platelets	IL-1, EGF, PDGF, TGFβ, HGF, PD-ECGF, RANTES, Groα
B-lymphocytes	TNFα, TNFβ, IL-3, G-CSF, M-CSF, GM-CSF, IFNγ, FasL
CD4 Th$_0$ cells	IL-2, IL-3, IL-4, IL-5, IL-6, IL-10, IL-13, IFNγ, TNFα, TNFβ, GM-CSF
CD4 Th$_1$	IL-2, IL-3, IFNγ, TNFα, TNFβ, GM-CSF
CD4 Th$_2$	IL-3, IL-4, IL-5, IL-6, IL-10, IL-13
Connective tissue cells	
Fibroblasts	IL-1α, IL-1β, IL-6, IL-8, IL-11, LIF, NGF, TNFα, G-CSF, bFGF, TGFβ, activin
Vascular smooth muscle cells	IL-1α, IL-1β, TNFα, NGF
Osteoblasts	IL-1α, IL-1β, IL-6, IL-8, GM-CSF, TGFβ, TNF
Chondrocytes	IL-1α, IL-1β, TNFα, IGF-1, TGFβ
Other cells	
Epithelial cells	IL-1α, IL-1β, IL-1ra, IL-7, IL-10, NGF, amphiregulin, TGFα, defensins
Vascular endothelial cells	IL-1α, IL-1β, IL-6, IL-8, TNFα, LIF, EPO
Placental cells	IL-1, IL-2, IL-6, IFNα, IFNβ, TNF
Oocytes	IL-1, IL-6, M-CSF
Granulosa cells	TNF
Anterior pituitary cells	IL-6, MIF

Abbreviations used: Gro, growth-regulated oncogene; MCAF, monocyte chemotactic and activating factor. For full list, see page xiii.

leukotriene (LT)B$_4$ by neutrophils [42,43]. Another mechanism which probably underlies the phenomenon of priming is receptor transmodulation. This is the regulation of the expression/affinity/post receptor coupling of one cytokine receptor by another cytokine. For example, IL-1 up-regulates the PDGF receptor on mouse osteoblasts, thus increasing the response of these cells to PDGF. Understanding the mechanisms of cell priming will be important in comprehending the interactions between cytokines and cells which constitute a cytokine network.

4.3.2   Cell populations producing cytokines

The various cytokines that exist in mammals and their general actions on cells have been explained, and the various specific actions of one important pro-inflammatory cytokine, IL-1, have been described. The newcomer to the field of cytokine biology often believes that cytokines are only produced by a select group of cells. However, nothing could be further from the truth. All mammalian cells, with the probable exception of the erythrocyte, can produce cytokines or contain cytokines. Even the non-nucleated platelet contains 'prepackaged' growth factors such as PDGF and PD-ECGF. Some cells, for example monocytes/macrophages, produce more cytokines than others. The pattern of cytokines produced by any one cell is an important guide to the behaviour of the cell and to its interactions with other cells. As will be described later in this chapter, the three types of CD4 lymphocytes — Th_o, Th_1 and Th_2 — are defined by the patterns of cytokines they produce [44]. The reader is referred to Table 4.11, which lists the cytokines produced by various cell populations. As is evident, many cell populations produce many cytokines. Inflammation can take place in all of the organs and tissues of the body and the resident cells in these organs and tissues can participate in this process.

4.3.3   Factors stimulating cytokine synthesis

Table 4.12 lists the agents that have been shown over the years to induce the synthesis of cytokines. A wide range of materials has been claimed to induce the synthesis of various cytokines. Some appear to be non-specific (e.g. silica particles). Others are specific, a good example being the induction of interferons by viral infection of cells. By far the best studied inducers of cytokine synthesis are LPS and endotoxin. As will be described in detail in this book, endotoxin is a mixture of bacterial surface proteins with LPS. LPS and the proteins of endotoxin are stable and ubiquitous components that are present as contaminants in most compounds. It is likely that some of the molecules described as cytokine inducers may be contaminated with endotoxin/LPS.

Many of the cytokines play roles in host defence either as: (i) growth factors for myeloid and lymphoid cell lineages; (ii) growth and activating

Table 4.12 Inducers of cytokine synthesis

Microbial components	Inflammatory mediators	Other components
LPS	C5a	Silica crystals
Peptidoglycan	Leukotrienes	Urate crystals
Muramyl dipeptide	Prostaglandins	PMA
Lipoarabinomannan	Platelet-activating factor	
Yeast cell walls	Products of other lipoxygenases	
Virus: haemagglutinins,	Fc receptor occupation	
double-stranded RNA	Cytokines	
plus many other components	Immune complexes	
(see Chapter 7)	Integrin receptor occupation	

factors for mature myeloid and lymphoid cells; (iii) growth and activating factors for mesenchymal cells and for the processes of wound healing; or (iv) inhibitory factors involved in the control and cessation of immune and inflammatory responses. Many of the inducers of these various cytokines are components from bacteria. However, there is very little understanding of the nature of such bacterial cytokine inducers. The following chapters (5 and 6) will outline the chemistry, biochemistry and mechanism of action of LPS as an inducer of cytokine synthesis. Chapter 7 will introduce the concept that bacteria produce a large range of cytokine-inducing components (mainly proteins) that play a role in controlling cytokine networks. In Chapter 9, the hypothesis that bacteria constituting the normal microflora can control local cytokine networks to block any inflammatory response to such bacteria will be explained. The formulation and testing of this hypothesis requires that the range of bacterial cytokine-inducing components is defined and their ability to induce pro- or anti-inflammatory cytokines and control cytokine receptor transmodulation is ascertained.

4.4   Cytokine receptors and cytokine-induced cell signalling

As has been explained, cytokines have no inherent activity and their capacity to produce selective biological effects is due to their binding to specific receptors on the surfaces of cells and the induction of specific intracellular signalling pathways leading to particular patterns of gene transcription. Early studies of cytokine receptors using radiolabelled ligands and protein cross-linking techniques revealed that such receptors were, in general, very avid binders of cytokines, with K_D values in the nano- to picomolar range (10^{-9} to 10^{-12} M). For readers not familiar with this nomenclature, consider the binding of a cytokine (C) to its receptor (R): $C + R \rightarrow CR$. For simple binding reactions the rate of association of C and R is equivalent to the concentrations of both species multiplied by an association or affinity constant which is measured in litres/mole. The higher the affinity between the cytokine and the receptor, the larger the size of this constant (K_A). However, it is usual to express affinity of protein–ligand interactions in terms of K_D, the reciprocal of KA, which is expressed in moles/litre. The smaller the value of K_D, the higher the affinity of binding between the receptor and its cytokine ligand. Another interesting finding from the early studies of cytokine receptors (mainly the IL-1, IL-2 and TNF receptors) was the low numbers of receptors on cells. For example, the numbers of IL-1, TNF or interferon receptors on different cell types varies from several hundred to several thousand. Many cytokine-responsive cells contain only small numbers of cytokine receptors. Presumably it is the high affinity of the interactions between cytokines and their receptors that makes it possible for cells to function with such a small number of receptors. As cells can express receptors for a large number of cytokines the capacity to respond to cytokines via only a few receptors is presumably of advantage since it lessens the cells' need to make these large and complex proteins.

During the past decade, a large number of the cytokine receptors have been cloned, their DNA sequenced and, in many cases, the proteins expressed. This has provided a growing database which has allowed the cytokine receptors to be categorized into different structural and functional classes (Table 4.10). To aid in the later discussions of the roles of various cytokines in inflammation and immunity, a brief description of each class of receptor and the intracellular signalling pathways it utilizes is provided. The mechanisms involved in intracellular signalling are complex and the reader should note that binding of cytokines to the various cytokine receptors results in the activation of many overlapping signal transduction pathways.

4.4.1   Immunoglobulin superfamily receptors

The immunoglobulin superfamily is a group of proteins, each containing the immunoglobulin domain (or domains) within their structure. The immunoglobulin domain must be a useful 'building block' as this superfamily contains such members as the immunoglobulins, T-cell receptor, MHC antigens, CD2, CD4, CD8, lymphocyte function-associated antigen (LFA)-3, ICAM, nerve cell adhesion molecule (NCAM), etc. Included in this family are a small number of cytokine receptors including those for IL-1, IL-6, CSF-1 and PDGF. The IL-1 receptors (types I and II) are the most unusual in this regard. The type I IL-1 receptor is composed of 569 amino acids divided into a 20-residue signal sequence, an extracellular region of 317 amino acids and a cytoplasmic domain of 210 amino acids. The extracellular domain is composed entirely of three immunoglobulin-like domains. It is fascinating to realize that the IL-1 receptor is unrelated to the receptors for any other cytokine. The intracellular domain is necessary for signalling but its sequence give no clue as to its mode of action. Thus, it is still not clear how binding of IL-1 to its receptor induces specific intracellular signalling and there has been much debate over this point. The reader interested in this receptor and its signalling is referred to [45]. The conclusion of studies conducted over the past decade is that IL-1 probably does not induce the stimulation of adenylate cyclase or the activation of protein kinase C (PKC) via the phosphoinositide pathway. A pathway of cell signalling which has attracted much attention in recent years involves the hydrolysis of cell membrane sphingomyelin with the release of ceramide. There is evidence that IL-1 acts, at least in part, via the generation of ceramide [46]. As will be described in Chapter 7, LPS is also thought to stimulate ceramide generation.

Cell signalling involves the phosphorylation and/or dephosphorylation [47,48] of serine, threonine or tyrosine residues on proteins. The enzymes which add phosphate groups fall into two main classes: tyrosine kinases and serine/threonine kinases. As will be described, cytokine receptors themselves may be protein kinases or they may be associated with kinases. To determine the kinases involved in IL-1 signalling, the proteins phosphorylated in IL-1-stimulated cells have been examined. It was subsequently shown that the EGF receptor and the small heat shock protein (hsp)27 were phosphorylated. One protein kinase known to be involved in the phosphorylation of the EGF receptor was the 42 kDa mitogen-activated protein (MAP) kinase — p42-MAPK. Bird and colleagues [49] subsequently found that IL-1 stimulated both 42 and 44 kDa MAP kinases. These kinases are involved in the activation of a range of transcription factors such as c-Jun, Elk-1, c-Myc, NF-IL-6 and activating transcription factor (ATF)-2. Further study of IL-1-mediated cell signalling has revealed that p54-MAPK, p40 (which phosphorylated hsp27) and an enzyme, casein kinase, are involved in IL-1 signalling. The reader should refer to Figure 6.8 to see similarities between IL-1 signalling and LPS signalling pathways. Further work is clearly needed to identify the signalling pathways which give rise to the multitude of biological activities of the IL-1 family. For a recent review of IL-1 signalling, see [45]. It is interesting that there is no apparent similarity in receptor structure or post-receptor events between the IL-1 and TNF receptors, despite the results of activation of cells by these two cytokines being very similar.

The other cytokine receptors which contain immunoglobulin domains may have additional non-immunoglobulin domain structures (e.g. the IL-6 receptor) and are therefore members of other receptor classes.

4.4.2   Receptor tyrosine kinases (RTKs)

One solution to the intracellular signalling problem has been the evolution of receptors with inbuilt kinase domains. One family of cell-surface receptors are the RTKs which possess tyrosine kinase domains. At the time of writing there are more than 50 such receptors for a wide range of agonists, including a number of growth factors such as PDGF, FGF, EGF and vascular endothelial growth factor (VEGF) (see Table 4.10). The RTKs consist of a single polypeptide chain which traverses the plasma membrane once only, the only exception to this rule being the receptor for insulin. These 50 or more receptors can be categorized into 14 subgroups. A number of these receptors, for example those for PDGF, FGF and VEGF, contain immunoglobulin domains. In contrast, the EGF receptor has no such domains, but has two cysteine-rich regions. For further details on the structure of RTKs the reader is referred to [50].

The interaction of cytokines with their appropriate RTK introduces a mode of molecular behaviour that is increasingly seen to be common to most cytokines and their receptors, namely the oligomerization of cytokine receptors. A number of the growth factors, for example PDGF, are homo- or heterodimers and are therefore able to form a complex with two receptors (Figure 4.8). It is now thought that receptor dimerization is important in signal transduction generally and this will become apparent during the description of cytokine receptor signalling. Oligomerization of RTKs enables the intracellu-

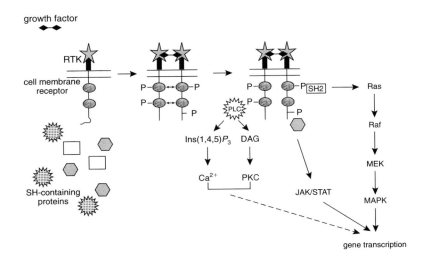

Figure 4.8 Interaction of a growth factor (e.g. PDGF) with its RTK
The dimeric growth factor dimerizes two receptors, allowing each member of the pair to phosphorylate its partner on selected tyrosine residues. These phosphorylated residues then become binding sites for proteins such as phospholipase C (PLC) containing Src-homology (SH)2 or SH3 domains. This results in the generation of signalling complexes, the activation of various kinases (MAPK, JAK, etc.) and the induction of selective gene transcription. Activation of PLC results in calcium release from intracellular stores and activation of PKC. For a full list of the abbreviations used, see page xiii.

lar domains to be autophosphorylated with each member of the pair phosphorylating the other member (a process termed trans-phosphorylation) on appropriate tyrosine residues. There are two consequences of this autophosphorylation. It can activate the kinase activity of the receptor and it also forms binding sites on the internal face of the receptor for phosphotyrosine-recognizing signalling molecules.

Many of the molecules involved in intracellular signalling are kinases, and with the cloning of many of these proteins the presence of conserved domains has been recognized. The major area of sequence similarity in the protein tyrosine kinases (PTKs) is their catalytic domain. This domain was found to share sequence similarity with the catalytic domain of the cellular oncogene c-Src and was therefore named the Src-homology (SH)1 domain. Two other domains, termed SH2 and SH3, are also found in PTKs. The SH2 domain binds with high affinity to selected phosphotyrosine-containing protein sequences. The SH3 domains bind to particular short peptide sequences containing proline residues. RTK autophosphorylation therefore produces SH2- and SH3-binding sites on intracellular proteins and such binding can result in the activation of the SH2/3-containing protein. A good example of this is provided by the enzyme phospholipase C-γ (PLC-γ), which hydrolyses phosphatidylinositol 4,5-bisphosphate [PtdIns(4,5)P_2] into diacylglycerol and inositol 1,4,5-trisphosphate [Ins(1,4,5)P_3]. Diacylglycerol is an activator of the PKC group of kinases and Ins(1,4,5)P_3 mobilizes Ca^{2+} from cellular stores. Other systems that can be activated by RTKs are the Ras pathways which can lead to the activation of MAP kinases, and the JAK/STAT pathway which will be dealt with in the next section (Figure 4.8).

4.4.3   Class I cytokine receptors

Many of the cytokines regulating the immune, inflammatory and haematopoi-etic cell systems are members of the class I cytokine receptors, which are also known as the haemopoietin receptors (Table 4.10). They are characterized by the presence of a conserved region of approximately 200 amino acids in their extracellular domain and two to three short conserved regions in the cytoplas-mic domain. In a similar manner to the RTKs, the binding of cytokines to type I receptors induces either homo- or heterodimerization. This results in tyrosine phosphorylation but, unlike the RTKs, this is indirect and requires the interaction of certain cytoplasmic tyrosine kinases known as Janus family (JAK) tyrosine kinases [51–54].

The possibility that cytokine receptors could contain more than one subunit was first shown for the IL-2 receptor, whereas the mechanism of action of the type I receptor subunits was first elucidated with the IL-6 receptor. To signal, the IL-6 receptor requires an additional membrane protein, gp130 [55]. It has now been established that the cytokines IL-6, IL-11, leukaemia inhibitory factor (LIF), oncostatin M (OM) and ciliary neurotroph-ic factor (CNTF) utilize this gp130 protein in order to induce intracellular signalling. This presumably accounts for the significant overlap in the biological activity of these cytokines. IL-6 binds to the IL-6 receptor and the ligand–receptor complex then associates with gp130, causing it to homodimer-ize. The IL-6 receptor has a short intra-cytoplasmic domain and does not signal. It is the dimerization of two gp130 proteins that initiates the intracellu-lar signal. IL-11 is also believed to induce homodimerization of gp130. In contrast to the homodimerization of gp130 induced by IL-6, the other cytokines in this group (LIF, OM and CNTF) form heterodimers between their receptors and gp130. Of interest is the finding that CNTF stimulates the formation of a heterodimeric complex between the LIF receptor and gp130 [56]. Other members of the type I receptor family, for example, IL-3, IL-5 and GM-CSF, use another common chain known as β_c. The cytokines IL-2, IL-4, IL-7, IL-9 and IL-15 use a different common signal transducer called γ_c. Having bound to the receptor and induced dimerization of the particular receptor–signalling protein complex, the intracytoplasmic domain of the dimerized receptor complex is phosphorylated on tyrosine residues by a family of bound tyrosine kinases — the JAK (or Jak) tyrosine kinases. There are currently four JAK kinases — JAK1, JAK2, JAK3 and TYK2. As described earlier, this phosphorylation induces binding sites for the SH2 binding domains of particular proteins. In this case the proteins are a group of latent

Table 4.13 Cytokine signalling controlled by STATs

Cytokine	JAK tyrosine kinase	STAT member
IL-2	JAK1, JAK3	STAT5
IL-4	JAK1, JAK3	STAT5, STAT6
IL-6/IL-11/LIF	JAK1, JAK2, TYK2	STAT1, STAT3
IL-12	JAK2, TYK2	STAT4
GM-CSF	JAK1, JAK2	STAT5
G-CSF	JAK1, JAK2	STAT3
IFNR I	JAK1, TYK2	STAT1, STAT2, STAT3
IFNR II	JAK1, JAK2	STAT1, STAT3
IL-10	JAK1, TYK2	STAT1

Abbreviation used: IFNR, interferon receptor.

cytosolic transcription factors known as signal transducers and activators of transcription (STATs). There are, at the time of writing, seven STAT proteins (STAT1–4, STAT5a, STAT5b and STAT6). The STATs bind to the dimerized receptor where they are phosphorylated by the JAKs. This allows them to dissociate and form homo- or heterodimers with other STATs, thus forming DNA-binding complexes which act as specific transcription activators.

By direct study of the STATs involved in cytokine signalling and by use of the knockout of specific STATs to look for functional consequences, it appears that some of the STATs may confer specificity on the signalling induced by certain cytokines (Table 4.13). Knockout of STAT4 (using the technique of homologous recombination), for example, results in mice which, although normal in appearance and fertile, have selective deficits in IL-12-controlled functions such as the production of Th$_1$ lymphocytes, lymphocyte proliferation and IL-12-stimulated production of IFNγ [57,58]. As will be explained later in this chapter, IL-12 plays a major role in the induction of acquired responses to infections and therefore it is important to understand the mechanisms underlying its signalling. Mice made deficient in STAT6 demonstrate impaired response to both IL-4 [59,60] and IL-13 [61]. It is not clear whether STAT4- or STAT6-deficient mice show any aberrant responses to infection. However, knockout of STAT1, which is involved in the signalling by the interferons, results in mice which are highly susceptible to infections by viruses and certain intracellular bacteria [62].

4.4.4 Class II cytokine receptors

The class II receptors are distantly related to the class I group and consist of the receptors for the interferons, IL-10 and tissue factor, a protein involved in blood coagulation. There are multiple forms of interferons and at least 18 proteins are involved in the binding to this class of receptor. These type II receptors are multimeric and binding of a cytokine leads to dimerization and the involvement of the JAK/STAT pathway [62a]. This pathway has been described in the previous section and is shown schematically in Figure 4.9.

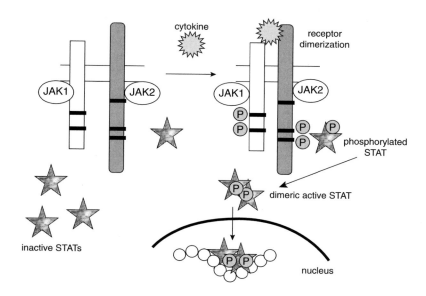

Figure 4.9 A simplified diagram showing the interaction of a cytokine receptor with the JAK/STAT signalling system
Oligomerization of the cytokine receptor results in the receptor chain-associated JAKs phosphorylating each other and also tyrosine residues on the cytoplasmic domains of the receptor. This creates phosphotyrosine docking sites for the SH2 domains of STAT proteins. The STAT proteins bind to these residues on the receptor chains and are phosphorylated by the JAKs on conserved tyrosine residues. The bound STATs then dissociate and dimerize forming homo- and heterodimers which translocate to the nucleus where they bind to DNA sites, regulating selected gene transcription. For example, many STATs bind to a TTCNNNGAA site. In addition it is likely that the STATs interact and cooperate with other transcriptional elements, allowing the multiple effects demonstrated by the cytokines binding type I and type II receptors.

4.4.5   Protein serine/threonine kinase receptors

The TGFβ superfamily consists of more than 30 structurally related proteins, which include three TGFβ molecules, the activins and inhibins and the BMPs. The receptors for TGFβ have received most attention and will be briefly described. There are three receptors: TGFβ type I receptor (TβR-I) is of molecular mass 53 kDa; TβR-II is a 75 kDa protein; and the type III receptor is a proteoglycan (also termed betaglycan) of greater than 200 kDa, containing a 100 kDa core protein. The type I and II receptors are signalling receptors, while the betaglycan does not signal, but plays a role in facilitating the interaction between the ligands and the signalling receptors.

TβR-I and -II are both receptors containing a serine/threonine kinase intracellular domain. Binding of the ligand to these receptors appears to induce a complex oligomeric assembly — possibly a heterotetramer between the type I and type II receptors. The nature of the signalling systems invoked by binding of the TGFβ family to its receptors has not been delineated. The interested reader is referred to [62b] for a current account of these receptors.

4.4.6   TNF receptor family

The overlap between the biological actions of IL-1 and TNFα has already been considered (see Table 4.3) and the IL-1 receptor has been described in Section 4.4.1. TNFα binds to two receptors (type I, TNFR1, 55 kDa; and type II, TNFR2, 75 kDa) [63,64], which belong to a family consisting of about a dozen members (see Table 4.5). This section will only briefly deal with the TNF receptors. Binding of TNF to the type I receptor triggers a wide range of cellular responses involved in inflammation and immunity. This appears to be due to the activation of the transcription factor nuclear factor (NF)κB. In most cells, NFκB exists as a cytoplasmic complex bound to an inhibitory protein, IκB. Activation of the type I receptor results in the phosphorylation and degradation of IκB and the migration of NFκB to the nucleus where it binds to specific κB elements and induces specific gene transcription [65]. TNFR1 occupancy also induces apoptosis in certain tumour cell lines. This involves the participation of an intracellular protease similar to the enzyme ICE (and Ced-3), described earlier in this chapter and also in Chapter 8. CrmA, a cowpox virus protein, described in detail in Chapter 8, is a potent inhibitor of TNF-induced apoptosis [66].

The TNFR1 signalling mechanism is still not fully defined. The intracellular segment of TNFR1 has no enzymic function and it is only recently that a protein has been found which interacts with the activated receptor. Both NFκB activation and apoptosis appear to be related to the binding of a 34 kDa protein termed TRADD (TNF receptor-associated death domain) [67]. Mice lacking TNFR1 were resistant to superantigen- and endotoxin-induced shock, but were deficient in their ability to deal with the intracellular bacterium *Listeria monocytogenes* [68]

The TNFR2 plays a role in immunity and is believed to also activate NFκB. Instead of TRADD this receptor seems to couple occupancy with two proteins, TNF receptor-associated factor (TRAF)1 and TRAF2 [69]. How these proteins couple receptor activation to cell activation is still not fully established, but a recent report suggests that SAPK/JNK (stress-activated protein kinase/c-Jun N-terminal kinase) is involved in the non-apoptotic NFκB-driven pathway of cell activation [70].

4.4.7   G-protein-coupled cytokine receptors

Approximately 30 chemokine genes have been discovered and all these proteins bind to the seven-strand G-protein-coupled receptors. As such receptors have been well documented in other areas of biology (e.g. G-protein-coupled receptors are involved in transducing a wide variety of signals, e.g. light, odorants, neurotransmitters, hormones, peptides, etc.), they will not be addressed here.

It is clear that cytokine signalling is an extremely complex area of biology and that similarities and differences between the receptors and intracellular signalling pathways utilized account for the wide spectrum of biological actions demonstrated by cytokines and also for the overlapping functions of many of these proteins. There are obvious exceptions to this, perhaps the biggest being the overlapping actions of IL-1 and TNF and the apparent major differences in the receptors used. Knowledge about cytokine receptors and receptor signalling will be invaluable in the development of therapeutics for diseases, particularly infectious diseases involving cytokines. The therapeutic potential of modulating cytokine synthesis or activity is reviewed in Chapter 8.

4.4.8 Soluble cytokine receptors

Many studies of the biological activities of cytokines have been hindered by the presence of inhibitory substances in biological fluids. It is now clear that some anomalous results are due to the presence of soluble forms of cytokine receptors [71,72]. A large number of cytokine receptors have now been shown to be released by cells, either as a result of proteolytic cleavage of the receptor at the cell surface, probably by cell-surface proteases, or by the production of an alternatively spliced soluble form of the surface receptor (Table 4.14). In Chapter 8 the strategy taken by several of the large-genome double-stranded

Table 4.14 Soluble human cytokine receptors

Receptor	Cell or tissue source
Proteolytically cleaved	
IL-1	Cell culture
IL-2	Cell, urine, serum
IL-4	Serum, urine
IL-6	Cell, serum
gp130	Serum
TNFRI	Serum, urine
TNFRII	Serum, urine
M-CSF	Cell culture
IFNα	Serum, urine
IFNγ	Cell culture
EGF	Cell culture
PDGF	Cell culture
EPO	Cell, serum
Alternative splicing	
LIF	Serum
IL-7	Cell culture
G-CSF	Cell culture
GM-CSF	Cell culture
EGF	Cell culture
VEGF	Cell culture

Abbreviation used: TNFR, tumour necrosis factor receptor. For full list, see page xiii.

DNA viruses is described. To inactivate host defence responses these viruses produce soluble forms of cytokine receptors. This may be the role that soluble receptors released from mammalian cells play — acting to inhibit cytokine-mediated effects by binding to and neutralizing cytokines. Of course, another possibility is that the soluble form of the receptor, if bound to its ligand, can interact with cytokine receptors forming an oligomeric structure and triggering a response. As reviewed in Chapter 6, this is what happens when LPS binds to the soluble form of CD14.

4.5   Cytokines in innate and acquired immunity

It is envisaged that since the appearance and diversification of large multicellular organisms (with multiple environments for bacteria to colonize) in the pre-Cambrian era (2.5 billion to 600 million years ago) there has been an evolutionary struggle between the Domain Bacteria and the Domain Eucarya. This struggle, it is assumed, has resulted in the development of the range of defence mechanisms that can be divided into the systems of innate immunity and acquired immunity. The innate systems were developed first, with the acquired systems of T-cell immunity being of relatively recent origin. Indeed, the majority of living species do not have T-cell immunity and rely for their survival on innate immune systems. A simplified and schematic diagram showing the major elements of the innate system of immunity is shown in Figure 4.10. The cellular elements of this system include the epithelial cells which form a barrier to bacterial entry into organisms, the phagocytes (monocytes, macrophages and neutrophils), natural killer cells and the blood-vessel endothelial cells. An additional cell which may play a role in innate immunity is the γδ T-lymphocyte which, unlike αβ T-cells, has a very restricted T-cell receptor (TCR) repertoire. These cells are exclusively found in association with epithelial cells and may function to monitor them for damage or changes due to infection. The epithelial cells, fixed-tissue macrophages and the endothelium are the cells which are always present in tissues and are therefore likely to be responsible for producing signals announcing the presence of bacteria. The cytokines which control the innate immune responses will be described in due course. In addition to cells and cytokines, the innate immune response utilizes a range of soluble components. These include the complement system, with its protease-driven cascade producing anaphylotoxins, opsonins and the membrane attack complex. The liver also produces a wide range of proteins called acute-phase proteins (Table 4.15), the functions of which are multifarious and include: control of haemostasis, protein transport, protease inactivation, bacterial opsonization and control of free radical and cytokine synthesis/action. In addition to these soluble factors, it is now established that epithelial cells and neutrophils produce a range of small, positively charged peptides which have antibiotic actions. These peptides are reviewed in some detail in Chapter 9. Such antibiotic peptides can

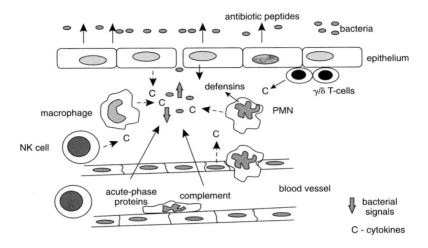

Figure 4.10 A simplified schematic drawing of the elements of the innate system of immunity
The first element of this system could be considered to consist of the antibiotic peptides produced by the epithelial cells which act as barriers to bacterial entry, and the antibiotic peptides (defensins) produced by PMNs. The epithelial cells and the fixed-tissue macrophages along with the endothelial cells of the blood vessels are the main cells able to respond to the invasion of the tissue by bacteria. Another cell population which may play a role in innate immunity is the γδ T-lymphocyte. Signals from the bacteria stimulate the production both of the antibiotic peptides and of cytokines. Bacteria can also directly activate (or can inhibit) the complement cascade. Production of cytokines by the epithelium, macrophages and endothelial cells induces multiple changes in the tissue and also stimulates the production of acute-phase reactants in the liver and causes the accumulation of leucocytes. Activation of complement also induces the formation of pro-inflammatory molecules such as the anaphylotoxins C3a and C5a.

kill Gram-positive and Gram-negative bacteria and can bind to and inactivate LPS. They have been shown to be induced by LPS, and the neutrophil peptides (called defensins) have been shown to have chemotactic properties, and thus are similar to chemokines in activity. This system of antibiotic peptides presumably overlaps, in terms of the biological and physiological consequences of their production, with the cytokines and the acute-phase proteins.

Table 4.15 Human acute-phase proteins and their cytokine regulators

Type 1 (induced by IL-1/TNF and IL-6/LIF/OM/IL-11)	Type 2 (induced by IL-6/LIF/OM/IL-11)
C-reactive protein	Fibrinogen
α_1-Acid glycoprotein	Haptoglobin
Serum amyloid A protein	α_1-Proteinase inhibitor
Serum amyloid P	α_1-Antichymotrypsin
Complement component C3	Caeruloplasmin
Complement factor B	

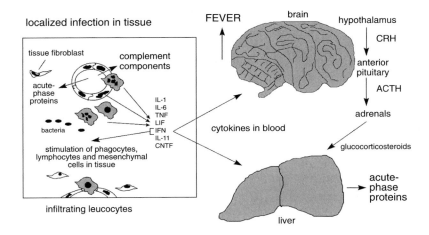

Figure 4.11 A schematic diagram showing the interactions involved in the acute-phase response
Local infiltration of tissues by bacteria releases signals that induce the formation of cytokines such as IL-1, TNFα, IL-6, IFNs, IL-11, OM and CNTF, which can enter the circulation. The cytokines IL-1, IL-6, IFN and TNF interact with the brain and produce fever. IL-1, TNF, IL-6, LIF, OM and IL-11 act as inducers of the synthesis of acute-phase proteins which have a wide range of biological actions including opsonization and inhibition of the actions of proteases and free radicals. The production of the acute-phase proteins also involves the hypothalamic-pituitary-adrenal axis. Abbreviations used: ACTH, adrenocorticotrophic hormone; CRH, corticotrophin-releasing hormone.

4.5.1   The involvement of cytokines in the acute-phase response

Infection, trauma, burns and other resolving stimuli produce what appears to be a stereotyped local and systemic network of cellular and humoral interactions which is known as the acute-phase response. A simplified diagram showing the major organs involved in this response is shown in Figure 4.11. In this chapter only the induction of the acute-phase response as a result of infection will be discussed.

Bacterial entry into tissues, past the epithelial barrier, results in the induction of the classic signs of inflammation — rubor, calor, tumour and possibly also dolor. The driving force for the induction of inflammation is thought to be LPS in Gram-negative infections. In Gram-positive infections it is less clear what components act to promote inflammatory changes. In truth, it is not clear what bacterial components are able to drive inflammatory processes and which actually do so in practice. As will be abundantly evident from reading Chapter 7, many bacterial components are potent stimulators of cytokine synthesis. Indeed, many of these molecules are more potent, and even more efficacious, than LPS. Defining the bacterial molecules which control the inflammatory response will be a major research goal and will constitute an important element of the cellular microbiology of infectious diseases.

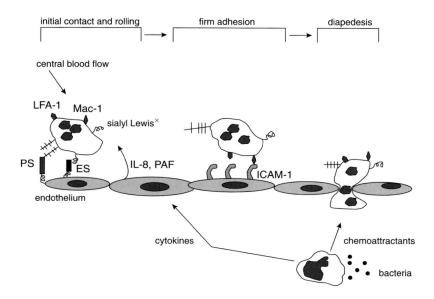

**Figure 4.12 The mechanism of neutrophil–endothelial interactions in inflamma-
tion and the role of cytokines in this process**
Activation of endothelial cells by bacteria induces the synthesis [or, in the case of P-selectin (PS),
surface expression] of three groups of proteins: selectins, ICAMs and VCAMs. Only the selectins
and ICAMs will be considered since the VCAMs only recognize β_1-integrins, which are not
expressed by neutrophils. The selectins [P and E (ES) on vascular endothelial cells (VECs) and L
on lymphocytes and neutrophils] bind, in a calcium-dependent manner, to cell-surface oligosaccharides such as sialyl Lewis^X. In addition, a newly emerging group of heavily glycosylated
proteins called sialomucins have been recognized to be involved in leucocyte adhesion. P-selectin
glycoprotein ligand (PSGL-1) is one member of this family present on neutrophils. The
neutrophil also expresses two β_2-integrins — LFA-1 and MAC-1 — which bind to ICAM-1. The
first step in neutrophil recruitment is the slowing down of these cells by binding of neutrophil
surface carbohydrate to P- and E-selectin. This results in the neutrophils rolling along the
endothelial surface and in doing so they are exposed to activating signals such as IL-8, platelet-
activating factor (PAF) and also direct cell-to-cell contact. Activation of the neutrophils results in
phosphorylation of the β_2-integrins, which produces a conformational change in these proteins
such that they can now bind to ICAM-1. This integrin–ICAM binding halts the neutrophils and, if
a sufficient gradient of chemoattractant exists across the vascular cell wall, the neutrophils will
undergo diapedesis and enter the tissue.

The major problem that the inflammatory response has solved is how to
get anti-bacterial leucocytes, which are flowing past the site of infection within
the blood vessels, into the infected tissue. This problem has been solved by
stimulating the induction or the activation of adhesion receptors on vascular
endothelial cells and circulating leucocytes (Figure 4.12). These mutually interacting sets of receptors initially slow and then halt the leucocytes by adherence
to the vascular endothelium, allowing these cells to undergo the process of
diapedesis which then enables them to enter the site of infection. As shown in
Figure 4.12 the induction of the synthesis of the various adhesion proteins on
the vascular endothelium (ICAMs, VCAMs and E-selectin) is controlled by

cytokines such as IL-1 and TNF. The leucocytes constitutively express a triad of counter receptors — the β_2-integrins — which, in unactivated leucocytes, are in a conformational state that does not bind to ICAMs and VCAMs. As the leucocytes slow down, due to binding of surface oligosaccharides to E-selectin, they become activated by other signals from the endothelium (or by diffusing through the endothelium) including chemokines such as IL-8 and lipid mediators such as platelet-activating factor (PAF). Leucocyte activation causes phosphorylation of the β_2-integrins, resulting in a conformational change in these proteins which enables them to bind with the ICAMs and VCAMs on the activated endothelial cell surface. This binding stops the leucocytes rolling along the blood-vessel wall and prepares the way for diapedesis [73]. Once leucocytes enter the tissue they can move to invading bacteria along gradients of host or bacterial chemoattractants. Once in the tissues the leucocytes have to remain there long enough to carry out their intended functions (phagocytosis, bacterial killing, tissue repair). It is now established that a second set of integrins — the β_1-integrins (which have a wider distribution than the β_2-integrins) — act to trap leucocytes in the connective tissues by binding to components of the extracellular matrix (collagen, fibronectin, laminin). The exception to this is the neutrophil which does not express β_1-integrins. Cells express these connective tissue-binding β_1-integrins as long as they are activated. Removal of the bacteria removes the stimulus and allows the leucocytes to decouple from the matrix components and either leave the area or apoptose.

The first leucocytes to enter a site of inflammation are polymorphonuclear leucocytes and these are generally followed by monocytes and lymphocytes. The C-X-C chemokines such as IL-8 are chemoattractants for neutrophils, while the C-C chemokines preferentially induce migration of monocytes. These small cytokines are vital to the correct mobilization of leucocytes from the blood, both in terms of the timing of cellular immigration and the populations of leucocytes recruited. IL-1 and TNF are probably the first cytokines to be produced and, acting alone or in combination with LPS (additively or synergistically), can induce the synthesis of other cytokines, including the chemokines (initially IL-8) and also IL-6. Other chemokines are then induced to attract monocytes and lymphocytes into the infected sites. Upwards of 30 chemokines have now been discovered and their selectivity for leucocytes is still the subject of active research.

Infiltrating leucocytes entering a site of infection will already be activated by exposure to chemokines and lipid mediators. Other activating signals present at the site of infection will include bacterial products, cytokines, complement components, acute-phase reactants, free radicals, lipid mediators, proteases and various cell receptors — all-in-all, a 'heady cocktail' for leucocytes. The relationship between cytokines and lipid mediators such as the prostanoids (products of COX II), leukotrienes (products of the action of 5-lipoxygenase) and other lipoxygenase products is not fully understood. The prostanoids and leukotrienes (generic title, eicosanoids; products of eicosa-tetraenoic acid or arachidonic acid) have generally been regarded as pro-inflammatory molecules, and the prostanoids have been, largely unknow-ingly, therapeutic targets for the past century. It was only in the 1970s that John Vane in London showed that non-steroidal anti-inflammatory drugs (NSAIDs) act by blocking the enzyme cyclo-oxygenase, which produces the prostanoids. Thus, the prostanoids, and later on the leukotrienes, were firmly set in the pantheon of pro-inflammatory mediators. However, it has been shown that prostanoids can inhibit the synthesis of pro-inflammatory cytokines such as IL-1 and TNF [73a]. In contrast, leukotriene B_4 has been claimed to stimulate IL-6 synthesis [73b]. Indeed the role of LTB_4 in inflam-mation is being rethought with the finding that this lipid is an activating signal for lipid-controlling transcription factors — the so-called peroxisome prolifer-ator-activated receptors (PPARs) [74]. While the interactions between the eicosanoids and PAF and the cytokines is probably very important in terms of controlling inflammatory events, the literature is still confusing and will not be discussed further. The interested reader is referred to [75] for a review of this literature.

Leucocytes are exposed to a variety of mediators on entering a site of infection and will, in turn, produce a range of mediators including cytokines, lipids, free radicals, enzymes, etc. These will all contribute to the final network of signals which controls the initiation, perpetutation and cessation of inflam-mation. Cells present in the tissue, including fibroblasts, vascular endothelial cells, fixed-tissue macrophages and epithelial cells, will also produce cytokines which will amplify the inflammatory signals within the lesion. With a simple infection, the cytokine-activated neutrophils and monocytes can ingest the opsonized bacteria and kill them. In the absence of further bacterial signals the leucocytes will apoptose and the tissues will heal without sequelae. Tissue healing will involve a range of growth factors such as EGF, TGFα, FGF, PDGF, and TGFβ.

The major unknown in inflammation is what switches it off. This may seem a facile question, but there are many important idiopathic human diseases in which the pathology is caused by chronic inflammation that does not 'switch off'. A good example of this is rheumatoid arthritis in which the synovial joints are permanently inflamed. It is envisaged that there will be as many cytokines involved in switching off inflammation as there are cytokines switching inflammation on and allowing it to perpetuate. The cytokines involved in a 'simple' acute-phase response to an infecting organism are delineated in Table 4.16. It is envisaged that these cytokines form a time-dependent network for controlling the local defence and repair systems in response to infection. The nature of the control of such networks will be discussed at the end of this chapter. It is not clear whether these acute-phase cytokine networks are stereotyped or whether they exhibit functional plasticity.

Table 4.16 Cytokines in acute-phase response to simple infection

Response	Cytokine
Initiation of inflammation	IL-1α
	IL-1β
	TNFα
	Cytokines interact and may synergize with each other and with bacterial factors such as LPS
Perpetuation of inflammation	Chemokines
	IL-8
	Groα
	Groβ
	MIP-1α
	MIP-1β
	MCP-1,2,3
	IL-6
	IFNs
Cessation of inflammation/controlling cytokines	IL-1ra
	IL-4
	IL-10
	IL-13
	TGFβ
Cytokines involved in growth and repair	Colony-stimulating factors
	EGF
	TGFα
	FGF
	PDGF
	TGFβ family

4.5.2   Acquired immunity and its interactions with innate immunity

As has been mentioned, the systems of innate and acquired immunity have evolved separately. The evolutionary history of innate immunity is lost in the aeons of time. However, acquired (adaptive or specific) immunity evolved about 400 million years ago and is found only in cartilaginous and bony fish, amphibians, reptiles, birds and mammals. The innate system of immunity has a limited repertoire of recognition receptors including sCD14, LBP (lipopolysaccharide-binding protein), the acute-phase proteins [C-reactive protein (CRP), stress-activated protein (SAP), mannose-binding protein (MBP)] and various receptors including mCD14, complement receptors (CD35) and scavenger receptors. These receptors and proteins and their role in cytokine production are discussed in more detail in the following two chapters. In contrast to this relatively inflexible recognition system, the acquired immune response with its T- and B-lymphocytes is breathtakingly adaptable. This adaptability is due to an amalgamation of combinatorial genetics, with the V, D and J elements of the immunoglobulin and T-cell receptor genes, and somatic mutation or junctional diversity which comes into play when these individual genetic elements are combined. The reader interested in the molecular genetics of the process known as 'generation of diversity' (GOD) in the specific immune system should refer to [76] for a current explanation. The mechanisms for introducing diversity into the genes encoding the variable region of the immunoglobulin molecule or the T-cell receptor are estimated to be able to generate 10^{18} receptors for antigens. Thus, GOD is a system for producing an astronomical number of molecular shapes in order that the myriad evolving molecules present in infectious organisms can always be 'recognized' by the immune system. It is important to realize that the process of generation of diversity does not depend on the presence of antigen. It is an inherent system that works continuously and without reference to external antigens.

The two recognition elements in the system of acquired immunity are the immunoglobulin receptor and the T-cell receptor which, in combination with cytokines, make a fully functional 'guardian system' for defence against exogenous parasites. The immunoglobulin receptors on B-cells are able to recognize conformational epitopes on any biomolecule, no matter what its molecular mass, while the T-cell receptor only recognizes peptides derived from protein antigens if presented in association with the polymorphic MHC proteins. Recognition of an infectious organism stimulates the clonal proliferation of B- and/or T-cells and, provided the infection is mastered, the immune system retains a memory of the antigens in long-lived antigen-specific memory cells. These ensure that subsequent exposure to the same antigen(s) results in a much more rapid response. This memory response, which can last for a lifetime, in the absence of the antigen [77], is one of the most fascinating tricks of acquired immunity and its mechanisms are still not fully understood [78].

4.5.2.1   T-cell subpopulations and infection

The cellular basis of the immune response has been shaped by the micro-organisms and parasites that have to be eliminated from the host (Figure 4.13). In addition to lymphocytes, the system of acquired immunity also requires the actions of antigen-presenting cells. The two main professional antigen-presenting cells are the dendritic cell and the macrophage. The macrophage is also a key cell in innate immunity. Lymphocytes can be subdivided into two main populations — B- and T-cells — and the T-cells are further subdivided into CD4 and CD8 cells on the basis of the presence of these cell-surface proteins. The CD8 cell is a cytolytic cell whose main role appears to be the killing of virally infected cells in an MHC class I-restricted manner. The key cytokines involved in CD8 lymphocyte function are TNF and IFNγ [79]. There is now evidence for the existence of two types of CD8 cells similar to CD4 T-cells [79]. CD8 cells are anti-viral in action but can also be involved in responses to intracellular bacteria such as *Listeria monocytogenes.* However, we will not deal with these cells in any detail in this chapter or in this book.

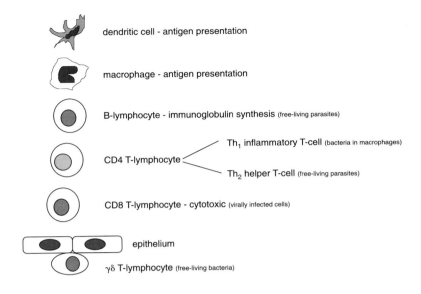

Figure 4.13 The cellular elements of the immune response
The particular micro-organisms dealt with by cells are shown in brackets.

Readers interested in learning more about the history of the CD8 lymphocyte should refer to [79] for a personal view.

Infectious micro-organisms and parasites can exist in a number of environments in the host. They can live on the external body surface, within connective tissue, or on the inside face of epithelia. Conversely, they can live within cells. Some bacteria, a good example being *Shigella* spp., can invade mucosal epithelial cells and survive and grow within these cells, eventually killing them. Other bacteria, for example *Mycobacterium tuberculosis* and *M. leprae*, live within the vacuolar apparatus of macrophages. The acquired immune system has developed to cope with these parasite survival strategies. Free-living bacteria and parasites, and any toxic molecules they produce, can be inhibited by the work-horse of immunology — the antibody molecule. Antibodies are the product of activated B-lymphocytes and exist as five different classes (IgG, IgM, IgA, IgD and IgE). IgG and IgM are antibodies involved in opsonizing bacteria, inhibiting toxins and activating the classic pathway of complement. IgA is involved in mucosal immunity and IgE is required for activating mast cells. IgE antibodies/mast cells are important in immunity to helminths and are also the cause of allergic conditions which afflict a large proportion of the population. Bacteria living inside cells are dealt with by CD8 T-lymphocytes and by natural killer (NK) cells. Bacteria living inside macrophages require that the macrophages be activated by antigen-specific T-lymphocytes to enable these resident bacteria to be killed.

The solution to the question of how acquired immunity coped with these various parasite strategies was the finding by Mosmann and colleagues that murine CD4 T-lymphocytes existed as two subsets distinguished by the network of cytokines they produced [80]. They named these distinguishable CD4 subsets, T-helper (Th)$_1$ and Th$_2$ lymphocytes. The Th$_1$ cells were characterized by their production of IL-2, TNFβ and IFNγ, while the Th$_2$ cells produced IL-4, IL-5, IL-6 and IL-13. Although IL-10 was originally described as a Th$_2$ cytokine, it has now been shown to be secreted by human Th$_1$ lymphocytes and macrophages [81]. The reason for these different 'patterns' of cytokine production was clarified when the nature of the immunological responses produced by these two CD4 subsets was defined. Th$_1$ cells were shown to be involved in cell-mediated immunity (CMI), such as delayed type hypersensitivity (DTH), which lies at the heart of a number of mankind's major chronic infectious and idiopathic diseases (tuberculosis, leprosy, rheumatoid arthritis). In contrast, Th$_2$ cells had a greater involvement in the production of humoral (antibody-mediated) immunity [82,83]. These differences are not absolute since Th$_1$ cells are involved in stimulating the synthesis of opsonizing IgG antibodies involved in the phagocytic removal of bacteria.

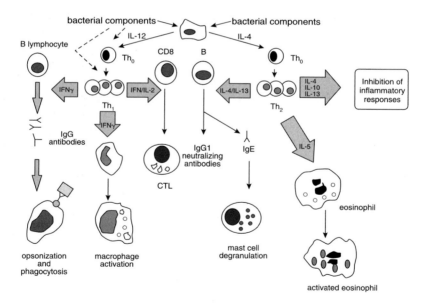

Figure 4.14 The role of bacterial components and cytokines in the workings of the Th$_1$ and Th$_2$ lymphocyte subpopulations
Bacterial components acting on various cell populations including macrophages and lymphocytes can induce the synthesis of IL-12 or IL-4. IL-12 acts to stimulate the differentiation of mature naive Th$_0$ cells into Th$_1$ lymphocytes which can then release IL-2 for autocrine stimulation and IFNγ to stimulate phagocyte function. Th$_2$ cells release IL-4, IL-5, etc. to activate mast cells or eosinophils and generate IgE antibodies. These cytokines can also act to inhibit inflammatory responses. Abbreviation used: CTL, cytotoxic T-lymphocyte.

Thus, it may be more appropriate to consider Th_1 lymphocytes as effectors of phagocyte-dependent responses and Th_2 lymphocytes as controllers of phagocyte-independent responses (Figure 4.14). This hypothesis, defining the division of labour in T-cell immunity, has been extremely productive in terms of scientific hypothesis testing. It is now believed that the two CD4 subsets produce cytokines that: (i) act as autocrine growth factors for the individual subsets; and (ii) act to inhibit the actions of the other subset. Thus, IL-2 is the growth factor for the Th_1 subset and IL-4 for the Th_2 subset. In terms of cross-control, IFNγ (a Th_1 cytokine) inhibits the proliferation of Th_2 cells, and IL-10, produced predominantly by Th_2 cells, inhibits the activation of Th_1 cells. This seems to provide a satisfying symmetry to T-cell functions. While these two CD4 subsets were discovered and extensively studied in the mouse there is now evidence that they also exist in man [84,85].

Is there evidence to support the role of Th_1 and Th_2 lymphocytes in natural infections? The most impressive data from animal models of infection is the response of different strains of mice to infection with the intracellular protozoal parasite *Leishmania major*. In this model, resistance or susceptibility to the parasite is dependent on the induction of a Th_1 or a Th_2 response respectively [86]. Thus, the production of the Th_2 cytokines, IL-4 and IL-10, by Leishmania-infected mice, was associated with a progressive form of disease [87]. Administration of neutralizing anti-IL-4 antibodies blocked these Th_2 responses and cured the mice of the lethal infection [88]. In contrast, administration of recombinant (r)IL-4 was able to stimulate Th_2 responses in mice which were normally resistant to this parasite [89].

Further evidence for the importance of these Th_1/Th_2 cytokines has come from the genetic knockout of specific cytokine genes. This literature on cytokine gene knockout is dealt with in detail in Chapters 7 and 9 and will not be discussed here; suffice to say that the abrogation of IL-2 and IL-10 results in animals which die due to colonic or enterocolonic inflammation in response to their normal microflora. Knockout of IL-4 lowers the production of Th_2 cytokines and renders animals susceptible to parasitic infections. Knockouts of IL-6 and members of the IFNγ family [cytokine, receptor or interferon regulatory factor (IRF-1)] result in increased susceptibility to intracellular bacterial infections with *M. tuberculosis, M. bovis, Listeria monocytogenes* and *L. major* [90].

Is there any evidence for a controlling role of Th_1/Th_2 subsets in human infections? The best studied condition is the chronic infection leprosy, caused by the organism *M. leprae*. This remains a common disease in certain areas of the world and is estimated to afflict 12–15 million people worldwide. The disease can exist in two forms — tuberculoid leprosy and lepromatous disease. Patients with tuberculoid leprosy exhibit blotchy red lesions with areas of loss of sensation on the face, trunk and extremities, due to growth of the organisms in the nerve sheaths. This form of the disease carries a better prognosis and in some cases is self-limiting. In patients with lepromatous leprosy there is extensive skin involvement with large numbers of bacteria being present in affected regions. Patients suffer gross deformities with substantial bone destruction. The differences in these two forms of leprosy are a reflection of differences in the Th_1/Th_2 responses of sufferers. Thus, tuberculoid leprosy patients demonstrate strong CMI responses to *M. leprae* that limit bacterial growth. In contrast, lepromatous leprosy patients are specifically unresponsive to this bacterium, allowing growth of the organism. When lesional tissue was collected from patients with both forms of this disease the tuberculoid patients demonstrated a Th_1 cytokine profile with IL-2 and IFNγ mRNA being detected. In contrast, biopsies from the lepromatous patients showed predominant expression of Th_2 cytokines (IL-4, IL-5 and IL-10) [91–93]. As antibodies are of no use in treating an infection due to bacteria living inside macrophages, the reason for the poor prognosis of the lepromatous leprosy patients is obvious. The only distinguishing features of Th_1/Th_2 cells are the cytokines they produce. However, a recent study has concluded that these cells also differ in their capacity to recognize vascular adhesion molecules. Thus P- and E-selectin (which recognize oligosaccharides on T-cells) mediate the recruitment of Th_1 cells into inflamed sites. However, Th_2 lymphocytes are unable to bind to P- or E-selectin. This suggests that these two CD4 subsets have differential tissue trafficking. The consequences of this are unclear [94].

4.5.2.2   Induction of Th_1/Th_2 lymphocytes

Th_1 and Th_2 lymphocytes are not derived from distinct bone marrow-derived lineages but develop from the same precursor cell population (often termed the Th_0 population), under the influence of environmental factors and in association with ubiquitous but undefined genetic factors [95]. This precursor is a mature naive $CD4^+$ T-lymphocyte that produces predominantly IL-2 on exposure to antigen. To date the most potent signals for the differentiation of this Th_0 lymphocyte precursor are cytokines. IL-12, the product of activated macrophages, dendritic cells and B-lymphocytes, is the key cytokine inducing the differentiation of Th_0 precursors into Th_1 cells [96–99]. LPS and undefined components from viruses, intracellular bacteria (mycobacteria and *Listeria*) and protozoa are able to stimulate IL-12 synthesis and thus induce Th_1-dominated responses (Figure 4.15). Antibody-mediated neutralization of IL-12 in animal models of Th_1-mediated diseases inhibited IFNγ synthesis and exacerbated disease symptoms. Thus, in animals infected with *L. monocytogenes,* neutralization of IL-12 led to a striking increase in susceptibility to sublethal doses of this organism [100]. IL-12 is important in primary responses to infection but does not appear to be so important in memory or secondary responses [99]. It has recently been reported that IL-12 activates cells by use of the transcription factor STAT4. Knock-out of STAT4 renders cells unresponsive to IL-12 and inhibits Th_1 responses [57,58].

IL-12 is a heterodimeric glycoprotein composed of two unrelated protein subunits of 35 (p35) and 40 (p40) kDa. Co-expression of both chains is

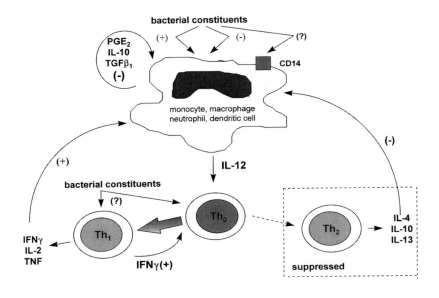

Figure 4.15 The possible interactions between bacterial constituents and IL-12
Bacterial constituents can stimulate macrophages and other cells to secrete IL-12. It is not clear what role CD14 plays in this process. In addition, it has not been determined if bacteria produce IL-12-inhibiting constituents. It is known that the measles virus can inhibit IL-12 synthesis. Bacterial constituents may interact directly with the various Th subsets. The IL-12 produced by macrophages stimulates the preferential differentiation of Th_0 cells into Th_1 cells.

necessary for biological activity. Examination of lesional tissues from patients with tuberculoid or lepromatous leprosy revealed that the former had 10-fold greater expression of the p35 and p40 IL-12 subunits than the latter, indicating that IL-12 is an important mediator of Th_1 cytokine responses in human infectious diseases [101].

The development of Th_2 cells from naive precursor CD4 Th_0 cells is induced by IL-4 [95]. This depends upon the transcription factor STAT6, and knockout of this factor renders animals unresponsive to IL-4 [59,60]. The source of the IL-4 is still not fully defined and may be the product of T-cells. The nature of the stimulatory signals inducing selective expression of IL-4 also needs to be more clearly defined.

4.5.2.3 What controls the balance of Th_1:Th_2 responses?
Inducing or maintaining the correct Th_1:Th_2 lymphocyte balance can mean the difference between life and death. This balance may lie at the heart of the huge increase in atopic disorders seen in the First World, as an intriguing report from a U.K./Japan collaboration suggests. Examination of children in Japan revealed a strong inverse association between delayed hypersensitivity to *M. tuberculosis* and atopy. Positive tuberculin responses predicted a lower incidence of asthma, lower serum IgE and Th_1-biased cytokine profiles. The

conclusion from this study was that immune responses to *M. tuberculosis* skewed the general immune response to a Th_1 state, repressing possible Th_2 atopic states. As the First World clears itself of tuberculosis this skewing effect disappears, leaving the way open for potential pathogenic Th_2 states, provided the genetic and environmental factors (e.g. air pollution) are present [102].

In addition to the cytokines IL-4 and IL-12, two other factors probably play a controlling role in the Th_1:Th_2 balance. The concentration of antigen appears to be able to influence this balance. Low concentrations of antigen preferentially induce Th_1 responses, while high-dose antigen promotes Th_2 responses [103,104]. The second factor important in determining the type of response is the role of co-stimulatory signals. The cell-surface proteins on antigen-presenting cells B7-1 and B7-2, which bind to CD28 on T-cells, are one of the best studied accessory factors in antigen presentation to T-cells and ensure that false-positive stimulation does not occur. Obviously, the development of Th_1 and Th_2 cells depends upon co-stimulatory factors. High levels of co-stimulation are reported to favour Th_2 responses [105,106]. In a world in which antibiotics have a rapidly decreasing efficacy, the capacity to modulate Th_1/Th_2 responses to specifically cope with infections appears an attractive proposition. It is predicted that pharmacological control of Th_1/Th_2 responses will become a major goal of the pharmaceutical industry.

4.5.3   Genetic control of cytokine responses in infection

It is certain that the responses of individuals to infection and their concomitant pattern of cytokine induction will owe something to genetic factors. Studies of adopted children have shown that they have a five fold greater chance of dying from a fatal infectious disease if a biological parent has succumbed to infection. Death of an adopted parent from infection carried no increased relative risk [107]. The finding that susceptibility to *L. major* is related to the synthesis of Th_1 or Th_2 cytokines has already been touched upon [108]. Experimental cerebral malaria in mice also appears to show a link between genetic susceptibility to the condition and cytokine synthesis [109]. Studies of the susceptibility of murine strains to various intracellular bacteria (e.g. *M. tuberculosis*) have identified a gene, *Nramp1*, which is involved in the resistance of animals to these bacteria. Indeed, a family of such genes has been defined and shown to be highly conserved [110].

In the past five or so years a number of reports have appeared of polymorphisms in the control elements of certain cytokine genes such as IL-1, IL-1ra, IL-2, IL-6 receptor and TNFα (reviewed in [111]). These polymorphisms relate to the kinetics of the induced synthesis of cytokines, for example, the synthesis of TNFβ in response to phytohaemagglutinin (PHA) or IL-1 in response to LPS [112,113]. Such polymorphic variants, which act in the control of cytokine synthesis, could obviously play a role in susceptibility to infection. The genetic influence of cytokine production and its role in the susceptibility to meningococcal sepsis has recently been examined [114]. The relatives of patients who either survived, or died, from meningococcal sepsis were studied to determine the capacity of their blood to produce the pro-inflammatory/protective cytokine TNFα and the anti-inflammatory/protective cytokine IL-10 in response to *Escherichia coli* endotoxin. Twenty six monozygotic twins were also examined. This study has claimed that there is a strong genetic influence on the production of both of these cytokines. In the monozygotic twins, 60% of the variation in the synthesis of TNFα and 75% of the variation in IL-10 production was genetically determined. Families that had a characteristic low level of TNF production had a 10-fold increased risk of a fatal outcome to sepsis, and those who had a high level of IL-10 production had a 20-fold increased risk. Families with both characteristics had the highest risk. This study therefore claims that the innate capacity to produce these diametrically opposed (in terms of bio-activity) cytokines contributes to susceptibility to fatal meningococcal disease. These findings have obvious ramifications for those attempting to treat septic shock by blocking TNF bioactivity.

It is clear that we now understand a great deal about the workings of the immune system at the genetic, molecular and cellular levels. However, one area of which we are still ignorant is the role that bacterial constituents play in controlling the immune response and the molecular and cellular levels at which they interact with immune cells. As will be described in Chapter 7 there are now many bacterial macromolecules and low-molecular-mass components that can induce cytokine synthesis and could therefore interfere with $Th_1:Th_2$ balance and function. Understanding the activity of these components, which can derive both from the normal microflora and from exogenous pathogens, is an urgent task if we are to truly understand how infections are controlled.

4.6   Cytokine networks: a synthesis

At the start of this chapter we introduced the concept defined by von Bertalanffy as General System Theory — a discipline which has its roots in mathematics, engineering, computer science and biology and which tries to make sense of complex systems and define their control mechanisms [4]. One key element of System Theory is information and its integrated control. Biological systems need enormous amounts of information processing to integrate their molecular, cellular, tissue and organismal homeostatic mechanisms. A key element in the control of systems is the concept of feedback control (Figure 4.16). A good example is the thermostat controlling domestic heating. Thus, the temperature in the house is monitored by the thermometer attached to the thermostat. This measures the ambient temperature and this is integrated by the thermostat which, depending on the temperature, will switch on or switch off the boiler. Feedback control can either be negative (i.e. inhibitory) or positive (i.e. stimulatory). Our experience is mostly of

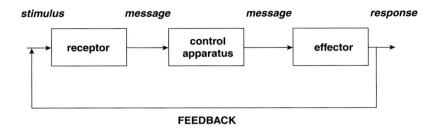

Figure 4.16 Schematic diagram of a feedback control system

negative feedback, but an example of biological positive feedback is the formation of a blood clot.

Feedback control mechanisms exist at all levels in biological systems. The complex metabolic pathways that adorn many laboratory walls are excellent examples of the workings of General System Theory. Classic endocrinology has, from its foundation, relied on the concepts of feedback control for its development, and the dissection of the complex hypothalamic–pituitary–adrenal/gonadal control systems is one of the triumphs of this discipline. The control of blood pressure, plasma ion balance, neuro-muscular communication, etc. are further examples of biological feedback control.

The discoverers and the protagonists of cytokine biology have largely come from non-physiological/endocrinological backgrounds and so the concepts of physiological feedback control with respect to cytokines have not developed as quickly as perhaps they should have. In the past decade the concept of cytokine control systems has led to the concept of the *cytokine network*. The common usage of the term network would refer to the complex interactions of cytokines with cytokine receptors, which can result in the sup-pression or synthesis of cytokines (both agonists and antagonists), the transmodulation of receptor affinities and the release of soluble receptors. We now appreciate that certain cytokines (IL-4, IL-10, TGFβ) act as physiological antagonists of the action of cytokines. A true cytokine antagonist exists in IL-1ra, and other examples may be found. In addition to modulating cytokine synthesis, cytokines also induce the production of low-molecular-mass mediators (eicosanoids, peptides, NO), proteases, cellular receptors, etc. (Figure 4.17). Some of these low-molecular-mass components have inhibitory actions on cytokine transcription or translation, a good example being the prostanoids which can inhibit the synthesis of certain cytokines by raising intracellular cyclic AMP [71]. It is this complex network of interacting elements that would be considered to be the fully interacting feedback control system in any tissue site (Figure 4.17).

Throughout this chapter examples have been given of the interaction of cytokines. Perhaps the best studied networks are those that are believed to lead

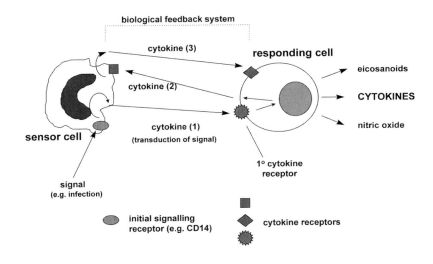

Figure 4.17 Simple schematic diagram which attempts to encapsulate the feedback systems which might exist in a cytokine network established during bacterially induced inflammation

Binding of the bacterial constituent (or more likely constituents) to a sensor cell (e.g. macrophage) would trigger the production of one or many cytokines. In this Figure only one cytokine is produced and this acts in a paracrine fashion to activate a nearby cell, the responding cell. This cell then produces cytokines which can feed back on to the sensor cell and induce more cytokine production. This mutual signalling can also lead to the production of other mediators (eicosanoids, etc.) which can feedback on to the producing cell in positive and negative ways. Receptor release is another method by which the network can be controlled as can receptor transmodulation in which the receptor affinity or number is altered. It is not clear if such networks have a hierarchy of cytokines or if the cytokines act in series or parallel.

to the maturation and establishment of Th$_1$ and Th$_2$ immunity in mice with their interacting cytokines inducing positive and negative biological signals. Is there any evidence for cytokine networks in man? Probably the best evidence has come from the study of cytokines involved in chronic inflammatory diseases. The cytokine profiles (networks) in the two major forms of leprosy have been described. Rheumatoid arthritis (RA), a chronic destructive inflammatory disease of the joints which may be due to an initial episode of infection, is one of the best studied diseases in terms of cytokine networks. Many cytokines, indeed 'too many' cytokines, are found in the joints of patients with RA. The presence of a Th$_1$ or Th$_2$ profile of cytokines in this disease has not been examined in detail and, in truth, very few T-cell cytokines are present in lesional tissues from individuals with established disease. A key observation in RA is that IL-1 is an important cytokine, its synthesis being dependent on the activity of TNFα. This observation was made using explant cultures of inflamed synovial membrane from patients with RA [115], and positioned TNFα at the 'top of the pile' in terms of cytokine induction. As a result, agents have been developed to block the action of TNFα, and these are now showing great promise in the treatment of RA (see Chapter 8). Thus, in RA it may be possible to view the induction of cytokines in terms of a cascade, with the cascade of cytokines being initiated by TNFα. Removal of this cytokine blocks the cascade and halts the disease. This suggests a cytokine interaction in which the cytokines are acting in series and not in parallel, by analogy with electrical circuits. Removal of one cytokine, in this case TNFα, halts the cytokine circuit in the same way that the removal of one electric bulb from the Christmas tree lights (wired in series) puts out all the lights. Of course, more evidence is required to establish that TNFα really is the cytokine that 'triggers' tissue pathology in RA.

Atherosclerosis is a common chronic disease of Western man which, although an inflammatory condition, is not immediately associated with immunological reactivity. However, it has recently been suggested that in atherosclerotic plaques there are important Th_1 network interactions. In such plaques there was strong expression of IFNγ but not IL-4, and IL-12 was also present. In some lesions there was also the expression of IL-10. Highly oxidized LDL (a potential pathogenic factor in atherosclerosis) was shown to stimulate human monocytes to produce IL-12, and this stimulation of a key Th_1-inducing cytokine was blocked by IL-10 [116]. Thus, in atherosclerosis there may be a cross-regulatory cytokine network controlling the degree of immunologically mediated injury.

Gram-negative sepsis is a lethal condition in which components of bacteria generate the rapid production of pro-inflammatory cytokines. The pathology of this condition is described in other chapters in this book (e.g. Chapter 6). The key pathological cytokines produced in patients with septic shock are believed to be TNFα, IL-1, IL-6 and IFNγ, and in experimental septic shock blockade of one or other of these cytokines can inhibit the lethal effects of LPS or of whole *E. coli*. In recent years the cytokine network interactions in septic shock have received considerable attention and it is now clear that the production of pro-inflammatory cytokines, which are believed to produce the pathology, is actually required for limiting and eliminating infectious organisms and contributing to host survival. As an example, administration of anti-TNF antibodies can worsen septic shock in animals and the outcome in patients [117]. This is reflected in the study of the genetics of the relatives of septic patients, which was described earlier. The understanding of the key parameters involved in the lethal effects of cytokine production in sepsis will involve understanding the network interactions. Recently, the balance sheet of cytokines has been studied in animals with mild, moderate or severe sepsis induced by caecal ligation and puncture. The cytokines IL-6, IL-10 and TNFα were measured in serum, peritoneal lavage fluid and in liver and lung tissues. The conclusion from this study was that in severe sepsis pro-inflammatory cytokines predominated (i.e. TNF and IL-6), but in less severe disease the anti-inflammatory cytokine IL-10 became predominant and could inhibit lethal symptoms. Indeed the lethal effects of moderate sepsis could be prevented by administration of IL-10 and exacerbated by anti-IL-10 antibodies [118]. This ties in with the finding that in human septic shock IL-10 is the principal inhibitory cytokine [119]. While these studies suggest that IL-10 acts to block the effects of TNF, it may in addition be responsible for blocking IL-12, which has also been shown to play a role in the pathology of septic shock [120,121].

An interesting, for want of a better word, micronetwork is the control of IL-9 synthesis by human T-cells in culture. IL-9 production by highly purified human peripheral blood T-cells, induced by anti-CD3 and phorbol 12-myristate 13-acetate (PMA), was initially shown to be mediated by IL-2, as blockade of the IL-2 receptor (IL-2R) inhibited IL-9 synthesis. This blocking antibody also inhibited the synthesis of IL-4, IL-6 and IL-10, and the replacement of IL-4 and IL-10 overcame the IL-2R-inhibiting activity. IL-10 synthesis was in turn inhibited by neutralizing IL-4, but IL-2 and IL-4 (and not IL-4 alone) were required to stimulate IL-10 synthesis. These results suggest that IL-9 synthesis in these purified T-cell cultures is controlled by a cytokine cascade with IL-2 inducing IL-4 synthesis, and a combination of IL-2 and IL-4 driving IL-10 synthesis, followed by IL-4 and IL-10 inducing IL-9 synthesis [122].

Prior to the discovery of cytokines, the control of physiological or pathological processes was seen in terms of individual molecules or simple regulatory circuits. The validity of this view is reflected in the findings that pharmacological antagonists of many individual molecules can have clear-cut effects on many bodily systems resulting in therapeutic benefit. The complexity of cytokine biology, with its agonists with overlapping activities, but controlled by distinct receptor families and different intracellular signalling pathways, suggests that a different level of control exists. It is not only the individual cytokine that is important but the pattern of cytokines acting on cells. These patterns of cytokines with their overlapping agonist/antagonist activities may provide a greater degree of stability than the simpler linear control systems with which biologists have hitherto been more familiar. Such stable networks can cope with multiple modulatory inputs without altering their overall command status. Of course, one down-side to this concept is the possibility of a chaotic switch in a pattern of cytokine synthesis leading to a new stable state which may be pathological in nature. It is possible that the rheumatoid joint with its idiopathic production of cytokines represents a chaotic switch from the pattern of cytokines that normally exists in joints to a new stable pattern which is RA.

It is likely that all normal tissues have particular cytokine patterns. At epithelial surfaces in contact with the bacteria constituting the normal microflora there will be cytokine networks which act as the pattern for health. As described in later chapters, it is proposed that such cytokine networks are controlled by the action of particular proteins produced by the normal microflora. In infections these cytokine patterns are disturbed, warning the body of infection, and it is the return of the set network in any particular tissue that is the goal of the innate and acquired immune systems. This concept will be described in more detail in Chapters 7 and 9.

4.7   Conclusions

Cytokine biology is still in its infancy and yet has led to a radical rethinking of our understanding of cellular communication in health and disease. It should be clear from this chapter, and this message will be made plain in subsequent chapters, that cytokines play a central role in the workings of innate and acquired responses to infections. Micro-organisms have the capacity to interact with cytokine networks for their own advantage. For example, a number of viruses have been shown to enter cells by binding to cytokine receptors. Thus, HIV binds to a chemokine receptor [31] and Herpes Simplex Virus-1 binds to a novel member of the TNF receptor family [32]. The capacity of viruses to control cytokine networks is discussed in more detail in Chapter 7. To bring home this message, a recent example of this ability comes from the report that measles virus suppresses cell-mediated immunity by inhibiting IL-12 synthesis. The cellular receptor for measles virus is CD46, a complement regulatory protein. Cross-linking of CD46 by other methods was also shown to inhibit IL-12 synthesis [123]. This is an interesting example of the networking of innate and acquired immunity. Another recent example of such networking is the finding that fusing the complement component C3d to hen egg lysozyme raised its immunogenicity by up to four log orders. This suggests that C3d is an adjuvant [124]. The final, and possibly most fascinating, example of the merging of innate and acquired immunity is the report that mice with a systemic lupus erythematosus (SLE)-like condition and patients with SLE produce a TGFβ–IgG complex, which is significantly more biologically active than free TGFβ in suppressing neutrophil antibacterial function and which may contribute to autoimmunity [125].

  Cytokine networks are immensely complex and can probably only be understood in terms of mathematical formulation. The reader interested in this subject should consult reference [126] for a recent review of the subject.

  In the next chapter the key stimulator of innate immunity, LPS, will be described in terms of its chemistry and structure, while its very interesting mechanism of action will be delineated in Chapter 6.

References

1.   Bernard, C. (1957) An Introduction to the Study of Experimental Medicine (translated by H.C. Breen), Dover Press, New York
2.   Bayliss, W.B. and Starling, E.H. (1902) On the causation of the so-called 'peripheral reflex section of the pancreas'. Proc. R. Soc. London Ser. B **69**, 325
3.   Cannon, W.B. (1932) The Wisdom of the Body. W.W. Norton, New York
4.   von Bertalanffy, L. (1968) General System Theory. Penguin, London
5.   Prigogine, I. (1986) Life and physics: new perspectives. Cell Biophysics **9**, 217–224
6.   Ibelgaufts, H. (1995) Dictionary of Cytokines. VCH, Weinheim

7. Jelkmann, W. (1992) Erythropoietin: structure, control of production, and function. Physiol. Rev. **72**, 449–489

8. Bomford, R. and Henderson, B. (eds.) (1989) Interleukin-1, Inflammation and Disease. Elsevier, North Holland

9. Meager, A. (1991) Cytokines. Open University Press, Buckingham

10. Aggarwal, B.B. and Gutterman, J.U. (eds.) (1992) Human Cytokines: A Handbook for Basic and Clinical Researchers. Blackwell, Boston

11. Nicola, N.A. (ed.) (1994) Guidebook to Cytokines and their Receptors. Oxford University Press, Oxford

12. Balkwill, F.R. (ed.) (1995) Cytokines — A Practical Approach. IRL Press, Oxford

13. Aggarwal, B.B. and Puri, R.K. (eds.) (1995) Human Cytokines: their Role in Disease and Therapy. Blackwell, Cambridge, MA

14. Henderson, B. and Bodmer, M.W. (eds.) (1996) Therapeutic Modulation of Cytokines. CRC Press, Boca Raton

15. Cohen, S., Bigazzi, P.E. and Yoshida, T. (1974) Similarities of T cell function in cell mediated immunity and antibody production. Cell Immunol. **12**, 150–159

16. Isaacs, A. and Lindemann, J. (1957) Virus interference. I. The interferon. Proc. R. Soc. London Ser. B **147**, 258–267

17. Dumonde, D.C., Wolstencroft, R.A., Panayi, G.S., Matthews, M., Morley, J. and Howson, W.T. (1969) Lymphokines: Non-antibody mediators of cellular immunity generated by lymphocyte activation. Nature (London) **224**, 38–42

18. Aarden, Burnner, T.K., Cerottini, J.C., Dayer, J.M., de Weck, A.L., Dinarello, C.A., Di Sabato, G., Farrar, J.J., Gery, I., Gillis, S. et al. (1979) Revised nomenclature for antigen-non-specific T cell proliferation and helper factors. J. Immunol. **123**, 2928–2929

19. Cohen, S. (1962) Isolation of a mouse submaxillary gland protein accelerating incisor eruption and eyelid opening in the new-born animal. J. Biol. Chem. **237**, 1555–1562

20. Henderson, B., Nair, S.P. and Coates, A.R.M. (1996) Review: molecular chaperones and disease. Inflamm. Res. **45**, 155–158

21. Dimmock, N.J. and Primrose, S.B. (1994) Introduction to Modern Virology, 4th edn. Blackwell, Oxford

22. Mims, C.A., Dimmock, N.J., Nash, A. and Stephen, J. (1995) Mim's Pathogenesis of Infectious Disease, 4th edn. Academic Press, London

22a. Kelner G.S., Kennedy, J., Bacon, J.B., Kleyensteuber, S., Largaespada, D.A., Jenkins, N.A., Copelend, N.G., Bazan, J.F., Moore, K.W., Schall, T.J. and Zlotnik, A. (1994) Lymphotactin: a cytokine that represents a new class of chemokines. Science **266**, 1395–1398

23. Finzel, B.C., Clancy, L.L., Holland, D.R., Muchmore, S.W., Watenpaugh, K.D. and Einspahr, H.M. (1989) Crystal structure of recombinant human interleukin-1β at 2.0Å resolution. J. Mol. Biol. **209**, 779–791

24. Clore, G.M., Wingfield, P.T. and Gronenborn, A.M. (1991) High-resolution three-dimensional structure of interleukin-1β in solution by three- and four-dimensional nuclear magnetic resonance spectroscopy. Biochemistry **30**, 2315–2323

25. Clore, G.M., Appella, E., Yamada, M., Matsushima, K. and Gronenborn, A.M. (1990) Three-dimensional structure of interleukin-8 in solution. Biochemistry **29**, 1689–1696

26. Baldwin, E.T., Weber, I.T., St Charles, R., Xuan, J.-C., Appella, A., Yamada, M., Matsushima, K., Edwards, B.E.P., Clore, G.M., Gronenborn, A.M. and Wlodawar, A. (1991) Crystal structure of interleukin-8: Symbiosis of NMR and crystallography. Proc. Natl. Acad. Sci. U.S.A. **88**, 502–506

27. Perutz, M. (1992) Protein Structure: New Approaches to Disease and Therapy. W.H. Freeman, New York

28. Barton, G.J., Ponting, C.P., Spraggon, G., Finnis, C., and Sleep D. (1992) Human platelet-derived endothelial cell growth factor is homologous to *Escherichia coli* thymidine phosphorylase. Protein Sci. **1**, 688–690

29. Ben-Rafael, Z. and Orvieto, R. (1992) Cytokines — involvement in reproduction. Fertil. Steril. **58**, 1093–1099

30. Ren, P., Lim, C.-S., Johnse, R., Albert, P.S., Pilgrim, D. and Riddle, D.L. (1996) Control of *C. elegans* larval development by neuronal expression of a TGF-β homolog. Science **274**, 1389–1391

31. Rucker, J., Samson, M., Doranz, B.J., Libert, F., Berson, J.F., Yi, Y., Smyth, R.J., Collman, R.G., Broder, C.C., Vassart, G., Doms, C.C. and Parmentier, M. (1996) Regions in β-chemokine receptors CCR5 and CCR2b that determine HIV-1 cofactor specificity. Cell **87**, 437–446

32. Montgomery, R.I., Warner, M.S., Lum, B.J. and Spear, P.G. (1996) Herpes simplex virus-1 entry into cells mediated by a novel member of the TNF/NGF receptor family. Cell **87**, 427–436

33. Miller, D.K. (1996) Cytokine convertase inhibitors. In Therapeutic Modulation of Cytokines (Henderson, B. and Bodmer, M.W., eds.), pp. 143–170, CRC Press, Boca Raton

34. Fantuzzi, G. and Dinarello, C.A. (1996) The inflammatory response in interleukin-1β-deficient mice: comparison with other cytokine-related knockout mice. J. Leukoc. Biol. **59**, 489–493

35. Dinarello, C.A. (1991) Interleukin-1 and interleukin-1 antagonism. Blood **77**, 1627–1652

36. Dinarello, C.A. (1994) The biological properties of IL-1. Eur. Cytokine Netw. **5**, 517–531

37. Henderson, B. (1994) Interleukin-1. In Textbook of Immunopharmacology, 3rd edn., (Dale, M.M., Foreman, J.C. and Fan, T.-P., eds.), pp. 193–199, Blackwell, London

38. Maemura, K., Kurihara, H., Morita, T., Oh-hashi, Y. and Yazaki, Y. (1992) Production of endothelin-1 in vascular endothelial cells is regulated by factors associated with vascular injury. Gerontology **38(suppl. 1)**, 29–35

39. Rothwell, N.J. and Luheshi, G. (1994) Pharmacology of interleukin-1 actions in the brain. Adv. Pharmacol. **25**, 1–20

40. Ferreira, S.H., Lorenzetti, B.B., Bristow, A.F. and Poole, S. (1988) Interleukin-1β is a potent hyperalgesic agent antagonized by a tripeptide analogue. Nature (London) **334**, 698–700

41. Follenfant, R.L., Nakamura-Craig, M., Henderson, B. and Higgs, G.A. (1989) Inhibition by neuro-peptides of interleukin-1β-induced prostaglandin-independent hyperalgesia. Br. J. Pharmacol. **98**, 41–43

42. Dahinden, C.A., Zingg, J., Maly, F.E. and de Weck, A.L. (1988) Leukotriene production in human neutrophils primed by recombinant human granulocyte/macrophage-colony stimulating factor and stimulated with the complement components C5a and FMLP as second signals. J. Exp. Med. **167**, 1281–1295

43. McColl, S.R., Krump, E., Naccache, P.H., Poubelle, P.E., Braquet, P., Braquet, M. and Borgeat, P. (1991) Granulocyte-macrophage colony-stimulating factor increases the synthesis of leukotriene B_4 by human neutrophils in response to platelet-activating factor. Enhancement of both arachidonic acid availability and 5-lipoxygenase activation. J. Immunol. **146**, 1204–1211

44. Abbas, A.K., Murphy, K.M. and Sher, A. (1996) Functional diversity of helper T lymphocytes. Nature (London) **383**, 787–793

45. Saklatvala, J. (1996) Signal transduction mechanisms of interleukin-1 and tumor necrosis factor. In Therapeutic Modulation of Cytokines (Henderson, B. and Bodmer, M. W., eds.), pp. 267–286, CRC Press, Boca Raton

46. Hannum, Y.A. (1996) Functions of ceramide in coordinating cellular responses to stress. Science **274**, 1855–1859

47. Johnson, L., Noble, M.E.M. and Owen, D.J. (1996) Active and inactive protein kinases: structural basis for regulation. Cell **85**, 149–158

48. Tonks, N.K. and Neel, B.G. (1996) From form to function: Signalling by protein tyrosine phosphatases. Cell **87**, 365–368

49. Bird, T.A., Sleath, P.R., DeRoos, P.C., Dower, S.K. and Virca, G.D. (1991) Interleukin-1 represents a new modality for the activation of extracellular signal-regulated kinases/microtubule-associated protein-2 kinases. J. Biol. Chem. **266**, 22661–22670

50. van der Greer, P., Hunter, T. and Lindberg, R.A. (1994) Receptor protein tyrosine kinases and their signal transduction pathways. Annu. Rev. Cell Biol. **10**, 251–337

51. Zemiecki, A., Harpur, A.G. and Wilks, A.F. (1994) MAP protein kinase tyrosine kinases: their role in cytokine signalling. Trends Cell Biol. **4**, 207–212

52. Schindler, C. and Darnell, J.E. (1995) Transcriptional responses to polypeptide ligands: the JAK-STAT pathway. Annu. Rev. Biochem. **64**, 621–651

53. Hill, C.S. and Treisman, R. (1995) Transcriptional regulation by extracellular signals: mechanisms and specificity. Cell **80**, 199–211

54. Leonard, W.J. (1996) STATs and cytokine specificity. Nature Med. **2**, 968–969

55. Hibi, M., Murakami, M., Saito, M., Hirano, T., Taga, T. and Kishimoto, T. (1990) Molecular cloning and expression of an IL-6 signal transducer. Cell **63**, 1149–1157

56. Taga, T. and Kishimoto, T. (1993) Cytokine receptors and signal transduction. FASEB J. **7**, 3387–3396

57. Thierfelder, W.E., van Deursen, J.M., Yamamoto, K., Tripp, R.A., Sarawar, S.R., Carson, R.T., Sangster, M.Y., Vignali, D.A.A., Doherty, P.C., Grosveld, G.C. and Ihle, J.N. (1996) Requirement for Stat4 in interleukin-12-mediated responses of natural killer and T cells. Nature (London) **382**, 171–174

58. Kaplan, M.H., Sun, Y.-L., Hoey, T. and Grusby, M.J. (1996) Impaired IL-12 responses and enhanced development of Th2 cells in Stat4-deficient mice. Nature (London) **382**, 174–177

59. Takeda, K., Tanaka, T., Shi, W., Matsumoto, M., Minami, M., Kashiwamura, S., Nakanishi, K., Yoshida, N., Kishimoto, T. and Akira, S. (1996) Essential role of Stat6 in IL-4 signalling. Nature (London) **380**, 627–630

60. Shimoda, K., van Deursen, J., Sanster, M.Y., Sarawar, S.R., Carson, R.T., Tripp, R.A., Chu, C., Quelle, F.W., Nosaka, T., Vignali, D.A.A., Doherty, P.C., Grosveld, G., Paul, W.E. and Ihle, J.N. (1996) Lack of IL-4-induced Th$_2$ response and IgE class switching in mice with disrupted Stat6 gene. Nature (London) **380**, 630–633

61. Takeda, K., Kamanaka, M., Tanaka, T., Kishimoto, T. and Akira, S. (1996) Impaired IL-13-mediated functions of macrophages in STAT-6-deficient mice. J. Immunol. **157**, 3220–3222

62. Durbin, J.E., Hackenmiller, R., Simon, M.C. and Levy, D.E. (1996) Targeted disruption of the mouse *Stat1* gene results in compromised immunity to viral disease. Cell **84**, 443–450

62a. Uze, G., Lutfalla, G. and Mogenson, K.E. (1995) α and β interferons and their receptors and their friends and relations. J. Interferon Cytokine Res. **15**, 3–26

62b. Miyazono, K. (1996) Signalling through protein serine/threonine kinase receptors. In Signal Transduction (Heldin, C.-H. and Purton, M., eds.), pp. 65–78, Chapman and Hall, London

63. Vandenabeele, P., Declercq, W., Beyaert, R. and Fiers, W. (1995) Two tumour necrosis factor receptors: structure and function. Trends Cell Biol. **5**, 392–399

64. Heller, R.A. and Kronke, M. (1994) Tumor necrosis factor receptor-mediated signalling pathways. J. Cell Biol. **126**, 5–9

65. Siebenlist, U., Franzoso, G. and Brown, K. (1994) Structure, regulation and function of NF-kappa B. Annu. Rev. Cell. Biol. **10**, 405–455

66. Tewari, M. and Dizit, V.M. (1995) Fas- and tumor necrosis factor-induced apoptosis is inhibited by the poxvirus crmA gene product. J. Biol. Chem. **270**, 3255–3260

67. Hsu, H., Xiong, J. and Goeddel, D.V. (1995) The TNF receptor 1-associated protein TRADD signals cell death and NF-kappa B activation. Cell **81**, 495–504

68. Rothe, J., Kesslauer, W., Lotscher, H., Lang, Y., Koebel, P., Kontgen, F., Althage, A., Zinkernagel, R., Steinmetz, M. and Bluethman, W. (1993) Mice lacking the tumour necrosis factor receptor I are resistant to TNF-mediated toxicity but highly susceptible to infection with *Listeria monocytogenes*. Nature (London) **364**, 798–802

69. Rothe, M., Wong, S.C., Henzel, W.J. and Goeddel, D.V. (1994) A novel family of putative signal transducers associated with the cytoplasmic domain of the 75kDa tumor necrosis factor receptor. Cell **78**, 681–692

70. Natoli, G., Costanzo, A., Ianni, A., Templeton, D.J., Woodgett, J.R., Balsano, C. and Levrero, M. (1997) Activation of SAPK/JNK by TNF-receptor 1 through a noncytotoxic TRAF2-dependent pathway. Science **275**, 200–203

71. Fernandez-Botran, R. (1991) Soluble cytokine receptors: their role in immunoregulation. FASEB J. **5**, 2567–2574

72. Heaney, M.L. and Golde, D.W. (1993) Soluble hormone receptors. Blood **82**, 1945–1948

73. Konstantopoulos, K. and McIntire, L.V. (1996) Effects of fluid dynamic forces on vascular cell adhesion. J. Clin. Invest. **98**, 2661–2665

73a. Scales, W. and Kunkel, S.L. (1989) Regulatory Interactions Between Interleukin-1, Inflammation and Disease (Bomford, R. and Henderson, B., eds.), pp. 163–172, Elsevier, North Holland

73b. Rola Pleszczynski, M.-R. and Stankova, J. (1992) Leukotriene B₄ enhances interleukin-6 (IL-6) production and IL-6 messenger RNA accumulation in human monocytes *in vitro*: transcriptional and posttranscriptional mechanisms. Blood **80**, 1004–1011

74. Devchand, P.R., Keller, H., Peters, J.M., Vazquez, M., Gonzalez, F.J. and Wahli, W. (1996) The PPARα-leukotriene B₄ pathway to inflammation control. Nature (London) **384**, 39–43

75. Henderson, B., Higgs, G.A., Pettipher, E.R. and Moncada, S. (1993) Fatty acid mediators in inflammation and immunology. In Clinical Aspects of Immunology, 5th edn. (Lachman, P. and Peters, D.K., eds.), pp. 395–410, Blackwell, London

76. Janeway, C.A. and Travers, P. (1997) Immunobiology; the Immune System in Health and Disease, 3rd edn., Current Biology, London

77. Slifka, M.K. and Ahmed, R. (1996) Long-term humoral immunity against viruses: revisiting the issue of plasma cell longevity. Trends Microbiol. **4**, 394–400

78. Ahmed, R. and Gray, D. (1996) Immunological memory and protective immunity: understanding their relation. Science **272**, 54–60

79. Dutton, R.W. (1996) The regulation of the development of CD8 effector T cells. J. Immunol. **157**, 4287–4292

80. Mosmann, T.R., Cherwinski, H., Bond, M.W., Giedlin, M.A. and Coffman, R.L. (1986) Two types of murine helper T cell clones. Definition according to profiles of lymphokine activities and secreted proteins. J. Immunol. **136**, 2348–2357

81. Sornasse, T., Larenas, P.V., Davis, K.A., de Vries, J.E. and Yssel, H. J. (1996) Differentiation and stability of T helper 1 and T helper 2 cells derived from naive human neonatal CD4+ T cells, analysed at the single cell level. Exp. Med. **184**, 473–483

82. Bottomly, K. (1988) A functional dichotomy in CD4+ T lymphocytes. Immunol. Today **9**, 268–273

83. Mosmann, T.R. and Coffman, R.L. (1993) T$_H$1 and T$_H$2 cells: different patterns of lymphokine secretion lead to different functional properties. Annu. Rev. Immunol. **7**, 145–173

84. Romagnani, S. (1991) Human TH1 and TH2 subsets: doubt no more. Immunol. Today **12**, 256–257

85. Romagnani, S. (1996) Understanding the role of Th1/Th2 cells in infection. Trends Microbiol. **4**, 470–473

86. Reiner, S.L. and Locksley, R.M. (1995) The regulation of immunity to Leishmania major. Annu. Rev. Immunol. **13**, 151–177

87. Heinzel, F.P., Sadick, M.D., Mutha, S.S. and Locksley, R.M. (1991) Production of interferon gamma, interleukin-2, interleukin-4 and interleukin-10 by CD4+ lymphocytes in vivo during healing and progressive murine leishmaniasis. Proc. Natl. Acad. Sci. U.S.A. **88**, 7011–7015

88. Sadick, M.D., Heinzell, F.P., Holaday, B.J., Pu, R.T., Dawkins, R.S. and Locksley, R.M. (1990) Cure of murine leishmaniasis with anti-interleukin 4 monoclonal antibody. Evidence for a T cell-dependent, interferon gamma-independent mechanism. J. Exp. Med. **171**, 115–127

89. Chatelain, R., Varkila, K. and Coffman, R.L. (1992) IL-4 induces a Th2 response in *Leishmania major*-infected mice. J. Immunol. **148**, 1182–1187

90. Kaufmann, S.H.E. and Ladel, C.H. (1994) Application of knockout mice to the experimental analysis of infections with bacteria and protozoa. Trends Microbiol. **2**, 235–242

91. Yamamura, M., Uyemura, K., Deans, R.J., Weinberg, K., Rea, T.H., Bloom. B.R. and Modlin, R.L. (1991) Defining protective responses to pathogens: cytokine profiles in leprosy lesions. Science **254**, 277–279

92. Salgame, P., Abrams, J.S., Clayberger, C., Goldstein, H., Convit, J., Modlin, R.L. and Bloom, B.R. (1991) Differing lymphokine profiles of functional subsets of human CD4 and CD8 T cell clones. Science **254**, 279–282

93. Yamamura, M., Wang, X.-H., Ohmen, J.D., Uyemura, K., Rea, T.H., Bloom, B.R. and Modlin, R.L. (1992) Cytokine patterns of immunologically mediated tissue damage. J. Immunol. **149**, 1470–1475

94. Austrup, F., Vestweber, D., Borges, E., Lohning, M., Brauer, R., Herz, U., Renz, H., Hallmann, R., Scheffold, A., Radbruch, A. and Hamann, A. (1997) P- and E-selectin mediate recruitment of T-helper-1 but not T-helper-2 cells into inflamed tissues. Nature (London) **385**, 81–83

95. Seder, R.A. and Paul, W. (1994) Acquisition of lymphokine-producing phenotype by CD4+ T cells. Annu. Rev. Immunol. **12**, 635–673

96. Gately, M.K., Desai, B.B., Wolitsky, A.G., Quinn, P.M., Dwyer, C.M., Podlaski, F.J., Familletti, P.C., Sinigaglia, F., Chizzonite, R., Gubler, U. and Stern, A.S. (1991) Regulation of human lymphocyte proliferation by a heterodimeric cytokine IL-12 (cytotoxic lymphocyte maturation factor). J. Immunol. **147**, 874–882

97. Hsieh, C., Macatonia, S.E., Tripp, C.S., Wolf, S.F., O'Garra, A. and Murphy, K.M. (1993) Development of Th1 CD4+ T cells through IL-12 produced by Listeria-induced macrophages. Science **260**, 547–549

98. Seder, R.A., Gazzinelli, R., Sher, A. and Paul, W.E. (1993) Interleukin-12 acts directly on CD4+ T cells to enhance priming for interferon-γ production and diminishes interleukin-4 inhibition of such priming. Proc. Natl. Acad. Sci. U.S.A. **90**, 10188–10192

99. Seder, R.A., Kelsall, B.L. and Jankovic, D. (1996) Differential roles for IL-12 in the maintenance of immune responses in infectious versus autoimmune diseases. J. Immunol. **157**, 2745–2748

100. Tripp, C.S., Kanagawa, S.O. and Unanue, E.R. (1995) Secondary responses to Listeria infection required IFN-γ but is partially-independent of IL-12. J. Immunol. **155**, 3427–3432

101. Sieling, P.A., Wang, X.-H., Gately, M.K., Oliveros, J.L., McHugh, T., Barnes, P.F., Wolf, S.F., Golkar, L., Yamamura, M., Yogi, Y., Uyemura, K., Rea, T.H. and Modlin, R.L. (1994) IL-12 regulates T helper type 1 cytokine responses in human infectious diseases. J. Immunol. **153**, 3639–3647

102. Shirakawa, T., Enomoto, T., Shimazu, S.-I. and Hopkin, J.M. (1997) The inverse association between tuberculin responses and atopic disorders. Science **275**, 77–79

103. Bretscher, P.A., Wei, G., Menon, J.N. and Bielenfeldt-Ohmann, H. (1992) Establishment of stable cell-mediated immunity makes 'susceptible' mice resistant to Leishmania major. Science **257**, 539–542

104. Hosken, N.A., Shibuya, K., Heath, A.W., Murphy, K.M. and O'Garra, A. (1995) The effect of antigen dose on CD4+ T helper cell phenotype, development in a T cell receptor-alpha-beta-transgenic model. J. Exp. Med. **182**, 1579–1584

105. Kuchroo, V.K., Prabhu Das, M., Brown, J.A., Ranger, A.M., Zamvil, S.S., Sobel, R.A., Weiner, H.L., Nabavi, N. and Glimcher, L.H. (1995) B7-1 and B7-2 costimulatory molecules activate differentially the Th1/Th2 developmental pathways: application to autoimmune disease therapy. Cell **80**, 707–718

106. Lenschow, D.J., Ho, S.C., Sattar, H., Rhee, L., Gray, G., Nabavi, N., Herold, K.C. and Bluestone, J.A. (1995) Differential effects of anti-B7-1 and anti-B7-2 monoclonal antibody treatment on the development of diabetes in the nonobese diabetic mouse. J. Exp. Med. **181**, 1145–1155

107. Sorensen, T.I.A., Nielsen, G.G., Andersen, P.K. and Teasdale T.W. (1988) Genetic and environmental influences on premature death in adult adoptees. N. Engl. J. Med. **318**, 727–732

108. Coffman, R.L., Varkila, K., Scott, P. and Chatelain, R. (1991) The role of cytokines in the differentiation of CD4+ T cell subsets in vivo. Immunol. Rev. **123**, 189–207

109. de Kossodo, S. and Grau, G.E. (1993) Profile of cytokine production in relation with susceptibility to experimental malaria. J. Immunol. **151**, 4811–4820

110. Barton, C.H., White, J.K., Roach, T.I. and Blackwell, J.M. (1994) NH2 terminal sequence of macrophage-expressed natural resistance-associated macrophage protein (Nramp) encodes a proline-rich putative Src homology 3-binding domain. J. Exp. Med. **179**, 1683–1687

111. Daser, A., Mitchison, H., Mitchison, A. and Muller, A. (1996) Non-classical MHC genetics of immunological disease in man and mouse. The key role of pro-inflammatory cytokine genes. Cytokine **8**, 593–597

112. Messer, G., Spengler, U., Jung, M.C., Honold, G., Blomer, K., Pape, G.R., Riethmuller, G.R. and Weiss, E.H. (1991) Polymorphic structure of the tumor necrosis factor (TNF) locus: an NeoI polymorphism in the first intron of the human TNF-beta gene correlates with a variant amino acid in position 26 and a reduced level of TNF-beta production. J. Exp. Med. **173**, 203–219

113. Pociot, F., Molvig, J., Wogensen, L., Worsaae, H. and Nerup, J. (1992) A TaqI polymorphism in the human interleukin-1 beta (IL-1 beta) gene correlates with IL-1 beta secretion in vitro. Eur. J. Clin. Invest. **22**, 396–402

114. Westenberg, R.G.J., Langermans, J.A.M., Huizinga, T.W.J., Elouali, A.H., Verweij, C.L., Boomsma, D.I. and Vandenbrouke, J.P. (1997) Genetic influences on cytokine production and fatal meningo-coccal disease. Lancet **349**, 170–173

115. Brennan, F.M., Chantry, D., Jackson, A., Maini, R.N. and Feldmann, M. (1989) Inhibitory effect of TNFα antibodies on synovial cell interleukin-1 production in rheumatoid arthritis. Lancet **2**, 244–247

116. Uyemura, K., Demer, L.L., Castle, S.C., Jullien, D., Berliner, J.A., Gately, M.K., Warrier, R.R., Pham, N., Fogelman, A.M. and Modlin, R.L. (1996) Cross-regulatory roles of interleukin (IL)-12 and IL-10 in atherosclerosis. J. Clin. Invest. **97**, 2130–2138

117. Abraham, E. and Raffin, T.A. (1994) Sepsis therapy trials: continued disappointment or reason for hope? JAMA **271**, 1876–1878

118. Walley, K.R., Lukacs, N.W., Standiford, T.J., Strieter, R.J. and Kunkel, S.L. (1996) Balance of inflammatory cytokines related to severity and mortality of murine sepsis. Infect. Immun. **64**, 4733–4738

119. Brandtzaeg, P., Osnes, L., Øvstebø, R., Joø, G.B., Westvik, Å.-B. and Kierulf, P. (1996) Net inflam-matory capacity of human septic shock plasma evaluated by a monocyte-based target cell assay: identification of interleukin-10 as a major functional deactivator of human monocytes. J. Exp. Med. **184**, 51–60

120. Wysocka, M., Kubin, M., Vieira, L.Q., Ozmen, L., Garotta, G., Scott, P. and Trinchieri, G. (1995) Interleukin-12 is required for interferon-γ-production and lethality in lipopolysaccharide-induced shock in mice. Eur. J. Immunol. **25**, 672–676

121. Heinzel, F.P., Rerko, R.M., Ahmed, F. and Hujer, A.M. (1996) IFN-γ-independent production of IL-12 during murine endotoxemia. J. Immunol. **157**, 4521–4528

122. Houssiau, F.A., Schandene, L., Stevens, M., Cambiaso, C., Goldman, M., van Snick, J. and Renauld, J.-C. (1995) A cascade of cytokines is responsible for IL-9 expression in human T cells. J. Immunol. **154**, 2624–2630

123. Karp, C.L., Wysocka, M., Wahl, L.M., Ahearn, J.M., Cuomo, P.J., Sherry, B., Trinchieri, G. and Griffin, D.E. (1996) Mechanism of suppression of cell-mediated immunity by measles virus. Science **273**, 228–231

124. Dempsey, P.W., Allison, M.E.D., Akkaraju, S., Goodnow, G.C. and Fearon, D.T. (1996) C3d of complement as a molecular adjuvant: bridging innate and acquired immunity. Science **271**, 348–350

125. Caver, T.E., O'Sullivan, F.X., Gold, L.I. and Gresham, H.D. (1996) Intracellular demonstration of active TGFβ$_1$ in B cells and plasma cells of autoimmune mice. J. Clin. Invest. **98**, 2496–2506

126. Henderson, B., Seymour, R. and Wilson, M. (1998) The cytokine network in infectious disease. J. Immunol. Immunopharmacol., in the press

5

Lipopolysaccharide: structure and function

"Endotoxins possess an intrinsic fascination that is nothing less than fabulous. They seem to have been endowed by Nature with virtues and vices in the exact and glamorous proportions needed to render them irresistible to any investigator who comes to know them".

Ivan L. Bennett, 1964 [1]

5.1   Introduction

The previous chapter has reviewed in detail the nature and activity of the local hormones known as cytokines. It is important to realize that the discovery of cytokines and our understanding of the interactions that occur between bacteria and cytokines rests full square on the complex Gram-negative bacterial cell-wall component lipopolysaccharide (LPS). The knowledge gained during the past 30 years of the physiology, pathology, cell biology and molecular biology of cytokines is largely based on the capacity of LPS to induce the transcription of cytokine genes. In this, and the following chapter, the chemistry and biology of LPS will be described and the central role of this molecule in the interactions between bacteria and host cytokines will be delineated.

Before embarking on this chapter it is important to clarify a problem of terminology. The terms lipopolysaccharide and endotoxin are employed interchangeably by scientists in many biomedical disciplines. These bacterial constituents have a myriad of biological actions, and the study of the cellular, physiological and pharmacological actions of LPS/endotoxin has generated many thousands of research papers. In this chapter we will employ the terminology introduced by Hitchcock and co-workers [2]. Thus, the term lipopolysaccharide "should be reserved for purified bacterial extracts which are reasonably free of detectable contaminants, particularly protein". Such

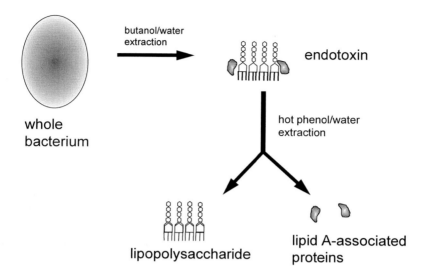

Figure 5.1 Schematic diagram showing how endotoxin, LPS and lipid A-associated (also known as endotoxin-associated) proteins are prepared from Gram-negative bacteria

Application to the whole bacteria of a butanol/water mixture extracts the LPS and proteins associated with the LPS. This complex, known as endotoxin, is separated into its constituent parts by the differential solubility of the LPS and the protein in phenol/water mixtures at 68°C.

preparations may be obtained using the Westphal extraction procedure as a starting point (see Figure 5.1). In contrast, the term endotoxin should be used to refer to "products of extraction procedures which result in macromolecular complexes of LPS, protein and phospholipid". Complexes of this type can be obtained by extraction of bacteria with trichloroacetic acid, butanol or EDTA (Figure 5.1). In this chapter, the emphasis will be on LPS. In Chapter 7 we will discuss the proteins associated with LPS in the bacterial membrane, and which form the complex known as endotoxin when Gram-negative bacteria are appropriately extracted. Such endotoxin-associated proteins are potent stimulators of inflammatory and immunological activity in their own right and are members of a growing population of bacterial proteins which have the capacity to modulate host cytokine synthesis.

5.2   Historical background to endotoxin and LPS research

5.2.1   Discovery and assay
One hundred and twelve years before the writing of this chapter, the Danish microbiologist Christian Gram discovered that bacteria differed in their ability to retain a Crystal Violet–iodine stain when exposed to organic solvents like ethanol or acetone. This staining technique has become known as the Gram

stain and its use has allowed microbiologists to divide bacteria into two broad classes — Gram-positive and Gram-negative. The structure of the cell envelopes of both classes of bacteria has been reviewed in Chapter 2. Briefly, the Gram-positive envelope is composed of two main structures, the cell wall and the cell membrane, the former consisting of a thick layer of peptidoglycan. In contrast, in Gram-negative bacteria, the peptidoglycan layer is much thinner (possibly a monolayer) and it is bounded on the outer surface by an outer membrane composed of phospholipids, in which are embedded a range of proteins. Within the three superkingdoms of life, this outer membrane is unique, in being asymmetric with the phospholipids attached to a complex amphiphilic molecule called LPS.

LPS/endotoxin, has been the subject of study for just over 100 years. As will be described in the next chapter, the discovery of LPS was the result of a search (which had been going on for more than two millennia) for the cause of fever. In the 19th century it was discovered that bacteria were a key cause of fever. It was Richard Pfeiffer, working in Berlin in the 1890s, who observed that cholera bacteria contained a toxin tightly anchored to the bacterial cell wall and named this toxin, endotoxin, to discriminate it from the soluble exotoxins released by bacteria [3]. It was other workers, particularly the Italian Eugenio Centanni, who demonstrated that bacteria contained pyrogenic (fever-inducing) substances [4–6]. Centanni showed that the pyrogenic material that he isolated from bacteria was not protein and was insensitive to heat. These are characteristic features of LPS. Contemporaneously, Coley, in New York, described the necrotizing activity of endotoxin (Coley's toxin) on tumours [7,8]. Coley's toxin, in the modern form of tumour necrosis factor (TNF)-α, has enjoyed a resurgence in cancer therapy in recent years [9].

Fever, the abnormal elevation of core body temperature, is today still one of the commonest indications of disease. However, in the pre-antibiotic age, fevers were the key manifestation of disease and their cause and effects were the subject of endless speculation [10]. As discussed by Atkins and Snell [10], by 1785 the concept that fever was caused by products of tissue destruction (inflammation) was under discussion. Scientific backing for the belief that inflammation and fever were linked came from the work of Billroth and Weber and this view was presciently expounded by the physician William Welch in 1888 (reviewed by Atkins [11]). Welch hypothesized that fever resulted from a disturbance of the normal central nervous system regulation of body temperature induced by microbes, or their products, acting on host cells to produce host products which he termed 'ferments' and which were the final inducers of fever. This hypothesis, that fever was due to molecules produced by the host (and as we now know them — cytokines) was far ahead of its time. It is ironic that just as the hypothesis that ferments (i.e. cytokines) control fever was postulated, the discovery was made of endotoxin, a ubiquitous bacterial component which is a potent cytokine inducer. The problems of removing or controlling for bacterial endotoxins was not solved until the 1950s and even at the time of writing (1997) endotoxin/LPS contamination is still one of the major confusing variables in much of biomedical research. It must be realized that Gram-positive bacteria also produce pyrogenic substances. Such constituents will not be considered in this chapter but will be discussed in detail in Chapter 7.

In the early part of this century the fevers that often accompanied the injection of therapeutic agents (so-called injection fevers) were ascribed to contaminating endotoxins and these substances were shown to be heat-stable and filterable agents which were presumably of bacterial origin. Hort and Penfold [12] pioneered these studies and introduced the rabbit pyrogen test for the assay of these substances. The most potent pyrogens seemed to be associated with Gram-negative bacteria, a finding that was later confirmed by the extensive studies of Seibert [13] who further developed the pyrogen assay, which was later incorporated in the United States Pharmacopoeia [14] and is still a standard test today for detecting endotoxin levels in pharmaceutical products. The rabbit pyrogen test involves measuring the rise in body temperature, in trained rabbits, evoked by intravenous injection of a sterile solution of the substance under investigation. In 1956 Bang [15] discovered that the endotoxin of a sea water *Vibrio* species, which was pathogenic for the horseshoe crab (*Limulus polyphemus*), caused fatal intravascular coagulation, and that this coagulation process could be replicated *in vitro*. Levin, Bang and co-workers [16–18] subsequently showed that coagulation was the result of an endotoxin-initiated reaction causing the enzymic conversion of a clottable protein of the blood cell (amoebocyte) of the crab (see review by Mikkelsen [19]). This ability of the haemolymph of this crab to clot in the presence of endotoxin became the basis of a variety of extremely sensitive (though not necessarily specific) assays for this bacterial constituent [20,21]. Both the rabbit pyrogen and Limulus amoebocyte assays have a number of disadvantages, and one of the authors (Stephen Poole) has pioneered the use of indirect pyrogen assays in which the induction of cellular cytokine synthesis is used to monitor and assay endotoxin/LPS [20,22]. More recent work on the clotting system of *L. polyphemus* has identified an additional protein, termed endotoxin-neutralizing protein, which is now being tested as a therapeutic agent for septic shock. The biological activity of this protein and its therapeutic potential is discussed in more detail in Chapter 8.

By the 1940s the wheel was coming round full circle and there was increasing interest in how endotoxin/LPS induced fever. In 1943, Menkin [23] reported on the isolation of a pyrogenic material from the euglobulin fraction of inflammatory exudates. He named this substance pyrexin and concluded that it was a peptide breakdown product of damaged tissue [24,25]. However, the heat stability of pyrexin suggested that it was contaminated with bacterial pyrogens [26]. It was the pioneering work of Beeson [27] and Bennett and Beeson [28] that finally established that the host was capable of producing endogenous pyrogens in response to exogenous pyrogens such as endotoxin.

The first endogenous pyrogen is now known throughout the world as IL-1 (reviewed in [10]). Thus, while it is rarely appreciated, it is clear that our current understanding of cytokines (reviewed in Chapter 4) has arisen directly from these pioneering studies on endotoxin-induced fever. The relationship between endotoxin and cytokines will be discussed in more detail in Chapters 6 and 7.

5.2.2   Evolution of endotoxin/LPS

The central theme of this book is the biology of the interactions between prokaryotic micro-organisms (specifically those of the superkingdom, Bacteria) and their eukaryotic hosts. Such interactions have to be examined in an evolutionary context. It is safe to assume that the evolution of prokaryotes and eukaryotes has been rich in mutual interactions and that such interactions have been responsible for at least part of the evolutionary drive. Indeed, it is now established that our own eukaryotic cells contain prokaryotes and even portions of prokaryotes. This hypothesis has been most vigorously championed by Lynn Margulis [29], who is best known for her proposal that the mitochondria in eukaryotic cells arose from a symbiotic relationship between early eukaryotic cells and bacteria. The Earth formed some 4.6 billion years ago, and fossil bacteria have been found in rocks 3.4–3.5 billion years old [29,30]. It is not known when the Gram-negative outer membrane with its complex LPS developed. Mikkelsen [19] has argued that the amoebocytes in the blood of the horseshoe crab (*L. polyphemus*), which, as described, degranulate and induce a gelling reaction to trap bacteria, represent one of the earliest known evolutionary responses to this bacterial component. However, Stephen Jay Gould [31] in his exegesis on the Burgess shale fossils claims that *L. polyphemus* is not a 'living fossil' — "the genus *Limulus* ranges back only some 20 million years, not 200 million". The evolutionary history of the development of antibacterial strategies is thus still confusing, and needs to be much more carefully and extensively studied to provide clues to the mutual interaction between prokaryotic and eukaryotic cells.

5.2.3   Structural studies of endotoxin

It is only within the past few years that the complete chemical structure of LPS has been elucidated and there is still much to learn about its three-dimensional structure and its structure in aqueous solution. This is important in understanding: (i) the role LPS plays on the bacterial cell surface (as essentially a two-dimensional structure); and (ii) the structural form that LPS adopts in protein-containing solutions such as plasma or extracellular fluid, which will obviously control its ability to induce inflammatory reactions. Before the structure of endotoxin/LPS could be determined it was vital that it be isolated in as pure a manner as possible. Pioneering work on the extraction of bacterial endotoxin was undertaken in the 1930s and 40s. Boivin and Mesrobeanu [32] in Paris developed the trichloroacetic acid extraction method with the

Enterobacteriaceae. Walter Morgan [33] at the Lister Institute in London developed a mild extraction procedure for endotoxin using diethylene glycol. Goebel, at the Rockefeller in New York, demonstrated that endotoxin could be extracted with pyridine/water mixtures [34]. However, the best known extraction procedure, and the one used by most workers at the present time, was that developed by Otto Westphal [35–37]. In this procedure, bacteria are extracted with phenol/water mixtures at a temperature of 65–68°C. When the solution is cooled the homogenous mixture separates into an upper (water) and a lower (phenol) layer with the LPS in the water phase. Galanos introduced a variant of this extraction methodology by using mixtures of phenol/chloroform/petroleum ether at room temperature [38]. It is now common to treat the LPS with a mixture of protease, DNase and RNase to remove contaminating macromolecules, the LPS being separated from the enzymes by ultracentrifugation. As will be described in Chapter 8, proteins co-extracted with LPS — the so-called lipid A-associated proteins (LAPs) or endotoxin-associated proteins (EAPs) — have potent pro-inflammatory properties in their own right and such proteins are often present in the commercially available endotoxin preparations used by most researchers [39]. Current views of the structure of LPS will be presented in Section 5.3 of this chapter.

While an increasing amount of information was gathered about the structure of LPS, the identification of the nature of the endotoxic activity still eluded discovery. It was not until the 1970s that it was shown convincingly that the endotoxic actions of LPS were due to the lipid A moiety [40,41], and this was only verified in the 1980s with the generation and biological testing of synthetic lipid A [42].

5.2.4   Immunochemistry of LPS/endotoxin

Much of our current understanding of the structure of LPS came from pioneering studies of bacterial immunochemistry. The 'polysaccharide' portion of LPS represents the species-specific antigens responsible for the specificity of antibacterial antibodies raised to Gram-negative bacteria. These heat-stable polysaccharide surface antigens are known as somatic or O antigens. Fritz Kaufmann and Bruce White classified the O antigens of *Salmonella* species [43]. The Kaufmann–White scheme subdivides the *Salmonellae* into some 30 groups on the basis of shared O antigens. A similar scheme was developed by Kaufmann and collaborators for *Escherichia coli.* Immunochemical analysis of LPS led to the discovery of the unique sugars present in the O antigens [sugars such as abequose, tyvelose and the eight-carbon 2-keto-3-deoxyoctulonate (KDO)] and their constitution in the whole molecule (reviewed in [44]). For further information on the history of LPS research the reader is referred to [45].

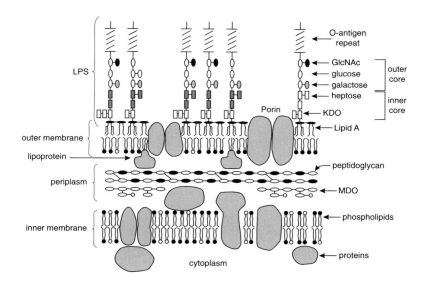

Figure 5.2 Structure of the Gram-negative outer membrane showing the proposed relationship between LPS and membrane proteins
Reproduced from [45a] with permission. Abbreviation used: MDO, membrane-derived oligosaccharide.

5.3   Synthesis and structure of LPS

Having provided a brief historical overview of LPS, the remainder of this chapter will describe the synthesis, structure and certain aspects of the biological activity of this fascinating bacterial component. Further coverage of the bioactivity of LPS will be given in the following chapter.

As described in Chapter 2, all bacteria are surrounded by an envelope which provides the shape and internal environment for metabolism and acts as a semipermeable filtering device working in both directions. In Gram-negative organisms this envelope contains a bilayered and asymmetrically organized membrane called the outer membrane (Figure 5.2). The outer leaflet of this membrane, which makes contact with the organisms' environment, contains the LPS. It has been estimated that the surface area of an *E. coli* cell is 6.7 μm^2 and that the 3.5×10^6 LPS molecules found in the cell cover an area of 4.9 μm^2. Thus, three-quarters of the bacterial surface consists of LPS with the remainder being composed of protein [46]. The LPS is postulated to form a very ordered structure on the bacterial surface, such as is depicted in Figure 5.3 [47]. The LPS molecule consists of three separate regions: lipid A, the core region and the O-specific side chain (Figure 5.4). The lipid A is responsible for most of the endotoxic (cytokine-inducing) capacity of LPS and the chemistry and biology of lipid A will be the focus of this chapter. However, the activity of lipid A is modified by the polysaccharide portion of LPS, and the chemistry and structure of this region will be briefly covered. The combination of a

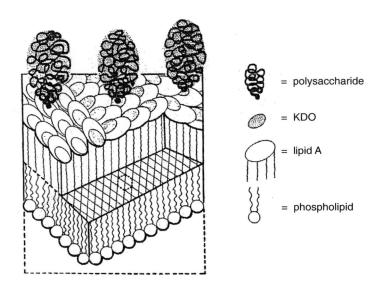

= polysaccharide

= KDO

= lipid A

= phospholipid

Figure 5.3 Schematic diagram of the outer membrane of a typical Gram-negative bacterium showing the relative relationships of the various components of the LPS molecule
Adapted from [47] with permission.

hydrophobic lipid and a hydrophilic polysaccharide makes LPS an amphipathic (or amphiphilic) molecule and this has repercussions in terms of the

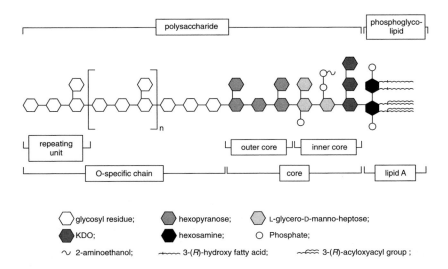

polysaccharide

phosphoglyco-lipid

repeating unit

outer core

inner core

O-specific chain

core

lipid A

◯ glycosyl residue; ⬡ hexopyranose; ⬡ L-glycero-D-manno-heptose;

⬢ KDO; ⬢ hexosamine; O Phosphate;

∿ 2-aminoethanol; ⟿ 3-(R)-hydroxy fatty acid; ⟿ 3-(R)-acyloxyacyl group ;

Figure 5.4 Schematic diagram of the basic structure of the LPS molecule
Reproduced with permission from Helander, M., Makela, P.H., Westphal, O. and Rietschel, E.T. (1996) Lipopolysaccharides, in Encyclopedia of Molecular Biology and Molecular Medicine, vol. 3 (R.A. Meyers, ed.), pp. 462–471, VCH Verlagsgesellschaft mbH, Weinheim

structure of LPS, its biological actions and its interaction with host-cell receptors and other proteins.

5.3.1   Structure and synthesis of lipid A

LPS is a highly polymorphic molecule, and variations in its chemical and physical properties are important in terms of both its biological properties in the bacterial cell wall and its ability to act as an endotoxin and induce cytokine synthesis. An example of this polymorphism can be seen clearly when LPS is fractionated by SDS/PAGE. Many LPSs exhibit what is termed a ladder pattern which is due to the very large range of molecular masses this molecule exhibits. It is therefore important to have some understanding of the chemical structure of LPS and, in particular, its endotoxic lipid A moiety. For readers not familiar with chemistry, the Figures provide all the required information.

The general structure of lipid A is provided in Figure 5.5 and at the outset it should be noted that lipid A is a phosphoglycolipid with a unique architecture. It is a disaccharide composed of either: (i) two residues of the amino sugar D-glucosamine; (ii) D-glucosamine and another amino sugar, 2,3-

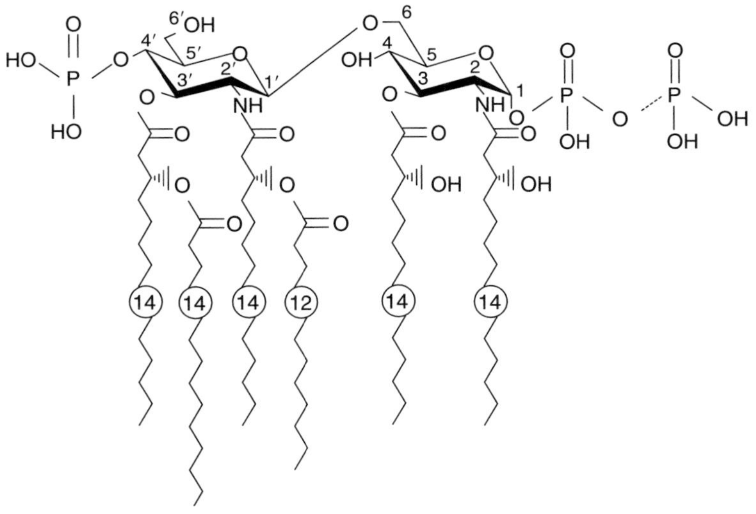

Figure 5.5 The structure of E. coli lipid A
The two D-glucosamine sugars are in the chair conformation and are linked by a β(1'→6) linkage. The terminal (reducing) saccharide has its carbon atoms numbered C-1 to C-6. The saccharide linked to the inner core of the LPS (the non-reducing) has its carbon atoms numbered C-1' to C-6'. Both sugars contain phosphate groups. One is α-linked to the glycosylic (anomeric) carbon at C-1 and the other is ester bound to the hydroxy group at position C-4'. Bound to the glucosamine molecules are four (R)-3-hydroxy fatty acids. Two are amide-linked at positions 2 and 2', and two are ester-linked at positions 3 and 3'. The hydroxy groups of some of these fatty acids are esterified with non-hydroxy fatty acids, creating the unique 3-acyloxyacyl structure. In E. coli lipid A (as shown), this esterification occurs at the 2' and 3' positions. In E. coli the length of the (R)-3-hydroxy fatty acids is 14 and the non-hydroxylated acyl chains can consist of 12 or 14 carbons. Other lipid A molecules have different patterns of acyl chains (see Table 5.1.).

diamino-2,3-dideoxy-D-glucose (DAG); or (iii) two DAG residues linked by a β(1→6) bond. As illustrated in Figure 5.5, the carbons in the sugars are numbered either 1, 2, etc., or 1′, 2′, and so on. Both sugars contain phosphate groups on the 1 and 4′ positions. Four sites on the disaccharide are linked to fatty acyl chains. The 3 and 3′ positions are linked to acyl chains via an ester (CO–OR) linkage, and at positions 2 and 2′ the acyl chains are amide (NH)-linked. The core oligosaccharide, composed of 2-keto-3-deoxyoctulonic acid (also termed 3-deoxy-D-*manno*-octulosonic acid; KDO) is linked to lipid A via the 6′ hydroxyl [48,49]. There is fine variation of the common architecture of lipid A, which accounts for the pronounced differences in biological activity of lipid A molecules. Such variations in structure result from: (i) the types of sugar present; (ii) the degree of phosphorylation; (iii) presence of phosphate substituents; and (iv) the presence of KDO. However, the major variable in terms of lipid A structure–function relationship is the nature, chain length, unsaturation, number and location of the acyl chains. The relationship between the structure of lipid A and its bioactivity will be reviewed in Section

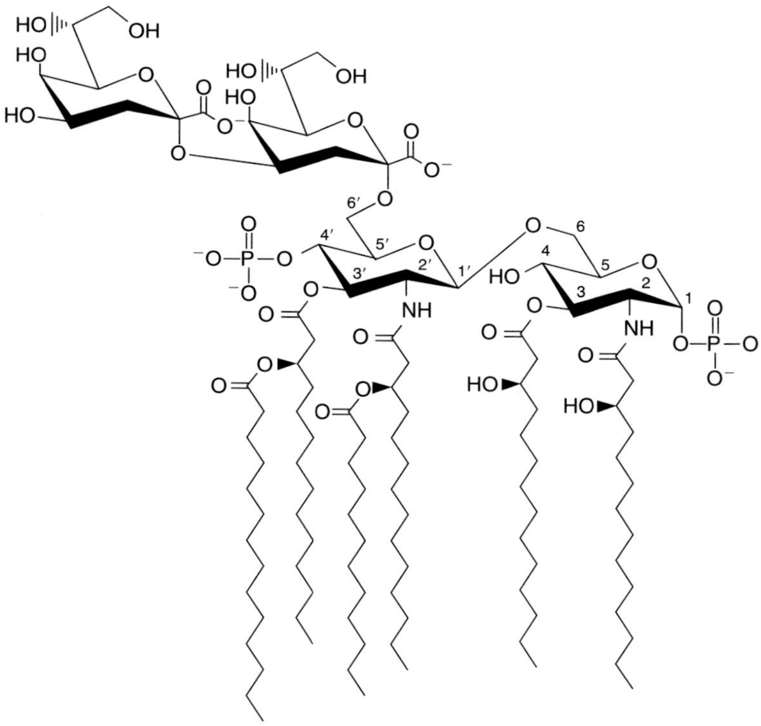

Figure 5.6 The structure of *E. coli* Re-LPS (KDO2–lipid A), which consists of the lipid A molecule shown in Figure 5.5 plus two KDO residues linked to the C-6′ position of GlcNII

5.3.1.4. In the next section the mechanism of the synthesis of lipid A in *E. coli* will be briefly reviewed.

5.3.1.1 Synthesis of lipid A

The minimal LPS structure compatible with bacterial survival and growth is termed Re-LPS, occurs in mutants lacking heptose, and consists of lipid A with two KDO sugars linked to the 6′ position [50]. The steps in the synthesis of this molecule (whose structure is shown in Figure 5.6) will be outlined. Given the importance of LPS, it is curious that it took so long to elucidate the biosynthetic pathway of lipid A. The delay lay in the complexity of lipid A and its unique structure. Just as the structure of lipid A was being defined in the early-to-mid 1980s, Raetz's group [51] in the States made the important discovery of a monosaccharide (2,3-diacylglucosamine 1-phosphate; Figure 5.7) in certain phosphatidylglycerol-deficient mutants of *E. coli.* This molecule, which was termed lipid X, was recognized as a monosaccharide substructure of lipid A, possibly an intermediate or breakdown product, and provided clues as to the sites of acylation on lipid A and the enzymic reactions involved in its biosynthesis. Nucleotide-linked sugars are the activated biosynthetic intermediates used in the synthesis of all oligosaccharides and polysaccharides. For example, uridine diphosphate glucose, commonly abbreviated to UDP-glucose or UDP-Glc, is the substrate for the synthesis of the mammalian storage polysaccharide, glycogen. To determine the putative nucleotide-sugar derivative of lipid X, the molecule UDP-2,3-diacylglucosamine was synthesized and incubated with a crude extract of wild-type *E. coli* in the presence of lipid X. This resulted in the generation of a disaccharide linked by a β(1′→6) linkage

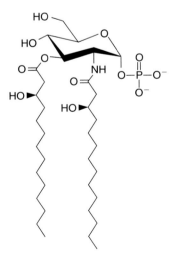

Figure 5.7 The structure of the monosaccharide 2,3-diacylglucosamine 1-phosphate, a molecule known as lipid X

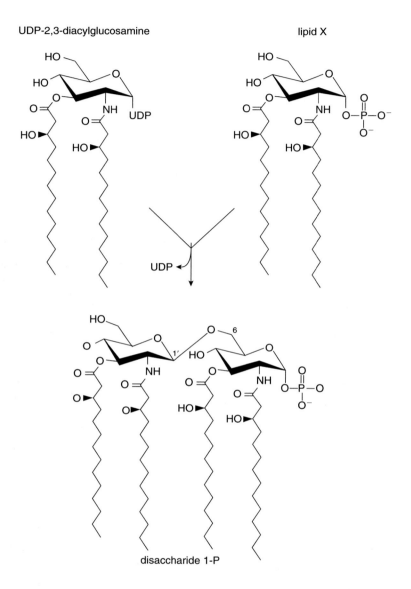

Figure 5.8 The enzymic reaction, catalysed by the enzyme disaccharide synthase, between lipid X and its UDP derivative, UDP-2,3-diacylglucosamine, to produce the lipid A disaccharide structure linked by a β(1→6) linkage

(termed lipid IV_A), demonstrating that lipid X and its UDP derivative were the precursors of lipid A (Figure 5.8) [52]. Lipid A is a glycophospholipid. In eukaryotic cells about 5% of the plasma-membrane lipids are glycolipids. These molecules have an asymmetric distribution, only being found in the non-cytoplasmic half of the plasma membrane. Glycolipids such as galacto-cerebroside and gangliosides have their acyl chains linked indirectly to the C_1

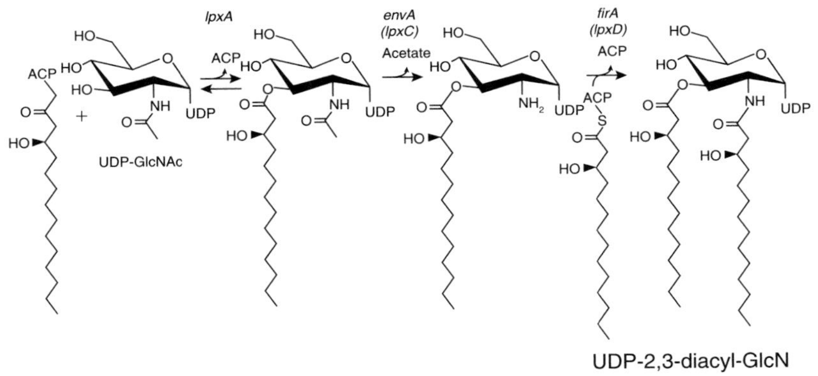

UDP-2,3-diacyl-GlcN

Figure 5.9 The synthesis of UDP-2,3-diacylglucosamine (UDP-2,3-diacyl-GlcN)
The starting materials for this synthesis are 3-hydroxy fatty acids linked to acyl carrier protein (ACP), and the saccharide, UDP-N-acetyl-glucosamine (UDP-GlcNAc). The first step is the acylation of the sugar at position C-3 by the enzyme UDP-GlcNAc O-acyltransferase which incorporates a C-14 fatty acid (myristic acid). The N-acetyl group at position C-2 is then removed by a deacetylase enzyme. A separate acylase enzyme then incorporates an ACP-linked fatty acid at the NH$_2$ group attached to the C-2 carbon. This intermediate can then be linked to lipid X to form the lipid A precursor, disaccharide 1-P. The abbreviations in italics (*lpxA*, *lpxC*, *lpxD*, *envA*, *firA*) refer to the genes encoding the enzymes indicated.

carbon, but not the C$_2$ and C$_3$ (and even C$_4$) positions as found in lipid A. Again, the structure of lipid A is unusual and this is reflected in the complex enzymology of its acylation (see Figure 5.9). The starting point for the generation of UDP-2,3-diacylglucosamine is, in fact, not UDP-glucosamine (UDP-GlcN), but UDP-N-acetyl-glucosamine (UDP-GlcNAc). The enzyme UDP-GlcNAc O-acyltransferase, which is selective for 3-hydroxymyristic acid (a saturated 14-C fatty acid), acylates UDP-GlcNAc at position 3. The N-acetyl group is then removed by a deacetylase, allowing incorporation, by a separate acylase, of an N-linked acyl chain, and thus forming one of the intermediates of lipid A synthesis — UDP-2,3-diacyl-GlcN [53–56]. Removal of the UDP from this molecule produces lipid X, and the enzyme disaccharide synthase couples lipid X and UDP-2,3-diacyl-GlcN at the 1→6 position, forming disaccharide 1-phosphate (Figure 5.8) [57]. Addition of another phosphate group at the C-4′ position by a specific kinase forms the molecule that is known as lipid IV$_A$ [58].

Lipid IV$_A$ is now the substrate for the enzyme KDO transferase, which uses the nucleotide sugar cytidine monophosphate (CMP)–KDO and couples KDO to the 6′ position. This same enzyme then couples a second CMP–KDO residue to the 4-OH of the innermost KDO [59]. It is only following this addition of the KDO residues to the 6′ position of lipid IV$_A$ that additional acylation reactions can occur. In *E. coli*, for which this enzymology and chemistry have been elucidated, additional acyltransferases complete the formation of lipid A by transferring laurate (12-carbon saturated fatty acid) and myristate to KDO–lipid IV$_A$, forming the acyloxyacyl units found in most lipid A molecules [60].

5.3.2   Structure–function relationship of the polysaccharide of LPS

Before considering the structure–function relationship of LPS (which is mainly that of lipid A), a brief description will be provided of the other two (polysaccharide) regions of this molecule, with regard to structure and relationship with cytokine-inducing action. Those readers interested in the molecular details of the structure of the inner core and O-antigens should refer to [61–63].

5.3.2.1   The core region

The core region of LPS can be divided into the inner core, which covalently bonds with lipid A, and the outer core which merges with the O-antigen. The synthesis of the inner core with its KDO residues has been briefly described during the discussion of the synthesis of the lipid A of Re-LPS, and it has been shown that the presence of KDO residues is required for the final acylation of lipid IV$_A$. Indeed, mutants lacking KDO, or the inhibition of KDO incorporation by bacteria, results in the accumulation of lipid IV$_A$ [64].

The inner core region of enterobacterial LPS is characterized by the unusual sugars heptose and KDO. All bacteria contain at least one KDO residue, although not all contain heptose. The chemical structure of the inner core of enterobacterial LPS is shown in Figure 5.10. The sugars in the inner core are often substituted with charged groups such as phosphate and pyrophosphate, resulting in a high negative charge density. This is likely to

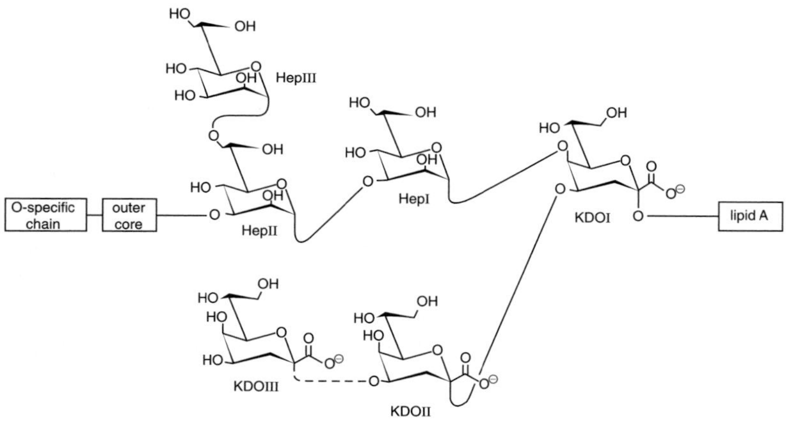

Figure 5.10 The inner core of the LPS of *Salmonella enterica*
The lipid A is linked to a KDO residue which forms a branched structure being attached to one or two KDO residues via an $\alpha(2{\to}4)$ linkage and to L-glycero-D-manno-heptose (Hep) at the C-5 position.

bring about the concentration of countercharges in this region, e.g. Ca^{2+} and Mg^{2+}, and such ion binding may play a structural role in stabilizing the outer membrane of the bacterium.

While the endotoxic activity of LPS resides within the lipid A domain, there is evidence that the polysaccharide moiety, the inner core in particular, can modulate LPS bioactivity to some degree. Thus, there are reports that polysaccharide-free lipid A is less potent than the KDO-containing Re-LPS in inducing IL-1 [65] or peptidoleukotriene [66] release from adherent mono-nuclear cells.

The outer core of enterobacterial LPS is composed of hexose (D-glucose, D-galactose, GlcNAc) sugars and is therefore also termed the hexose region. In *Salmonella* spp. and *E. coli*, the outer core is a branched pentasaccharide. There is little evidence that the outer core contributes to the cytokine-modulating activity of LPS.

5.3.2.2 O-specific side-chain

The unique chemical nature of lipid A has already been described. The outer core is also unique among polysaccharides, in that it is a carbohydrate polymer composed not of monosaccharides or disaccharides but of repeating units of oligosaccharides which can consist of up to eight residues. In addition, various unusual sugars, including abequose, tyvelose and rhamnose, are found in these repeating oligosaccharides. In wild-type *Salmonella typhimurium*, the O-antigen repeating unit has the following structure: abequose-α(1→3)-mannose-α(1→4)-rhamnose-β(1→3)-galactose.

This oligosaccharide unit is assembled on a lipid carrier, undecaprenol phosphate (equivalent to dolichol phosphate used in eukaryotic systems), a 55-carbon compound comprising of 11 isoprenoid units with a phosphate linked at the terminus. This lipid-linked tetrasaccharide is assembled inside the bacterium and then passes through the cell membrane to link to the growing LPS O-antigen [67].

The structure of the oligosaccharide repeating units can vary in terms of number and nature of the sugars, their ring form, sequence, substitution and type of linkage. This gives rise to an enormous amount of structural variation and accounts for the known serological variation in O-antigens.

The O-antigens can be regarded as the public face of the Gram-negative bacterium — a sea of polysaccharide chains. What role does this coating have? Perhaps the most important function, in terms of pathology, is the capacity of the O-antigens to activate complement through the alternative pathway [68]. Subtle differences in the chemistry and chirality of the O-antigens can modify the bacterium's capacity to activate the complement cascade [69,70]. There are also reports that the O-antigen polysaccharide can regulate the endotoxic activity of lipid A [66,71]. In addition, the O-antigens can function as receptors for bacteriophages [72] and in the interaction of nitrogen-fixing bacteria with their leguminous host [73].

5.3.3   Structure–function relationships of lipid A

As stated previously, it is now firmly established that the endotoxic actions of LPS are due to the lipid A moiety. Studies conducted during the past decade have established the structure–function relationships of lipid A, largely in terms of the capacity of this molecule to directly (in *in vitro* assays), and indirectly (in *in vivo* assays involving measurement of lethality, pyrogenicity or the skin response known as the Schwartzman reaction) induce cytokine synthesis. Discussion will be limited to this literature. Such studies have utilized both synthetic lipid A agonists and lipid A molecules derived from various bacteria or bacterial mutants. This has led to the synthesis of chemically stable lipid A molecules with potent antagonist action against LPS. Such LPS antagonists may have therapeutic potential in infections and particularly in Gram-negative septic shock. The structure and therapeutic potential of lipid A antagonists is discussed in more detail in Chapter 8.

    A growing number of lipid A molecules whose complete structure has been elucidated are now available for analysis. Table 5.1 provides information on the acyl chain composition of some of these glucosamine disaccharide structures. Comparing the acylation patterns shown in Table 5.1, it is clear that one can divide the lipid A molecules into those that have an asymmetric acylation pattern (e.g. *E. coli*) and those that have a symmetric pattern of fatty acid residues (e.g. *Neisseria meningitidis*). The hexa-acyl diphosphoryl lipid A (DPLA) molecules found in *E. coli* and *Salmonella* strains are identical. This lipid A has the highest endotoxic properties and is considered to be the standard 'toxic' lipid A [81]. The LPS of *Salmonella minnesota* 595 is unique in that it contains a lipid A with seven acyl chains. The lipid A of *Rhodobacter*

Table 5.1 The chemical structures of various lipid A molecules

Bacterial source	No. of fatty acids	Fatty acyl group			
		R1	R2	R3	R4
E. coli	6	$C_{14}OC_{14}$	$C_{12}OC_{14}$	OHC_{14}	OHC_{14} [74]
E. coli (msbB mutant)	5	C_{14}	$C_{12}OC_{14}$	OHC_{14}	OHC_{14} [75]
S. typhimurium	6	$C_{14}OC_{14}$	$C_{12}OC_{14}$	OHC_{14}	OHC_{14} [76]
S. minnesota (R595)	7	$C_{14}OC_{14}$	$C_{12}OC_{14}$	OHC_{14}	$C_{16}OC_{14}$ [77]
N. meningitidis	6	OHC_{12}	$C_{14}OC_{12}$	OHC_{12}	$C_{14}OC_{12}$ [78]
C. violaceum	6	OHC_{14}	$C_{14}OC_{14}$	OHC_{14}	$C_{14}OC_{14}$ [79]
R. sphaeroides	5	OHC_{10}	$\Delta^7\text{-}C_{14}OC_{14}$	OHC_{10}	$3kC_{14}$ [80]

Refer to Figure 5.5 for the general structure of lipid A. For all structures the backbone sugar is a glucosamine disaccharide with a $\beta(1{\rightarrow}6)$ linkage and the reducing end sugar has an α anomeric configuration (at carbon 1). The 1 and 4′ positions are phosphorylated and fatty acyl groups occupy the 3′ (R1) 2′ (R2), 3 (R3) and 2 (R4) positions. Abbreviations used: OHC_{10}, 3-hydroxy-decanoate; OHC_{14}, 3-hydroxytetradecanoate; $C_{14}OC_{14}$, tetradecanoyloxytetradecanoate; $C_{16}OC_{14}$, hexadecanoyloxytetradecanoate; $\Delta^7\text{-}C_{14}OC_{14}$, Δ^7-tetradecenoyloxytetradecanoate; $3kC_{14}$, 3-keto or 3-oxo tetradecanoate; *C. violaceum*, *Chromobacterium violaceum*; *N. meningitidis*, *Neisseria meningitidis*; *R. sphaeroides*, *Rhodobacter sphaeroides*.

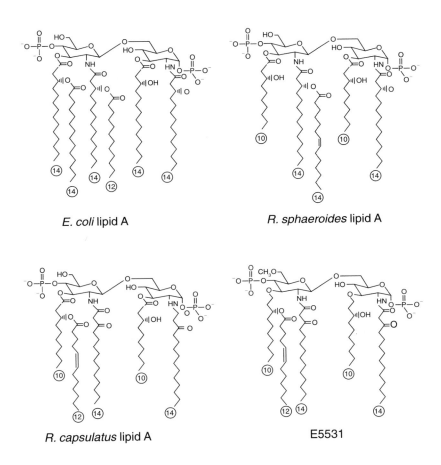

Figure 5.11 Comparison of the structures of the potent lipid A agonist from E. coli and the antagonist lipid As from R. sphaeroides and R. capsulatis
Compare these structures with E5531, a synthetic lipid A antagonist.

sphaeroides is currently the focus of much interest as this molecule has LPS antagonistic activity. This is also true of the LPS made by the *E. coli* mutant *msb*B (Table 5.1). The penta-acyl DPLA of *R. sphaeroides* is now considered to be the model 'non-toxic' lipid A. The structures of the *E. coli* and the *R. sphaeroides* lipid A molecules are shown in Figure 5.11.

As has been discussed, a proportion of bacteria examined (e.g. *Pseudomonas diminuta*) contain lipid A molecules with a DAG disaccharide rather than a glucosamine disaccharide [82]. Others contain a heterodisaccharide composed of DAG and glucosamine (e.g. the LPS of *Campylobacter jejuni*). LPS containing DAG in the backbone has been reported to be both endotoxic (e.g. *Campylobacter jejuni* [83]) and non-toxic [84]. The reasons for these differences await fuller analysis of the structures of these lipid A molecules. Indeed, Nature has further surprises up her sleeve. LPS is

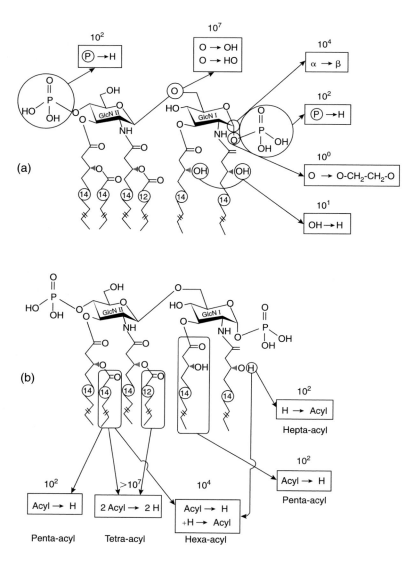

Figure 5.12 Schematic representation of the effects of altering the structure of E. coli lipid A on its endotoxic activities
Each region highlighted shows the nature of chemical changes that have been made to the lipid A and the factor (10^x) by which such a change decreases the activity of the analogue. For example, exchanging the phosphate group at C-4′ for a hydrogen atom decreases the activity of this analogue by 100-fold compared with the parent structure. Breaking the β(1→6) linkage to produce two monosaccharides results in almost complete loss of endotoxic activity (by a factor of 10^7). In (a) the modifications have been made to the hydrophilic portion of the lipid A. In (b) the modifications have been introduced to the hydrophobic part of this molecule. Redrawn with permission from [49].

considered to be an obligatory component of the outer membrane of Gram-negative bacteria and, as has been discussed, mutants lacking this amphiphile do not survive. However, it has recently been established that the common water-borne Gram-negative bacterium *Sphingomonas paucimobilis* (formerly *Pseudomonas paucimobilis*) lacks LPS [85]. In the place of LPS, this organism produces two different amphiphilic glycosphingolipids, which appear to have similar physicochemical and biological properties to LPS. The glycosphingolipids of this bacterium are only weakly active in stimulating cytokine production, and activity is blocked by neither CD14 (see Chapter 6) or lipid A partial structures with LPS antagonist properties [86]. The role of CD14 in the activity of LPS will be considered in detail in the next chapter.

The reader should now have some grasp of the structure of lipid A, and in the next part of this section the role of the individual parts of the molecule in the expression of endotoxic activity will be described. The key parts of the structure which have been examined include: (i) the requirement for a disaccharide structure; (ii) the phosphorylation of the sugars; (iii) stereochemistry of the anomeric carbon; and (iv) the number and length of the acyl chains. The importance of these structures in the overall endotoxic activity of lipid A is shown diagrammatically in Figure 5.12 and the reader is referred to [49,87,88] for recent reviews of the literature.

5.3.3.1 Requirement for disaccharide structure

Although there are some discrepant reports in the literature, the general consensus is that monosaccharide partial structures of lipid A, such as the intermediate lipid X or the synthetic compound lipid Y (2-tetradecanoyloxy-lipid A), are much weaker endotoxic agonists than the disaccharide lipid A [89]. In terms of stimulating human mononuclear cells to synthesize and release either IL-1 or IL-6, lipid X was inactive [79,90] in comparison with synthetic lipid A (Table 5.2). Even the presence of IFNγ, which synergizes with LPS, was unable to boost the cytokine-stimulating activity of lipid X. Indeed, as can be seen from Table 5.2, lipid X appeared to inhibit IL-6 production induced by LPS, suggesting that this monosaccharide had some antagonistic properties. Thus, it is clear that a disaccharide structure is required for cytokine induction. However, no comparative studies of the cytokine-

Table 5.2 Comparison of LPS and lipid X in stimulating IL-6 synthesis by human peripheral blood leucocytes

Substance	IL-6 synthesis (units/ml)			
	Without IFNγ	With IFNγ	Without IFNγ	With IFNγ
LPS (1 ng/ml)	18710	28157	20704	33787
Lipid X (1000 ng/ml)	0	0	0	0
LPS + lipid X	5099	9922	5653	6743

Results of two separate experiments are shown. Each value represents the mean of triplicate estimations. Standard deviations were less than 20%. The LPS used was prepared from *Salmonella abortus equi*. Results are taken from [90] with permission.

inducing activity of di-glucosamine, di-DAG and heterodimers of both sugars have been performed in order to define the optimum endotoxic disaccharide structure.

5.3.3.2 Sugar phosphorylation

Phosphorylation of the sugars at positions C-1 and C-4′ is important for the endotoxic activity of lipid A. This is best exemplified in a comparative study of synthetic *E. coli* hexa-acyl lipid A molecules in which either or both phosphates were removed and the resulting compounds were tested in a range of *in vivo* and *in vitro* assays. Removal of the C-4′ phosphate caused a significant reduction in activity, although this analogue was more active than that in which the C-1 phosphate group was removed. When both phosphate groups were removed the lipid A analogue formed was essentially inactive [91]. Rietschel's group also showed that the phosphate groups are required for induction of IL-1 synthesis [49,79,92].

5.3.3.3 Stereochemistry of the anomeric carbon

Hexoses such as the glucosamine in lipid A have an asymmetric or chiral centre at C-1 (the so-called anomeric carbon) and can exist as two stereoisomers — α-D-glucosamine and β-D-glucosamine. Thus, if one refers to the drawing of lipid A in Figure 5.6, the phosphate group at C-1 is below the sugar ring, and the molecule is therefore the α-anomeric form. If the phosphate were to be above the sugar ring then this would be the β-anomer. The stereochemistry of this anomeric site is claimed to be important in the overall activity of lipid A [49].

5.3.3.4 Acyl chains

It turns out that the major structural variable in terms of lipid A bioactivity is the number and length of the acyl chains that the molecule possesses. The importance of the acyl chains was first suggested by the studies of Tanamoto and colleagues [93]. The key determinants of activity are the 3-acyloxyacyl groups, which are in positions C-2′ and C-3′ in *E. coli* lipid A. Detailed comparisons have been made between *E. coli* lipid A (which has two 3-acyloxyacyl groups at position C-2′ and C-3′), *Sal. minnesota* lipid A (which has three 3-acyloxyacyl groups at positions C-2′, C-3′ and C-2) and lipid IV$_A$ (which has no acyloxyacyl groups). In *in vivo* tests of lethality, pyrogenicity and the Schwartzman reaction, the lipid A from *E. coli* was significantly more active than the *Sal. minnesota* lipid A, which was, in turn, more active than the lipid IV$_A$. The only exception was the induction of lethality in galactosamine-treated mice where all three molecules showed similar potency (reviewed in [87]). In *in vitro* tests of cytokine induction, the *E. coli* lipid A was generally more potent than the other two analogues [79,89,94]. Comparative studies of *E. coli* lipid A, a lipid A analogue lacking the 3-acyloxy group at position C-3′ and lipid IV$_A$ (in which both the C-2′ and C-3′ acyloxy groups are missing) revealed a hierarchy of biological potency with a 10-fold loss of pyrogenic activity with each missing 3-acyloxy group [95,96]. Similar results were found with cytokine induction *in vitro* [79,97]. As has been stated, some of the synthetic lipid A analogues demonstrated inhibitory (possible antagonistic) activity when co-cultured with LPS. This inhibitory activity also showed a dependence on structure. For example, lipid IV$_A$, which has four acyl chains, was a potent inhibitor of human monocyte pro-inflammatory cytokine induction. In contrast, a bisacyl lipid A showed minimal activity, revealing the importance of the number of acyl chains for activity [79,97–100]. Most studies of LPS are performed using human and rodent cells and it is well known that mice and rats are relatively resistant to endotoxin, whereas human and rabbits, for example, respond to nanogram/kg amounts of LPS. It is therefore interesting to note that, with human cells, lipid IV$_A$ had very weak agonist activity but was a potent inhibitor of LPS [97–99]. In contrast, with murine cells, lipid IV$_A$ still had some significant agonist activity [95,101,102]. Major differences in the response of human and rodent cells to LPS will be discussed in more detail in Chapter 6.

Further evidence of the importance of the acyloxyacyl groups in lipid A came from use of the neutrophil enzyme acyloxyacyl hydrolase, which removes the myristate and laurate at the 3-acyloxy positions in lipid A. Such enzymically de-acylated lipid As lose endotoxic activity and display antagonistic behaviour [103,104]. A recently described *E. coli* mutant (*msb*B), which produces a non-myristoylated LPS (i.e. lacking the 3-acyloxy group at position 3′), is the only viable non-temperature-sensitive lipid A mutant yet described. As expected, the LPS from this mutant was 10^3–10^4 times less active than wild-type *E. coli* LPS in stimulating E-selectin expression by human endothelial cells and TNFα production by monocytes [75].

The finding that modifications of lipid A, particularly the number of the acyl chains, can result in the production of molecules with apparent antagonist actions has galvanized the biotechnological/pharmaceutical industry and those interested in blocking the life-threatening condition of Gram-negative septic shock. A non-toxic LPS from *R. sphaeroides*, which has become the focus of much attention, was first described in the early 1980s. It was shown to be 10^4 times less toxic in the galactosamine-sensitized mouse and about 10^2 times less pyrogenic in the rabbit than the LPS from *Salmonella abortus-equi* [105]. The structure of this penta-acyl lipid A is shown in Figure 5.11 [106,107]. This non-toxic antagonistic LPS has been shown to be active *in vivo* [107] and has been used as the basis for the generation of a synthetic stable LPS antagonist E5531 [108–113]. The pharmacology of *R. sphaeroides* LPS and of E5531 is described in more detail in Chapter 8.

5.3.4   Lipid A structure and bacterial lifestyle

An enormous amount of information has been gathered during the past decade concerning the relationship between the structure of lipid A and its capacity to stimulate mammalian cells to produce cytokines. However, the fundamental role that lipid A plays is in anchoring the LPS to the other half of the bacterial outer membrane. As Gram-negative organisms evolved long before mammals, it is unlikely that the biological activity of lipid A in mammals was a driving force in the shaping of this molecule. The major variation in lipid A structure appears to be the number and nature of the acyl chains. An obvious question, and one that does not appear to have been seriously addressed, is whether the nature of the lipid A acyl chains is related to the composition of the phospholipids in the other half of the bacterial outer membrane. In bacteria, phospholipids may constitute 10% of the dry weight of the cell and their only known role is as components of cell membranes. The cell membranes in the common gut bacterium *E. coli* contain only three phospholipids in significant amounts: phosphatidylethanolamine (75–85%); phosphatidylglycerol (10–20%); and cardiolipin (5–15%). The fatty acid composition of these lipids is equally simple, with three major species: palmitate ($C_{16:0}$); palmitoleate ($C_{16:1}\Delta^9$) and *cis*-vaccenate ($C_{18:1}\Delta^{11}$). Compare this with the plasma membrane of an erythrocyte which has a more complex composition containing: phosphatidylcholine, phosphatidylethanolamine, phosphatidylserine, phosphatidic acid, sphingomyelin, glycolipids and cholesterol. Membranes are dynamic structures, and organisms respond to environmental changes, particularly temperature, by altering the size and saturation of the acyl chains. Could the differences in the acylation of lipid A molecules reflect the environmental temperature or other environmental variables of the bacteria producing them? *R. sphaeroides*, a photosynthetic organism, will live at temperatures much lower than, for example, would *E. coli*. Thus, its membranes would be likely to be more fluid. However, until the acyl chain composition of a greater number of bacterial lipid A molecules is known, the answer to this question is in abeyance.

5.4   Macromolecular structure of LPS

The structure–function relationships of lipid A molecules, both natural and synthetic, have just been described, but it is important to remember that the multicellular organism responds to LPS and not to free lipid A. As has been described on a number of occasions in this chapter, LPS (and lipid A) is an amphiphile (that is, it contains a hydrophilic and a hydrophobic domain) and, like all such molecules (phospholipids are good examples), it forms micellar structures above the critical micellar concentration. This capacity, it is assumed, has consequences for the biological effects of LPS and will be briefly discussed in this section. Indeed, as will be described in the next chapter, mammals have evolved systems that can deal with both individual LPS molecules and also with LPS aggregates.

The most common arrangement for phospholipids to assume is lamellar, which is the conformation found in lipid bilayers. Other more complex

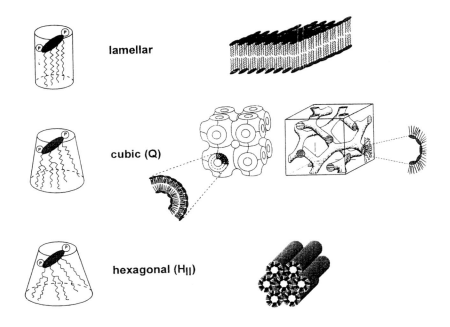

lamellar

cubic (Q)

hexagonal (H_{II})

Figure 5.13 Above the critical micellar concentration, LPS and lipid A form supramolecular aggregates which may assume a variety of three-dimensional forms including lamellar and non-lamellar [cubic (Q) or inverted hexagonal (HII)] structures
In this Figure the relationship between the molecular shape of lipid A (in terms mainly of its acyl chain packing) and the three-dimensional supramolecular structure formed is demonstrated. Reproduced with permission from [115].

structures such a non-lamellar (cubic) or inverted hexagonal are possible, depending on the structure and conformation of the amphiphile. As lipid A has a wide range of acyl chain compositions, it is believed that it can exist in these various forms of supramolecular structure (Figure 5.13). In a recent study from Rietschel's group it was found that the LPS from *Rhodobacter capsulatis, Rhodopseudomonas viridis* and *Rhodospirullum fulvum* had a lamellar structure, while the bisphosphorylated lipid A from *E. coli* and *Sal. minnesota* had a cubic structure and the lipid A of *Rhodocyclus gelatinosa* an inverted hexagonal structure [114,115]. Correlating the supramolecular structure with biological activity suggests that LPS/lipid A, which assumes a lamellar conformation, lacks biological activity. Only those LPS/lipid A molecules that form cubic or inverted hexagonal structures had bio-activity [115]. As can be seen from Figure 5.13 these active forms of lipid A have a conical shape. Brandenburg et al. [115] argue that this conical shape, with its slightly higher cross-section of the hydrophobic as opposed to the hydrophilic segment of the molecule, may cause a strong disturbance in the lamellar target-cell membrane and trigger a cellular response. However, it is likely that the supramolecular structure of the lipid A will depend upon its binding to CD14, LPS-binding protein and other serum proteins, and also on the nature of the cellular receptor on the target cell. Thus, it is obvious that much more work is needed to define the importance of the macromolecular structure of LPS to its final biological activity.

5.5 Conclusions

The LPS molecule is extremely complex and has not yielded all its secrets. This molecule is a potent stimulator of a wide variety of biological effects in mammals. These effects were believed to be due to the LPS alone. However, since the late 1980s, it has become evident that the biological actions of LPS in mammals is dependent upon a number of host proteins with, in fact, the LPS being a reasonably inactive moiety in the absence of such host proteins. The next chapter will review the literature on such LPS-interacting host molecules and their role in the biology and pharmacology of LPS.

References

1. Bennett, I.L. (1994) Introduction: approaches to the mechanisms of endotoxin action. J. Endotoxin Res. **1**, 2–3
2. Hitchcock, P.J., Leive, L., Makela, P.H., Rietschel, E.T., Strittmatter, W. and Morrison, D.C. (1986) Lipopolysaccharide nomenclature — past, present and future. J. Bacteriol. **166**, 699–701
3. Pfeiffer, R. (1892) Untersuchungen uber das Choleragift. Fortschr. Med. **11**, 393–412
4. Kanthack, A.A. (1892) Acute leucocytosis produced by bacterial products. Brit. Med. J. **1**, 1301–1303
5. Centanni, E. (1894) Untersuchungen uder das Infektions fieber-das Fubergift der Becterien. Deut. Med. Wochschr. **20**, 148
6. Westphal, O. (1974) Bacterial endotoxins. Int. Archs. Allergy Appl. Immunol. **49**, 1–43
7. Coley, W.B. (1893) The treatment of malignant tumors by repeated injection of Erysipelas. Am. J. Med. Sci. **105**, 487–511
8. Coley, W.B. (1898) The treatment of inoperable sarcoma with the mixed toxins of erysipelas and *Bacillus prodigiosus*: immediate and final results of one hundred and forty cases. J. Am. Med. Assoc. **31**, 389–395
9. Balkwill, F.R. and Fiers, W. (eds.) (1989) Biological response modifiers. Cancer Surveys **8**, no. 4
10. Atkins, E. and Snell, E.S. (1965) Fever. In The Inflammatory Process (Zweifach, B.W., Grant, L. and McCluskey, R.T., eds.), pp. 495–534, Academic Press, New York
11. Atkins, E.A. (1989) Fever: historical aspects. In Interleukin-1, Inflammation and Disease (Bomford, R. and Henderson, B., eds.), pp. 3–15, Elsevier, North Holland
12. Hort, E. and Penfold, W. (1912) Micro-organisms and their relation to fever. J. Hyg. **12**, 361–390
13. Seibert, F.B. (1925) The cause of many febrile reactions following intravenous injections. Am. J. Physiol. **71**, 621–651
14. Welch, H., Calvery, H.D., McClosky, W.T. and Price, C.W. (1943) Method of preparation and test for bacterial pyrogens. J. Am. Pharmacol. Assoc. **32**, 65–69
15. Bang, F.B. (1956) A bacterial disease of *Limulus polyphemus*. Bull. Johns Hopkins Hosp. **98**, 325–351
16. Levin, J. and Bang, F.B. (1964a) The role of endotoxin in the extracellular coagulation of *Limulus* blood. Bull. Johns Hopkins Hosp. **115**, 265–274

17. Levin, J. and Bang, F.B. (1964) A description of cellular coagulation in the *Limulus*. Bull. Johns Hopkins Hosp. **115**, 337–345
18. Levin, J. and Bang, F.B. (1968) Clottable protein in Limulus: its localization and kinetics of its coagulation by endotoxin. Thromb. Diath. Haemorrh. **19**, 186–197
19. Mikkelsen, M. (1988) The Secret in the Blue Blood. Science Press, Beijing
20. Poole, S. (1991) Pyrogen testing of polypeptide and protein drugs. In Polypeptide and Protein Drugs — Production, Characterization and Formulation (Hider, R.C. and Barlow, D., eds.), pp. 146–153, Ellis Horwood, New York
21. Hurley, J.C. (1995) Endotoxaemia: methods of detection and clinical correlates. Clin. Microbiol. Rev. **8**, 268–292
22. Poole, S., Thorpe, R., Meager, A., Hubbard, A. and Gearing, A.J.H. (1988) Detection of pyrogen by cytokine release. Lancet **i**, 130
23. Menkin, V. (1943) Chemical basis of injury in inflammation. Arch. Pathol. **36**, 269–288
24. Menkin, V. (1952) Studies on the crystallization of pyrexin, the pyrogenic factor of inflammatory exudates. Arch. Intern. Pharmacodyn. **89**, 229–236
25. Menkin, V. (1955) Pyrexin, the pyrogenic factor of inflammatory exudates and its relation to some bacterial pyrogens. J. Lab. Clin. Med. **46**, 423–442
26. Bennett, I.L. and Beeson, P.B. (1953) Studies on the pathogenesis of fever. II Characterization of fever-producing substances from polymorphonuclear leukocytes from the fluid of sterile exudates. J. Exp. Med. **98**, 493–508
27. Beeson, P.B. (1948) Temperature-elevating effect of a substance obtained from polymorphonuclear leukocytes. J. Clin. Invest. **27**, 524
28. Bennett, I.L. and Beeson, P.B. (1953) Studies on the pathogenesis of fever. I. The effect of injection of extracts and suspensions of uninfected rabbit tissues upon the body temperature of normal rabbits. J. Exp. Med. **98**, 477–492
29. Margulis L. (1993) Symbiosis in Cell Evolution, 2nd edn., W.H. Freeman, New York
30. Maynard-Smith, J. and Szathmary, E. (1995) The Major Transitions in Evolution, W.H. Freeman, Oxford
31. Gould, S.J. (1989) Wonderful Life: The Burgess Shale and the Nature of History, Hutchinson Radius, London
32. Boivin, A. and Mesrobeanu, L. (1935) Recherche sur les antigenes somatique et sur les endotoxines des bacteries. Revue. Immun. **1**, 553–569
33. Morgan, W.T.J. (1937) Studies in immunochemistry. II. The isolation and properties of a specific antigenic substance from *B. dysenteriae* (shiga) Biochem. J. **31**, 2003–2021
34. Goebel, W.F., Binkley, F. and Perlman, E. (1945) Studies on the Flexner group of dysentery bacilli. J. Exp. Med. **81**, 315–330
35. Westphal, O., Luderitz, O. and Bister, F. (1952) Uber die Extraktion von Bacterien mit Phenol/Wasser. Z. Naturforsch. **7b**, 148–156
36. Westphal, O., Luderitz, O., Eichenberger, E. and Keiderling, W. (1952) Uber bakterielle Reizstoff I. Mitt.: Reindarstellung eines Polysaccharid-Pyrogens aus *Bacterium coli*. Z. Naturf. **7b**, 536–548
37. Westphal, O. and Jann, K. (1965) Bacterial lipopolysaccharide. Extraction with phenol-water and further applications of the procedure. Meth. Carbohyd. Chem. **5**, 83–91
38. Galanos, C., Luderitz, O. and Westphal, O. (1969) A new method for extraction of R-lipopolysaccharide. Eur. J. Biochem. **9**, 245–250
39. Hitchcock, P.J. and Morrison, D.C. (1984) The protein component of bacterial endotoxins. In Handbook of Endotoxin Vol. 1 Chemistry of Endotoxin (Rietschel, E.T., ed.), pp. 339–374, Elsevier, North Holland
40. Rietschel, E.T., Galanos, C., Tanaka, A., Ruschmann, E., Luderitz, O. and Westphal, O. (1971) Biological activities of chemically modified endotoxins. Eur. J. Biochem. **22**, 218–224
41. Galanos, C., Rietschel, E.T., Luderitz, O., Westphal, O., Kim, Y.B. and Watson, D.W. (1972) Biological activities of lipid A complexed with bovine-serum albumin. Eur. J. Biochem. **31**, 230–233

42. Imoto, M., Yoshimura, H., Sakaguchi, N., Kusumoto, S. and Shiba, T. (1985) Total synthesis of *Escherichia coli* lipid A. Tetrahedron Lett. **26**, 1545
43. Kaufmann, F (1971) Serological Diagnosis of Salmonella Species, Munksgaard, Copenhagen
44. Luderitz, O., Staub, A.M. and Westphal, O. (1966) Immunochemistry of O- and R-antigens of Salmonella and related Enterobacteriaciae. Bacteriol. Rev. **30**, 193–255
45. Westphal, O., Westphal, U. and Sommer, T (1978) The history of pyrogen research. In Microbiology — 1977 (Schlessinger, D., ed.), pp. 221–238, American Society for Microbiology, Washington
45a. Raetz, C.R. (1993) Bacterial endotoxins: extraordinary lipids that activate eukaryotic signal transduction. J. Bacteriol. **175**, 5745–5753
46. Nikaido, H. and Vaara, M. (1987) Outer membrane. In *Escherichia coli* and *Salmonella typhimurium*. Cellular and Molecular Biology (Neidhardt, C., Ingraham, J.L., Brooks, E., Low, K., Magasanik, B., Schaechter, M. and Umbarger, H.E., eds.), pp. 7–22, American Society for Microbiology, Washington
47. Labischinski, H., Barnickel, G., Bradaczek, D., Rietschel, E.T. and Giesbrecht, P. (1985) High state order of isolated bacterial lipopolysaccharide and its possible contribution to the permeation barrier property of the outer membrane. J. Bacteriol. **162**, 9–20
48. Takayama, K. and Qureshi, N. (1992) Chemical structure of lipid A. In Bacterial Endotoxic Lipopolysaccharides, vol. 1, Molecular Biochemistry and Cellular Biology (Morrison, D.C. and Ryan, J.L., eds.), pp. 43–60, CRC Press, Boca Raton
49. Rietschel, E.T., Kirikae, T., Schade, F.U., Mamat, U., Schmidt, G., Loppnow, H., Ulmer, A.J., Zahringer, U., Seydel, U., Di Padova, F., Schreier, M. and Brade, H. (1994) Bacterial endotoxin: molecular relationships of structure to activity and function. FASEB J. **8**, 217–225
50. Raetz, C.H.R. (1990) Biochemistry of endotoxins. Annu. Rev. Biochem. **59**, 129–170
51. Takayama, K., Qureshi, N., Mascagni, P., Nashed, M.A., Anderson, L. and Raetz, C.R.P. (1993) Fatty acyl derivatives of glucosamine 1-phosphate in *Escherichia coli* and their relationship to lipid A. Complete structure of a diacyl GlcN-1-P found in a phosphatidylglycerol-deficient mutant. J. Biol. Chem. **258**, 7379–7385
52. Ray, B.L., Painter, G. and Raetz, C.R.H. (1984) The biosynthesis of gram-negative endotoxin: formation of lipid A disaccharides from monosaccharide precursors in extracts of *Escherichia coli*. J. Biol. Chem. **259**, 4852–4859
53. Anderson, M.S., Bulawa, C.E. and Raetz, C.R.H. (1985) The biosynthesis of gram-negative endotoxin: formation of lipid A precursors from UDP-GlcNAc in extracts of *Escherichia coli*. J. Biol. Chem. **260**, 15536–15541
54. Anderson, M.S. and Raetz, C.R.H. (1987) Biosynthesis of lipid A precursors in *Escherichia coli*: a cytoplasmic acyltransferase that converts UDP-N-acetylglucosamine to UDP-3-0-(R-hydroxymyristoyl)-N-acetylglucosamine. J. Biol. Chem. **262**, 5159–5169
55. Anderson, M.S., Bull, H.S., Galloway, S.M., Kelly, T.M., Mohan, S., Radika, K. and Raetz, C.R.H. (1993) UDP-N-acetylglucosamine acyltransferase of *Escherichia coli*: the first step of endotoxin biosynthesis is thermodynamically unfavourable. J. Biol. Chem. **268**, 19858–19865
56. Kelly, T.M., Stachula, S.A., Raetz, C.R.H. and Anderson, M.S. (1993) The *firA* gene of *Escherichia coli* encodes UDP-3-0-(R-3-hydroxymyristoyl)-α-D-glucosamine N-acetyltransferase: the third step of endotoxin biosynthesis. J. Biol. Chem. **268**, 19866–19874
57. Radika, K. and Raetz, C.H.R. (1988) Purification and properties of lipid A disaccharide synthase of *Escherichia coli*. J. Biol. Chem. **263**, 14859–14867
58. Ray, B.L. and Raetz, C.R.H. (1987) The biosynthesis of gram-negative endotoxin: a novel kinase in *Escherichia coli* membranes that incorporates the 4′ phosphate of lipid A. J. Biol. Chem. **262**, 1122–1128
59. Belunis, C.J. and Raetz, C.R.H. (1992) Biosynthesis of endotoxins: purification and catalytic properties of 3-deoxy-D-*manno*-octulosonic acid transferase from *Escherichia coli*. J. Biol. Chem. **267**, 9988–9997

60. Brozek, K.A. and Raetz, C.R.H. (1990) Biosynthesis of lipid A in *Escherichia coli*: acyl carrier protein-dependent incorporation of laurate and myristate. J. Biol. Chem. **265**, 15410–15417

61. Holst, O. and Brade, H. (1992) Chemical structure of the core region of lipopolysaccharides. In Bacterial Endotoxic Lipopolysaccharides, vol.1. Molecular Biochemistry and Cellular Biology (Morrison, D.C. and Ryan, J.L., eds.), pp. 135–170, CRC Press, Boca Raton

62. Stutz, P.L. and Unger, F.M. (1992) Chemical synthesis of core structures. In Bacterial Endotoxic Lipopolysaccharides, vol. 1. Molecular Biochemistry and Cellular Biology (Morrison, D.C. and Ryan, J.L., eds.), pp. 171–204, CRC Press, Boca Raton

63. Rietschel, E.T., Brade, L., Lindner, B. and Zahringer, U. (1992) Biochemistry of lipopolysaccharides. In Bacterial Endotoxic Lipopolysaccharides, vol.1. Molecular Biochemistry and Cellular Biology (Morrison, D.C. and Ryan, J.L., eds.), pp. 3–41, CRC Press, Boca Raton

64. Goldman, R.C., Doran, C.C. and Capobianco, J.O. (1988) Analysis of lipopolysaccharide biosynthesis in *Salmonella typhimurium* and *Escherichia coli* using agents which specifically block incorporation of 3-deoxy-D-*manno*-octulosonate. J. Bacteriol. **170**, 2185–2192

65. Haeffner-Cavaillon, N., Caroff, N. and Cavaillon, J.-M. (1989) Interleukin-1 induction by lipopolysaccharide: structural requirement of the 3-deoxy-D-*manno*-2-octulosonic acid (Kdo). Mol. Immunol. **26**, 485–494

66. Luderitz, T., Brandenburg, K., Seydel, U., Roth, A., Galanos, C. and Rietschel, E.T. (1989) Structural and physicochemical requirements of endotoxins for the activation of arachidonic acid metabolism in mouse peritoneal macrophages *in vitro*. Eur. J. Biochem. **179**, 11–16

67. Jann, K. and Jann, B. (1984) The structure and biosynthesis of O-antigens. In Chemistry of Endotoxin (Rietschel, E.T., ed.), pp. 138–147, Elsevier, North Holland

68. Tomlinson, S. (1993) Complement defence mechanisms. Curr. Opin. Immunol. **5**, 83–89

69. Grossman, N. and Leive, L. (1984) Complement activation via the alternative pathway by purified *Salmonella* lipopolysaccharide is affected by its structure but not its O-antigen length. J. Immunol. **132**, 376–385

70. Makela, P.H., Hovi, M., Saxen, H., Muutiala, A. and Rhen, M. (1990) Role of LPS in the pathogenesis of salmonellosis. In Cellular and Molecular Aspects of Endotoxin Research (Nowotny, A., Spitzer, J.Y. and Ziegler, E.J., eds.), pp. 537–544, Elsevier, North Holland

71. Morrison, D.C., Vukajlovich, S.W., Goodman, S.A. and Wollenweber, H.-W. (1985) Regulation of lipopolysaccharide biologic activity by polysaccharide. In The Pathogenesis of Bacterial Infections. Bayer Symp. VIII. (Jackson, G.G. and Thomas, H., eds.), pp. 68–76, Springer-Verlag, Berlin

72. Lindberg, A.A., Wollin, R., Bruse, G., Ekwall, E. and Svenson, S.B. (1983) Immunology and immunochemistry of synthetic and semisynthetic *Salmonella* O-antigen-specific glycoconjugates. Acta Chirurgica Scandinavica Symp. Ser. **231**, 83–90

73. Wolpert, J.S. and Albersheim, P. (1976) Host symbiont interactions I. The lectins of legumes interact with the O-antigen-containing lipopolysaccharides of their symbiont rhizoba. Biochem. Biophys. Res. Commun. **70**, 729–737

74. Imoto, M., Kusumoto, S., Shiba, T., Rietschel, E.T., Galanos, C. and Luderitz, O. (1985) Chemical structure of *Escherichia coli* lipid A. Tetrahedron Lett. **25**, 2667–2671

75. Somerville, J.E., Cassiano, L., Bainbridge, B., Cunningham, M.D. and Darveau, R.P. (1996) A novel *Escherichia coli* lipid A mutant that produces an antiinflammatory lipopolysaccharide. J. Clin. Invest. **97**, 359–365

76. Takayama, K., Qureshi, N. and Mascagni, P. (1983) Complete structure of lipid A obtained from the lipopolysaccharides of the heptoseless mutant of *Salmonella typhimurium*. J. Biol. Chem. **258**, 12801–12803

77. Qureshi, N., Mascagni, P., Ribi, E. and Takayama, K. (1985) Monophosphoryl lipid A obtained from lipopolysaccharides of *Salmonella minnesota* R595. Purification of the dimethyl derivative by high performance liquid chromatography and complete structural determination. J. Biol. Chem. **260**, 5271–5278

78. Kulshin, V., Zahringer, U., Lindner, B., Frasch, C.E., Tsai, C., Dmitirev, B.A. and Rietschel, E.T. (1992) Structural characterization of the lipid A component of pathogenic *Neisseria meningitidis*. J. Bacteriol. **174**, 1793–1800

79. Loppnow, H., Brade, H., Durrbaum, I., Dinarello, C.A., Kusumoto, S., Rietschel, E.T. and Flad, H.-D. (1989) IL-1 induction-capacity of defined lipopolysaccharide partial structures. J. Immunol. **142**, 3229–3238

80. Qureshi, N., Honovich, J.P., Hara, H., Cotter, R.J. and Takayama, K. (1988) Location of fatty acids in lipid A obtained from lipopolysaccharide of *Rhodospeudomonas sphaeroides* ATCC 17023. J. Biol. Chem. **263**, 5502–5504

81. Qureshi, N. and Takayama, K. (1990) Structure and function of lipid A. In The Bacteria, vol. 9 (Iglewski, B.J. and Clark, V.L., eds.), pp. 319–330, Academic Press, San Francisco

82. Kasai, N., Arata, S., Mashimo, J., Akiyama, Y., Tanaka, C., Egawa, K. and Tanaka, S. (1987) *Pseudomonas diminuta* LPS with a new endotoxic lipid A structure. Biochem. Biophys. Res. Commun. **142**, 972–978

83. Moran, A.P., Rietschel, E.T., Kusunen, T.U. and Zahringer, U. (1991) Chemical characterization of *Campylobacter jejuni* lipopolysaccharides containing N-acetylneuramic acid and 2,3-diamino-3,4-dideoxy-D-glucose. J. Bacteriol. **173**, 618–626

84. Roppel, J., Mayer, H. and Weckesser, J. (1975) Identification of a 2,3-diamino-2,3-dideoxyhexose in the lipid A component of lipopolysaccharides of *Rhodopseudomonas viridis* and *Rhodopseudomonas palustris*. Carbohydr. Res. **40**, 31–40

85. Kawahara, K., Seydel, U., Matsuura, M., Danbara, H., Rietschel. E.T. and Zahringer, U. (1991) Chemical structure of glycosphingolipids isolated from *Sphingomonas paucimobilis*. FEBS Lett. **292**, 107–110

86. Krziwon, C., Zahringer, U., Kawahara, K., Weidemann, B., Kusumoto, S., Rietschel, E.T., Flad, H.-D. and Ulmer, A.J. (1995) Glycosphingolipids from *Sphingomonas paucimobilis* induce monokine production in human mononuclear cells. Infect. Immun. **63**, 2899–2905

87. Takada, H. and Kotani, S. (1992) Structure–function relationships of lipid A. In Bacterial Endotoxic Lipopolysaccharides, vol.1. Molecular Biochemistry and Cellular Biology (Morrison, D.C. and Ryan, J.L., eds.), pp. 108–134, CRC Press, Boca Raton

88. Zahringer, U., Lindner, B. and Rietschel, E.T. (1994) Molecular structure of lipid A. The endotoxic centre of bacterial lipopolysaccharide. Adv. Carbohydr. Chem. Biochem. **50**, 211–276

89. Takahashi, I., Kotani, S., Takeda, H., Tsujimoto, M., Ogawa, T., Shiba, T., Kusumoto, S., Yamamoto, M., Hasegawa, A., Kisu, M., Nishijima, M., Amano, F., Akamatsu, Y., Harada, K., Takada, S., Okamura, H. and Tamura, T. (1987) Requirement of a properly acetylated $\beta(1\rightarrow6)$-D-glucosamine disaccharide bisphosphate structure for efficient manifestation of full endotoxic and associated bioactivities of lipid A. Infect. Immun. **55**, 57–68

90. Wang, M.-H., Flad, H.-P., Feist, W., Brade, H., Kusumoto, S., Rietschel, E.T. and Ulmer, A.J. (1991) Inhibition of endotoxin-induced interleukin-6 production by synthetic lipid A partial structures in human peripheral blood mononuclear cells. Infect. Immun. **59**, 4655–4664

91. Kotani, S., Takada, H., Tsujimoto, M., Ogawa, T., Takahashi, I., Ikeda, T., Otsuka, K., Shimauchi, H., Kasai, N., Mashimo, J., et al. (1985) Synthetic lipid A with endotoxic and related biological activities comparable to those of a natural lipid A from an *Escherichia coli* Re-mutant. Infect. Immun. **49**, 225–237

92. Loppnow, H., Brade, L., Brade, H., Rietschel, E.T., Kusumoto, S., Shiba, T. and Flad, H.-D. (1986) Induction of human interleukin-1 by bacterial and synthetic lipid A. Eur. J. Biochem. **16**, 1263–1267

93. Tanamoto, K., Galanos, C., Luderitz, O., Kusumoto, S. and Shiba, T. (1984) Mitogenic activities of synthetic lipid A analogues and suppression of mitogenicity of lipid A. Infect. Immun. **44**, 427–433

94. Takahashi, I., Kotani, S., Takada, H., Shiba, T. and Kusumoto, S. (1988) Structural requirements of endotoxic lipopolysaccharides and bacterial cell walls in induction of interleukin-1. Blood Purification **6**, 188–206

95. Galanos, C., Lehmann, V., Luderitz, O., Rietschel, E.T., Westphal, O., Brade, H., Brade, L., Freudenberg, M.A., Hansen-Hagge, T., Luderitz, T., et al. (1984) Endotoxic properties of chemically-synthesized lipid A part structures: comparison of synthetic lipid A precursor and synthetic analogues with biosynthetic lipid A precursor and free lipid A. Eur. J. Biochem. **140**, 221–227

96. Rietschel, E.T., Brade, L., Schade, U., Galanos, C., Freudenberg, M.A., Luderitz, O., Kusumoto, S. and Shiba, T. (1987) Endotoxic properties of synthetic pentaacyl lipid A precursor Ib and a structural isomer. Eur. J. Biochem. **169**, 27–32

97. Feist, W., Ulmer, A.J., Musehold, J., Brade, H., Kusumoto, S. and Flad, H.-D. (1989) Induction of tumor necrosis factor alpha release by lipopolysaccharide and defined lipopolysaccharide partial structures. Immunobiology **179**, 293–307

98. Wang, M.-H., Feist, W., Herzbeck, H., Brade, H., Kusumoto, S., Rietschel, E.T., Flad, H.-D. and Ulmer, A.J. (1990) Suppressive effect of lipid A partial structures on lipopolysaccharide or lipid A-induced release of interleukin-1 by human monocytes. FEMS Microbiol. Immunol. **64**, 179–186

99. Kovach, N.L., Yee, E., Munford, R.S., Raetz, C.R.H. and Harlan, J.M. (1990) Lipid IV_A inhibits synthesis and release of tumor necrosis factor induced by lipopolysaccharide in human whole blood ex vivo. J. Exp. Med. **172**, 77–84

100. Golenbock, D.T., Hampton, R.Y., Qureshi, N., Takayama, K. and Raetz, C.R.H. (1991) Lipid A-like molecule that antagonizes the effects of endotoxins on human monocytes. J. Biol. Chem. **266**, 19490–19498

101. Loppnow, H., Libby, P., Freudenberg, M.A., Krauss, J.H., Weckesser, J. and Mayer, H. (1990) Cytokine induction by lipopolysaccharide (LPS) corresponds to lethal toxicity and is inhibited by non-toxic Rhodobacter capsulatis LPS. Infect. Immunol. **58**, 3743–3750

102. Tanamoto, K.-I. (1995) Chemically detoxified lipid A precursor derivatives antagonize the TNF-α-inducing actions of LPS in both murine macrophages and a human macrophage cell line. J. Immunol. **155**, 5391–5396

103. Pohlman, T.H., Munford, R.S. and Harlan, J.M. (1987) Deacylated lipopolysaccharide inhibits neutrophil adherence to endothelium induced by lipopolysaccharide in vitro. J. Exp. Med. **165**, 1393–1402

104. Erwin, A.L., Mandrell, R.E. and Munford, R.S. (1991) Enzymatically-deacylated Neisseria lipopolysaccharide (LPS) inhibits murine splenocyte mitogenesis induced by LPS. Infect. Immun. **59**, 1881–1887

105. Strittmatter, W., Weckesser, J., Salimath, P.V. and Galanos, C. (1983) Nontoxic lipopolysaccharide from Rhodopseudomonas sphaeroides ATCC 17023. J. Bacteriol. **155**, 153–158

106. Salimath, P.V., Weckesser, J., Strittmatter, W., Mayer, H. (1983) Structural studies on the non-toxic lipid A from Rhodopseudomonas sphaeroides. Eur. J. Biochem. **136**, 195–200

107. Takayama, K., Qureshi, N., Beutler, B. and Kirkland, T.N. (1989) Diphosphoryl lipid A obtained from Rhodopseudomonas sphaeroides ATCC 17023 blocks production of cachectin in macrophages by lipopolysaccharide. Infect. Immun. **57**, 1336–1338

108. Qureshi, N., Hofman, J., Takayama, K., Vogel, S.N. and Morrison, D.C. (1996) Diphosphoryl lipid A from Rhodobacter sphaeroides: a novel lipopolysaccharide antagonist. In Novel Therapeutic Strategies in the Treatment of Sepsis (Morrison, D.C. and Ryan, J.L., eds.), pp. 111–131, Marcel Dekker, New York

109. Christ, W.J., Kawata, T., Hawkins, LD., Asano, O., Kobayashi, S. and Rossignol, D.P. (1992) Anti-endotoxin compounds and related molecules and methods. U.S. Patent application no. 935050

110. Christ, W.J., McGuinness, P.D., Asano, O., Wang, Y., Mullarkey, M.A., Perez, M., Hawkins, L.D., Blythe, T.A., Dubue, G.R. and Robidoux, A.L. (1994) Total synthesis of the proposed structure of Rhodobacter sphaeroides lipid A resulting in the synthesis of new potent lipopolysaccharide antago-nists. J. Am. Chem. Soc. **116**, 3637–3638

111. Rose, J.R., Christ, W.J., Bristol, J.R., Kawata, T. and Rossignol, D.P. (1995) Agonistic and antago-nistic activities of bacterially-derived Rhodobacter sphaeroides lipid A: comparison with activities of synthetic material of the proposed structure and analogs. Infect. Immun. **63**, 833–839

112. Christ, W.J., Asano, O., Robidoux, A.L.C., Perez, M., Wang, Y., Dubuc, G.R., Gavin, W.E., Hawkins, L.D., McGuiness, P.D., Mullarkey, M.A., et al. (1995) E5531, a pure endotoxin antagonist of high potency. Science **268**, 80–83

113. Kawata, T., Bristol, J.R., Rose, J.R., Rossignol, D.P., Christ, W.J., Asano, O., Dubuc, G.R., Gavin, W.E., Hawkins, L.D., Lewis, M.D., et al. (1996) Specific lipid A analog which exhibits exclusive antagonism of endotoxin. In Novel Therapeutic Strategies in the Treatment of Sepsis. (Morrison, D.C. and Ryan, J.L., eds.), pp. 171–186, Marcel Dekker, New York

114. Brandenburg, K., Mayer, H., Koch, M.J.H., Weckesser, J., Rietschel, E.T. and Seydel, U. (1993) Influence of the supramolecular structure of free lipid A on its biological activity. Eur. J. Biochem. **218**, 555–563

115. Brandenburg, K., Seydel, U., Schromm, A.B., Loppnow, H., Koch, M.J.H. and Rietchel, E.T. (1996) Conformation of lipid A, the endotoxic center of bacterial lipopolysaccharide. J. Endotox. Res. **3**, 173–178

6

Lipopolysaccharide and cytokines: a tangled web

"The gram-negative bacteria…display lipopolysaccharide…in their cell walls and these macromolecules are read by our tissues as the very worst of bad news. When we sense lipopolysaccharide we are likely to turn on every defence at our disposal;… cells believe that it signifies the presence of gram-negative bacteria and they will stop at nothing to avoid this threat."

Lewis Thomas (1974)
The Lives of a Cell:
Notes of a Biology Watcher.
Viking Press, New York

6.1   Introduction

Chapter 4 described the nature and interactions of cytokines, while in Chapter 5 the reader was introduced to the complex molecule known as endotoxin or lipopolysaccharide (LPS). In this chapter, these two areas of investigation will be brought together. The history of endotoxin research will be briefly sketched out, and the symbiotic relationship between endotoxin and cytokine research will be revealed. LPS has the capacity to stimulate the synthesis of many, many cytokines and the reasons for this will be explored. In the past decade, the mechanisms by which LPS stimulates host cells to produce cytokines have been partially uncovered and, in what appears a bizarre and paradoxical evolutionary development, it has been found that LPS is, by itself, relatively inactive, and is only able to show its potent biological effects with help from the host, through proteins such as LPS-binding protein (LBP) and CD14. Such studies are also suggesting that LPS is but one of the bacterial motifs recognized by multicellular organisms to warn them of bacterial invasion. The host has also developed an armamentarium of proteins for neutralizing LPS and Gram-negative bacteria, and so a complex cellular and molecular network of interactions occurs when LPS enters multicellular organisms. These interactions are even more interesting when one considers the ease with which it is possible to render animals resistant to endotoxin/LPS. It should also be borne in mind that there is a very wide variation between species in their response to LPS. For example, mice and rats are relatively resistant to LPS. In contrast, both the coprophagic rabbit and the non-coprophagic human are exquisitely sensitive to LPS. While much of the interaction between LPS and host organisms has been made clear since the pioneering work of the 1950s, there is still much to understand. This will be highlighted in the following chapter in which it is shown that LPS is simply one of dozens of bacterial constituents which have the capacity to induce the synthesis and secretion of cytokines.

6.2   Historical introduction

Cytokines and LPS — LPS and cytokines — there has been the most fascinating relationship, defined during the past 30–40 years, between this bacterial cell wall constituent and the large family of local hormones which we now call cytokines. It was the discovery of endotoxin which snuffed out the nascent studies of the endogenous fever-controlling substances (cytokines?) which had been proposed by William Welch in 1888 [1]. Contamination of biological material with the ubiquitous endotoxin proved a major trial to those researchers attempting to understand how fever was controlled. It was not until the 1940s that the Russian scientist Menkin, working in the U.S.A., kick-started the work on LPS which eventually resulted in the recognition of the first cytokine — IL-1. The experimental protocol used by Menkin, and replicated in various forms by many who followed, was to inject inflammatory substances into the peritoneal cavity of rabbits and harvest the inflammatory exudate, which is rich in leucocytes, in particular polymorphonuclear neutrophils (PMNs). Menkin then demonstrated that this inflammatory exudate contained a factor, or factors, able to induce fever, neutrophilia (increased numbers of circulating neutrophils) and tumour necrosis. This activity was given the name 'pyrexin' [2]. Unfortunately, Menkin was unable to disprove the hypothesis that pyrexin was nothing more than LPS. It was not until the late 1940s that Paul Beeson in the U.S.A., taking care to exclude the biological effects of endotoxin, showed that sterile saline extracts of rabbit neutrophils (also termed granulocytes because of their prominent intracellular granules) contained a pyrogenic substance which was distinct from endotoxin [3]. As this molecule appeared to come from the predominant cell in the peritoneal washout of the rabbit — the granulocyte — this activity was termed 'granulocyte pyrogen' [3]. Although he did not know it at the time, Beeson had initiated cytokine biology. Table 6.1 shows how many times IL-1 was rediscovered after this 'initial discovery'. Further papers from Beeson, and co-worker Bennett [4,5], confirmed the presence of a heat-labile pyrogen in

Table 6.1 History of the nomenclature of the first cytokine, interleukin-1

Date	Experimental finding/event	Synonym for IL-1
1948	Peritoneal exudate cells release pyrogen	Granulocyte pyrogen
1955	Endotoxin induces pyrogenic protein in rabbits	Endogenous pyrogen (EP)
1969	Material from leucocytes induces acute-phase response	Leucocyte endogenous mediator (LEM)
1972	Macrophages produce lymphocyte-stimulating protein	Lymphocyte-activating factor (LAF)
1976	Monocyte factor stimulates synovial cell PGE$_2$ synthesis	Mononuclear cell factor (MCF)
1977	Leucocyte-produced pyrogen	Leucocytic pyrogen (LP)
1979	International Lymphokine Workshop	Interleukin-1
1980	Pig synovial lining factor stimulating cartilage breakdown	Catabolin
1980	Human synovial lining factor stimulating chondrocytes	Synovial factor (SF)
1983	Factor stimulating bone resorption	Osteoclast-activating factor (OAF)
1984	IL-1 gene cloned and two distinct genes discovered	
1985/6	IL-1 receptor identified	
1990	Third member of IL-1 family (IL-1ra) cloned and expressed	

In addition to the main historically important terms for IL-1 shown above, IL-1 has been given a variety of additional names, including: B-cell differentiation factor (BDF); B-cell-activating factor (BAF); epidermal cell-derived thymocyte-activating factor (ETAF); corneal epithelial cell-derived thymocyte-activating factor (CETAF); fibroblast-activating factor (FAF); mitogenic protein (MP); neutrophil-releasing activity (NRA); proteolysis-inducing factor (PIF); ornithine decarboxylase-inducing factor (ODC); serum amyloid A-inducer (SAA); thymocyte-activating factor (TAF), etc.

acute inflammatory exudates, in disrupted leucocytes and in inflamed skin. Uninflamed 'control' tissues and organs were negative. Two important controls for LPS contamination were used in these studies. The first was heat stability. As LPS is very heat stable, if the pyrogenic activity was abrogated by heating it could not be due to LPS contamination. The second was the use of endotoxin-tolerant animals, which have a very much lowered biological response to LPS, but not to other pyrogens. Tolerance to LPS is a fascinating and still poorly understood phenomenon which can be induced by daily injections of this bacterial component. In contrast, Beeson and Bennett could not induce tolerance to the endogenous pyrogen. The activity described in these studies was termed 'leucocytic pyrogen'. Menkin's 'pyrexin' was tested using these bioassay techniques and was shown to be contaminated with endotoxin [4,5].

In more physiologically based experiments, Grant and Whalen [6] showed that rabbits given endotoxin rapidly exhibited a pyrogen in their blood, and this was subsequently named 'endogenous pyrogen' (EP). This study of the chemical physiology of fever was subsequently championed by Elisha Atkins, one of the major pioneers of pyrogen research. For example, Atkins and Wood [7] injected rabbits with a bacterial pyrogen — typhoid vaccine — and showed that this was rapidly cleared from the blood, but was replaced by an endogenous pyrogenic substance in amounts that related to the degree of fever at the time of sampling. Injections of rabbits with a whole range of bacterial components, viruses, or components capable of inducing hypersensitivity, resulted in the generation of EP activity (reviewed in [8,9]). Thus, endotoxin was not the only bacterial component able to induce the formation of EP activity. However, this finding, as with many others in the field of EP research, was forgotten or overlooked in subsequent decades.

The source of EP was studied by making extracts of various rabbit tissues and it was shown that, if sufficient tissue was used, many different organs had the capacity to induce fever. However, it appeared that the monocyte was a very effective producer of EP [10] and, contrary to the original belief, PMNs (granulocytes) did not release much of this pyrogenic activity. Kampschmidt et al. [11] showed that material extracted from leucocytes could elicit granulocytosis and the synthesis of the acute-phase response. He thus coined another name — 'leucocyte endogenous mediator' (LEM) to describe this activity.

Attempts were made during the 1950s and 60s to purify EP, and by the early 1970s it had been established that it was a single protein of 14–15 kDa [12]. Murphy, using a sequential protein purification scheme with large amounts of rabbit exudate, isolated a homogeneous protein of 14 kDa with a pI of 7.3 [13]. Thus, endogenous pyrogen was isolated to homogeneity, but this important finding was largely ignored. Charles Dinarello has been the major figure in endotoxin/EP research since the 1970s. Using human monocytes as a source of EP, his group showed that there were two different forms of EP with different isoelectric points [14]. Later, Murphy and colleagues also found a second form of rabbit EP [15].

In the 1960s immunologists discovered that macrophages and lymphocytes released a number of proteins with important biological actions. These included activities such as migration-inhibitory factor (MIF) and macrophage-activating factor (MAF), and such proteins were given the generic title, lymphokines, by Dudley Dumonde working in London [16]. In the early 1970s Gery and colleagues [17,18] reported that activated human or murine leucocytes produced a factor which acted in synergy with T-cell mitogens, such as phytohaemagglutinin (PHA) or concanavalin A (con A) to cause thymocyte proliferation. This factor was named lymphocyte-activating factor (LAF). The source of this factor was demonstrated to be the macrophage, and LPS was found to be a major stimulator of its production [19]. By the 1970s a number of groups of workers were investigating the biochemical nature and mechanism of action of factors that can be produced by LPS-activated macrophages. Those interested in the nature of fever and the actions of LPS were pursuing EP, those interested in the acute-phase response were seeking LEM, and immunologists interested in the control of lymphocyte proliferation and the mechanisms of antigen presentation were hunting LAF. During the 1970s and early 1980s a number of other macrophage-derived mediators were discovered. These included mononuclear cell factor [20], catabolin [21], synovial factor [22], osteoclast-activating factor [23] and hepatocyte-activating factor [24]. This proliferation of poorly defined factors seemed to be a repeat of the experience of the 1960s when immunologists discovered the lymphokines — proteins that were still ill-defined in the 1970s. Fortunately, by the late 1970s and early 1980s there was increasing evidence that these various factors, many induced by LPS, were one-and-the-same molecule(s). For example, in 1977 Kampschmidt's group reported that EP was inseparable, in terms of its biological actions, from LEM [25]. The key experiment, however, was the demonstration that purified EP could produce 'lymphocyte activation' and that LAF and EP were in fact the same molecule [26,27]. These experiments were reported at around the time that the term interleukin-1 was suggested to describe the biological activities of LAF [28], and initially there was some scepticism about the proposal that EP and IL-1 were the same molecule (reviewed in [29]).

These initial uncertainties about the consanguinity of IL-1 and EP, and indeed all the other factors named in this section, were swept away when IL-1 was cloned, and it was shown that IL-1α and IL-1β (as these molecules were then called) had an enormously wide range of biological actions encompassing those reported for EP, LEM, LAF, catabolin, etc., etc. Thus, we can see that research into the biological effects of LPS led to the discovery of the first cytokine, IL-1, and the discovery that this molecule had a multitude of actions resulted in intense search, which remains unabated to this day, for further cytokines [30]. The number of cytokines produced by human cells must now be in the hundreds, and the study of these proteins has, in turn, shed fresh light on the cellular and molecular actions of LPS.

6.3   Biological activity of LPS

Injection of LPS or its biologically active moiety, lipid A, into animals or human volunteers induces a multitude of effects upon all the major organs and tissues of the body (Table 6.2). These effects are clearly seen in patients with Gram-negative septic shock, who exhibit the major signs and symptoms of fever, hypotension, diffuse intravascular coagulation (DIC) and metabolic derangements. It is this pattern of biological effects caused by LPS which is presumed to result in the deaths of a large proportion of patients with septic (or endotoxin) shock [31]. In experimental animals, endotoxin-induced mortality can usually be inhibited by neutralizing one of the triad of pro-

Table 6.2 Biological actions of LPS/lipid A

In vitro effects of LPS

Induces all cell populations to produce some cytokines

Associated with above, induces synthesis of arachidonic acid metabolites by a variety of cells

Induces platelet-activating factor synthesis

In synergy with cytokines, induces NO synthesis

Induces vascular endothelial cell adhesion protein synthesis (ICAM, VCAM, E-selectin)

Directly/indirectly activates leucocyte integrins

Alters coagulation balance in cultured endothelial cells by multiple actions

Up-regulates CD14 expression

Up-regulates synthesis of antibiotic peptides

Acts on PMNs to cause lysosomal degranulation

Decreases PMN chemotaxis

Increases PMN adhesion

Stimulates phagocytosis

Enhances free radical production

Mitogenic for B-lymphocytes

Inhibits connective tissue cell matrix synthesis

Inhibits fibroblast proliferation

Stimulates bone resorption

In vivo effects of LPS

Pyrogenic

Induces lethal toxicity (due to following actions)

Induces hypotension due to action both on cardiac output and on vascular tone

Alters coagulation/fibrinolysis balance of vasculature

Induces disseminated intravascular coagulation and thrombosis

Activates factor XII (Hageman factor)

Causes platelet aggregation

Causes hypoglycaemia

Induces Schwartzmann reaction in skin

Activates alternative complement pathway

Acts as immunogen activating B-cells

Induces tolerance to itself

As Lewis Thomas observed: "when we sense LPS we are likely to turn on every defence at our disposal". An enormous number of LPS effects on cells, organs, tissues and whole organisms have been reported. Most of these are likely to be due to the induction of cytokine synthesis with the subsequent induction of other mediators (e.g. prostanoids, leukotrienes, nitric oxide, etc.). This table highlights only a selection of the actions of LPS.

inflammatory cytokines — IL-1, IFNγ or TNFα [32,33] — or by administering an anti-inflammatory cytokine such as IL-10 [34], thus exemplifying the role of cytokine networks in controlling endotoxin/LPS toxicity. It is not within the scope of this book to describe the biology of LPS. However, it should be noted that this molecule has a truly amazing spectrum of activities, many of which are related to cytokine induction. Examples of the actions of LPS include: inducing release of vasoactive substances, activation of the alternative pathway of complement, activation of the coagulation cascade, B-cell mitogenesis, and modulation of steroid and peptide hormone secretion and control (Table 6.2). This ability of LPS to affect diverse biological systems has led to its popularity as an experimental tool in the various biological disciplines, and many thousands of papers have appeared since the 1960s on the actions of this molecule [35,36]. It is now clear that most of the actions of LPS are due to the induction of cytokine synthesis. Precisely how LPS induces cytokine synthesis remains controversial, as will be described in the next section.

Another interesting facet of the biology of LPS is its capacity for inducing self-tolerance. Thus, a single, or repeated, exposure to LPS (depending on the dose) can render animals, or man, tolerant to LPS. It is also possible to make cultured cells tolerant to LPS. Endotoxin tolerance is selective for LPS, and animals or cells unresponsive to LPS will respond to cytokines or other cytokine-inducing components from bacteria. The mechanism of endotoxin tolerance has still not been defined and it remains another of the mysteries of this fascinating molecule [37].

In the next section, the nature of the interaction between LPS and cells will be considered and the vexing question of the nature of the LPS receptor reviewed.

6.4   LPS receptors

The remainder of this chapter will be concerned with the mechanism(s) by which LPS can stimulate cells to produce cytokines. In this section we will consider the manner in which LPS selectively binds and activates cells. If LPS was like any other biological agonist it would bind to a receptor on the external face of the plasma membrane of the target cell, thereby inducing the production of an intracellular message. The consequence of this would be the transcription of selected genes. Surprisingly, LPS does not appear to obey these conventional rules of agonist–receptor interaction. First, the cell-surface signal-transducing receptor(s) for LPS has still not been absolutely identified. Secondly, the nature of the intracellular signalling pathways induced by LPS are many and varied. Thirdly, LPS is itself relatively inactive and requires several host proteins to allow it to exhibit full agonist potency. The problems that researchers have had in defining the exact nature of the LPS receptor may be related to the growing belief that it is one of the classes of bacterial molecules that bind to pattern-recognition receptors, a concept introduced by Janeway [38]. It is possible that pattern-recognition receptors may employ different biological rules to 'normal' receptors.

6.4.1   The search for the cell-surface LPS receptor

Many hundreds of proteins exist on the outer surface of the average eukaryotic cell. Thus, finding the one protein that acts as a receptor for a particular agonist is akin to finding the proverbial needle in the haystack. However, a number of methods have been developed to identify specific receptors on the cell surface. The classic biochemical methodology relies on having a method of recognizing the agonist. The simplest way of doing this is to label the agonist with a radioactive, fluorescent or chromogenic tag. The labelled agonist can then be incubated with the cell containing the putative receptor, and the complex between the tagged agonist and the receptor isolated and identified. To aid this process the agonist and receptor can be covalently linked by a protein cross-linking agent. The receptor agonist can also be linked to an affinity matrix and this can be used to isolate the receptor from preparations of plasma membranes of cells which respond to the agonist.

This combination of tagging and protein cross-linking was used by David Morrison's group [39–41] to identify LPS receptors on the surface of leucocytes. This approach used a photoactivatable disulphide-reducible ^{125}I-linked cross-linking reagent, 2-*p*-azidosalicylamido-1,3′-dithiopropionate, to label the LPS. This complex was incubated with cells and the cell-surface proteins binding to LPS were covalently cross-linked to keep the complex together during the processes of isolation and identification of the receptor. This technology identified a receptor on a range of rodent cells, including macrophages, T- and B-lymphocytes, trophoblasts and various cell lines which, depending on the techniques of PAGE used, had an estimated molecular mass of 80 or 73 kDa. This latter molecular mass was taken as being the correct one. The receptor recognized both smooth and rough mutants of LPS (i.e. those respectively lacking or containing outer-core oligosaccharides), suggesting that it bound to the lipid A moiety of the molecule. In a survey of mononuclear cells from a variety of species, it was found that species such as chickens and frogs, which are not susceptible to LPS, did not contain this 73 kDa receptor [42]. It should be noted, at this point, that other workers have suggested that cells contain receptors recognizing the polysaccharide portion of LPS [43–45].

A curious finding was that the putative 73 kDa LPS receptor was present on the surface of cells from the C3H/HeJ LPS-unresponsive strain of mouse, and at levels comparable to those in the congenic non-LPS-sensitive strain [46]. The finding of this receptor on C3H/HeJ mice implied that it was functional, at least as far as binding of LPS was concerned. A 73 kDa LPS receptor has also been reported to exist on the plasma membranes of human monocytes, lymphocytes, PMNs and platelets [47]. Using a similar cross-linking technology to Morrison's group, Dziarski reported that a 73 kDa receptor also existed on cells, that was capable of binding peptidoglycan [48–50], lipoteichoic acid, heparin and certain sulphated heparinoids [51].

Evidence for the biological importance of the 73 kDa receptor was adduced from the generation of hamster IgM monoclonal antibodies (mAbs) to this protein [52,53]. These antibodies were designated 3D7 and 5D3. Both mAbs recognized the 73 kDa putative LPS receptor in ELISAs, and both competed with LPS. Binding of 3D7 to the 73 kDa protein was prevented by treatment of the latter with periodate, whereas binding of 5D3 was inhibited by treating the 73 kDa protein preparation with proteinase K. This was interpreted as 3D7 recognizing a carbohydrate epitope, and 5D3 a peptide epitope.

When incubated with macrophages, 5D3 mimicked LPS and induced C3H/HeN macrophages to exhibit tumour cell cytotoxicity. This could be enhanced by IFNγ. It was of note that 5D3 did not stimulate the same activity in the LPS-unresponsive C3H/HeJ strain of mouse [53]. 5D3 was also reported to induce nitric oxide production [54] and induce cytolytic activity against virally infected cells [55]. Surprisingly, in light of the findings just reported, 5D3 has been claimed to protect D-galactosamine-sensitized mice against the toxic effects of both LPS and TNF [56].

Thus far, the hypothesis being tested is that LPS-responsive cells contain a 73 kDa receptor which binds and transduces the LPS signal. The finding that mAbs to this receptor have, what appear to be, appropriate actions supports the hypothesis. It seems that LPS binds to cells much as any other cellular agonist would do. It is, of course, surprising that the nature of this receptor has not been defined, particularly as mAbs are available. However, this hypothesis has recently been challenged. Dziarski [57], who was responsible for showing that other bacterial macromolecules, such as peptidoglycan, bound to the 73 kDa putative LPS receptor, has now provided strong evidence that this LPS-, peptidoglycan- and lipoteichoic acid-binding receptor is nothing more than cell-bound serum albumin. The albumin is suggested to bind strongly to an inducible albumin-binding protein. Dziarski also suggests that the failure of the antibodies 3D7 and 5D3 to immunoprecipitate the LPS receptor from cell membranes is related to the fact that these IgM molecules activate cells by non-specific binding to plasma-membrane proteins. However, this does not explain why 5D3 did not activate cells from C3H/HeJ mice [57].

Dziarski's data seem to rule out the existence of a 73 kDa receptor. However, more recent data from Rietschel's group [58], using plasma-membrane proteins which have been Western-blotted and then incubated with LPS or lipid A (plus specific antibodies to these components) have reported the presence of an 80 kDa receptor on the surfaces of monocytes and vascular endothelial cells. In the absence of serum proteins, LPS bound to several plasma-membrane proteins of 90, 65, 60 and 35–50 kDa. When the LPS was incubated with Western-blotted proteins in the presence of serum, an additional band of 80 kDa was seen. Lipid A only bound to the 35–50 kDa proteins in the absence of serum and to the 80 kDa protein in its presence. The factors in serum which mediated binding to the 80 kDa protein were LBP and soluble CD14 (to be described later). Hexa-acyl lipid A bound to this protein, while tetra-acyl lipid A only bound in the presence of serum, and compound 606 (a bisacylated structure: see Chapter 5 for details of lipid A structure) did not bind. This is in line with the known structure–function relationships of lipid A, which have been described in detail in the previous chapter. This 80 kDa protein was not serum albumin. However, it should be borne in mind that the protocol used to define this protein involves boiling it in the detergent, sodium dodecyl sulphate (SDS), and therefore it is denatured when it binds LPS and lipid A. The obvious question therefore is: does this protein bind lipid A and LPS when it is in the native state?

Other cell-surface LPS receptors, with a wide range of molecular masses, have been reported. The complete list of LPS/lipid A receptors is provided in Table 6.3. Using a similar cross-linking methodology to that used by Morrison and Dziarski, Kirkland and colleagues [59] failed to find a 70–80 kDa receptor on the pre-B-cell line 70Z/3, but did find binding proteins of 18 and 25 kDa. Cross-linking of these proteins with LPS was saturable and was blocked by a variety of LPS and lipid A molecules, suggesting some degree of specificity. A 38 kDa LPS-binding surface protein present on murine macrophages, lymphocytes, splenocytes and the 70Z/3 cell line has been reported by Morrison's group [60]. LPS binding to this protein could not be antagonized by excess purified lipid A, but could be inhibited by Re-LPS which, as described in

Table 6.3 Cell-surface proteins acting as putative LPS receptors

Protein	Reference	Comment
73/80 kDa	[39–41,57]	Probably serum albumin
73 kDa-binding peptidoglycan, LTA, etc.	[48–51,57]	Probably serum albumin
CD14	[114]	Not a signalling receptor
CD18	[67,69]	Not clear if it is a signalling receptor
Scavenger receptor	[66]	Probably not a signalling receptor
80 kDa	[58]	Binds LPS and relevant lipid A analogues in presence of serum. Not albumin, but not clear if a signalling receptor
90 kDa	[58]	Proteins bind to LPS in absence of serum
65 kDa	[58]	
60 kDa	[58]	
35–50 kDa	[58]	
18 kDa	[59]	Binding suggests some degree of specificity
25 kDa	[59]	
38 kDa	[60]	May bind to LPS inner core
69 kDa	[65]	May be a cytokine
95 kDa	[62]	
96 kDa	[63,64]	

Chapter 5, is lipid A plus two KDO residues at position C-6'. The conclusion is that this 38 kDa receptor recognizes the inner core of LPS. The possibility that cell-surface receptors can recognize the polysaccharide moieties of LPS has been proposed [43–45] and there is one report in the literature that regions of the core oligosaccharide in LPS can induce human monocytes to synthesize IL-1 [61]. Ligand blotting, using a similar methodology as employed by Rietchel's group [58], has identified a 95 kDa lipid A-binding protein in macrophages [62] and a 96 kDa protein in murine erythrocytes, macrophages and lymphocytes [63,64].

The most recent candidate for the cell-surface LPS receptor is a 69 kDa protein which has been identified with a polyclonal antibody to the N-terminal 20-amino-acid sequence of a cytokine known as soluble immune response suppressor. Mutant cell lines, unresponsive to LPS, lacked this receptor [65].

Two established cell-surface receptors have also been shown to bind LPS. The first is the scavenger (acetylated low-density lipoprotein) receptor, which can bind to lipid IV_A at the C-1 position (see Chapter 5), thus allowing the entry and dephosphorylation of this molecule. However, the scavenger receptor appears not to be a signalling receptor, as acetylated lipid IV_A totally blocks lipid IV_A binding without affecting LPS-induced cytokine synthesis [66]. The second receptor is the family of proteins known as β_2-integrins, which are heterodimers found on the surface of leucocytes and composed of a common chain (CD18) and three unique chains (CD11a/b/c). Wright and Jong [67] reported that all three β_2-integrins were capable of binding to LPS, although the finding that CD18-deficient cells responded to LPS suggested that these were not signalling receptors [68]. However, this view has had to be revised with the discovery that transfection of CD11c/CD18 into the, normally LPS-unresponsive, Chinese hamster ovary (CHO) cell line results in these cells becoming responsive to LPS [69]. The reader is referred to the following reviews [70,71] for further information about LPS receptors.

6.4.2   Does the LPS receptor exist?

The literature on the LPS receptor is extremely confusing, particularly in a world in which the identification, cloning and expression of receptors is becoming commonplace. It is not clear why the cell-surface receptor or receptors for LPS have not been identified and the literature suggests that there may be unprecedented heterogeneity in the cell-surface proteins which bind LPS. One obvious reason for the difficulties encountered in conclusively detailing the LPS receptor is the enormous heterogeneity of the LPS molecule itself, with the prospects of significant non-specific cell-surface interactions which would confuse investigators. During the past six years two proteins have shed much light on the nature of the interaction of LPS with eukaryotic cells. The first is a serum protein and acute-phase reactant called LBP and the second is a cell-surface protein, but one also found in serum, called CD14. The next two sections will describe these two proteins and their interactions in binding of LPS to cells, and the subsequent cell signalling produced by LPS binding.

6.4.3   Lipopolysaccharide-binding protein

Measurements of the binding of LPS to high-density lipoproteins in sera revealed that such binding was significantly slowed in acute-phase sera. This was due to the binding of the LPS to protein(s) present in such sera but absent from normal sera. Fractionation of sera identified a 60 kDa glycoprotein as the LPS-binding protein, and it was named, appropriately, LBP [72] and was shown to bind to the lipid A region of LPS [73]. LBP has a high-affinity site which binds to the lipid A region of LPS with a dissociation constant (K_D) in the nanomolar range forming a 1:1 complex [73]. Cloning of the lapine and human cDNAs for LBP revealed a mature protein containing 452 amino acids with four cystines and five potential glycosylation sites. The protein is highly conserved with the human and rabbit LBPs sharing 69% sequence homology [74]. LBP also shares 44% sequence homology with an antibacterial protein produced by PMNs called bactericidal/permeability-increasing protein (BPI) [74,75]. LBP and BPI also show sequence similarity to two other proteins involved in binding lipids — cholesterol ester-transfer protein (CETP) and phospholipid-transfer protein (PLTP) [76,77]. It can be concluded from this that LBP is one of a group of related proteins that bind amphipathic molecules (including LPS) and are involved in their transport in an aqueous environment. LBP is predominantly produced by the liver, as a 55 kDa polypeptide which is released as a glycoprotein of 60 kDa [78]. Extrahepatic synthesis of LBP has also been reported [79]. The concentration of LBP in normal serum is in the range 12–22 µg/ml and this can increase up to 200 µg/ml within 24 h of the induction of the acute-phase response [80,81]. Like other acute-phase proteins, the synthesis of LBP is under the control of cytokines and steroid hormones [82].

What are the consequences of the binding of LBP to LPS? Work from the Scripp's Institute has shown that LBP enhances the responsiveness of both rabbit and human macrophages to LPS by up to 1000-fold. LBP also enhances the activity of synthetic lipid A, but not of lipid A partial structures such as lipid IV$_A$. However, LBP was not able to enhance the responses to other cell activators such as phorbol myristate or heat-killed *Staphylococcus aureus*. An interesting finding was the increased stability of TNF mRNA in macrophages exposed to LPS–LBP complexes. Antibodies to CD14 inhibited the activity of LPS–LBP complexes, indicating the importance of CD14 in the cellular activation produced by such complexes [83,84]. LBP also enhanced the capacity of LPS to: (i) induce NO [85]; (ii) induce vascular leucocyte adhesiveness [86]; and (iii) activate arachidonate oxidation [87]. In addition, LBP enhanced the LPS-induced release of soluble 75 kDa TNF receptor from human monocytes, and BPI inhibited such release, suggesting an interesting control circuit for monocytes [88]. If LBP is removed from serum by immunoadsorption, the serum loses its ability to activate LPS [74]. The role of LBP in the pathogenesis of experimental septic shock has been identified by the use of LBP-neutralizing polyclonal IgG antibodies. Administration of such antibodies inhibited the symptoms of shock by two proposed mechanisms: (i) prevention of binding of LPS to LBP; and (ii) increasing LPS clearance [81,89].

The LPS–LBP complex is not in itself any more active than free LPS. The function of the LBP appears to be the transfer of LPS to membrane-bound or soluble CD14 (Figure 6.1) [90–94]. LBP has two domains — one involved in LPS binding and another which enhances the interaction of LPS with CD14. Studies of the homologous LPS-binding protein, BPI, had limited the LPS-binding domain of BPI to the N-terminus. Thus, a 25 kDa proteolytically produced N-terminal fragment of BPI [95] and a truncated mutant, consisting of the first 199 amino acids (BPI$^{1-199}$) [96], had similar abilities to bind LPS. A similar fragment of LBP was prepared and shown to be able to bind LPS with comparable affinity to the intact molecule [92,97]. However, this fragment of LBP was unable to form a complex with LPS which was able to stimulate monocyte cytokine synthesis [96], and it could not catalyse the transfer of flu-oresceinated LPS to a CD14-transfected cell line [92]. Indeed, the truncated mutant could antagonize the binding of LPS to CD14 and the LPS-dependent induction of monocyte TNF synthesis. In this respect the truncated LBP resembles PLTP, which has been shown to bind LPS but transfer it not to CD14 but to high-density lipoprotein, thus acting as an inhibitor of LPS bio-

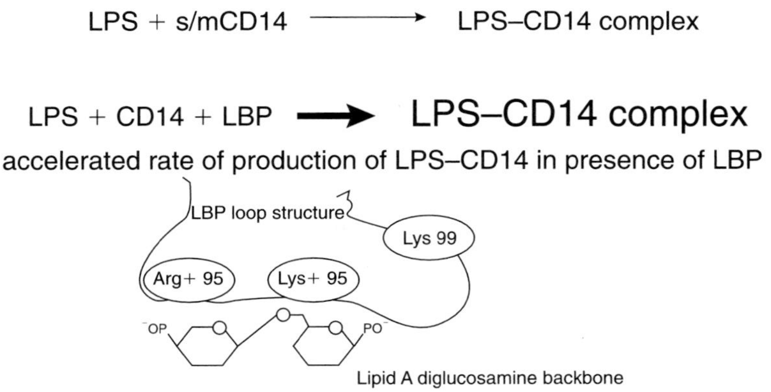

Figure 6.1 The mechanism of action of LBP
This protein acts to increase the rate at which LPS binds to either soluble (s) or membrane-bound (m) CD14. Mutational analysis of LBP has revealed that the basic residues Arg-94, Lys-95 and, to a lesser extent, the lysine at position 99, are involved in the binding of this protein to LPS. A conceivable model for this interaction (shown in the bottom diagram) is that the positively charged residues in LBP are in a loop structure which interacts with the negatively charged phosphates on the diglucosamine backbone of lipid A. Confirmation of this interaction will require crystallographic studies.

activity [98]. Site-directed mutation of LBP has identified as the lipid A-binding domain the positively charged residues Arg-94 and Lys-95, with some contribution from residue 99. Deletion of residues 94 and 95 resulted in a protein with no ability to bind LPS, transfer it to CD14 or stimulate cell activity. Interestingly, if residues 94/95 were replaced by alanine, the LPS-binding activity of the mutant LBP was decreased markedly, but the ability of LBP to transfer LPS aggregates to membrane CD14 was unaffected. The proposed interaction between LBP and LPS is shown schematically in Figure 6.1 [98a].

Structural data defining how LBP binds LPS are not yet available. However, the X-ray crystallographic structure of the *Limulus* anti-LPS factor (also called endotoxin neutralizing protein), an LPS-binding protein, has been derived. The LPS-binding site is proposed to be an amphipathic loop with similarity to amino acid residues 86–104 of BPI and to the LPS-binding peptide antibiotic polymyxin B [99]. Indeed, LBP has a sequence with very similar structure [74], as described above. In spite of the similarities in sequence and structure between LBP and BPI, the former transfers LPS to CD14 while the latter manifestly does not do so. The simplest assumption is that LBP contains a domain or domains lacking in BPI required for such transfer. However, if such domains exist they are not easily recognized.

Another serum factor which acts to integrate the interaction of LPS and CD14 is septin, which is found in normal human plasma [100]. Less attention has been paid to this protein than to the other LPS-binding proteins described earlier. The reader should note that there is also a class of intracellular proteins called septins, and these should not be confused with Wright's nomenclature for this plasma protein.

6.4.4 CD14: an LPS receptor or a quasi-receptor?

LBP can opsonize LPS-bearing particles, such as Gram-negative bacteria or cells which have bound LPS. The finding that erythrocytes coated with LPS bound to macrophages in the presence of LBP [101] suggested that the LBP had a cell-bound receptor, and this was subsequently identified as CD14 [102]. These initial studies suggested that, for cell activation, CD14 was required to interact with an LPS–LBP complex. This is now known not to be the case, and the primary role of LBP is, in fact, to function as a lipid (i.e. LPS)-transfer protein, increasing the rate at which LPS interacts with CD14 rather than forming a stable complex with LPS (Figure 6.1).

CD14 was originally identified as a myeloid differentiation antigen present on mature cells, particularly monocytes, but absent on myeloid precursors [103]. This protein is often used as a marker for monocytes. Peripheral blood monocytes have an estimated 5×10^4 CD14 receptors per cell [104], and PMNs have a tenth of this number [86]. B-cells have been claimed to be CD14-positive (e.g. [105]), but it is generally accepted that normal B-lymphocytes do not express this protein. The human and murine genes for CD14 were cloned in the late 1980s [106,107], and the human gene is located on chromosome 5 in a region containing several genes for growth factors or growth factor receptors, such as IL-3, GM-CSF and PDGF [106]. The primary amino acid sequence similarity between the murine, lapine (rabbit) and human CD14s is high. For example, the rabbit CD14 gene shares 73% and 64% sequence similarity with human and murine CD14, respectively [107].

Proteins can associate with cellular membranes in a variety of ways. Many intracellular proteins attach to membranes via a variety of lipid structures. With certain proteins, like CD14, present on the outer surface of the plasma membrane, the attachment is via a complex glycolipid known as glycosyl phosphatidylinositol (GPI; Figure 6.2) [108,109]. The fatty acyl chains of GPI form part of the phospholipid bilayer, and the terminal phosphoethanolamine unit of the GPI is linked to the C-terminus of the protein. Thus, GPI-linked proteins are tethered to the cell by a flexible leash and the whole of the protein is external to the cell. The consequence of this is that such GPI-anchored proteins cannot be involved in intracellular signalling as they have no signalling domain within the cell. This point will be discussed in more detail in a later section of this chapter.

Direct binding studies suggested that monocyte cell-surface CD14 had the capacity to bind LPS [93]. However, does this binding event have any role to play in the biological actions of LPS? The first piece of evidence for a role for CD14 in the biological activity of LPS was the finding that antibodies to CD14 could block the LPS-induced activation of leucocytes in whole blood. In the

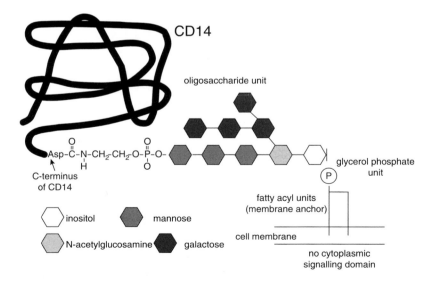

Figure 6.2 The chemical structure of the GPI linkage of CD14

presence of such antibodies, LPS-induced TNFα synthesis was blocked [102]. Other workers have shown the importance of CD14 in the induction of various pro-inflammatory cytokines, including IL-1, IL-6 and IL-8 [110–112] The reader is referred to some recent reviews which describe this early work in more detail [113,114]. The antibodies used in such studies are listed in reference [115]. A key study in identifying the role of CD14 in LPS-induced cell activation involved the transfection of the murine pre-B-cell line 70Z/3 with human CD14 cDNA. The non-transfected cell responds to LPS, by expressing surface IgM, but only at high LPS concentrations. Transfection with human CD14 reduced the amount of LPS required to stimulate IgM expression by up to 10000-fold [116]. This intriguing finding suggests that the non-transformed cell has a low-affinity LPS receptor, but that addition of CD14 either provides the cell with a distinct high-affinity receptor or produces a complex between CD14 and the low-affinity receptor which now acts as a high-affinity receptor. This behaviour is analogous to certain cytokine receptors such as the type I receptors discussed in Chapter 4. A transgenic mouse expressing high levels of human CD14 on blood leucocytes has been produced and is reported to be hypersensitive to LPS compared with the non-transgenic control [117].

The evidence to date suggests that surface-bound CD14 on myelomono-cytic cells is responsible, in some as yet undiscovered fashion, for recognizing and transducing the biological signal for LPS. There have been few studies of the receptor properties of membrane-bound CD14. Kirkland and co-workers [91] have examined the binding of LPS to the monocytic cell line, THP-1, and to stably transfected CHO cells. LPS binding occurred rapidly and relatively independently of temperature, over the range 10–37°C. The apparent dissociation constant (K_D) of CD14 for LPS was an estimated 3×10^{-8} M. To put this in context, the K_D for binding of human IL-1 to the human type I IL-1 receptor is in the range 5×10^{-9}–5×10^{-10} M [118]. Thus, the apparent receptor affinity of LPS for cell-bound CD14 is probably within the lower range for cytokine receptors, but is high for agonist–receptor interactions in general. The stoichiometry of binding of LPS to membrane CD14 is not clear. For soluble CD14, the stoichiometry of binding with LPS is 1:1 [94].

Without exception, studies of the interaction of bacteria with CD14 have utilized endotoxin or LPS. However, it has recently been reported that whole E. coli can interact with PMNs in an LBP/CD14-dependent manner, and that bacteria, on a weight basis, were >10-fold more active than isolated LPS and that as few as one bacterium per 20–200 leucocytes was able to stimulate PMNs [94a]. This suggests that, in conditions such as septic shock, the bacteria may be a more important target than previously thought.

6.4.4.1   Regulation of expression of CD14

A common characteristic of plasma-membrane receptor proteins is that their numbers and affinities can be modulated by the ligand or by other factors.

Cytokine receptor transmodulation and its role in cytokine network control has been discussed in Chapter 4. The control of CD14 expression during monocyte maturation induced by 1,25-dihydroxy-vitamin D_3 has suggested that gene transcription is the major controlling factor, and that the 5'-upstream region of the CD14 gene contains binding sites for the transcription factor Sp1 — a member of the zinc-finger group of transcriptional control proteins — and that this is the major regulatory site for this gene [119,120]. There is evidence of differential expression of CD14 among resident macrophage populations. For example, human alveolar macrophages express low levels of CD14, while peritoneal macrophages express relatively much higher levels [121–123]. This may reflect different networks of CD14-inducers in these different microenvironments.

What effect does LPS have on the expression of CD14? As with many aspects of the biology of LPS, the answer to this question is controversial. A number of workers have reported that LPS stimulation of (i) whole blood [124], (ii) peripheral blood monocytes [125,126], (iii) alveolar macrophages [127] and (iv) monocytic cell lines [128,129] caused an increase in the expression of CD14 on the cell surface. In contrast, other workers have reported that LPS reduced CD14 expression on monocytes [130] and macrophages [131]. This may reflect differences in experimental variables such as the source of LPS and its contamination with biologically active proteins, LPS concentration, cell type, state of monocyte differentiation, etc. In a recent study, some of these parameters have been examined using human monocytes and monocyte-derived macrophages. These cells were exposed to different concentrations of LPS for different periods of time, and the membrane-bound mCD14, soluble sCD14 and CD14 mRNA measured. During the first 3 h of LPS exposure no changes were noted. Between 6 and 15 h of exposure, there was a small decrease in CD14 mRNA and in mCD14, but sCD14 transiently increased. This early change is similar to the decreased CD14 expression reported by Bazil and Strominger [130] and Wright [131]. With prolonged exposure (48 h) to LPS (>1 ng/ml) there were significant increases in CD14 mRNA, and in mCD14 and sCD14, and cells bound increased amounts of fluoresceinated LPS. Of particular interest was the finding that lipoteichoic acid and a cell-wall extract of *Staph. aureus* were also able to up-regulate CD14 expression. This study clearly shows the capacity of bacterial components to up-regulate the transcription of the gene for CD14, and suggests the reasons for the discrepant results in the literature [132]. Intraperitoneal injection of LPS into CB6 mice resulted in the appearance of CD14 in the serum (normal murine serum contains undetectable levels of CD14) and in the up-regulation of expression of mCD14 in many of the body tissues, including heart, lung, liver, uterus, etc. Expression was found both in the myeloid cells (e.g. liver Kupffer cells) and in epithelial cells [133]. It is established that cells expressing mCD14 can bind bacteria [134]. Thus, as suggested by Fearns et al. [133], the up-regulation of mCD14 in the liver may be part of the clearance mechanism for Gram-negative bacteria.

The major result of the binding of LPS to CD14 is the induction of cytokines, mainly those with pro-inflammatory activity. What role, if any, do cytokines play in the regulation of membrane or soluble CD14. A number of reports have suggested that cytokines such as IL-1β, IL-2, IL-3, IL-5, IL-6, TNFα, GM-CSF and TGFβ have no effects on the expression of CD14 on human macrophages [135–138]. Other workers have reported that IL-6 [139,140] and TNFα [131] can up-regulate CD14 expression. Human PMNs cultured in the presence of TNFα, G-CSF, GM-CSF or formyl peptide have been reported to rapidly double their expression of CD14. As this increase took place within 20 min of the addition of such molecules, the process presumably was the result of the insertion of preformed stores of CD14 into the plasma membrane [86]. In contrast, infusion of recombinant human (rh)G-CSF, but not rhGM-CSF, was reported to increase CD14 expression on peripheral blood PMNs [141].

Two cytokines down-regulate the expression of mCD14 by acting at the level of gene transcription. IL-4 was the first cytokine to be shown to act in this manner [137]. The second cytokine reported to do so is IL-13, a product of T-lymphocytes and mast cells, which shares structural and functional similarities with IL-4. IL-13 decreases the pro-inflammatory activity of macrophages, reducing production of inflammatory cytokines in response to IFNγ and LPS [142]. This cytokine has been shown to dose-dependently inhibit the expression of CD14 on human monocytes, while concurrently increasing expression of the $β_2$-integrin protein CD11b. As has been alluded to, the down-regulation of cell-surface CD14 can occur by stimulation of the shedding of this protein. However, in the case of IL-13, shedding was not increased. Rather, IL-13 decreased the level of CD14 transcripts in monocytes, suggesting that it was acting to inhibit the transcription of the CD14 gene [143]. It is likely that the inhibitory effect of IL-13 on monocyte CD14 expression explains why this cytokine can inhibit monocyte pro-inflammatory cytokine synthesis. The range of interactions between LPS, cytokines and CD14 is shown schematically in Figure 6.3.

These finding may have implications for the replication of HIV-1, the virus responsible for AIDS. Monocytes are important in HIV infection as they act as a reservoir for the persistence of the virus in the host [144]. LPS is a potent stimulator of HIV-1 expression in monocytes [145] via a CD14-dependent mechanism [146]. IL-13 has been shown to suppress HIV-1 replication in monocytes [147] and this may possibly relate to the ability of this cytokine to inhibit the transcription of CD14.

Membrane-bound CD14 may have functions in addition to acting as the receptor for LPS. For example, antibodies to CD14 can interfere with the adhesion of leucocytes to cytokine-activated vascular endothelial cells [148,149], and antigen-driven T-cell proliferation can also be blocked by

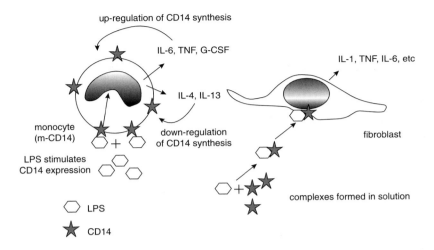

up-regulation of CD14 synthesis

IL-6, TNF, G-CSF

IL-1, TNF, IL-6, etc

IL-4, IL-13

monocyte
(m-CD14)

down-regulation
of CD14 synthesis

fibroblast

LPS stimulates
CD14 expression

complexes formed in solution

LPS

CD14

Figure 6.3 The relationships between LPS, soluble or membrane-bound CD14 and cellular cytokine synthesis

LPS can bind directly to membrane-bound CD14 on myeloid cells and induce the transcription of cytokines. With cells which do not express CD14 (e.g. most types of fibroblasts, epithelial cells, vascular endothelial cells, etc.) the LPS has to bind to soluble CD14 to form a complex which then interacts with the cells to induce cytokine transcription. The binding of LPS to CD14 induces the synthesis of various cytokines. These cytokines can, in turn, either stimulate, or inhibit, the synthesis of CD14. LPS itself is able to up-regulate the synthesis of CD14.

certain anti-CD14 mAbs [150]. Anti-CD14 mAbs also inhibited the spontaneous production of IL-6 which occurred when rheumatoid synovial fibroblasts were co-cultured with human monocytes, suggesting that the monocyte–fibroblast interactions resulting in cellular activation were CD14-dependent. Such inhibition was not related to LPS levels in the cultures since polymyxin B had no effect on IL-6 synthesis [151]. This finding may have relevance for the treatment of RA, a disease which is currently untreatable except at the symptomatic level.

The reader should not be left with the impression that LPS can only activate monocytes and macrophages via a CD14-dependent mechanism. Human monocytes will respond to LPS in, for example, the presence of saturating concentrations of neutralizing antibodies to CD14 if the LPS concentration is sufficiently high. Thus, as we will show in the next section on non-myeloid cells, CD14 sensitizes monocytes to LPS [152]. This capacity to stimulate human monocytes by a CD14-independent pathway is, surprisingly, absent in human neonates whose monocytes can only be stimulated in a CD14-dependent manner as shown by the fact that in the presence of saturating concentrations of anti-CD14 antibodies the neonatal monocytes fail to respond to LPS. This CD14-independent response to LPS is also dependent on an, as yet, unidentified plasma factor which is absent in neonatal plasma [153]. The consequence to the neonate of the absence of this factor is not clear, nor is it known when this plasma factor is produced in childhood develop-ment. The answers to both questions are obviously important in understanding how children respond to bacterial, particularly Gram-negative, infections.

6.4.4.2 Soluble CD14

It is now well established that cell-surface receptors can be shed. With cytokine receptors, for example, shedding is believed to be a regulatory mechanism to inhibit cytokine activity. However, it is rare to find measurable levels of such receptors in biological fluids such as blood. The presence of soluble (s) CD14 in cultures of mononuclear cells was first described in the mid-1980s, at a time when the function of this protein had not been determined [154]. Since then, CD14 has been found in both serum and urine, and in the former is present at relatively high concentrations, of the order of 3–6 µg/ml [155,156]. To put this circulating level of CD14 in context, the con-centration of immunoglobulin (Ig)D in serum is 30 µg/ml and that of IgE is only 0.05 µg/ml. The source of this sCD14 is not entirely clear. As has been discussed, CD14 can be shed from monocytes in response to stimuli such as LPS and IFNγ, but it is not known if this is the major physiological pathway for generating the high steady-state level of this protein in blood. Serum sCD14 exists as a number of isoforms of molecular mass 53, 50, 46 and 43 kDa [157]. The release of CD14 from cells can be due to the action of proteases or to the action of a phospholipase [158]. What role does this large amount of LPS receptor play in the blood? The finding that sCD14 could inhibit the LPS-induced respiratory burst in cultured human monocytes [159] suggested that the sCD14 could act to bind and inactivate the small amounts of LPS that continually enter into the blood from the gut, mouth, etc. As will be discussed later in this section and in Chapter 8, sCD14 has been proposed as a therapeu-tic agent for LPS-induced pathology (e.g. septic shock). However, it is now becoming clear that the major role of sCD14 is probably to allow cells that do not express GPI-anchored CD14 to respond to low levels of LPS. Frey et al. [160] and Pugin and co-workers [90] were the first to show that sCD14 confers on both vascular endothelial and epithelial cells a dramatically increased sensitivity to LPS. These findings have been confirmed by a number of other workers [161–163] and it has also been shown that sCD14 confers LPS sensitivity on vascular smooth muscle cells [164]. Indeed, sCD14 has also been shown to confer sensitivity on monocytes from patients with paroxysmal nocturnal haemoglobinuria which lack GPI-anchored proteins including CD14 [165].

The capacity of both soluble and membrane-associated CD14 to render cells more sensitive to LPS suggests that the inhibition of this protein could have therapeutic benefit. Indeed, it has been reported that sCD14 is elevated in patients with sepsis [166,167]. Neutralizing antibodies to CD14, as discussed, can block cellular responses to LPS. Administration of neutralizing anti-CD14

IgG$_1$ mAbs to primates with experimental septic shock has been reported to be clinically beneficial [168]. One of the truly surprising aspects of the LPS/LBP/CD14 story is that CD14 itself can also inhibit LPS-induced cytokine synthesis by leucocytes [169] and can protect mice from experimental endotoxic shock [170]. The therapeutic potential of anti-CD14 and CD14 is discussed in more detail in Chapter 8.

6.4.5 Interactions between LPS, LBP, CD14 and the CD14 receptor

This chapter has discussed the individual parts of the complex molecular machinery that has evolved to enable the bacterial surface component, LPS, to stimulate eukaryotic cells to produce cytokines. In this section the interacting pathways leading to the binding of LPS to its final signalling receptor, whatever that may be, will be described.

LBP has been described in detail in a previous section. A key question was whether this molecule needed to be complexed with LPS/CD14 to allow LPS to express its full endotoxic activity. It is now clear that this is not the case and

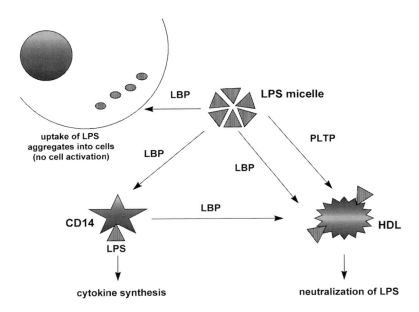

Figure 6.4 The interaction of LPS, LBP, CD14, other lipid-binding serum proteins, and HDL

LBP acts as a transfer protein for LPS, in transporting LPS monomers from micellar structures to either CD14 (thus activating cells) or to high-density lipoproteins (HDLs). LBP can also transfer LPS from CD14 to HDL. Another recently recognized activity of LBP is its ability to stimulate the uptake of aggregates of LPS into phagocytes without activating these cells [173]. Other serum lipid-binding proteins, such as cholesterol ester-transfer protein (CETP) and phospholipid-transfer protein (PLTP), can bind to LPS. The latter protein can also transfer LPS to HDL and, in this manner, act as an inhibitor of LPS.

that LBP acts, as an activator of LPS activity, by catalysing the transfer of LPS from micelles containing this molecule to either sCD14 or mCD14. However, it plays no further role in the response of CD14-bearing cells to LPS [90,94,171–175]. LBP is simply a lipid-transfer protein which, in addition to transferring LPS from LPS micelles to CD14 (thus promoting LPS-induced cell activation), can also transfer LPS (from micelles) to high-density lipoproteins (HDLs), forming a biologically inactive complex. LBP can also transfer LPS from CD14 to HDL, again forming a biologically inactive complex [171]. In addition, LBP has been shown to catalyse the uptake of LPS aggregates into phagocytic cells without their being activated. Thus, with this ability to transfer LPS to HDL or into phagocytes, LBP could act both as an activator and as an inhibitor of LPS action. This would depend on the kinetics of transfer of LPS between CD14 and HDL, and on the relative uptake of LPS aggregates versus their solubilization by LBP and transfer to CD14 (Figure 6.4).

Thus, it is clear that the eukaryotic organism has a Janus-like interaction with LPS (Figure 6.5). The response of the multicellular host to LPS will depend on the balance of a range of host-derived factors. To achieve maximum response, LPS, after release from the bacterium, must interact with CD14 in a reaction catalysed by LBP. As we have seen, LBP can transfer aggregates of LPS into phagocytes without activating such cells. A number of other proteins and peptides can interact with LPS to inhibit its biological activity. At epithelial surfaces and in phagocytes are a large number of so-called antibiotic peptides, some of which can bind and inactivate LPS. These host-defence peptides will be discussed in more detail in Chapter 9. Other proteins, either cell-derived like BPI or present in the serum (PLTP and CETP), can bind LPS but not transfer it to CD14, thus acting as inhibitors of LBP. These various pro- and anti-inflammatory molecules are shown schematically in Figure 6.5.

In the remaining part of this section the interaction between LPS, CD14 and the ultimate cell-surface LPS receptor will be considered. Douglas Golenbock's group in Boston has described a fascinating study of the interaction of LPS and lipid A analogues with CD14 [176]. As discussed in Chapter 5, the lipid A analogues, lipid IV$_A$ and the *Rhodobacter sphaeroides* lipid A (RSLA), act as antagonists of LPS when tested on human monocytes. In contrast, with murine mononuclear cells lipid IV$_A$ was an agonist while RSLA was an antagonist. The simplest explanation for these differences in the responses of cells to these lipid A analogues is that they reflect molecular differences in the CD14 of the two species. An alternative explanation is that they reflect differences in post-CD14-dependent cellular mechanisms. To discriminate between these two possibilities the cDNAs encoding human or murine CD14 were cross-transfected, and each combination of cell and receptor was tested with both lipid A analogues. The results of this study were surprising. Human cells transfected with mouse CD14 recognized both lipid A analogues as antagonists. Mouse cells transfected with human CD14 responded to lipid

INHIBITION OF INFLAMMATION

BPI, CETP, PLTP, antibiotic peptides, HDL [LBP, CD14]

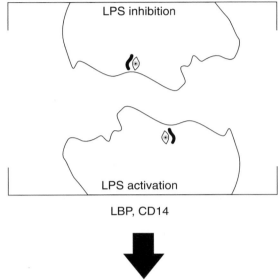

LPS inhibition

LPS activation

LBP, CD14

INFLAMMATION

Figure 6.5 The 'Janus system' of LPS activation/inactivation by host factors
LPS is relatively inactive, but interaction with the host-derived factors LBP and CD14 dramatically increases its biological activity. This is one face of the interaction of LPS with the host. The
opposite face shows the many host factors that are employed to inhibit the biological actions of
LPS. These include bactericidal/permeability-increasing protein (BPI), a number of host antibiotic
peptides, PLTP, CETP and HDL. In addition, LBP can limit the actions of LPS by transfer to HDL
or stimulation of uptake into phagocyte, and excess sCD14 can compete with mCD14, thus
inhibiting the activity of LPS.

IV$_A$. Normal hamster phagocytes responded to both lipid A analogues. This
study clearly demonstrates that CD14 is not acting as a conventional receptor,
as it cannot discriminate between the different lipid A structures. Thus, the
actual receptor for LPS must be an additional lipid A-recognizing protein that
functions as the true signalling receptor for LPS.

The interaction of LPS with CD14 has now been studied in some detail. It
has been established by limited proteolysis and LPS-protection studies that the
LPS-binding site on CD14 resides in the N-terminus between amino acids 57
and 64 [177–179]. Expression of the N-terminal 1–152 region of CD14 in
CHO cells revealed that this truncated protein could act as a fully functional

'receptor' [115]. A number of mAbs to CD14 can inhibit the response of cells to LPS. Two mAbs (MEM-18 and 3C10) have been shown to recognize the N-terminal functional domain of CD14. MEM-18 has been reported to bind to the LPS-binding domain of CD14, but 3C10 recognizes a different part of this protein between residues 7 and 14 [180]. Site-directed mutagenesis was used to replace residues 7–10 in CD14 with alanine and this mutant [$CD14_{(7-10)A}$] was shown to bind LPS normally, but to be impaired in its capacity to activate cells. The simplest explanation to account for the data is that residues 7–10 in CD14 are responsible for signalling to some adjacent membrane protein that the CD14 has complexed with LPS (Figure 6.6).

This brings us back to the key question: what is the identity of the LPS receptor(s) and how does it interact with LPS? The evidence suggests that this interaction with LPS is indirect and via CD14. This may be similar to the situation with a growing number of cytokine receptors [181]. This literature has been reviewed in Chapter 4. A good example is the IL-6 receptor. Upon binding of IL-6, this receptor becomes associated with a signal-transducing receptor known as gp130, which dimerizes to form a signalling element. This protein has no IL-6-binding capacity, but serves as a signal transducer for IL-6 and also for a number of other cytokines, including LIF, OSM and IL-11 [182]. In the case of LPS, the signal-transducing receptor(s) probably has some affinity for LPS, accounting for the activity seen at high concentrations of LPS

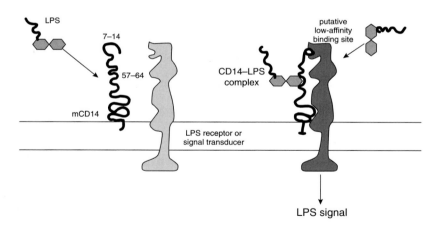

Figure 6.6 Schematic diagram suggesting the nature of the interaction between LPS, CD14 and the putative LPS receptor

LPS binds to residues 57–64 in CD14, and site-directed mutagenesis has suggested that residues 7–14 are important in CD14 signalling. This implies that residues 7–14 interact with the true LPS receptor (or more correctly the LPS–CD14 receptor) to induce intracellular signalling. This 'true' receptor may have two binding sites – a high-affinity site for LPS–CD14 and a lower-affinity site for LPS. The diagram shows that the CD14 oligomerizes with the 'true' LPS receptor. A possible alternative is that the binding of the LPS to CD14 is sufficient to oligomerize a second activating protein, in a manner similar to the binding of IL-6 to the IL-6 receptor causing dimerization of gp130 (see Chapter 4).

when CD14 binding is abrogated (Figure 6.6). The finding of multiple LPS-binding cell-surface proteins (Table 6.3) suggests parallels with the cytokines that bind to type I receptors (Chapter 4). Such receptors utilize a number of signalling proteins such as gp130, LIF-R, and the common β or γ subunits that undergo homo- or heterodimerization when the cytokines bind to target cells.

The nature of the cell-surface machinery required in order to allow cells to bind and respond to LPS is obviously complex. Is there a satisfying explanation for this complexity? As will be discussed in more detail in the final chapter, CD14 is distinct from most other receptors in that it can recognize a wide range of bacterial macromolecules and has recently been classified as an example of a pattern-recognition receptor [183]. The immune system, which could also be classified as a pattern-recognition network, requires multiple receptors to be triggered before lymphocytes are activated. Thus, for CD4 T-lymphocytes to recognize antigen, fragments of the antigen bind to MHC class II, and this complex binds to the T-cell receptor. This binding is helped by the interaction of the CD4 protein with MHC class II. However, to trigger the T-cell into clonal expansion, another receptor–co-receptor pair must be formed — namely the binding of CD28 on the T-cell to B7-1/B7-2 (CD80/CD86) on the antigen-presenting cell. It is believed that the requirement for the co-stimulatory signal via CD28/B7-1/2 ensures that only 'professional' antigen-presenting cells can initiate T-cell responses. Could the same safety net be produced by the co-stimulatory actions of CD14? In addition, it is possible that cell activation may depend on the rate of binding of LPS to the ultimate receptor rather than on the concentration of the agonist. This may explain the requirement for LBP.

6.5   LPS-induced transmembrane signalling

Agonist–receptor binding is usually followed by intracellular signalling events involving phosphorylation/dephosphorylation of proteins, which leads to the generation of a specific transcription-activating complex or complexes and a specific pattern of gene activation. The key to the selectivity of individual cell-stimulating agonists (hormones, cytokines, eicosanoids, etc.) is the pattern of intracellular protein phosphorylation, which is controlled by a large number of intracellular kinases and phosphatases [184]. These patterns of kinase/phosphatase activity are responsible for regulating all the major variables of cell behaviour, including growth, motility, differentiation and division. The number of reported sequences encoding distinct kinases is now around 400 [185], revealing the enormous complexity of the systems known as signal transduction or intracellular signalling. These intracellular signal transduction pathways are another complicated network of interacting signals mimicking to some extent the network of mediators (cytokines, lipid mediators, etc.) which exist external to the cell. To date, the studies of LPS cell signalling have focused on kinases and their activation.

The early history of LPS-induced intracellular signalling has been reviewed by Dolph Adams [186]. There is currently a large and highly confusing literature on LPS-induced intracellular signalling. This literature will be briefly reviewed in this section. The importance of certain signalling pathways in LPS-induced pathology has been defined by use of selective inhibitors. For example, tyrphostins, which are synthetic inhibitors of tyrosine kinases, have been shown to inhibit LPS-induced lethality in mice [187]. A range of tyrphostins were examined in this study, and the most active at inhibiting LPS-induced monocyte TNF synthesis *in vitro* were tested *in vivo* where they proved effective. The inhibitory effect of the tyrphostins correlated with the inhibition of LPS-induced tyrosine phosphorylation of the 42 kDa mitogen-activated protein kinase (MAPK), a kinase family that will be discussed later in this section. Lisofylline [(R)-1-5-(5-hydroxyhexyl)-3,7-dimethylxanthine], a metabolite of the rheological/cytokine-inhibitory agent pentoxyfylline, is an inhibitor of the synthesis of the intracellular signalling molecule, phosphatidic acid (PA) [188]. Lisofylline has also been shown to protect mice against experimental endotoxic shock [189]. These *in vivo* studies suggest that tyrosine kinase-dependent and/or inositol-dependent signalling pathways are important in the pathophysiology induced by excessive levels of LPS.

One of the earliest signal-transduction pathways discovered was the G-protein-linked phospholipase C-inositol phosphate pathway, in which cleavage of membrane-bound PtdIns(4,5)P_2 by phospholipase C produces Ins(1,4,5)P_3, which can release calcium from the endoplasmic reticulum, and diacylglycerol, which activates the PKC group of protein kinases [190]. LPS activation of monocytes does result in the activation of phospholipase C and the formation of Ins(1,4,5)P_3 [191,192], but this is not accompanied by the expected rise in intracellular calcium levels [193,194]. Translocation of PKC within the cell often accompanies the activation of this family of enzyme isoforms [195]. LPS has been reported to translocate the PKC isoform PKCβ in LPS-responsive macrophages from C3H/HeN mice, but not in macrophages from the LPS-unresponsive C3H/HeJ mouse, suggesting that the translocation of this kinase is important in LPS signalling [196]. Activation of PKC is normally mediated by diacylglycerol, but Shapira and co-workers [197] have reported that LPS-induced activation of human monocytes was not accompanied by significant elevations in intracellular diacylglycerol, and that an inhibitor of diacylglycerol kinase did not block the LPS-induced activation of monocytes. Recent studies suggest that LPS causes the activation of a calcium-independent isoform of PKC [198] and that this may be due to the capacity of LPS to activate phosphatidylinositol 3-kinase in monocytes [199], a kinase that can activate calcium-independent isoforms of PKC.

Another approach to understanding the role of protein kinases in LPS-induced cell signalling is to identify intracellular proteins that are phosphorylated in response to LPS. A major phosphorylated protein in LPS-stimulated murine macrophages is a 65 kDa species (pp65). The phosphorylation of pp65 closely corresponds to the macrophage's capacity to secrete cytokines [200]. This protein has recently been identified as the murine homologue of the human transformation-induced protein, L-plastin which contains a unique series of domains that can bind calcium, calmodulin and actin. Analysis of the sites of phosphorylation has shown that this protein is a substrate for cyclic AMP-dependent protein kinase A, creatine kinase II and PKC [201]. Of interest, in light of the discussion to follow, was the failure to find a specific motif in pp65 for MAPKs, which have been shown to be involved in LPS-induced signalling.

The MAPKs are a group of protein kinases that are activated, not by the addition of one phosphate group to serine, threonine or tyrosine residues, but by dual phosphorylation of threonine and tyrosine in response to a large array of extracellular stimuli. For example, in the yeast *Saccharomyces cerevisiae,* many of the cellular functions are controlled by MAPK signal transduction pathways. The role of MAPKs in cytokine signalling has been discussed in Chapter 4. A well-studied example is the kinase HOG1, which controls the yeasts' response to extracellular osmolarity [202]. The role of tyrosine kinases in LPS pathophysiology has already been discussed. Weinstein and co-workers [203,204] were the first to show that LPS can stimulate macrophage protein tyrosine phosphorylation, and identified MAPK as well as other proteins as targets of this phosphorylation. Such phosphorylation is CD14-dependent and stimulates MAPKs of 42 and 44 kDa (designated p42 and p44) [205,206]. LPS-induced activation of stably transfected 70Z/3 cells expressing human CD14 [207,208], or of human neutrophils [209], resulted in rapid protein tyrosine phosphorylation of a 38 kDa MAPK which is related to the HOG1 MAPK of *S. cerevisiae.* It is not only myeloid cells that respond in this manner to LPS. Cultured vascular endothelial cells have also been reported to have their p44, p42, p41 and p38 MAPKs phosphorylated on tyrosine. This LPS-induced activation required the presence of sCD14 [210,211].

The MAPKs, also known as extracellular-signal-regulated kinases (ERKs), function to activate transcription factors such as Jun, which can then combine with the Fos protein to produce the active transcription factor, AP1. However, the MAPKs are at the distal end of a complex cascade of kinases. Thus, the MAPKs are phosphorylated by a MAP kinase kinase (such as MEK1) which, in turn, is phosphorylated by a MAP kinase kinase kinase (Raf-1 or MEKK1). LPS has been reported to activate MEK1 [212] and Raf-1 [213]. Another group of enzymes that are related to the MAPKs are the stress-activated protein kinases (SAPKs)/c-Jun N-terminal kinases (JNKs) [214]. SAPK/JNK phosphorylation of c-Jun induces gene transcription as described. LPS has been shown to activate SAPK/JNK [215,216]. Weinstein and colleagues have recently suggested a scheme to incorporate the interaction of these kinases in LPS-stimulated cells (Figure 6.7).

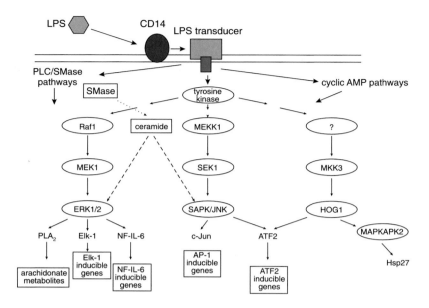

Figure 6.7 LPS-induced intracellular signal transduction cascade
Binding of LPS to CD14 and the subsequent activation of the LPS–CD14 'receptor' is proposed
to result in the activation of multiple cell signalling pathways. In this scheme, the LPS activates a
cellular tyrosine kinase which then sets off a network of kinase interactions, leading to the
induction of the transcription of multiple genes. LPS can also activate other cellular pathways.
One of the most interesting is the activation of membrane sphingomyelinases to produce
ceramide, which is then involved in the activation of certain of the MAP kinases and also the
stress-activated protein kinases. Abbreviations used: AP-1, activator protein 1; ATF, activating
transcription factor; ERK, extracellularly regulated kinase; HOG, high-osmolarity glycerol protein
kinase; hsp27, heat-shock protein of 27 kDa; JNK, c-Jun N-terminal kinase; MAPKAPK2,
mitogen-activated protein kinase-activated protein kinase 2; MEK1, MAPK/ERK kinase 1; MEKK,
mitogen-activated protein kinase/ERK kinase kinase; MKK, mitogen-activated protein kinase
kinase; NF-IL-6, nuclear factor IL-6 (IL-6 transcription factor); PLA$_2$, phospholipase A$_2$; PLC,
phospholipase C; SAPK, stress-activated protein kinase; SEK, stress-activated protein kinase/ERK
kinase; SMase, sphingomyelinase.

LPS has also been reported to cause the phosphorylation of two cytosolic
proteins of 36 and 38 kDa (p36 and p38 respectively), which are not related to
the MAPKs. This LPS-induced phosphorylation can be prevented by lipid IV$_A$
and, at low concentrations of LPS, mAbs against CD14 can also block.
However, at higher LPS concentrations these mAbs fail to prevent phosphorylation of p36/p38. Of interest is the finding that inhibitors of
ADP-ribosylation inhibit LPS-induced phosphorylation of p36/p38 and can
inhibit the synthesis of TNFα and IL-6, but not IL-1 [217]. These results
support the earlier claim of Hauschildt and colleagues that LPS induces ADP
ribosylation [218]. As is discussed in Chapter 7, a number of the ADP-ribosylating bacterial exotoxins can modulate the synthesis of cytokines, suggesting
that this process may play a key role in the control of the cytokine network.

Workers at SmithKline Beecham in the U.S.A. have taken another approach to LPS-induced cell signalling by using compounds known to inhibit LPS-induced cytokine synthesis as probes for kinases. As will be discussed in more detail in Chapter 8, using this approach has identified that these compounds interact with the p38 MAPK (reviewed in [219]).

The phosphoinositide pathway of intracellular signalling, which uses $Ins(1,4,5)P_3$ and the lipid diacylglycerol as second messengers, has been mentioned briefly. An analogous pathway has recently been discovered which uses ceramide as a second messenger. In this pathway a sphingomyelin-specific form of phospholipase C (a sphingomyelinase; SMase) hydrolyses sphingomyelin in the plasma membrane to release ceramide [220]. The actions of ceramide have not been fully elucidated, but there is evidence that this soluble lipid can activate a protein phosphatase [221] and a protein kinase (termed ceramide-activated protein kinase; CAPK) [222]. CAPK can phosphorylate Raf-1, linking ceramide signalling to the MAPK signalling pathway [223]. Of interest is the suggestion that the ceramide pathway is a signal for apoptosis [224].

In the past two years attention has focused on this ceramide pathway as an additional mechanism for cell stimulation by LPS. Thus, exogenous SMase, or analogues of ceramide which are cell-permeable, have been shown to mimic cellular responses associated with LPS, such as induction of IL-6 synthesis [225] and activation of MAPKs [226]. One of the most intriguing suggestions is that part of the lipid A has structural similarity to ceramide and that LPS may act by mimicking ceramide [227,228]. The basis of this suggestion was the finding that LPS activated CAPK, but did not induce the production of ceramide. Further evidence for the role of ceramide in LPS-induced cell activation has come from the study of the LPS-unresponsive C3H/HeJ mouse strain. While macrophages from an LPS-responsive strain of mouse (C3H/OuJ) responded to cell-permeable analogues of ceramide, macrophages from the LPS-unresponsive strain failed to be activated by such analogues [229]. The capacity to respond to LPS has been linked genetically to the *Lps* locus on chromosome 4 in the mouse [230]. Thus, the possibility exists that this *Lps* locus contains genes that control the regulation of the ceramide signalling system. In a recent study designed to compare the response of cells to LPS or ceramide, it was found that ceramide did not mimic all the actions of LPS on macrophages, suggesting that ceramide only activates a subset of the LPS-induced signalling pathways [231].

The final signal transduction pathway that will be discussed is the newly discovered JAK/STAT pathway. This involves Janus family tyrosine kinases (JAKs), which interact with latent cytosolic transcription factors known as 'signal transducers and activators of transcription' (STATs). These, in turn, consist of six proteins (STAT1– STAT6; see also Chapter 4). Binding of a specific cytokine to its receptor results in the recruitment and activation of a distinct pattern of STAT proteins, which become activated by the JAK tyrosine kinases and form homodimers and heterodimers, which, in turn, translocate to the nucleus and promote the transcription of cytokine-inducible genes [232]. The role of LPS in regulating this novel pathway has not been elucidated, but is an exciting area for research. Targeted disruption of the STATs has, as expected, shown profound changes in homozygotes. For example, knockout of STAT4 results in animals with selective defects in IL-12-dependent functions, including defects in Th$_1$ lymphocyte differentiation, lymphocyte proliferation and IL-12-dependent IFNγ production [233,234]. Knockout of STAT6 causes defective IL-4 signalling and defects in Th$_2$ responses [235,236], and with abrogation of STAT1 animals have impaired immune responses to intracellular bacteria and viruses [237,238]. For the reader with only a limited background knowledge in cell signalling we recommend [239].

In this abbreviated overview of the large and complex literature on LPS-induced cell signalling, it is obvious that LPS can activate multiple intracellular signal transduction pathways resulting in the switching on of multiple genes in responsive cells. Two factors contribute to the complexity of this literature and are often not taken into account in the published experimental studies. The first is that LPS rapidly induces the production of a range of potent cell-modulating cytokines, which can act on cells to produce additional cell-signalling events and confuse the interpretation of the data. The second confusing factor is the purity of the LPS used. Unless care is taken, LPS is con-taminated by bioactive proteins known as lipid A-associated proteins or endotoxin-associated proteins. These will be discussed in more detail in Chapter 8. Vogel and co-workers [240] have shown that contaminating proteins in LPS can stimulate IL-1β, TNFα and TNF receptor synthesis and tyrosine phosphorylation of three 41–47 kDa proteins in LPS-unresponsive murine macrophages. Removing these proteins results in the loss of respon-siveness of the LPS-unresponsive macrophages. Thus, unless care is taken to remove such proteins, much of the reported data on the intracellular signalling pathways induced by LPS may be due either to contaminating proteins or to synergy between LPS and such proteins. A random survey of the literature has revealed that only a fraction of the papers on LPS-induced cell signalling state that they use protein-free LPS preparations, and a number use Sigma LPS which contains large amounts of protein.

6.6   Conclusions

LPS is a complex biomolecule that still retains many of its mysteries. The study of this molecule has been responsible for the discovery of the cytokines — probably the most important molecules in maintaining the *milieu interieur*. We now understand that LPS, to induce its full agonist function, requires host accessory molecules like LBP and CD14. However, it is surprising that in 1997, with so many receptors being cloned and expressed, the nature of the

LPS receptor or receptors is still not known. This failure to determine the true LPS receptor may be related to the fact that LPS requires LBP and CD14 for cell activation and thus may be a prototypic system, the nature of which has yet to be discovered.

In the next chapter the discussion will move on from LPS to the discoveries made in the 1990s that bacteria produce a wide range of molecules, in addition to LPS, which can cause mammalian cells to produce cytokines. The possibility that these additional cytokine-modulating molecules may form part of a network of bacterial–host interactions will be considered in detail in Chapter 9.

References

1.   Welch, W.H. and Atkins, E.A. (1989) Fever: historical aspects. In Interleukin-1, Inflammation and Disease (Bomford, R., and Henderson, B., eds.), pp. 3–15, Elsevier, North Holland

2.   Menkin, V. (1944) Special articles. Chemical basis of fever. Science 100, 337–338

3.   Beeson, P.B. (1948) Temperature-elevating effect of a substance obtained from polymorphonuclear leukocytes. J. Clin. Invest. 27, 524

4.   Bennett, I.L. and Beeson, P.B. (1953) Studies on the pathogenesis of fever. I. The effect of injection of extracts and suspensions of uninfected rabbit tissues upon the body temperature of normal rabbits. J. Exp. Med. 98, 477–492

5.   Bennett, I.L. and Beeson, P.B. (1953) Studies on the pathogenesis of fever. II. Characterization of fever-producing substances from polymorphonuclear leukocytes from the fluid of sterile exudates. J. Exp. Med. 98, 493–508

6.   Grant, R. and Whalen, W.J. (1953) Latency of pyrogen fever. Appearance of a fast-acting pyrogen in the blood of febrile animals and in plasma incubated with bacterial pyrogen. Am. J. Physiol. 173, 47–54

7.   Atkins, E.A. and Wood, W.B. (1955) Studies on the pathogenesis of fever. II. Identification of an endogenous pyrogen in the blood stream following the injection of typhoid vaccine. J. Exp. Med. 102, 499–516

8.   Atkins, E.A. (1960) The pathogenesis of fever. Physiol. Rev. 40, 580–646

9.   Atkins, E.A. and Snell, E.S. (1965) Fever. In The Inflammatory Process (Zweifach, B.W., Grant, L. and McCluskey, R.T., eds.), pp. 495–534, Academic Press, New York

10. Bodel, P. and Atkins, E. (1967) Release of endogenous pyrogen by human monocytes. N. Engl. J. Med. 276, 1002–1008

11. Kampschmidt, R.F., Long, R.D. and Upchurch, H.F. (1972) Neutrophil releasing activity in rats injected with endogenous pyrogen. Proc. Soc. Exp. Biol. Med. 139, 1224–1226

12. Bodel, P. (1970) Studies on the mechanism of endogenous pyrogen production. I. Investigation of new protein synthesis in stimulated human blood leukocytes. Yale J. Biol. Med. 43, 145–163

13. Murphy, P.A., Chesney, P.J. and Wood, W.B. (1974) Further purification of rabbit leukocyte pyrogen. J. Lab. Clin. Med. 83, 310–322

14. Dinarello, C.A., Goldin, N.P. and Wolff, S.M. (1974) Demonstration and characterization of two distinct human leukocytic pyrogens. J. Exp. Med. 139, 1369–1381

15. Cebula, T.A., Hanson, D.F., Moore, D.M. and Murphy, P.A. (1979) Synthesis of four endogenous pyrogens by rabbit macrophages. J. Lab. Clin. Med. 94, 95–105

16. Dumonde, D.C., Wolstencroft, R.A., Panayi, G.S., Matthew, M., Morley, J. and Howson, W.T. (1969) Lymphokines: non-antibody mediators of cellular immunity generated by lymphocyte activation. Nature (London) 224, 38–39

17. Gery, I., Gershon, R.K. and Waksman, B.H. (1971) Potentiation of cultured mouse thymocyte responses by factors released by peripheral leukocytes. J. Immunol. 107, 1778–1780

18. Gery, I., Gershon, R.K. and Waksman, B.H. (1972) Potentiation of the T lymphocyte response to mitogens I. The responding cell. J. Exp. Med. **136**, 128–142

19. Gery, I. and Waksman, B.H. (1972) Potentiation of the T-lymphocyte response to mitogens. II. The cellular source of potentiating mediator(s). J. Exp. Med. **136**, 143–155

20. Dayer, J.-M., Robinson, D.R. and Krane, S.M. (1977) Prostaglandin production by rheumatoid synovial cells: stimulation by a factor from human mononuclear cells. J. Exp. Med. **145**, 1399–1404

21. Saklatvala, J. (1981) Characterization of catabolin, the major product of pig synovial tissue that induces resorption of cartilage proteoglycan *in vitro*. Biochem. J. **199**, 705–714

22. Meats, J.E., McGuire, M.B. and Russell, R.G.G. (1980) Human synovium releases a factor which stimulates chondrocyte production of PGE and plasminogen activator. Nature (London) **286**, 891–892

23. Horton, J.E., Raisz, L.G., Simmons, H.A., Oppenheim, J.J. and Mergenhagen, S.E. (1972) Bone resorbing activity in supernatant fluid from cultured human peripheral blood leukocytes. Science **177**, 793–795

24. Ritchie, D.G. and Fuller, G.M. (1983) Hepatocyte-stimulating factor: a monocyte-derived acute phase regulatory protein. Annu. N.Y. Acad. Sci. **408**, 490–502

25. Merriman, C.R., Pulliam, C.A. and Kampschmidt, R.F. (1977) Comparison of leukocytic pyrogen and leukocytic endogenous mediator. Proc. Soc. Exp. Biol. Med. **154**, 224–227

26. Rosenwasser, L.J., Dinarello, C.A. and Rosenthal, A.S. (1979) Adherent cell function in murine T-lymphocyte antigen recognition. IV. Enhancement of murine T-cell antigen recognition by human leukocytic pyrogen. J. Exp. Med. **150**, 709–714

27. Murphy, P.A., Simon, P.L. and Willoughby, W.F. (1980) Endogenous pyrogens made by rabbit peritoneal exudate cells are identical with lymphocyte activating factors made by rabbit alveolar macrophages. J. Immunol. **124**, 2498–2501

28. Aarden, L.A., Burnner, T.K., Cerottini, J.C., Dayer, J.M., de Weck, A.L., Dinarello, C.A., Di Sabato, G., Farrar, J.J., Gery, I., Gillis, S. et al. (1979) Revised nomenclature for antigen non-specific T cell proliferation and helper factors. J. Immunol. **123**, 2928–2929

29. Dinarello, C.A. (1979) Was the original endogenous pyrogen interleukin-1? In Interleukin-1, Inflammation and Disease (Bomford, R. and Henderson, B., eds.), pp. 17–28, Elsevier Press, North Holland

30. Ibelgaufts, H. (1995) Dictionary of Cytokines. VCH, Weinheim

31. Morrison, D.C. and Ryan, J.L. (1987) Endotoxins and disease mechanisms. Annu. Rev. Med. **38**, 417–432

32. Beutler, B., Milsark, I.W. and Cerami, A.C. (1985) Passive immunization against cachectin/tumor necrosis factor protects mice from lethal effects of endotoxin. Science **229**, 869–871

33. Kohler, J., Heumann, D., Garotta, G., Le, R.D., Bailat, S., Barras, C., Baumgartner, J.D. and Glauser, M.P. (1993) IFN-gamma involvement in the severity of gram-negative infections in mice. J. Immunol. **151**, 916–921

34. Berg, D.J., Kuhn, R., Rajewsky, K., Muller, W., Menon, S., Davidson, N., Grunig, G. and Rennick, D. (1995) Interleukin-10 is a central regulator of the response to LPS in murine models of endotoxic shock and the Shwartzman reaction but not endotoxin tolerance. J. Clin. Invest. **96**, 2339–2347

35. Doran, J.E. (1992) Biological effects of endotoxin. Curr. Stud. Hematol. Blood Transfus. **59**, 66–99

36. Manthey, C.L. and Vogel, S.N. (1994) Interactions of lipopolysaccharide with macrophages. Immunol. Ser. **60**, 63–81

37. Zeigler-Heitbrock, H.W. (1995) Review article: molecular mechanism in tolerance to lipopolysaccharide. J. Inflamm. **45**, 13–26

38. Janeway, C.A. (1992) The immune system evolved to discriminate infectious non-self from non-infectious self. Immunol. Today **13**, 11–16

39. Lei, M.-G. and Morrison, D.C. (1988) Specific endotoxic lipopolysaccharide binding proteins on murine splenocytes. I. Detection of lipopolysaccharide-binding sites on splenocytes and splenocyte subpopulations. J. Immunol. **141**, 996–1005

40. Hunt, J.S., Soares, M.J., Lei, M.-G., Smith, R.N., Wheaton, D., Atherton, R.A. and Morrison, D.C. (1988) Products of lipopolysaccharide-activated macrophages (tumor necrosis factor-α, transforming growth factor-β) but not lipopolysaccharide modify DNA synthesis by rat trophoblast cells exhibiting the 80 kDa lipopolysaccharide-binding protein. J. Immunol. **143**, 1606–1613

41. Lei, M.-G., Stimpson, S.A. and Morrison, D.C. (1991) Specific endotoxic lipopolysaccharide-binding receptors on murine splenocytes. III. Binding specificity and characterization. J. Immunol. **147**, 1925–1932

42. Roeder, D.J., Lei, M.-G. and Morrison, D.C. (1989) Endotoxic lipopolysaccharide-specific binding proteins on lymphoid cells of various animal species: association with endotoxin susceptibility. Infect. Immun. **57**, 1054–1058

43. Haeffner-Cavaillon, N., Chaby, R., Cavaillon, J.M. and Szabo, L. (1982) Lipopolysaccharide receptor on rabbit peritoneal macrophages. J. Immunol. **128**, 1950–1954

44. Haeffner-Cavaillon, N., Cavaillon, J.M., Etievant, M., Lebbar, S. and Szabo, L. (1985) Specific binding of endotoxin to human monocytes and mouse macrophages: serum requirements. Cell Immunol. **91**, 119–131

45. Warner, S.J.C., Savage, N. and Mitchell, D. (1985) Characteristics of lipopolysaccharide interaction with human peripheral blood monocytes. Biochem. J. **232**, 379–383

46. Flebbe, L.M., Chapes, S.K. and Morrison, D.C. (1990) Activation of C3H/HeJ macrophage tumoricidal activity and cytokine release by R-chemotype lipopolysaccharide preparations. J. Immunol. **145**, 1505–1511

47. Hailing, J.L., Hamill, D.R., Lei, M.-G. and Morrison, D.C. (1992) Identification and characterization of lipopolysaccharide-binding proteins on human peripheral blood cell populations. Infect. Immun. **60**, 845–852

48. Dziarski, R. (1991) Demonstration of peptidoglycan binding sites on lymphocytes and macrophages by photoaffinity cross-linking. J. Biol. Chem. **266**, 4713–4718

49. Dziarski, R. (1991) Peptidoglycan and lipopolysaccharide bind to the binding site on lymphocytes. J. Biol. Chem. **266**, 4719–4725

50. Dziarski, R. (1992) Letter to the editor. J. Immunol. **148**, 1590–1591

51. Dziarski, R. and Gupta, D. (1994) Heparan, sulfated heparinoids, and lipoteichoic acids bind to the 70 kDa peptidoglycan/lipopolysaccharide receptor protein on lymphocytes. J. Biol. Chem. **269**, 2100–2110

52. Bright, S.W., Chen, T.-Y., Flebbe, L.M., Lei, M.-G. and Morrison, D.C. (1990) Generation and characterization of hamster–mouse hybridomas secreting monoclonal antibodies with specificity for lipopolysaccharide receptor. J. Immunol. **145**, 1–7

53. Chen, T.-Y., Bright, S.W., Place, J.L., Russell, S.W. and Morrison, D.C. (1990) Induction of macrophage-mediated tumor cytotoxicity by a hamster monoclonal antibody with specificity for lipopolysaccharide receptor. J. Immunol. **145**, 8–12

54. Green, S.J., Chen, T.-Y., Crawford, R.M., Nacy, C.A., Morrison, D.C. and Meltzer, M.S. (1992) Cytotoxic activity and production of toxic nitrogen oxides by macrophages treated with IFN-γ and monoclonal antibodies against the 73 kDa lipopolysacharide receptor. J. Immunol. **149**, 2069–2075

55. LeBlanc, P.A. (1994) Activation of macrophages for cytolysis of virally-infected cells by monoclonal antibody to the 73 kDa lipopolysaccharide receptor. J. Leukoc. Biol. **55**, 262–264

56. Morrison, D.C., Silverstein, R., Bright, S.W., Chen, T.-Y., Flebbe, L.M. and Lei, M.-G. (1990) Monoclonal antibody to mouse lipopolysacchride receptor protects mice against the lethal effects of endotoxin. J. Infect. Dis. **162**, 1063–1068

57. Dziarski, R. (1994) Cell-bound albumin is the 70-kDa peptidoglycan-, lipopolysaccharide- and lipoteichoic acid-binding protein on lymphocytes and macrophages. J. Biol. Chem. **269**, 20431–20436

58. Schletter, J., Brade, H., Brade, L., Kruger, C., Loppnow, H., Kusumoto, S., Rietschel, E.T., Flad, H.-D. and Ulmer, A.J. (1995) Binding of lipopolysaccharide to an 80-kilodalton membrane protein of human cells is mediated by soluble CD14 and LPS-binding protein. Infect. Immun. **63**, 2576–2580

59. Kirkland, T.N., Virca, G.D., Kuus-Reichel, T., Multer, F.K., Kim, S.Y., Ulevitch, R.J. and Tobias, P.S. (1990) Identification of lipopolysaccharide-binding proteins in 70Z/3 cells by photoaffinity cross-linking. J. Biol. Chem. **265**, 9520–9525

60. Lei, M.-G., Qureshi, N. and Morrison, D.C. (1993) Lipopolysaccharide (LPS) binding to 73-kDa and 38-kDa surface proteins on lymphoreticular cells: preferential inhibition of LPS binding to the former by *Rhodopseudomonas sphaeroides* lipid A. Immunol. Lett. **36**, 245–250

61. Lebbar, S., Cavaillon, J.-M., Caroff, M., Ledur, A., Brade, H., Sarfati, R. and Haeffner-Cavaillon, N. (1986) Molecular requirement for interleukin 1 induction by lipopolysaccharide-stimulated human monocytes: involvement of the heptosyl-2-keto-3-deoxyoctulosonate region. Eur. J. Immunol. **16**, 87–91

62. Hampton, R.Y., Golenbock, D.T. and Raetz, C.R.H. (1988) Lipid A binding sites in membranes of macrophage tumor cells. J. Biol. Chem. **263**, 14802–14807

63. Kirikae, T., Inada, K., Hirata, M., Yoshida, M., Kondo, S., and Hisatsune, K. (1988) Identification of Re lipopolysaccharide-binding protein on murine erythrocyte membrane. Microbiol. Immunol. **32**, 33–44

64. Kirikae, T., Kirikae, F., Schade, U.F., Yoshida, M., Kondo, S., Hisatsune, K., Nishikawa, S.-I. and Rietschel, E.T. (1991) Detection of lipopolysaccharide-binding proteins on membrane of murine lymphocyte and macrophage-like cell lines. FEMS Microbiol. Immunol. **76**, 327–336

65. Fukuse, S., Maeda, T., Webb, D.R. and Devens, B.H. (1995) A 69-kDa membrane protein associated with lipopolysaccharide (LPS)-induced signal transduction in the human monocytic cell line THP-1. Cell. Immunol. **164**, 248–254

66. Hampton, R.Y., Golenbock, D.T., Penman, M., Krieger, M. and Raetz, C.R.H. (1991) Recognition and plasma clearance of endotoxin by scavenger receptors. Nature (London) **352**, 342–344

67. Wright, S.D. and Jong, M.T.C. (1986) Adhesion-promoting receptors on human macrophages recognize *Escherichia coli* by binding to lipopolysaccharide. J. Exp. Med. **164**, 1876–1888

68. Wright, S.D., Detmers, P.A., Aida, Y., Adamowski, R., Anderson, D.C., Chad, Z., Kabbash, L.G. and Pabst, M.J. (1990) CD18-deficient cells respond to lipopolysaccharide *in vitro*. J. Exp. Med. **144**, 2566–2571

69. Ingalls, R.R. and Golenbock, D.T. (1995) CD11c/CD18, a transmembrane signalling receptor for lipopolysaccharide. J. Exp. Med. **181**, 1473–1479

70. Lei, M.-G., and Chen, T.-Y. (1992) Cellular membrane receptors for lipopolysaccharide. In Bacterial Endotoxic Lipopolysaccharides Vol.1. Molecular Biochemistry and Cellular Biology (Morrison, D.C. and Ryan, J.L., eds.), pp. 254–267, CRC Press, Boca Raton

71. Morrison, D.C., Lei, M.-G., Kirikae, T. and Chen, T.-Y. (1993) Endotoxin receptors on mammalian cells. Immunobiology **187**, 212–226

72. Tobias, P., Soldau, K. and Ulevitch, R. (1986) Isolation of a lipopolysaccharide-binding acute phase reactant from rabbit serum. J. Exp. Med. **164**, 777–793

73. Tobias, P.S., Soldau, K. and Ulevitch, R.J. (1989) Identification of a lipid A binding site in the acute phase reactant lipopolysaccharide binding protein. J. Biol. Chem. **264**, 10867–10871

74. Schumann, R.R., Leong, S.R., Flaggs, G.W., Gray, P.W., Wright, S.D., Mathison, J.C., Tobias, P.S. and Ulevitch, R.J. (1990) Structure and function of lipopolysaccharide binding protein. Science **249**, 1429–1433

75. Tobias, P.S., Mathison, J.C. and Ulevitch, R.J. (1988) A family of lipopolysaccharide binding proteins involved in responses to gram-negative sepsis. J. Biol. Chem. **263**, 13479–13481

76. Gray, P.W., Flaggs, G., Leong, S.R., Gumina, R.J., Weiss, J., Ooi, C.E. and Elsbach, P. (1989) Cloning of the cDNA of a human neutrophil bactericidal protein. Structure and functional correlations. J. Biol. Chem. **264**, 9505–9509

77. Day, J.R., Albers, J.J., Lofton-Day, C.E., Gilbert, T.L., Ching, A.F.T., Grant, F.J., O'Hara, P., Marcovina, S.M. and Adolphson, J.L. (1994) Complete cDNA encoding human phospholipid transfer protein from human endothelial cells. J. Biol. Chem. **269**, 9388–9391

78. Ramadori, G., Meyer zum Buschenfelde, K.-H., Tobias, P.S., Mathison, J.C. and Ulevitch, R.J. (1990) Biosynthesis of lipopolysaccharide binding protein in rabbit hepatocytes. Pathobiology **58**, 89–94

79. Su, G.L., Freeswick, P.D., Geller, D.A., Wang, Q., Shapiro, R.A., Wan, Y.-H., Billiar, T.R., Tweardy, D.J., Simmons, R.L. and Wang, S.C. (1994) Molecular cloning, characterization, and tissue distribution of rat lipopolysaccharide binding protein. J. Immunol. **153**, 743–752

80. Tobias, P.S. and Ulevitch, R.J. (1993) Lipopolysaccharide binding protein and CD14 in LPS-dependent macrophage activation. Immunobiology **187**, 227–232

81. Gallay, P., Heumann, D., Le Roy, D., Barras, C. and Glauser, M.P. (1994) Mode of action of anti-lipopolysaccharide-binding protein antibodies for prevention of endotoxic shock in mice. Proc. Natl. Acad. Sci. U.S.A. **91**, 7922–7926

82. Grube, B.J., Cochrane, C.G., Ye, R.D., Ulevitch, R.J. and Tobias, P.S. (1994) Cytokine and dexamethasone regulation of lipopolysaccharide binding protein (LBP) expression in human hepatoma (HepG2) cells. J. Biol. Chem. **269**, 8477–8482

83. Martin, T.R., Mathison, J.C., Tobias, P.S., Leturcq, D.J., Moriarty, A.M., Maunder, R.J. and Ulevitch, R.J. (1992) Lipopolysaccharide binding protein enhances the responsiveness of alveolar macrophages to bacterial lipopolysaccharide. J. Clin. Invest. **90**, 2209–2219

84. Matthison, J.C., Tobias, P.S., Wolfson, E. and Ulevitch, R.J. (1992) Plasma lipopolysaccharide (LPS)-binding protein: a key component in macrophage recognition of Gram-negative LPS. J. Immunol. **149**, 200–206

85. Corradin, S.B., Mauel, J., Ulevitch, R.J. and Tobias, P.S. (1992) Enhancement of murine macrophage binding of and response to bacterial lipopolysaccharide (LPS) by LPS-binding protein. J. Leukoc. Biol. **52**, 363–368

86. Wright, S.D., Ramos, R.A., Hermanowski-Vosatka, A., Rockwell, P. and Detmers, P.A. (1991) Activation of the adhesive capacity of CR3 on neutrophils by endotoxin dependence on lipopolysaccharide binding protein and CD14. J. Exp. Med. **173**, 1281–1286

87. Surette, M.E., Palmantier, R., Gosselin, J. and Borgeat, P. (1993) Lipopolysaccharides prime whole human blood and isolated neutrophils for the increased synthesis of 5-lipoxygenase products by enhancing arachidonic acid availability: involvement of the CD14 antigen. J. Exp. Med. **178**, 1347–1355

88. Leeuwenberg, J.F.M., Dentener, M.A. and Buurman, W.A. (1994) Lipopolysaccharide LPS-mediated soluble TNF receptor release and TNF expression by monocytes. J. Immunol. **152**, 5070–5076

89. Gallay, P., Heumann, D., Le Roy, D., Barras, C. and Glauser, M.P. (1993) Lipopolysaccharide-binding protein as a major plasma protein responsible for endotoxin shock. Proc. Natl. Acad. Sci. U.S.A. **90**, 9935–9938

90. Pugin, J., Schurer-Maly, C.C., Leturcq, D., Moriarty, A., Ulevitch, R.J. and Tobias, P.S. (1993) Lipopolysaccharide (LPS) activation of human endothelial and epithelial cells is mediated by LPS binding protein and CD14. Proc. Natl. Acad. Sci. U.S.A. **90**, 2744–2748

91. Kirkland, T.N., Finley, F., Leturcq, D., Moriarty, A., Lee, J.-D., Ulevitch, R.J. and Tobias, P.S. (1993) Analysis of lipopolysaccharide binding by CD14. J. Biol. Chem. **268**, 24818–24823

92. Han, J., Mathison, J., Ulevitch, R. and Tobias, P. (1994) LPS binding protein, truncated at Ile-197, binds LPS but does not transfer LPS to CD14. J. Biol. Chem. **269**, 8172–8175

93. Tobias, P.S., Soldau, K., Kline, L., Lee, J.D., Kato, K., Martin, T.P. and Ulevitch, R.J. (1993) Crosslinking of lipopolysaccharide to CD14 on THP-1 cells mediated by lipopolysaccharide binding protein. J. Immunol. **150**, 3011–3021

94. Hailman, E., Lichtenstein, H.S., Wurfel, M.M., Miller, D.S., Johnson, D.A., Kelley, M., Busse, L.A., Zukowski, M.M. and Wright, S.D. (1994) Lipopolysaccharide (LPS)-binding protein accelerates the binding of LPS to CD14. J. Exp. Med. **179**, 269–277

94a. Katz, S.S., Chen, K., Chen, S., Doerfler, M.E., Elsbach, P. and Weiss, J. (1996) Potent CD14-mediated signalling of human leukocytes by *Escherichia coli* can be mediated by interaction of whole bacteria and host cells without extensive prior release of endotoxin. Infect. Immun. **64**, 3592–3600

95. Weiss, J., Elsbach, P., Shu, C., Castillo, J., Grinna, L., Horwitz, A. and Theofan, G. (1992) Human bactericidal/permeability-increasing protein and a recombinant NH2-terminal fragment cause killing of serum-resistant Gram-negative bacteria in whole blood and inhibit tumor necrosis factor release induced by bacteria. J. Clin. Invest. **90**, 1122–1130

96. Gazzano-Santoro, H., Meszaros, K., Birr, C., Carroll, S.F., Theofan, G., Horwitz, A.H., Lim, E., Aberle, S., Kasler, H. and Parent, J.B. (1994) Competition between rBP123, a recombinant fragment of bactericidal/permeability-increasing protein, and lipopolysaccharide binding protein (LBP) for binding to LPS and Gram-negative bacteria. Infect. Immun. **62**, 1185–1191

97. Theofan, G., Horwitz, A.H., Williams, R.E., Liu, P.-S., Chan, I., Birr, C., Carroll, S.F., Meszaros, K., Parent, J.B., Kasler, H., Aberle, S., Trown, P.W. and Gazzano-Santoro, H. (1994) An amino-terminal fragment of human lipopolysaccharide-binding protein retains lipid A binding but not CD14-stimulatory activity. J. Immunol. **152**, 3623–3629

98. Hailman, E., Albers, J.J., Wolfbauer, G., Tu, A.Y. and Wright, S.D. (1996) Neutralization and transfer of lipopolysaccharide by phospholipid transfer protein. J. Biol. Chem. **271**, 12172–12178

98a. Lamping, N., Hoess, A., Yu, B., Park, T.C., Kirschning, C.-J., Pfeil, D., Reuter, D., Wright, S.D., Herrmann, F. and Schumann, R.R. (1996) Effects of site-directed mutagenesis of basic residues (Arg94, Lys95, Lys99) of lipopolysaccharide (LPS)-binding protein on binding and transfer of LPS and subsequent immune cell activation. J. Immunol. **157**, 4648–4656

99. Hoess, A., Watson, S., Siber, G.R. and Liddington, R. (1993) Crystal structure of an endotoxin-neutralizing protein from the horse-shoe crab. Limulus anti-LPS factor at 1.5Å resolution. EMBO J. **12**, 3351–3356

100. Wright, S.D., Ramos, R.A., Patel, M. and Miller, D.S. (1992) Septin: a factor in plasma that opsonizes lipopolysaccharide-bearing particles for recognition by CD14 on phagocytes. J. Exp. Med. **176**, 719–727

101. Wright, S.D., Tobias, P.S., Ulevitch, R.J. and Ramos, R.A. (1989) Lipopolysaccharide (LPS) binding protein opsonizes LPS-bearing particles by a novel receptor on macrophages. J. Exp. Med. **170**, 1231–1241

102. Wright, S.D., Ramos, R.A., Tobias, P.S., Ulevitch R.J. and Mathison, J.C. (1990) CD14, a receptor for complexes of lipopolysaccharide (LPS) and LPS binding proteins. Science **249**, 1431–1433

103. Hogg, N. and Horton, M.A. (1986) Myeloid antigens: new and previously defined clusters. In Leukocyte Typing III, White Cell Differentiation Antigens (McMichael, A.J., ed.), pp. 576–602, Oxford University Press, Oxford

104. Van Voorhis, W.C., Steinman, R.M., Hair, L.S., Luban, J., Witmer, M.D., Koide, S. and Cohn, Z.A. (1983) Specific anti-mononuclear phagocyte monoclonal antibodies. Applications to the purification of dendritic cells and the tissue localization of macrophages. J. Exp. Med. **158**, 126–145

105. Zeitler-Heitbrock, H.W.L., Pechumer, J., Petersman, I., Durieux, J.J., Vita, N., Labeta, M.O. and Strobel, M. (1994) CD14 is expressed and functional in human B cells. Eur. J. Immunol. **24**, 1937–1940

106. Goyert, S.M., Ferrero, E., Rettig, W.J., Yenamandra, A.K., Obata, F. and LeBeau, M.M. (1990) The CD14 monocyte differentiation antigen maps to a region encoding growth factors and receptors. Science **239**, 497–500

107. Matsuura, K., Setoguchi, M., Nasu, N., Higuchi, Y., Yoshida, S., Akizuki, S. and Yamamoto, S. (1989) Nucleotide and amino acid sequences of the mouse CD14 gene. Nucleic Acids Res. **17**, 2132

108. Haziot, A., Chen, S., Ferrero, E., Low, M.G., Silber, R. and Goyert, S.M. (1988) The monocyte differentiation antigen, CD14, is anchored to the cell membrane by a phosphatidylinositol linkage. J. Immunol. **141**, 547–552

109. Simmons, D.L., Tan, S., Tenen, D.G., Nicholson-Weller, A. and Seed, B. (1989) Monocyte antigen CD14 is a phospholipid-anchored membrane protein. Blood **73**, 284–289

110. Heumann, D., Gallay, P., Barras, C., Zaesch, P., Ulevitch, R.J., Tobias, P.S., Glauser, M.-P. and Baumgartner, J.D. (1992) Control of lipopolysaccharide (LPS) binding and LPS-induced tumor necrosis factor secretion in human peripheral blood monocytes. J. Immunol. **148**, 3505–3512

111. Couturier, C., Haeffner-Cavaillon, N., Caroff, M. and Kazatchkine, M.D. (1991) Binding sites for endotoxins (lipopolysaccharides) on human monocytes. J. Immunol. **147**, 1899–1904

112. Dentener, M.A., Bazil, V., Von Asmuth, E.J.U., Ceska, M. and Buurman, W.A. (1993) Involvement of CD14 in lipopolysaccharide-induced tumor necrosis factor-α, IL-6 and IL-8 release by human monocytes and alveolar macrophages. J. Immunol. **150**, 2885–2891

113. Kielian, T.L. and Blecha, F. (1995) CD14 and other recognition molecules for lipopolysaccharide: a review. Immunopharmacology **29**, 187–205

114. Ulevitch, R.J. and Tobias, P.S. (1995) Receptor-dependent mechanisms of cell stimulation by bacterial endotoxin. Annu. Rev. Immunol. **13**, 437–457

115. Viriyakosol, S. and Kirkland, T.N. (1996) The N-terminal half of membrane CD14 is a functional cellular lipopolysaccharide receptor. Infect. Immun. **64**, 653–656

116. Lee, J.D., Kato, K., Tobias, P.S., Kirkland, T.N. and Ulevitch, R.J. (1992) Transfection of CD14 into 70Z/3 cells dramatically enhances the sensitivity to complexes of lipopolysaccharide (LPS) and LPS binding protein. J. Exp. Med. **175**, 1697–1705

117. Ferrero, E., Jiao, D., Tsuberi, B.Z., Tesio, L., Rong, G.W., Haziot, A. and Goyert, S.M. (1993) Transgenic mice expressing human CD14 are hypersensitive to lipopolysaccharide. Proc. Natl. Acad. Sci. U.S.A. **90**, 2380–2384

118. Slack, J., McMahan, C.J., Waugh, S., Schooley, K., Spriggs, M.K., Sims, J.E. and Dower, S.K. (1993) Independent binding of interleukin-1α and interleukin-1β to type I and type II interleukin-1 receptors. J. Biol. Chem. **268**, 2513–2524

119. Martin, T.R., Mongovin, S.M., Tobias, P.S., Mathison, J.C., Moriarty, A.M., Leturcq, D.J. and Ulevitch, R.J. (1994) The CD14 differentiation antigen mediates the development of endotoxin responsiveness during differentiation of mononuclear phagocytes. J. Leukoc. Biol. **56**, 1–9

120. Zhang, D.-E., Hetherington, C.J., Gonzalez, D.A., Chen, H.-M. and Tenen, D.G. (1994) Regulation of CD14 expression during monocytic differentiation induced with 1α,25-dihydroxyvitamin D$_3$. J. Immunol. **153**, 3276–3284

121. Passlick, B., Flieger, D. and Ziegler-Heitbrock, H.W.L. (1989) Identification and characterization of a novel monocyte subpopulation in human peripheral blood. Blood **74**, 2527–2534

122. Andreesen, R., Brugger, W., Scheibenbogen, C., Kreutz, M., Leser, H.-G., Rehm, A. and Lohr, G.W. (1990) Surface phenotype analysis of human monocyte to macrophage maturation. J. Leukoc. Biol. **47**, 490–497

123. Ziegler-Heitbrock, H.W.L. and Ulevitch, R.J. (1993) CD14: cell surface differentiation receptor and differentiation marker. Immunol. Today **14**, 121–125

124. Marchant, A., Duchow, J., Delville, J.-P. and Goldman, M. (1992) Lipopolysaccharide induces upregulation of CD14 molecule on monocytes in human whole blood. Eur. J. Immunol. **22**, 1663–1665

125. Brugger, W., Reinhardt, D., Galanos, C. and Andreesen, R. (1991) Inhibition of in vitro differentiation of human monocytes to macrophages by lipopolysaccharide (LPS): phenotypic and functional analysis. Int. Immunol. **3**, 221–227

126. Birkenmaier, C., Hong, Y.S. and Horn, J.K. (1992) Modulation of the endotoxin receptor (CD14) in septic patients. J. Trauma **32**, 473–479

127. Kielian T.L., Ross, C.R., McVey, D.S., Chapes, S.K. and Blecha, F (1995) Lipopolysaccharide modulation of a CD14-like molecule on porcine alveolar macrophages. J. Leukoc. Biol **57**, 581–586

128. Ikewaki, N., Tamauchi, H. and Inoko, H. (1993) Modulation of cell surface antigens and regulation of phagocytic activity mediated by CD11b in the monocyte-like cell line U937 in response to lipopolysaccharide. Tissue Antigens **42**, 125–132

129. Zeigler-Heitbrock, H.W.L., Weber, C., Aepfelbacher, M., Ehlers, M., Schutt, C. and Haas, J.G. (1994) Distinct patterns of differentiation induced in the monocytic cell line Mono Mac 6. J. Leukoc. Biol. **55**, 73–80

130. Bazil, V. and Strominger, J.L. (1991) Shedding as a mechanism of down-modulation of CD14 on stimulated human monocytes. J. Immunol. **147**, 1567–1574

131. Wright, S.D. (1991) CD14 and the immune response to lipopolysaccharide. Science **252**, 1321–1322

132. Landmann, R., Knopf, H.-P., Link, S., Sansano, S., Schumann, R. and Zimmerli, W. (1996) Human monocyte CD14 is upregulated by lipopolysaccharide. Infect. Immun. **64**, 1762–1769

133. Fearns, C., Kravchenko, V.V., Ulevitch, R.J. and Loskutoff, D.J. (1995) Murine CD14 gene expression in vivo: extramyeloid synthesis and regulation by lipopolysaccharide. J. Exp. Med. **181**, 857–866

134. Jack, R.S., Grunwald, U., Stelter, F., Workalemahu, G. and Schutt, C. (1995) Both membrane-bound and soluble forms of CD14 bind to Gram-negative bacteria. Eur. J. Immunol. **25**, 1436–1441

135. Landmann, R., Wesp, M. amd Obrecht, J.P. (1990) Cytokine regulation of the myeloid glycoproteins CD14. Pathobiology **59**, 131–135

136. Landmann, R., Ludwig, C., Obrist, R. and Obrecht, J.P. (1991) Effects of cytokines and lipopolysaccharide on CD14 antigen expression in human monocytes and macrophages. J. Cell. Biochem. **47**, 317–329

137. Launer, R.P., Goyert, S.M., Geha, R.S. and Vercelli, D. (1990) Interleukin-4 downregulates the expression of CD14 in normal human monocytes. Eur. J. Immunol. **20**, 2375–2381

138. Ruppert, J., Friedrichs, D., Xu, H. and Peters, J.H. (1991) IL-4 decreases the expression of the monocyte differentiation marker CD14, paralleled by an increasing accessory potency. Immunobiology **182**, 449–464

139. Ikewaki, N. and Inoko, H. (1991) Induction of CD14 antigen on the surface of U937 cells by an interleukin-6 autocrine mechanism after culture with formalin-fixed Gram-negative bacteria. Tissue Antigens **38**, 117–123

140. Labeta, M.O., Durieux, J.-J., Spagnoli, G., Fernandez, N., Wijdenes, J. and Herrmann, R. (1993) CD14 and tolerance to lipopolysaccharide: biochemical and functional analysis. Immunology **80**, 415–423

141. Hansen, P.B., Kjaersgaard, E., Johnsen, H.E., Gram, J., Pedersen, M., Nicolajsen, K. and Hansen, N.E. (1993) Different membrane expression of CD11b and CD14 on blood neutrophils following in vivo administration of growth factors. Br. J. Haemat. **85**, 50–56

142. de Waal Malefyt, R., Figdor, C.G., Huijbens, R., Mohan-Peterson, S., Bennett, B., Culpepper, J., Dang, W., Zurawski, G. and de Vries, J.E. (1993) Effects of IL-13 on phenotype, cytokine production, and cytotoxic function of human monocytes. J. Immunol. **151**, 6370–6377

143. Cosentino, G., Soprana, E., Thienes, C.P., Siccardi, A.G., Viale, G. and Vercelli, D. (1995) IL-13 down-regulates CD14 expression and TNF-α secretion in normal human monocytes. J. Immunol. **155**, 3145–3151

144. Ho, D.D., Roger, M.D., Pomerantz, R.J. and Kaplan, J.C. (1987) Pathogenesis of infection with human immunodeficiency virus. N. Engl. J. Med. **317**, 278–283

145. Pomerantz, R.J., Feinberg, M.B., Trono, D. and Baltimore, D. (1990) Lipopolysaccharide is a potent monocyte/macrophage-specific stimulator of human immunodeficiency virus type I expression. J. Exp. Med. **172**, 253–260

146. Bagasra, O., Wright, S.D., Seshamma, T., Oakes, J.W. and Pomerantz, R.J. (1992) CD14 is involved in control of human immunodeficiency virus type I expression in latently infected cells by lipopolysaccharide. Proc. Natl. Acad. Sci. U.S.A. **89**, 6285–6290

147. Montaner, L. J., Doyle, A.G., Collin, M., Herbein, G., Illei, P., James, W., Minty, A., Caput, D., Ferrara, P. and Gordon, S. (1993) Interleukin-13 inhibits human immunodeficiency virus type I production in primary blood-derived human macrophages in vitro. J. Exp. Med. **178**, 743–747

148. Beekhuizen, H., Blokland, I., Corsel-van Tilburg, A.J., Koning, F. and van Furth, R. (1991) CD14 contributes to the adherence of human monocytes to cytokine-stimulated human endothelial cells. J. Immunol. **147**, 3761–3767

149. Worthen, G.S., Avdi, N., Vukajlovich, S. and Tobias, P.S. (1992) Neutrophil adherence induced by lipopolysaccharide in vitro. J. Clin. Invest. **90**, 2526–2535

150. Lue, K.-H., Lauener, R.P., Winchester, R.J., Geha, R.S. and Vercelli, D. (1991) Engagement of CD14 on human monocytes terminates T cell proliferation by delivering a negative signal to T cells. J. Immunol. **147**, 1134–1138

151. Chomarat, P., Riossan, M.C., Pin, J.J., Banchereau, J. and Miossec, P. (1995) Contribution of IL-1, CD14 and CD13 in the increased IL-6 production induced by in vitro monocyte-synoviocyte interactions. J. Immunol. **155**, 3645–3652

152. Lynn, W.A., Liu, Y. and Golenbock, D.T. (1993) Neither CD14 nor serum is absolutely necessary for activation of mononuclear phagocytes by bacterial lipopolysaccharide. Infect. Immun. **61**, 4452–4461

153. Cohen, L., Haziot, A., Shen, D.R., Lin, X.-Y., Sia, C., Harper, R., Silver, J. and Goyert, S.M. (1995) CD14-independent response to LPS requires a serum factor which is absent from neonates. J. Immunol. **155**, 5337–5342

154. Maliszewski, C.R., Ball, E.D., Graziano, R.F. and Fanger, M.W. (1985) Isolation and characterization of My23, a myeloid cell-derived antigen reactive with the monoclonal antibody AML-2-23. J. Immunol. **135**, 1929–1936

155. Bazil, V., Horejsi, V., Baudys, M., Kristova, H., Strominger, J., Kostka, W. and Hilgert, I. (1986) Biochemical characterization of a soluble form of the 53 kDa monocyte surface antigen. Eur. J. Immunol. **16**, 1583–1589

156. Grunwald, U., Kruger, C., Westermann, J., Lukowsky, A., Ehlers, M. and Schutt, C. (1994) An enzyme linked immunoassay for the quantification of solubilized CD14 in biological fluids. J. Immunol. Methods **155**, 225–232

157. Stelter, F., Pfister, M., Bernheiden, M., Jack, R.S., Bufler, P., Engelmann, H. and Schutt, C. (1996) The myeloid differentiation antigen CD14 is N- and O-glycosylated. Contributions of N-linked glycosylation to different soluble CD14 isoforms. Eur. J. Biochem. **236**, 457–464

158. Bufler, P., Stiegler, G., Schuchmann, M., Hess, S., Kruger, C., Stelter, F., Eckerskorn, C., Schutt, C. and Engelmann, H. (1995) Soluble lipopolysaccharide receptor (CD14) is released via two different mechanisms from human monocytes and CD14 transfectants. Eur. J. Immunol. **25**, 604–610

159. Schutt, C., Schilling, T., Grunwald, U., Schonfeld, W. and Kruger, C. (1992) Endotoxin-neutralizing capacity of soluble CD14. Res. Immunol. **143**, 71–78

160. Frey, E.A., Miller, D.S., Jahr, T.G., Sundan, A., Bazil, V., Espevik, T., Finlay, B.B. and Wright, S.D. (1992) Soluble CD14 participates in the response of cells to lipopolysaccharide. J. Exp. Med. **176**, 1665–1671

161. Arditi, M., Zhou, J., Dorio, R., Rong, G.W., Goyert, S.M. and Kim, K.S. (1993) Endotoxin-mediated endothelial cell injury and activation: role of soluble CD14. Infect. Immun. **61**, 3149–3156

162. Haziot, A., Rong, G.-W., Silver J. and Goyert, S.M. (1993) Recombinant soluble CD14 mediates the activation of endothelial cells by lipopolysaccharide. J. Immunol. **151**, 1500–1507

163. Noel, R.F., Sato, T.T., Mendez, C., Johnson, M.C. and Pohlman, T.H. (1995) Activation of human endothelial cells by viable or heat-killed Gram-negative bacteria requires soluble CD14. Infect. Immun. **63**, 4046–4053

164. Loppnow, H., Steltner, F., Schonbeck, U., Schluter, C., Ernst, M., Schutt, C. and Flad, H.-D. (1995) Endotoxin activates human vascular smooth muscle cells despite lack of expression of CD14 mRNA or endogenous membrane CD14. Infect. Immun. **63**, 1020–1026

165. Golenock, D.T., Bach, R.R., Lichtenstein, H., Juan, T.S., Tadavarthy, A. and Moldow, C.F. (1995) Soluble CD14 promotes LPS activation of CD14-deficient PNH monocytes and endothelial cells. J. Lab. Clin. Med. **125**, 662–671

166. Landmann, R., Zimmerli, W., Sansano, S., Link, S., Hahn, A., Glauser, M.P. and Calandra, T. (1995) Increased circulating soluble CD14 is associated with high mortality in gram-negative septic shock. J. Infect. Dis. **171**, 639–644

167. Landmann, R., Reber, A.M., Sansano, S. and Zimmerli, W. (1996) Function of soluble CD14 in serum from patients with septic shock. J. Infect. Dis. **173**, 661–668

168. Leturcq, D.J., Moriarty, A.M., Talbott, G., Winn, R.K., Martin, T.R. and Ulevitch, R.J. (1996) Antibodies against CD14 protect primates from endotoxin-induced shock. J. Clin. Invest. **98**, 1533–1538

169. Haziot, A., Rong, G.-W., Bazil, V., Silver, J. and Goyert, S.M. (1994) Recombinant soluble CD14 inhibits LPS-induced tumor necrosis factor-α production by cells in whole blood. J. Immunol. **152**, 5868–5876

170. Haziot, A., Rong, G.W., Lin, X.-Y., Silver, J. and Goyert, S.M. (1995) Recombinant soluble CD14 prevents mortality in mice treated with endotoxin (lipopolysaccharide) J. Immunol. **154**, 6529–6532

171. Wurfel, M.M., Hailman, E. and Wright, S.D. (1995) Soluble CD14 acts as a shuttle in the neutralization of lipopolysaccharide (LPS) by LPS-binding protein and reconstituted high density lipoprotein. J. Exp. Med. **181**, 1743–1754

172. Tobias, P.S., Soldau, K., Gegner, J.A., Mintz, D. and Ulevitch, R.J. (1995) Lipopolysaccharide binding protein-mediated complexation of lipopolysaccharide with soluble CD14. J. Biol. Chem. **270**, 10482–10488

173. Gegner, J.A., Ulevitch, R.J. and Tobias, P.S. (1995) Lipopolysaccharide (LPS) signal transduction and clearance: dual roles for LPS binding protein and membrane CD14. J. Biol. Chem. **270**, 5320–5325

174. Yu, B. and Wright, S.D. (1996) Catalytic properties of lipopolysaccharide (LPS) binding protein. Transfer of LPS to soluble CD14. J. Biol. Chem. **271**, 4100–4105

175. Hailman, E., Vasselon, T., Kelley, M., Busse, L.A., Hu, M.C.-T., Lichtenstein, H.S., Detmers, P.A. and Wright, S.D. (1996) Stimulation of macrophages and neutrophils by complexes of lipopolysaccharide and soluble CD14. J. Immunol. **156**, 4384–4390

176. Delude, R.L., Savedra, R., Zhao, H., Thieringer, R., Yamamoto, S., Fenton, M.J. and Golenbock, D.T. (1995) CD14 enhances cellular responses to endotoxin without imparting ligand-specific recognition. Proc. Natl. Acad. Sci. U.S.A. **92**, 9288–9292

177. Juan, T.S.-C., Kelley, M.J., Johnson, D.A., Busse, L.A., Hailman, E., Wright, S.D. and Lichtenstein, H.S. (1995) Soluble CD14 truncated at amino acid 152 binds lipopolysaccharide (LPS) and enables cellular response to LPS. J. Biol. Chem. **270**, 1382–1387

178. McGinley, M.D., Narhi, L.O., Kelley, M.M., Davy, E., Robinson, J., Rhode, M.F., Wright, S.D. and Lichtenstein, H.S. (1995) CD14: physical properties and identification of an exposed site that is protected by lipopolysaccharide. J. Biol. Chem. **270**, 5213–5218

179. Todd, S.-C., Hailman, E., Kelley, M.J., Busse, L.A., Davy, E., Empig, C.J., Nahri, L.O., Wright, S.D. and Lichtenstein, H.S. (1995) Identification of a lipopolysaccharide binding domain in CD14 between amino acids 57 and 64. J. Biol. Chem. **270**, 5219–5224

180. Todd, S.-C., Hailman, E., Kelley, M.J., Wright, S.D. and Lichtenstein, H.S. (1995) Identification of a domain in soluble CD14 essential for lipopolysaccharide (LPS) signalling but not LPS binding. J. Biol. Chem. **270**, 17237–17242

181. Nicola, N.A. (1994) Guidebook to Cytokines and their Receptors. Oxford University Press, Oxford

182. Murukami, H., Hibi, M., Nakagawa, N., Nakagawa, T., Yasukawa, K., Yamanishi, K., Taga, T. and Kishimoto, T. (1993) IL-6-induced homodimerization of gp130 and associated activation of a tyrosine kinase. Science **260**, 1808–1810

183. Pugin, J., Heumann, D., Tomasz, A., Kravchenko, W., Akamatsu, Y., Nishijima, M., Glauser, M.P., Tobias, P.S. and Ulevitch, R.J. (1994) CD14 is a pattern recognition receptor. Immunity **1**, 509–516

184. Hardie, G. and Hanks, S (1995) The Protein Kinase Facts Book. Academic Press, London

185. Johnson, L.N., Noble, M.E.M. and Owen, D.J. (1996) Active and inactive protein kinases: Structural basis for regulation. Cell **85**, 149–158

186. Adams, D.O. (1992) LPS-initiated signal transduction pathways in macrophages. In Bacterial Endotoxic Lipopolysaccharides Vol.1. Molecular Biochemistry and Cellular Biology (Morrison, D.C. and Ryan, J.L., eds.), pp. 285–309, CRC Press, Boca Raton

187. Novogrosky, A., Vanichkin, A., Patya, M., Gazit, A., Osherov, N. and Levitzki, A. (1994) Prevention of lipopolysaccharide-induced lethal toxicity by tyrosine kinase inhibitors. Science **264**, 1319–1322

188. Thompson, N.T., Bonser, R.W. and Garland, L.G. (1991) Receptor-coupled phospholipase D and its inhibition. Trends Pharmacol. Sci. **12**, 404–408

189. Bursten, S.L., Harris, W.E. and Rice, G.C. (1996) Selective inhibition of phosphatidic acid synthesis: a novel approach to the treatment of sepsis and the systemic inflammatory response syndrome. In Novel Therapeutic Strategies in the Treatment of Sepsis (Morrison, D.C. and Ryan, J.L., eds.), pp. 199–226, Marcel Dekker, New York

190. Hardie, D.G. (1990) Biochemical Messengers: Hormones, Neurotransmitters and Growth Factors. Chapman and Hall, London

191. Prpic, V., Wciel, J.E., Somers, S.D., Diguiseppi, J., Gomas, S.L., Pizzo, S.V., Hamilton, T.A., Herman, B. and Adams, D.O. (1987) Effect of bacterial lipopolysaccharide on the hydrolysis of phosphatidylinositol-4,5-bisphosphate in murine peritoneal macrophages. J. Immunol. **139**, 526–533

192. Chang, Z.L., Novotney, A. and Suzuki, T. (1990) Phospholipase C and A_2 in tumoricidal activation of murine macrophage-like cell lines. FASEB J. **4**, A1753

193. Drysdale, P.E., Yapundich, R.A., Shin, M.L. and Shin, H.S. (1987) Lipopolysaccharide-mediated macrophage activation: the role of calcium in the generation of tumoricidal activity. J. Immunol. **139**, 951–956

194. Hurme, M., Viherluoto, J. and Nordstrom, T. (1992) The effect of calcium mobilization on LPS-induced IL-1β production depends on the differentiation stage of the monocytes/macrophages. Scand. J. Immunol. **36**, 507–511

195. Bosca, L., Marquez, C. and Martinez, C. (1991) B cell triggering by bacterial lipopeptide involves both translocation and activation of the membrane-bound form of protein kinase C. J. Immunol. **147**, 1463–1469

196. Shinji, H., Akagawa, K.S. and Yoshida, T. (1994) LPS induces selective translocation of protein kinase C-β in LPS-responsive mouse macrophages, but not in LPS-unresponsive mouse macrophages. J. Immunol. **153**, 5760–5771

197. Shapira, L., Yakashiba, S., Champagne, C., Amar, S. and Van Dyke, T.E. (1994) Involvement of protein kinase C and protein tyrosine kinase in lipopolysaccharide-induced TNFα and IL-1β production by human monocytes. J. Immunol. **153**, 1818–1824

198. Liu, M.K., Herrera-Velit, P., Brownsey, R.W. and Reiner, N.E. (1994) CD14-dependent activation of protein kinase C and mitogen-activated protein kinases (p42 and p44) in human monocytes treated with bacterial lipopolysaccharide. J. Immunol. **153**, 2642–2650

199. Herrera-Velit, P. and Reiner, N.E. (1996) Bacterial lipopolysaccharide induces the association and coordinate activation of p53/56*lyn* and phosphatidylinositol 3-kinase in human monocytes. J. Immunol. **156**, 1157–1165

200. Shinomiya, H., Hirata, H. and Nakano, M. (1991) Purification and characterization of the 65-kDa protein phosphorylated in murine macrophages by stimulation with bacterial lipopolysaccharide. J. Immunol. **146**, 3617–3625

201. Shinomiya, H., Hagi, A., Fukuzumi, M., Mizobuchi, M., Hirata, H. and Utsumi, S. (1995) Complete primary structure and phosphorylation site of the 65-kDa macrophage protein phosphorylated by stimulation with bacterial lipopolysaccharide. J. Immunol. **154**, 3471–3478

202. Davis, R.J. (1994) MAPKs: new JNK expands the group. Trends Biochem. Sci. **19**, 470–473

203. Weinstein, S.L., Gold, M.R. and DeFranco, A.L. (1991) Bacterial lipopolysaccharide stimulates protein tyrosine phosphorylation in macrophages. Proc. Natl. Acad. Sci. U.S.A. **88**, 4148–4152

204. Weinstein, S.L., Sanghera, J.S., Lemke, K., DeFranco, A.L. and Pelech, S.L. (1992) Bacterial lipopolysaccharide induces tyrosine phosphorylation and activation of mitogen-activated protein kinases in macrophages. J. Biol. Chem. **267**, 14955–14962

205. Weinstein, S.L., June, C.H. and De Franco, A.L. (1993) Lipopolysaccharide-induced protein tyrosine phosphorylation in human macrophages is mediated by CD14. J. Immunol. **151**, 3829–3838

206. Liu, M.K., Herrera-Velit, P., Brownsey, R.W. and Reiner, N.E. (1994) CD14-dependent activation of protein kinase C and mitogen-activated protein kinases (p42 and p44) in human monocytes treated with bacterial lipopolysaccharide. J. Immunol. **153**, 2642–2652

207. Han, J., Lee, J.D., Tobias, P.S. and Ulevitch, R.J. (1993) Endotoxin induces rapid protein tyrosine phosphorylation in 70Z/3 cells expressing CD14. J. Biol Chem. **268**, 25009–25014

208. Han, J., Lee, J.D., Bibbs, L. and Ulevitch, R.J. (1994) A MAP kinase targeted by endotoxin and hyperosmolarity in mammalian cells. Science **265**, 808–811

209. Nick, J.A., Avdi, N.J., Gerwins, P., Johnson, G.L. and Worthen, G.S. (1996) Activation of a p38 mitogen-activated protein kinase in human neutrophils by lipopolysaccharide. J. Immunol. **156**, 4867–4875

210. Arditi, M., Zhou, J., Torres, M., Durden, D.L., Stins, M. and Ki, K.S. (1995) Lipopolysaccharide stimulates the tyrosine phosphorylation of mitogen-activated protein kinases p44, p42, and p41 in vascular endothelial cells in a soluble CD14-dependent manner. J. Immunol. **155**, 3994–4003

211. Schumann, R.R., Pfeil, D., Kirschning, C., Scherzinger, G., Karawajew, L. and Hermann, F. (1996) Lipopolysaccharide induces the rapid tyrosine phosphorylation of the mitogen-activated protein kinases erk-1 and p38 in cultured human vascular endothelial cells requiring the presence of soluble CD14. Blood **87**, 2806–2814

212. Geppert, T.D., Whitehurst, C.E., Thompson, P. and Beutler, B. (1994) Lipopolysaccharide signals activation of tumor necrosis factor biosynthesis through the Ras/Raf-1/MEK/MAPK pathway. Mol. Med. **1**, 93–97

213. Reimann, T.D., Bucher, D., Hipskind, R.A., Krautwald, S., Lohmann-Matthes, M.-L. and Baccarini, M. (1994) Lipopolysaccharide induces activation of the Raf-1/MAP kinase pathway. J. Immunol. **153**, 5740–5746

214. Kyriakis, J., Banerjee, P., Nikolakaki, E., Dai, T., Rubie, E., Ahmad, M., Avruch, J. and Wodgett, J.R. (1994) The stress-activated protein kinase subfamily of c-jun kinases. Nature (London) **369**, 156–160

215. Hambleton, J., Weinstein, S.L., Lem, L. and De Franco, A.L. (1996) Activation of c-Jun N-terminal kinase in bacterial lipopolysaccharide-stimulated macrophages. Proc. Natl Acad. Sci. U.S.A. **93**, 2774–2778

216. Sanghera, J.S., Weinstein, S.L., Aluwalia, M., Girn, J. and Pelech, S.L. (1996) Activation of multiple proline-directed kinases by bacterial lipopolysaccharide in murine macrophages. J. Immunol. **156**, 4457–4465

217. Heine, H., Ulmer, A.J., Flad, H.-D. and Hauschildt, S. (1995) Lipopolysaccharide-induced change of phosphorylation of two cytosolic proteins in human monocytes is prevented by inhibitors of ADP-ribosylation. J. Immunol. **155**, 4899–4908

218. Hauschildt, S.H., Scheipers, P. and Bessler, W.G. (1994) Lipopolysaccharide-induced changes of ADP-ribosylation of a cytosolic protein in bone-marrow-derived macrophages. Biochem. J. **297**, 17–20

219. Lee, J.C. and Young, P.R. (1996) Role of CSBP/p38/RK stress response kinase in LPS and cytokine signalling mechanisms. J. Leukoc. Biol. **59**, 152–157

220. Kolesnick, R. and Dolde, D.W. (1994) The sphingomyelin pathway in tumor necrosis factor and interleukin-1 signalling. Cell **77**, 325–328

221. Dobrowsky, R.Y. and Hannum, Y.A. (1993) Ceramide-activated protein phosphatase: partial purification and relationship to protein phosphatase 2A. Adv. Lipid Res. **25**, 91–104

222. Liu, J., Mathias, S., Yang, Z. and Kolesnick, R.N. (1994) Renaturation and tumor necrosis factor-α stimulation of a 97-kDa ceramide-activated protein kinase. J. Biol. Chem. **269**, 3047–3052

223. Yao, B., Zhang, Y., Delikat, S., Mathias, S., Basu, S. and Kolesnick, R. (1995) Phosphorylation of Raf by ceramide-activated protein kinase. Nature (London) **378**, 307–310

224. Kolesnick, R. and Fuks, Z. (1995) Ceramide: A signal for apoptosis or mitogenesis? J. Exp. Med. **181**, 1949–1952

225. Laulederkind, S.J.F., Bielawska, A., Raghow, R., Hannun, Y.A. and Ballou, L.R. (1995) Ceramide induces interleukin-6 gene expression in human fibroblasts. J. Exp. Med. **182**, 599–604

226. Raines, M.A., Kolesnick, R.N. and Golde, D.W. (1993) Sphingomyelinase and ceramide activate mitogen-activated protein kinase in myeloid HL-60 cells. J. Biol. Chem. **268**, 14572–14575

227. Joseph, C.K., Wright, S.D., Bornmann, W.G., Randolph, J.T., Kumar, E.R., Bittman, R., Liu, J. and Kolesnick, R.N. (1995) Bacterial lipopolysaccharide has structural similarities to ceramide and stimulates ceramide-activated protein kinase in myeloid cells. J. Biol. Chem. **269**, 17606–17610

228. Wright, S.D. and Kolesnick, R.N. (1995) Does endotoxin stimulate cells by mimicking ceramide? Immunol. Today **16**, 297–302

229. Barber, S.A., Perera, P.-Y. and Vogel, S.N. (1995) Defective ceramide responses in C3H/HeJ (*Lpsd*) macrophages. J. Immunol. **155**, 2303–2305

230. Watson, J., Kelly, K., Largen, M. and Taylor, B.A. (1978) The genetic mapping of a defective LPS response gene in C3H/HeJ mice. J. Immunol. **120**, 422–424

231. Barber, S.A., Detore, G., McNally, R. and Vogel, S.N. (1996) Stimulation of the ceramide pathway partially mimics lipopolysaccharide-induced responses in murine peritoneal macrophages. J. Immunol. **64**, 3397–3400

232. Schindler, C. and Darnell, J.E. (1995) Transcriptional responses to polypeptide ligands: The JAK-STAT pathway. Annu. Rev. Biochem. **64**, 621–651

233. Thierfelder W.E., van Deursen, J.M., Yamamoto, K., Tripp, R.A., Sarawar, S.R., Carson, R.T., Sangster, M.Y., Vignali, D.A.A., Doherty, P.C., Grosveld, G. and Ihle, J.N. (1996) Requirement for STAT4 in interleukin-12-mediated responses of natural killer cells. Nature (London) **382**, 171–174

234. Kaplan, M.H., Sun, Y.-L., Hoey, T. and Grusby, M.J. (1996) Impaired IL-12 responses and enhanced development of Th2 cells in STAT-4-deficient mice. Nature (London) **382**, 174–177

235. Takeda K., Tanaka, T., Shi, W., Matsumoto, M., Minami, M., Kashiwamura, S.-I., Nakanishi, K., Yoshida, N., Kishimoto, T. and Akira, S. (1996) Essential role of STAT6 in IL-4 signalling. Nature (London) **380**, 627–630

236. Shimoda, K. van Deursen, J., Sangsterm, M.V., Sarawar, S.R., Carson, R.T., Tripp, R.A., Chu, C., Quelle, F.W., Nosaka, T., Vignali, D.A.A., Doherty, P.C., Grosveld, G., Paul, W.E. and Ihle, J.N. (1996) Lack of IL-4-induced Th2 response and IgE class switching in mice with disrupted *Stat6* gene. Nature (London) **380**, 630–633

237. Meraz, M.A., White, J.M., Sheenan, K.C.F., Bach, E.A., Rodig, S.J., Dighe, A.S., Kaplan, D.H., Riley, J.K., Greenlund, A.C., Campbell, D., Carver-Moore, K., DuBois, R.N., Clark, R., Aguet, M. and Schreiber, R.D. (1996) Targeted disruption of the *Stat1* gene in mice reveals unexpected physiological specificity in the JAK-STAT signalling pathway. Cell **84**, 431–442

238. Durbin, J.E., Hackenmiller, R., Simon, M.C. and Levy, D.E. (1996) Targeted disruption of the mouse Stat1 gene results in compromised immunity to viral disease. Cell **84**, 443–450

239. Heldin, C.-H. and Purton, M. (eds) (1996) Signal Transduction. Chapman and Hall, London

240. Manthey, C.L., Perera, P.-Y., Henricson, B.E., Hamilton, T.A., Qureshi, N. and Vogel, S.N. (1994) Endotoxin-induced early gene expression in C3H/HeJ (*Lpsd*) macrophages. J. Immunol. **153**, 2653–2663

7

Bacteria and cytokines: beyond lipopolysaccharide

"There could be no fairer destiny for any...theory than that it should point the way to a more comprehensive theory in which it lives on, as the limiting case."

Albert Einstein

7.1   Introduction

As has been highlighted in Chapters 5 and 6, research into the interactions between bacteria and cytokines has been dominated by endotoxin and LPS. The potent and multifarious actions of LPS, and proteins associated with this molecule in the bacterial membrane, have been an inspiration to two generations of biologists, and with the upsurge in interest in the cellular and molecular basis of sepsis and septic shock in the 1970s and 80s has come an understanding of the complex interactions that occur between LPS and other host factors (LBP, CD14, etc.) in the induction of cytokine synthesis. These interactions have been reviewed in detail in Chapter 6. As we will describe in more detail in Chapter 9, CD14 may be more than a pre-receptor for LPS. It is now suggested that this protein is not a conventional receptor, recognizing one, or a set, of structurally related ligands, but is a pattern-recognition receptor, able to bind to a certain class of bacterial constituents (complex 'lipocarbohydrates' and carbohydrates). In doing so it warns the host that it has been invaded and concurrently induces the acute-phase response by stimulating cytokine gene expression [1,2]. CD14 may only be one of a family of structurally unrelated bacterial-recognition receptors, which include the scavenger receptors on macrophages [3,4], receptors on B-cells able to recognize bacterial DNA [5], and possibly complement receptors such as CD46 which recognizes and binds measles virus and in doing so inhibits cellular IL-12 synthesis (see Chapter 4). The role of these additional 'pattern-recognition' receptors in inducing cytokine synthesis has not been investigated in any detail.

This fascination with LPS has overshadowed studies of the capacity of other bacterial constituents to induce cytokine synthesis. However, over the past decade a growing number of reports have revealed that bacteria contain many constituents, including lipids, carbohydrates, proteins, lipoproteins and glycoproteins, which also have the capacity to induce the transcription of cytokine genes. Among the most active of these bacterial 'constituents' are the exotoxins, some of which are able to induce myelomonocytic cells to synthesize pro-inflammatory cytokines at pico- to atomolar concentrations (10^{-12}–10^{-18} M). Many of these molecules clearly do not function by binding to CD14 and so it must be postulated that eukaryotic cells have a spectrum of receptors for bacterial constituents and exported proteins which, when occupied, can induce cytokine gene expression. As has been reviewed in Chapter 4, it is clear that cytokines can be, at the very least, divided into those that induce inflammatory mechanisms (pro-inflammatory cytokines such as IL-1, IL-8, TNFα) and those that have the capacity to inhibit inflammatory and immune mechanisms (anti-inflammatory cytokines such as IL-1ra, IL-4, IL-10, IL-13, TGFβ). LPS is universally regarded as a pro-inflammatory signal, and yet it can also induce the synthesis of anti-inflammatory cytokines such as IL-1ra [6] and the IL-10 receptor [7]. IL-10 has recently been claimed to be the major deactivator of human monocytes in human sepsis [8]. A growing number of bacterially derived proteins, again including exotoxins, have been reported to have the capacity to either inhibit LPS-induced cytokine synthesis or induce the synthesis of anti-inflammatory cytokines, suggesting that bacteria may have the capacity to control the cytokine networks of the host organism.

Chapter 2 has outlined the microbiology of multicellular organisms and the case has been made that most, if not all, multicellular organisms are symbionts, composed of host cells and a normal microbial flora. The startling statistic that the average human body contains 10^{13} eukaryotic cells, but also harbours 10^{14} bacteria [9], raises the obvious question of how we cope with such a huge number of potentially pro-inflammatory organisms. The authors believe that it is the presence of the various cytokine-inducing molecules produced by all bacteria which enables harmony to be established between the variety of bacteria that constitute the normal microflora and the epithelial cells of the host. This hypothesis has been discussed in recent review articles [10–13] and will be fully developed in this chapter and in the concluding chapter (Chapter 9). Such ideas are totally in line with the growing realization that a continuum exists at the molecular level between micro-organisms and eukaryotic cells. Much of this interaction between the prokaryotic and eukaryotic worlds is part of the competition between pathogenic organisms and the host (see recent reviews [14–16]). This area of overlap between the kingdoms Bacteria and Eukarya makes up the recently emerging discipline of

'Cellular Microbiology' [14], which is largely fired by a need to understand the molecular and cellular mechanisms of bacterial virulence. Our own view is that understanding how the host organism copes with its own normal microflora will provide answers to key questions about bacterial pathogenesis. The starting point for this chapter will be a consideration of how viruses have evolved cytokine-modulating proteins which allow them to evade host defences and which may also protect the host from its own defence mechanisms. This will be followed by a selective review of cytokine gene transgenics and the findings that certain transgenics are lethal due to the normal commensal bacteria inducing host pathology. This will be followed by an overview of the current literature on cytokine-inducing bacterial constituents, which will attempt to discriminate between those components that are part of the CD14 recognition system, those that are pro-inflammatory and those that may act to inhibit or control cytokine networks for the benefit of bacteria and host. Chapter 9 will attempt to integrate this information into a new paradigm of the homeostatic mechanisms that control the interactions between prokaryotic and eukaryotic cells in the mammal.

7.2   Viral control of the cytokine network

The common cold, influenza, measles, mumps, rubella, hepatitis B, polio, yellow fever, rabies — and the list of human infections caused by viruses goes on and on. Host defences against viral infections involve both the adaptive and acquired arms of the host defence system (Figure 7.1). The former includes monocytes, neutrophils, natural killer cells, the complement system and the family of cytokines known as interferons. In addition, the cellular populations described will produce a veritable panoply of cytokines. The acquired immune system involves the participation of CD4 Th_2 'helper' lymphocytes and B-lymphocytes to produce neutralizing antibodies, Th_1 'inflammatory' lymphocytes and the CD8 cytotoxic T-lymphocyte. The clonal expansion of these various lymphocyte populations involves the growth-promoting effects of a range of cytokines such as IL-2 and IL-4. Thus cytokines clearly play a major role in the host defence processes combatting viral infections and it therefore should not have come as a surprise that, in the early 1990s, reports appeared showing that certain viral genomes encoded proteins able to 'inhibit' specific cytokines. At the present time the viruses which have been found to produce such anti-cytokine proteins belong to four viral families: poxviruses, herpesviruses, adenoviruses and baculoviruses. These are all double-stranded DNA viruses with large genomes in the range 85–260 kDa [17]. It is now established that viruses employ three mechanisms for inhibiting cytokine networks: (i) inhibition of the synthesis or release of cytokines; (ii) antagonism of cytokine–receptor interactions; and (iii) functional antagonism (Figure 7.2 and Table 7.1). The literature has been extensively reviewed in recent years [18–24]. As described in more detail in Chapter 8, monocytes contain a unique

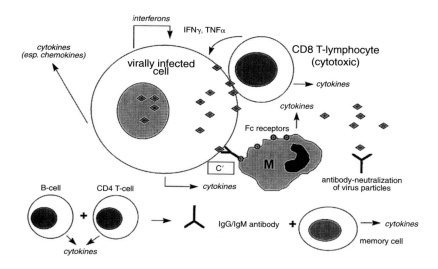

Figure 7.1 The host-defence mechanisms used by mammals to protect against viral infections

This involves the generation of a specific immune response to the virus via CD4, CD8 and B-lymphocytes. Specifically, sensitized CD8 T-cells can recognize viral antigens presented in the context of class I MHC antigens and can kill such cells through the release of perforin and granzymes. They also release TNFα and IFNγ, which enhances MHC expression, and activate macrophages. CD4 and B-cells can collaborate to induce specific anti-viral antibodies. Such antibodies can neutralize cell-free viral particles. Antibodies binding to cell-surface viruses stimulate Fc-bearing macrophages to kill infected cells by a mechanism known as antibody-dependent cellular cytotoxicity. Complement [C'] can also be activated by antibody–antigen complexes and can play a role in anti-viral defences. The attraction of macrophages and lymphocytes to virally infected cells involves the generation of various chemokines. As can be seen from this diagram all cell populations involved in anti-viral defences produce a wide range of cytokines.

cysteine proteinase which cleaves the 31 kDa pro-form of IL-1β at Asp[116]-Ala[117] to produce the 17 kDa active form of this cytokine [25]. This proteinase has been termed pro-IL-1β-converting enzyme (ICE) and the cloned active form of the enzyme is a cysteine proteinase composed of two subunits of molecular masses 10 and 20 kDa with the larger subunit containing the active site [26]. Of significance is the finding that ICE is homologous to the *ced*-3 gene of the nematode worm *Caenorhabditis elegans*, a gene encoding a cysteine proteinase essential for the control of apoptosis in this organism [27]. In the mammal, ICE is one of at least six related enzymes (ICE, ICE_{relII}/TX/Ich-2, ICE_{relIII}, Nedd 2, CPP32, mch2), which are now known to be involved in the control of apoptosis [28].

Vaccinia virus contains a gene, *crm*A, which encodes a serpin inhibitor (a member of a class of proteins best known for inhibiting serine proteases), Crm (cytokine response modifier) A. This viral protein inhibits ICE by forming a tight complex with an equilibrium constant of inhibition of $<4 \times 10^{-12}$ M

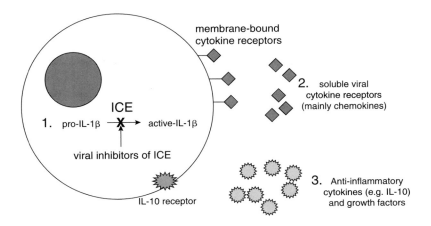

Figure 7.2 The three mechanisms used by viruses to inhibit the activity of cytokines

(1) The use of inhibitors of pro-interleukin-1β converting enzyme (ICE), which catalyses the proteolysis of the inactive 31 kDa precursor protein to the 17 kDa active cytokine; (2) the synthesis of soluble cytokine receptors (mainly chemokine receptors) to mop up cytokines; and (3) the synthesis of anti-inflammatory cytokines such as IL-10.

Table 7. 1 Viral proteins which inhibit host defence mechanisms

Virus	Viral protein	Host homologue	Function
Cowpox	CrmA	Serpin-type protease inhibitor	Inhibits ICE/ blocks IL-1β synthesis
Baculovirus	p35	Protease inhibitor	Inhibits ICE/ blocks IL-1β synthesis
Vaccinia	B15R	IL-1β receptor	Inhibits IL-1β activity
HVS	ORF78	Soluble IL-8 receptor	Inhibits IL-8 activity
HCMV	US28	Soluble IL-8 receptor	Blocks chemokine activity
Myxoma	T7	Soluble IFNγ receptor	Blocks IFNγ activity
SFV	T2	Soluble TNF receptor	Blocks TNF activity
EBV	BCRF1	IL-10	Anti-inflammatory
EHV2	IL-10-like protein	IL-10	Anti-inflammatory
Poxviruses	EGF, TGFα	Cell growth promoters	
KSHV	v-chemokines	MIP-I, MIP-II	?
KSHV	v-IL-6	IL-6	Inhibits apoptosis?
KSHV	v-IRF	IRF	?

Abbreviations used: EBV, Epstein–Barr virus; EHV2, equine herpesvirus type 2; HCMV, human cytomegalovirus; HVS, herpesvirus saimiri; IRF, interferon regulatory factor; KSHV, Kaposi's sarcoma-associated herpesvirus; MIP, macrophage inflammatory protein; SFV, Shope fibroma virus.

[29,30]. Thus, cells infected with vaccinia virus cannot produce the active IL-1β which would be utilized to induce local inflammation and activate the

acute-phase response. Mutation of the *crm*A gene, to produce an inactive serpin, results in a vaccinia virus which has a lower rate of replication, and there is a local increased inflammatory response at sites of viral infection. If the modified virus is injected intracranially, there is an increased rate of death compared with the wild-type virus. One interpretation of these data is that the virus has evolved the *crm*A gene to limit the response of the host to its own inflammatory system [31–33]. This would imply a very intimate evolutionary relationship between the virus and the host, but, as we peer more closely at microbial host interactions, such intimacy appears, increasingly, to be the order of the day.

The suggestion that apoptosis (of virally infected cells) could be an anti-viral strategy was first propounded by Clouston and Kerr [34] and the experimental evidence in support of this hypothesis was subsequently provided by Clem and Miller [35]. The finding that a growing number of viruses encode specific inhibitors of apoptosis supports this idea. Baculoviruses, which infect insects, encode two apoptosis inhibitors, p35 and inhibitor of apoptosis (IAP). The former is believed to act like CrmA as a direct protease inhibitor, whereas IAP may act upstream to prevent activation of apoptotic proteases [36,37]. Both baculovirus genes can function in heterologous systems to prevent apoptosis. Thus, both p35 and IAP block apoptosis in mammalian cells induced by overexpression of ICE, and p35 can block Ced-3-induced apoptosis in *C. elegans* [38–40]. If these baculovirus genes are mutated to produce inactive proteins then the infected insect cells respond to infection by undergoing apoptosis, drastically reducing viral replication and titre [35,41,42].

The second set of gene products expressed by these double-stranded DNA viruses are soluble forms of cytokine receptors (Table 7.1). Proteins with the capacity to bind IL-1, TNFα, IFNγ and various chemokines [IL-8, macrophage inhibitory protein (MIP-1α) RANTES, stromal cell-derived factor (SDF)-1] have been described. It has been demonstrated that: (i) certain chemokines (RANTES, MIP-1α, MIP-1β) suppress the replication of HIV-1; and (ii) HIV-1 enters cells via binding to a C-X-C chemokine receptor known as fusin or LESTR (leucocyte-derived 7-transmembrane domain receptor — CXCR4). This receptor does not bind RANTES, MIP-1α or MIP-1β [43–45]. These findings highlight the importance of chemokines in host anti-viral strategies and may explain the number of chemokines that viruses can inhibit with their soluble receptors. In addition, hepatitis B virus contains an envelope protein with no homology to the IL-6 receptor, but which binds IL-6 [46]. We have already described how the vaccinia virus contains the gene *crm*A encoding an inhibitor of IL-1β convertase. This virus contains another gene which encodes the gene product B15R, which has 30% identity with the type II non-signalling IL-1 receptor [47]. This protein binds with high affinity to IL-1β but does not bind to IL-1α [48,49] giving an evolutionary view of how this virus regards the IL-1 family members. It is of interest that the type II IL-

1 receptor does not transduce a signal and has been termed a decoy receptor, which probably acts to regulate the action of IL-1 at the cell surface [50]. When the gene encoding B15R was disrupted, the mutant virus produced increased morbidity, again supporting the hypothesis that these cytokine-modulating proteins have as much a role in limiting host tissue damage as they have in improving virus survival.

One of the most striking findings in the study of viral 'cytokine genes' has been the number of receptor-like proteins they produce that can inhibit the activity of chemokines. The biological activity of chemokines has been described in Chapter 4 and, to reiterate, members of this cytokine 'family', which is one of the largest and now includes about 30 proteins, are promoters of leucocyte movement to inflammatory sites [51,52]. Thus, they are fundamental in the induction of localized inflammation, and their inhibition would have serious consequences for the host. Viruses such as human cytomegalovirus and herpesvirus saimiri contain genes which encode soluble forms of receptors for a range of chemokines including IL-8 [20]. An interesting response to the knock-out of a chemokine gene (MIP-1α: chemotactic for monocytes, basophils and eosinophils) has recently been reported. Mice infected with coxsackie virus develop a serious, sometimes fatal, inflammation of the heart, and those animals infected with influenza virus develop pneumonia. Transgenic mice lacking the gene for MIP-1α, when infected with these viruses, failed to show serious inflammatory pathology of the heart or lungs, yet were still able to cope with the viruses and resolve infections [53]. This again raises the question of whether the viral inhibition of a host defence molecule benefits the virus or the host.

The third method developed by viruses to, apparently, combat host defence systems is the production of anti-inflammatory cytokines. The only example of this strategy, to date, is the presence of genes in Epstein–Barr virus (EBV) and equine herpes virus type 2 (EHV2) for the anti-inflammatory cytokine IL-10. The human IL-10 molecule consists of 160 amino acids and has a molecular mass of approximately 18 kDa. This molecule has a major inhibitory effect on monocytes and Th$_1$ lymphocytes and its administration to animals can inhibit a number of experimental conditions including lethal endotoxin shock (reviewed in [54]). In the early 1990s it was discovered that the human (h) and murine (m) IL-10 genes exhibited marked homology to an open reading frame in the EBV genome, BCRF-1 (*Bam*HIC fragment rightward reading frame) [55–57]. Surprisingly, the protein sequences of hIL-10 and BCRF-1 are 84% identical. Comparison of the mouse, human and viral sequences show that the human and viral sequences are the most closely related at the protein level, while mIL-10 and hIL-10 are significantly more similar in their DNA sequences. Such sequence relationships suggest that the mIL-10 and the hIL-10 genes evolved from a common ancestor. The amino acid sequence of BCRF-1, showing such identity with the human protein, suggests that the ancestor of BCRF-1 may have been a captured, processed host IL-10 gene. The sheep virus EHV2 also encodes a viral homologue of IL-10 [58]. Viral IL-10 appears to play a crucial role during the early transformation of B-cells by EBV [59], and transformed B-cells produce large amounts of human IL-10 [60]. The viral IL-10 product of the BCRF-1 gene shares the cytokine synthesis inhibitory activity of murine IL-10, although it has a lower specific activity and this may be important in suppressing the anti-viral activities of IFNγ and macrophages. However, viral IL-10 lacks the activities of mammalian IL-10s in co-stimulating thymocyte and mast cell proliferation and in inducing class II MHC expression on B-cells [57,61]. These are obviously activities which may militate against virus survival and, if BCRF-1 is a captured human gene, have presumably been evolved out of the original gene structure.

Kaposi's sarcoma-associated herpes virus (KSHV) is related to the EBV and herpes virus. This virus has recently been found to encode four cytokine-like proteins with sequence similarity to MIP-I and II, IL-6 and interferon regulatory factor (IRF) [61a]. It is not immediately clear what these various gene products would do if secreted from virally infected cells. The authors of this paper [61a] have suggested that the IL-6-like protein may have anti-apoptotic activity.

These viral gene products have obvious roles to play in modulating host cytokine networks, and as these proteins interact directly with cytokine networks the term introduced to describe these molecules — virokines — is appropriate.

In addition to having genes for cytokine-like molecules and cytokine-inhibitory molecules, viruses also contain genes encoding other proteins which can modulate host immune and inflammatory responses to viruses. For example, vaccinia virus encodes a 3β-hydroxysteroid dehydrogenase that synthesizes steroids and could damp-down inflammation [62]. Human cytomegalovirus and herpes virus saimiri express an IgG Fc receptor on the surface of infected cells [63,64]. As the class I MHC antigens signal that cells are virally infected, a number of strategies have been developed to mask this event. These include inhibition of the transcription of class I genes [65] and prevention of class I expression at the cell surface [66]. These MHC-modulating mechanisms have been reviewed [67] (see Table 7.1).

7.3   Cytokine gene knockouts and the response to the normal microflora

Cytokine induction by bacterial products is of obvious importance in the pathology of infectious diseases caused by viruses, bacteria and parasites. The textbook by Janeway and Travers [68] gives a current traditional overview of the role of cytokines in such infections. This is seen as a simple linear interaction in which products of the infecting micro-organism (e.g. LPS or lipoteichoic acids) interact with host cells and induce cytokine synthesis. The

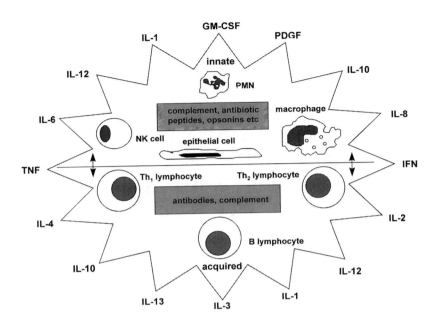

Figure 7.3 Schematic diagram illustrating the cells and mediators involved in innate and acquired immune responses and some of the major cytokines which drive and control such responses

cytokines then activate host defence systems which subsequently inhibit and finally get rid of the infection. The relationship of cytokines to the various cells involved in innate and acquired immunity is shown diagrammatically in Figure 7.3. Of course, if the infectious organism cannot be defeated, the continued production of cytokines can then be the cause of tissue pathology. This is the situation in tuberculosis and leprosy (see Chapters 2 and 4). It is becoming increasingly clear that this simple linear control sequence (bacteria-cytokine-inflammatory response-resolution of infection) is far too simplistic and that infectious organisms are able to modify it. The previous section has introduced the idea that viruses can exert significant control over cytokine-driven host immune defences to, at first sight, promote growth of the virus. However, the available evidence also suggests that the host can also benefit from the anti-cytokine actions of the viral gene products. Bacteria can also subvert host defence mechanisms such as complement activation [15], but less is known of their abilities to modulate cytokines. Although the molecular and cellular details are not known, it is clear that certain bacteria, mycobacteria being a good example, can control the nature of the T-cell response (i.e. Th_1 anti-inflammatory, macrophage-activating, T-lymphocyte or Th_2 helper T-lymphocyte) that they induce. In turn, this T-cell response is dependent on the pattern of cytokines produced. Induction of the inappropriate T-cell response (in the case of mycobacteria this would be a Th$_2$ response) results in the failure to kill bacteria and the continuance of the infection [69]. In this chapter we will also introduce the concept that the commensal microflora can control host cytokine network interactions at epithelial surfaces. Part of the evidence in support of this radical idea has already been reviewed in the previous section on virokines. In this section, the literature on cytokine gene knockouts will be described and the consequences of ablating certain cytokines on the hosts' response to its own commensal microflora will be discussed.

7.3.1   Cytokine gene knockouts and infections

At the time of writing a large number of cytokine gene knockouts have been reported (Table 7.2). The rationale for knocking out cytokine genes is that it will provide information about the role of the gene product in the functioning of the whole animal. However, the consequences of knocking out individual cytokines are difficult to predict and certain unusual results have now been reported in the literature. For example, the ablation of the activity of the growth-promoting cytokine, TGFα, produces mice in which the major phenotypic change is the presence of wavy hair and whiskers and a propensity to develop corneal inflammation. However, the expected changes — failure of embryonic development or impairment of wound healing in the adult — failed to materialize [70,71]. The ability of these TGFα knockout mice to develop normally may be due to the action of an overlapping cytokine, EGF, which

Table 7.2 Cytokine gene knockouts which compromise antibacterial defence mechanisms

Cytokine inactivated	Phenotypic changes
IL-1β	Responsive to endotoxin but not responsive to certain inflammatory stimuli (e.g. turpentine)
ICE	Resistant to endotoxin shock
IL-1ra	More susceptible to lethal endotoxaemia but less susceptible to *Listeria* infections, has role in development
IL-2	Ulcerative colitis in response to commensal gut microflora
IL-4	Deficient in Th$_2$ responses
IL-6	More susceptible to *Listeria monocytogenes*, lowered acute-phase response poor liver regeneration
NF-IL-6	As for IL-6 knockout
IL-7	Early lymphocyte expansion impaired
IL-10	Lethal enterocolitis
TGFβ	Multifocal inflammatory wasting condition
IFNγ	Susceptible to intracellular bacteria
G-CSF	Defects in granulopoiesis
GM-CSF	Normal haematopoiesis but show pulmonary pathology
Lymphotoxin	Abnormal development of peripheral lymphoid organs
MIP-1α	Inhibition of virally induced inflammation

can bind to TGFα receptors. This is an example of functional redundancy in the cytokine network. In contrast, knockout of the cytokine, TGFβ — which is one of the cytokines showing significant anti-inflammatory activity — results in animals that survive to term and then rapidly develop a diffuse, lethal, inflammatory syndrome characterized by massive mononuclear infiltration of the heart, lungs and gastrointestinal tract [72,73]. This is in spite of the fact that there are about 30 proteins in the TGFβ superfamily.

Certain cytokine or cytokine receptor knockouts result in susceptibility to microbial infections. For example, mice deficient in active IFNγ remained healthy in the absence of pathogens but were susceptible to normally sublethal doses of the intracellular bacterium *Mycobacterium bovis* [74]. Knockout of NF-IL-6, the major transcriptional control element of the IL-6 gene, renders mice highly susceptible to infection with *Listeria monocytogenes* [75]. Mice lacking the 55 kDa TNFα receptor (TNF-R55, TNFR1) showed the same response to this intracellular bacterium [76]. Knockout of the TNF-R55 receptor [76] or of the cysteine protease ICE [77] renders mice insensitive to the effects of endotoxin. However, if the gene for IL-1β is ablated, animals respond normally to LPS [78,79]. This discrepancy between the IL-1β knockout mice and the ICE knockouts may be due to the fact that in the latter there is also a deficiency in production of IL-1α [77,80]. IL-1β knockout mice, however, do show impaired inflammatory responses to the tissue-damaging agent turpentine [81] and an impaired contact hypersensitivity response to trinitrochlorobenzene [79]. Knockout of the gene encoding IL-6 rendered mice deficient in the local inflammatory response to turpentine [82], but failed to alter the pathological responses animals could mount to TNFα [83] or LPS [82]. As described in Chapters 4–6, IL-1, IL-6 and TNFα are the major endogenous pyrogens responsible for the fever response to bacterial products. Thus, to further complicate the issue of the role of pro-inflammatory cytokines in the acute-phase response, the injection of LPS or IL-1β into IL-6-deficient mice failed to produce fever although the same treatment caused fever in wild-type mice. Intracerebroventricular injection of IL-6 into IL-6 knockouts induced fever, but similar injection of IL-1 was without effect [84]. This suggests that centrally produced IL-6 is an obligatory factor in the fever response and that IL-6 acts downstream from both peripheral and central IL-1β. The role of IL-6 in combatting infection will be discussed later in this chapter.

The genetic ablation of individual pro-inflammatory cytokines appears to have confused, rather than clarified, the role that these potent molecules play in the acute-phase response and in infections. For example, it is clear in terms of the evolution of the pox virus that IL-1β is an important cytokine. Yet its ablation appears to have little effect in terms of the organisms' response to endotoxin. Obviously, it is now important to determine how animals, in which pro-inflammatory cytokines such as IL-1β and IL-6 have been ablated, respond to various infectious micro-organisms, and also to determine if lack of these or other cytokines has any effects on the response of the organism to its own commensal microflora. The effect of ablating individual cytokines on the capacity of mice to cope with infectious micro-organisms is detailed in Table 7.2.

Two cytokine knockouts have provided important clues to the nature of the interactions that occur between the commensal microflora and the host. Knockout of the cytokine, IL-2 — a cytokine which acts as a growth factor for the Th_1 subset of T-lymphocytes [85] — had no effect on the embryonic development of mice or on the early development of T-lymphocytes. However, animals showed abnormalities in T-cell responses and had high levels of IgG [86]. Further examination of the homozygous knockouts revealed that they developed severe inflammation of the colon with significant infiltration of the colonic mucosa with T- and B-lymphocytes. Indeed, the pathology resembled that of the chronic idiopathic human disease, ulcerative colitis. Homozygous knockouts also produced autoantibodies to antigens of the colonic tissues. The most obvious source of this immune responsiveness was the presence of intestinal pathogens. However, when animals were examined for the presence of a range of likely bacterial and viral pathogens, none were found. The clue to the underlying pathology came when animals were bred and maintained under gnotobiotic conditions. Such germ-free animals failed to develop inflammation of the colon. Thus, the source of the inflammatory stimuli in these animals was the normal gut microflora [87]. Interestingly, a child with IL-2 deficiency was reported to have problems with bacterial infections, although the possibility that the response was to the commensal microflora was not explored [88]. A proportion of mice in which the α-chain of the T-cell receptor was genetically ablated also developed an ulcerative colitis-like condition. Knockout animals with disease symptoms had a decreased IL-2, but an increased IL-4 and IFNγ production in mesenteric lymph nodes compared with homozygous knockouts free of disease symptoms [89].

The second cytokine knockout to show unexpected responses was that involving the ablation of IL-10. Unlike IL-2, which is produced by Th_1 lymphocytes, IL-10 is largely a product of Th_2 lymphocytes and, as has been described earlier in this chapter, inhibits the actions of Th_1 lymphocytes and macrophages. Knockout of this important anti-inflammatory cytokine resulted in homozygotes developing chronic enterocolitis. It has not been established if animals kept under germ-free conditions fail to develop inflammation, but inflammation was much reduced when animals were kept under extra clean (i.e. specific pathogen-free) conditions which may lower the bacterial load in the colon [90].

7.3.2   An explanation for gut inflammation in cytokine knockouts

How can one account for the inexplicable chronic gut inflammation in mice in which the genes for IL-2 or IL-10 have been inactivated? We have utilized these findings to construct a hypothesis which, we believe, provides an explanation for our ability to harbour the enormous number of bacteria which live

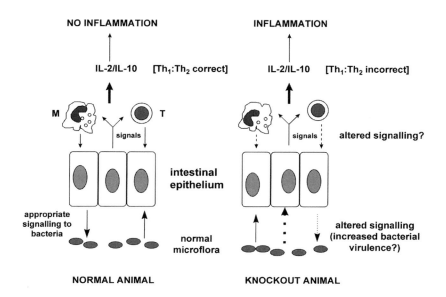

Figure 7.4 Diagram attempting to explain the mechanism by which mice, in which IL-2 or IL-10 have been inactivated by homologous recombination (KO mice), develop chronic colitis or enterocolitis
This model assumes that there is mutual signalling between the normal microflora and the intestinal epithelial cells with which these bacteria are in contact. This may be through soluble signals and/or cell-to-cell contact. This signalling is transmitted through the epithelial cell population, either as a failure to produce pro-inflammatory cytokines or as a positive signal(s), which regulates the actions of the myeloid and lymphoid cell populations present in the submucosa. This signalling presumably acts via IL-2/IL-10 and this may produce the 'correct' balance of $Th_1:Th_2$ lymphocytes, which, in turn, limits overt inflammation. In the absence of IL-2/IL-10, signalling in various parts of this complex system may go awry. For example, down-regulatory signals to the epithelial cells coming from myeloid or lymphoid cells may alter signalling to the normal microflora resulting in the up-regulation of virulence factor synthesis or the down-regulation of bacteriokine synthesis. In addition there may be imbalances in the $Th_1:Th_2$ ratio which contributes to inflammation.

on and in the human body. The basis of this hypothesis is shown schematically in Figure 7.4. The hypothesis states that in all tissues exposed to the normal microflora there are sets of interacting mediators — one set coming from the bacteria and the other from the host tissues — which regulate the activity of both the prokaryotic and eukaryotic cells such that inflammatory and immunological reactivity is minimized. The mediators produced by host tissues are the local hormones which we term cytokines. They may also include antibiotic peptides, some of which are claimed to have cytokine-like activity. The down-regulatory mediators produced by the bacteria will be described in detail in the next section of this chapter and we have termed them 'bacteriokines' [13]. Thus, in the colon we envisage that there is a constant cross-talk between epithelial cells and the normal gut microflora which causes a particular network of cytokine interactions in the colonic tissues. We propose that this cytokine network would be mirrored by and an equivalent set of bacteriokines produced by the normal microflora. However, when one of the key cytokines, IL-2 or IL-10, is missing, the correct cytokine network cannot be maintained. The consequence of this is either: (i) failure to generate a self-perpetuating non-inflammatory cytokine network in the colonic tissues, possibly due to an imbalance in Th_1:Th_2 interactions (e.g. dysregulation in the synthesis of IL-4 or IL-12, or both); or (ii) failure of the colon to produce signals which prevent the bacteria in the normal gut microflora from producing pro-inflammatory signals. We have proposed that the epithelial cells are key to the control of inflammation, and Eckman and co-workers have also proposed a role for these cells [91]. However, our hypothesis does not rule out a role for other cell populations (e.g. mononuclear cells, or $\alpha\beta$ or $\gamma\delta$ T-lymphocytes) in the maintenance of tissue homeostasis in response to the normal microflora.

To reiterate, we propose that there is an obligatory interaction between organisms of the normal microflora and host tissues to allow both cell kingdoms to live in harmony. If the existence of this interacting system is confirmed, it will alter our perception about infectious bacteria and cytokines. This hypothesis will be discussed briefly at the end of this chapter and in detail in Chapter 9. Indirect support for this hypothesis has come from the study of viruses and their ability to produce cytokine-modulating proteins which, as previously stated, have been termed virokines. The phenotypes of the IL-2 and IL-10 gene knockout mice have provided direct evidence that alteration in host tissue cytokine networks can induce a state of reactivity to the normal gut microflora. However, to fully support the hypothesis it must be shown that bacteria do produce bacteriokines. In the next section, the evidence will be produced to demonstrate that bacteria have an enormous potential to control cytokine networks and that they produce anti-inflammatory molecules able to modulate host tissue responses. We believe that it is the ability of bacteria to modulate host cytokine networks that can allow such organisms to live in harmony with — or kill — the host.

7.4   Cytokine-stimulating molecules of bacteria other than LPS

It is only since 1990 that the number of bacterial constituents able to stimulate mammalian, mainly human, cells to produce cytokines has begun to be appreciated. While it is the aim of this book to show that bacteria and host cells interact in a completely novel way to maintain homeostasis, and that such interactions employ cytokine-inducing bacterial constituents, it must not be forgotten that we have very little understanding of the mechanisms infectious bacteria use to induce cytokine synthesis, apart from the obvious factor, endotoxin. Thus, in chronic infectious diseases such as leprosy, tuberculosis, syphilis and brucellosis, in which there is cytokine-driven inflammatory and destructive tissue pathology (and in which endotoxin is absent), we know very little about the bacterial constituents that are able to induce such cytokine synthesis. Most attention, in terms of bacterial pathophysiology, has concentrated on various exotoxins, which are not produced by many of the bacteria causing these chronic diseases.

In this section, the bacterial constituents which have recently been shown capable of inducing cytokine production will be divided into three categories: (i) those which act or are believed to act via CD14 (mainly carbohydrates); (ii) those molecules (the majority) which have been shown to induce pro-inflammatory cytokine synthesis and which, in consequence, may be viewed as virulence factors (modulins); and (iii) those which appear to have the capacity to regulate the activity of pro-inflammatory cytokines or the synthesis of such proteins. It is important at the outset to raise the obvious problem of LPS contamination in the studies to be reviewed. As has been discussed in Chapters 5 and 6, certain bacterial species, *E. coli* being a good example, produce extremely potent LPS molecules able to stimulate cellular cytokine synthesis at nano- to picomolar concentrations. If small amounts of LPS were to contaminate the bacterial constituent under study. this could give rise to a 'false-positive' result; i.e. an inactive bacterial molecule would appear to have

Table 7.3 Controls for LPS contamination in biological experiments

1. LPS is resistant to heating to $>100\,°C$ and, if biological activity is still present after such treatment, the active moiety may be LPS. Conversely, loss of biological activity implies that the active moiety is not LPS.

2. LPS, not being proteinaceous, is resistant to proteolytic digestion. However, it is worth using a number of proteases as not all will inhibit activity. This control obviously only works if the unknown compound is a protein.

3. The cyclic peptide antibiotic polymyxin B (PB) binds to and inactivates the LPS of a number of Gram-negative bacteria. If activity is blocked by PB then the activity is almost certainly due to LPS. If activity is not blocked the possibility exists that the LPS contaminant is not inhibited by PB and the PB sensitivity of the contaminating LPS needs to be checked.

4. The activity of LPS, at least at low concentrations, is due to binding to CD14. If the activity of the substance under test is not blocked by anti-CD14 antibodies, and the activity of contaminating LPS species is, then activity is not due to LPS

5. The C3H/HeJ mouse strain is resistant to LPS. If cells from this strain are activated by the bacterial components under test, then activity is not due to LPS. It should be note that murine and human cells respond differently to certain lipid A analogues (see Chapter 7 for details).

6. A variant of (4) and a control much used in early studies of LPS was to tolerize animals to LPS and then check the activity of the molecule under test. If it stimulated animals then it was unlikely that activity was due to LPS. It is also possible to tolerize cells in culture.

the capacity to stimulate cytokine synthesis. In addition, contamination with LPS could produce synergistic interactions which would not be representative of LPS or the molecule under study. Thus, it is important that any study of the cytokine-inducing activity of bacterial constituents should be controlled for LPS contamination. These control procedures are reasonably simple and are highlighted in Table 7.3. It is obviously important to assay the bacterial constituent under examination for the presence of LPS using some form of the *Limulus* amoebocyte lysate (LAL) assay [92]. The most obvious control, and that which most researchers use, is the inclusion in the assay of polymyxin B — a cyclic peptide antibiotic which binds to the lipid A region of certain, but not all, LPS molecules [93], inhibiting biological activity. Polymyxin B-induced inhibition of activity certainly means that LPS is present, but the converse is not necessarily true. Failure to inhibit may simply mean that the LPS present does not bind to polymyxin B. LPS is an extremely stable molecule and is resistant to heating (up to 140–160°C) and to proteolytic digestion. Therefore, if the activity of the bacterial constituent or fraction survives such rigorous treatment it is very likely due to LPS contamination. As described in detail in Chapter 6, LPS interacts with CD14 to produce its full biological activity. The binding of LPS to CD14 can be blocked by a number of antibodies and this knowledge can be used to determine if activity is due to LPS contamination. A final control used by many researchers utilizes the fact that the C3H/HeJ mouse strain is unable to respond to LPS, possibly due to a defect in intracellular signalling via ceramide [94]. Thus, if the bacterial molecule under study can stimulate macrophages from this LPS-resistant strain, then activity cannot be due to LPS contamination. In the studies to be reviewed, one or other, or indeed all, of these controls have been used to check that the activity recorded is not simply the biological response to contaminating LPS.

7.4.1 Bacterial molecules which stimulate cytokine synthesis via CD14

There is growing evidence that carbohydrates and carbohydrate-containing molecules of bacteria (including Gram-negative, Gram-positive and Mycobacteria) have the capacity to induce cytokine synthesis. This evidence will be reviewed and the recent studies suggesting that such carbohydrates act by binding to CD14 discussed.

7.4.1.1 Cell-surface polysaccharides

As described in Chapter 3, capsular polysaccharides play important roles in bacterial virulence by, for example, hindering phagocytosis or inhibiting complement activation and complement-mediated killing. They are relatively poor immunogens and may lower the host's antibody response to the bacterium. There are now a number of reports that polysaccharides from the cell walls and capsules of various bacteria can stimulate cytokine synthesis.

Purified capsular polysaccharides from two serotypes (5 and 8) of *Staphylococcus aureus* have been shown to stimulate the release of IL-1β, IL-6, IL-8 and TNFα from human peripheral blood mononuclear cells (HPBMCs), IL-8 from the human epithelial KB cell line, and IL-6 and IL-8 from human endothelial cells [95]. However, cytokine release was only induced at polysaccharide concentrations ⩾10 μg/ml. Binding of the polysaccharides was demonstrated to occur in a dose-dependent, saturable fashion and was enhanced by calcium ions but inhibited by serum. The results of competitive binding assays suggested that both polysaccharides bind to the same receptor. The *Streptococcus mutans* serotype f polysaccharide (a rhamnose/glucose polymer) has been shown to stimulate the release of TNFα from human monocytes at a concentration of 25 μg/ml. Although the polysaccharide bound to both CD11b and CD14 on the monocyte cell surface, only binding to CD14 elicited cytokine release. However, the presence of heat-inactivated human serum inhibited cytokine production. The blood component responsible for inhibition was shown to be mannan-binding protein. This component forms a complex with the polysaccharides which then binds to the C1q receptor prior to uptake by the monocyte [96].

Takahashi et al. [97] have shown that the serotype-specific polysaccharides from the oral bacterium *Actinobacillus actinomycetemcomitans* induce IL-1 release from murine macrophages at concentrations ranging from 12.5 to 100 μg/ml. The polysaccharide group-specific antigens of another Gram-negative oral organism, *Porphyromonas gingivalis,* stimulated the release of IL-1β from HPBMCs at a significantly lower concentration — 100 ng/ml [98]. The mucoid exopolysaccharide of *Pseudomonas aeruginosa*, an alginate containing mannuronic and guluronic acid residues, which is involved in the pathology of cystic fibrosis, stimulated the release of IL-1 from mouse macrophages, but only at relatively high concentration (>10 μg/ml). On a weight basis, however, this exopolysaccharide was more potent than the organisms's LPS. Surprisingly, addition of polymyxin B resulted in enhanced production of IL-1 [99]. It is not known whether this polysaccharide contributes to the lung inflammation found in patients with cystic fibrosis.

The relationship between the chemical structure of defined polysaccharides and cytokine synthesis has been reported by Otterlei et al. [100] who showed that certain polysaccharides were as potent as (albeit fairly inactive) *E. coli* LPS, and concluded that β(1→4)-linked polyuronic acids probably bind to the same receptor as LPS.

At least one study has shown that polysaccharides are able to induce cytokine production *in vivo* [101]. Injection of neonatal rats with type III or group-specific polysaccharides of Group B streptococci resulted in increased levels of circulating TNFα, with the peak level induced by the group-specific polysaccharide being almost three times greater than that induced by the type III polysaccharide.

7.4.1.2 Peptidoglycan

A brief description of the structure of the Gram-negative and Gram-positive cell walls has been presented in Chapter 2 and the key structural role of peptidoglycan has been discussed. Gram-positive cell walls contain large amounts of peptidoglycan with associated teichoic and lipoteichoic acids. The role of peptidoglycan as an inducer of pathology is an area of active study. As reviewed by Schwab [102,103], peptidoglycan has long been recognized to have potent immunomodulatory actions. In terms of the ability of this structurally important cell-wall component to induce cytokine synthesis, attention has largely focused on the synthetic muramyl dipeptide, N-acetyl-muramyl-L-alanyl-D-isoglutamine (MDP), the smallest common structural unit of peptidoglycans. MDP and its analogues can stimulate the synthesis of a variety of cytokines by a range of cell populations and the literature to support this has been well reviewed [104,105]. However, MDP is a synthetic molecule and, since the amount produced by bacterial degradation *in vivo* and able to interact with host cells is unclear, the literature on this molecule will not be discussed.

There have been few studies of the cytokine-stimulating ability of peptidoglycan. This is probably due to the complications involved in dealing with this material which is generally insoluble and which, depending on the method of preparation and dissolution, has a wide range of molecular masses and structures. Purified cell walls from a number of Gram-positive bacteria have been reported to stimulate the release of IL-6 and TNFα from HPBMCs at a concentration of 100 ng/ml [106]. Cytokine release was unaffected by polymyxin B or by the presence of anti-CD14 mAbs, except for in the case of *Streptococcus pyogenes*. Fractionation of the cell walls revealed, in each case, that most of the cytokine-stimulating activity was attributable to the peptidoglycan components.

A soluble peptidoglycan (125 kDa) obtained from the supernatant of cultures of *Staph. aureus* grown in the presence of penicillin, was able to induce release of IL-1 and IL-6 from HPBMCs in a dose-dependent manner over the range 1–30 µg/ml [107]. As was found in the case of peptidoglycan from *Strep. pyogenes* [106], anti-CD14 monoclonal antibodies were able to block peptidoglycan-induced cytokine release.

In a comparative study of the capacity of whole *Staph. aureus, Staph. epidermidis* and their respective peptidoglycans to stimulate TNFα release from human monocytes, it was demonstrated that, while whole cells and the peptidoglycans were able to stimulate release of this cytokine, the former were more effective stimuli than the isolated cell-wall components [108]. Thus, 10^7 staphylococcal cells, corresponding to 0.1 µg of peptidoglycan, produced the same amount of TNF as 1–10 µg of purified peptidoglycan. The explanation for these differences in activity could relate to degradation of the peptidoglycan during isolation, to differences in steric interaction between whole-cell peptidoglycan and isolated material, or to synergistic interactions between peptidoglycan and other cell-wall components in whole bacteria. Indeed, degradation of the peptidoglycan by enzymes or by sonication decreased its capacity to induce TNF synthesis.

Monocytes are not the only cells which respond to peptidoglycans by releasing cytokines. Lichtman et al. [109] have shown that rat Kupffer cells can be induced to release IL-1 and TNFα by exposure to peptidoglycan-polysaccharide (PP) from *Strep. pyogenes*. However, the potency of the PP was quite low, requiring μg/ml concentrations to induce detectable amounts of cytokine and 100 μg/ml for maximum release. LPS from *E. coli* was a far more potent inducer of cytokine release from the Kupffer cells and it is interesting to compare the mechanism of cytokine induction by these two components. Cytokine induction is preceded by endocytosis in both cases, as cytochalasin B blocked cytokine release by PP and LPS. However, microtubule function appears to be more important in LPS-stimulated cytokine release as this was reduced considerably by addition of colchicine. Since nisoldipine, a calcium channel blocker, inhibited cytokine production by both components, this would imply that calcium is the second messenger with both agonists. It is of interest that taxol, an anti-cancer agent and microtubule stabilizer, has a similar profile of activity to LPS when added to myelomonocytic cells [110].

7.4.1.3 Peptidoglycan fragments

Peptidoglycans released *in vivo* will be degraded by host enzymes and such breakdown products may in turn have cytokine-inducing activity. Two reports have shown that N-acetylglucosaminyl-1,6-anhydro-N-acetylmuramyl-L-alanyl-D-isoglutamyl-*m*-diaminopimelyl-D-alanine [G(Anh)MTetra], a naturally occurring breakdown product of peptidoglycan, at concentrations as low as 50 ng/ml, induces the synthesis of IL-1β, IL-6 and G-CSF by human monocytes [111,112]. The synthesis of all three cytokines can be blocked by inhibitors of protein kinase C, but not by inhibitors of protein kinase A or tyrosine kinases. Interestingly, the transcriptional control of these three cytokines showed distinct differences. Using run-on transcription assays, G(Anh)MTetra markedly increased IL-1β transcription, whereas it had much less effect on the transcription rate of the gene for IL-6, and the increased level of mRNA for G-CSF was shown to be due to stabilization of the mRNA transcripts. Using the protein synthesis inhibitor cycloheximide, it was demonstrated that IL-6 mRNA expression depended on the synthesis of new protein, but that this was not the case for IL-1β or G-CSF.

Bordetella pertussis, the causative organism of whooping cough, releases a low-molecular-mass (921 Da) peptidoglycan fragment called tracheal cytotoxin (TCT), which damages airway epithelium and contributes to disease pathology. Addition of TCT to respiratory epithelial cells resulted in the rapid accumulation of intracellular IL-1α but without release of this cytokine. It has been proposed that this up-regulation of intracellular IL-1 may contribute to epithelial cell pathology in pertussis [113]. This is but one of a growing number of examples where bacterial toxins also show the capacity to induce cytokine synthesis, and will be dealt with in more detail in section 7.4.2.3.2.

Water-soluble peptidoglycan fragments from *Staph. epidermidis* have also been reported to stimulate the proliferation of spleen mononuclear cells from various strains of mice and the synthesis of various cytokines including IL-1 and GM-CSF [114].

7.4.1.4   Teichoic acids

Anionic polymers such as teichoic and lipoteichoic acids (LTAs) are major components of the walls of Gram-positive bacteria. Teichoic acids consist of chains of glycerol, ribitol, mannitol or sugars linked by phosphodiester bonds, and are attached to muramic acid residues in the peptidoglycan. D-Alanine or L-lysine are common substituents of the chains, which usually contain approximately 40 residues. LTAs consist of chains of glycerol phosphate, with D-alanine and sugar substituents, attached to a glycolipid (or diglyceride) in the cytoplasmic membrane. Both teichoic acids and LTAs are antigenic and often constitute the major somatic antigens of Gram-positive bacteria. LTAs are amphiphilic molecules with a number of biological activities. In some ways, they can be regarded as the Gram-positive equivalent of lipopolysaccharides, although the latter have a much greater potency and range of biological activities.

Riesenfeld-Orn et al. [115] reported that LTA from pneumococci stimulated human monocytes to release IL-1 but not TNF. In contrast, the LTA from *Strep. faecalis* stimulated murine mononuclear cells to release both cytokines with deacylation of the LTA abrogating this cytokine-stimulating activity [116]. In comparative studies of the LTAs from a range of Gram-positive bacterial species, Bhakdi and colleagues [117] reported that there were major differences in their capacities to stimulate human monocyte pro-inflammatory cytokine synthesis (specifically IL-1β, IL-6, TNFα). LTAs from several enterococcal species were capable of stimulating the release of similar amounts of all three cytokines to those induced by a crude preparation of *E. coli* LPS, although the potency of the LTA was significantly lower than that of the LPS. In contrast, LTAs from organisms such as *Staph. aureus*, *Strep. mutans* and *Leuconostoc mesenteroides* were inactive. In agreement with the findings of Tsutsui et al. [116], deacylation abolished this activity. Addition of polycations (poly-L-arginine or poly-L-lysine) to monocytes stimulated with LTA abolished the stimulatory activity, although the polycations had no effect on the cytokine-stimulating activity of LPS. Analysis of the kinetics of cytokine stimulation revealed that very short exposure of cells to LTA (5–30 min) was sufficient to trigger cytokine production, suggesting that the cellular receptor for LTA is distinct from that of LPS and is perhaps the macrophage scavenger receptor [118]. This can also be inferred from the report that LTA stimulation of cytokine production by human monocytes is not inhibited by blockade of CD14 [106]. While Bhakdi and co-workers [117] failed to find that LTA from *Staph. aureus* stimulated cytokine production, Standiford et al. [119] reported that LTA from this organism and from *Strep. pyogenes* stimulated human monocytes to produce IL-8. In a comparative study of LPS, peptidoglycan and teichoic acid (the latter two components from *Staph. epidermidis*), it was found that the teichoic acid could stimulate the production of IL-1β, IL-6 and TNFα, although only at concentrations of 10–100 μg/ml [120]. Thus, the conclusion from the literature is that the teichoic acids are weak stimulators of cytokine synthesis, requiring microgram-per-ml concentrations for activity. Nevertheless, evidence for the possible involvement of LTA-induced cytokine release in an infectious process comes from the study of Danforth et al. [121] who investigated the expression of MIP-1α — a macrophage-activating and chemotactic cytokine — in endocardial samples from patients with acute *Staph. aureus* endocarditis. Cell-associated MIP-1α expression was detected immunohistochemically in neutrophils, macrophages and fibroblasts. Furthermore, HPBMCs treated with LTA *in vitro* stimulated the release of MIP-1α.

Gram-positive bacteria can produce a septic shock-like condition and as many people die each year from Gram-positive as from Gram-negative sepsis. LPS is obviously a very potent inducer of cytokine synthesis. Peptidoglycans and teichoic acids are believed to be the Gram-positive equivalent of LPS, causing cytokine-induced shock in patients with Gram-positive sepsis. However, the relatively weak cytokine-inducing activity of peptidoglycan and teichoic acids is a problem in implicating these molecules in pathology. It has been shown, nevertheless, that peptidoglycan and teichoic acids synergize *in vivo* in inducing shock in rats [122] and this may be one mechanism by which these relatively inactive components induce pathology. Of course, we have to look at other Gram-positive bacterial components as the inducers of the shock-like state. A recent paper from Sam Wright's group [123] has reported the extraction, from the commercially available LTA of *Staph. aureus*, of a minor component (which is not a teichoic acid) able to stimulate human monocyte cytokine synthesis in a CD14-dependent manner. This molecule binds CD14 and blocks the binding of LPS to CD14.

An interesting role for teichoic acids in oral health has been suggested from the finding that LTA from the common oral bacterium *Strep. sanguis* stimulates human gingival fibroblasts to produce the epithelial cell mitogen, hepatocyte growth factor (scatter factor). The LTA and IL-1 interacted synergistically in the induction of scatter factor synthesis. This has raised the suggestion that oral bacterial LTA could promote oral epithelial growth [124].

7.4.1.5 Mycobacterial lipoarabinomannan

Mycobacterium tuberculosis is able to induce cytokine synthesis by human monocytes [125], but the nature of the cytokine-inducing molecules is still being defined. In the genus *Mycobacterium*, the peptidoglycan layer is covered by lipid-rich layers so that up to 60% of the dry weight of the cell wall may consist of lipids, rendering it extremely hydrophobic. A variety of lipids, gly-colipids and lipoproteins have been isolated from mycobacterial cell walls and several of these contain mycolic acids, these being unique to the mycobacteria, nocardiae and corynebacteria. Most of the studies of cytokine-inducing components have concentrated on lipoarabinomannan (LAM), a major cell-wall component of *Mycobacterium* spp., which exhibits immunoregulatory and anti-inflammatory effects that favour the survival of the mycobacteria. These effects include suppression of T-lymphocyte proliferation through interference with antigen processing [126], inhibition of macrophage activation by IFNγ [127,128] and scavenging of oxygen-derived free radicals [129]. Moreno and co-workers [126,130] were the first to report that LAM had the capacity to stimulate human blood monocytes and activate murine peritoneal macrophages to release TNFα. Maximal release was found at a concentration of 10 μg/ml with significant release at 100 ng/ml. A surprising finding was that polymyxin B, a well known inhibitor of LPS activity, bound to and inactivated LAM, suggesting a similarity in structure between LAM and LPS. Treatment of the LAM with dilute alkali significantly diminished the TNF-stimulating activity, suggesting that the *o*-acyl groups may be responsible for stimulation of cytokine synthesis [130]. In a more detailed study, Barnes et al. [131] showed that LAM stimulated HPBMCs to transcribe mRNA for cytokines normally thought of as being macrophage products — IL-1α and β, IL-6, IL-8, GM-CSF, TNF and IL-10. In contrast, the LAM did not stimulate the transcription of cytokines normally associated with lymphocytes — IFNγ, IL-2, IL-3 or IL-4 — although the whole bacterium *M. tuberculosis* was capable of stimulating the transcription of these cytokine genes. Lipomannans and phosphatidylinositol mannoside were also capable of stimulating the same cytokine profile as the LAM. Deacylation of the LAM almost totally inhibited its capacity to induce cytokine synthesis showing that the activity was associated with the phosphatidylinositol portion of the molecule. While LAM is clearly seen as an inducer of pro-inflammatory signals there is a recent report that it can induce the synthesis of the inflammation-modulating cytokine, TGFβ [132], and this may be important in the immunosuppression that occurs in patients with tuberculosis [133].

The mechanism by which LAM, or its structural analogues, stimulates cytokine synthesis is becoming clearer. It has been demonstrated that LAM and LPS activate human monocyte IL-6 gene expression by an identical pathway involving NF-κB and NF-IL-6 [125,134]. There is evidence that the phagocytosis of mycobacteria is dependent on binding to mannose and receptors for complement and the leucocyte β_2-integrins [135]. However, it is likely that LAM acts via the CD14 receptor [2,136], at least in its interactions with monocytes. The recent report that LAM is chemotactic for T-cells in serum-free conditions strongly suggests that T-lymphocytes have a separate receptor for LAM [137].

Interestingly, LAM preparations from virulent or attenuated strains of mycobacteria differ markedly in their biological actions, with the former, in contrast to the latter, being unable to stimulate murine resident peritoneal macrophages to release TNFα [138]. The LAM from an avirulent mycobacterium, strain H37Ra, was able to stimulate the transcription of the immediate early-response genes c-*fos*, JE and KC and the production of TNFα. In contrast, LAM from a virulent strain, Erdman, failed to trigger these genes and the production of TNFα [139,140]. This capacity to produce LAM, which is incapable of stimulating a macrophage response, is now seen as a key determinant in mycobacterial virulence, and the structural basis of the differences is under investigation.

7.4.1.6 Role of CD14
It is now clear that many of the carbohydrates emanating from bacteria can stimulate cell activity in a CD14-dependent manner (Table 7.4). The potential mechanism of activation of cells via the cell-bound or soluble forms of CD14, and the biological role of CD14 in controlling cytokine networks, has been reviewed in Chapter 6 and is discussed in detail in Chapter 9.

7.4.2   Bacterial cytokine-inducing molecules inducing pro-inflammatory cytokine synthesis
One of the surprising findings of the past decade has been the number of bacterial molecules which have the capacity to stimulate eukaryotic cells to produce cytokines. Indeed very few scientists have appreciated this sea change in bacterial cytokine induction. We have already summarized the data on those molecules (predominantly carbohydrates) which stimulate cells by binding to CD14. There are a large number of bacterial molecules, of all biochemical classes, which stimulate cytokine synthesis by receptor-driven pathways independent of CD14. At first sight the accumulated data suggest that these cytokine-inducing bacterial components play some role in bacterial pathology. Indeed, the authors of this monograph have suggested that the bacterial cytokine-stimulating components are part of the global bacterial virulence mechanism and may form a new class of virulence factor which, in line with

Table 7.4 Carbohydrates stimulating cytokine synthesis via CD14	
Carbohydrate	**Reference**
Lipoarabinomannan	[2,136,141,142]
Mannuronic acid polymers (*Pseudomonas* spp.)	[143]
Rhamnose-glucose polymers (*Streptococcus mutans*)	[96]
Staphylococcus aureus soluble peptidoglycan	[107]
Gram-positive insoluble cell walls	[2]
Minor component of lipoteichoic acid	[123]
Chitosans	[144]

the other bacterial virulence families (i.e. adhesins, aggressins, impedins, invasins; see Chapters 2 and 3), we have named 'modulins'. This term signifies that the capacity of the bacterial factor to induce cellular cytokine synthesis 'modulates' the function of the cell, tissue, organ or organism, and the altered cellular behaviour contributes to pathology. LPS and the components present in endotoxin would be members of this virulence family. However, until we know precisely what pattern of cytokines, cytokine inhibitors, cytokine-modulating mediators, soluble cytokine receptors, cytokine-transcription/ translation, etc., any particular bacterial component induces, it is almost impossible to define the nature of the said component. In this section the bacterial molecules that induce pro-inflammatory cytokine synthesis will be discussed, but this should not be taken to imply that each of the molecules described is a modulin, functioning as a mediator of tissue pathology.

7.4.2.1 Lipids

Apart from LPS, few bacterial lipids have been investigated for their cytokine-stimulating ability. Nevertheless, an uncharacterized, protein-free, lipid/polyol isolated from the membranes of *Mycoplasma fermentans* has been shown to stimulate the release of IL-6 and TNFα from murine macrophages at concentrations as low as 100 pg/ml. Periodate treatment of the extract significantly decreased its cytokine-stimulating activity, demonstrating the importance of the polyol moiety of this molecule in cell activation [145]. A purified membrane preparation from *M. fermentans,* consisting of a dipalmitoyl- and a stearoyl-palmitoyl-glycerodiphosphatidylcholine, has also been shown to stimulate human monocytes to secrete TNFα [146].

7.4.2.2 Proteins

The story emerging from the study of cytokine-inducing bacterial proteins is that the bacterial genome encodes a very large number of proteins, lipoproteins and glycoproteins which have the capacity to control host cytokine synthesis.

7.4.2.2.1 Lipoproteins

Lipoproteins are found in the bacterial cytoplasmic membrane and are also common constituents of the cell wall of both Gram-negative and Gram-positive bacteria. Indeed, the Braun lipoprotein of *E. coli,* which is responsible for anchoring the outer membrane to the peptidoglycan layer, is one of the most abundant proteins in this organism. There are several reports of the cytokine-inducing ability of lipoproteins. A hydrophobic 48 kDa membrane lipoprotein from *M. fermentans* stimulated the release of IL-1β and TNF from human peripheral blood monocytes over the dose range 10 ng/ml to 1 μg/ml [147]. The TNF-stimulating ability of this lipoprotein was abrogated by proteinase treatment, whereas a lipoprotein lipase removed some, but not all, of its cytokine-inducing ability. Anti-CD14 mAbs had no effect on the cytokine-inducing ability of this lipoprotein.

A membrane preparation extracted from *Mycoplasma arginini* and consisting of five lipoproteins was a potent stimulator (in the concentration range 0.5–50 ng/ml) of the release of IL-1, TNFα and IL-6 from human monocytes. Most of the IL-1 (approximately 80%) remained cell-associated. Addition of polymyxin B had no effect on cytokine release [148].

Cytokine-stimulating ability is not confined to the lipoproteins of mycoplasmas, as relatively low concentrations of a surface lipoprotein (OspA) from *Borrelia burgdorferi*, the causative agent of Lyme disease, has been shown to stimulate the induction of mRNA for IL-1, IL-6, IL-12, IFNβ and TNFα in macrophages from LPS-responsive (BALB/c) and LPS-unresponsive (C3H/HeJ) mice [149] and to induce IL-12 synthesis by monocytes/macrophages [150]. As described in Chapter 4, IL-12 is an important cytokine in acquired immunity able to induce the development of Th_1 T-lymphocytes. *B. burgdorferi* does not contain LPS. Lipid modification of OspA protein is essential for its cytokine-stimulating activity [151,152] and it has been established that the OspA protein and lipoproteins from the causative organism of syphilis, *Treponema pallidum*, activate human monocytes by a pathway distinct from that produced by LPS [153]. These outer-membrane lipoproteins can be used as a vaccine to protect mice from *B. burgdorferi* infection and pathology [154]. The role of IL-12 production in experimental Lyme disease in mice has been assessed by administering a neutralizing mAb. This caused a reduction in arthritis and in the Th_1-type immune response, showing the importance of induced IL-12 in this experimental lesion [155].

The importance of the lipid in these lipoproteins has been emphasized by the finding that the synthetic lipopeptide, *N*-palmitoyl-*S*-[2,3-bis(palmitoyloxy)-2*RS*-propyl]-(*R*)-cysteinyl-alanyl-glycine (Pam₃Cys-Ala-Gly), an analogue of the N-terminus of a bacterial lipoprotein, stimulated the synthesis of IL-1, IL-6 and TNFα by murine macrophages [156]. Shimizu et al. [157] have demonstrated that the cytokine-inducing potency of a number of synthetic peptides (which are analogues of the N-terminus of the *E. coli* lipoprotein) is affected by structural differences in the glycerol moiety. The synthetic lipopeptide CGP 31362, a structure derived from a Gram-negative lipopeptide, stimulates LPS-unresponsive murine macrophages to release TNFα by a process involving protein tyrosine phosphorylation and MAP kinases [158]. (See Chapters 4, 6 and 8 for further information on the role of MAP kinases in the stimulation of host cells by bacterial constituents.)

7.4.2.2.2 Glycoproteins

There are increasing reports of the presence of glycoproteins in prokaryotes and a number of cell-wall glycoproteins from *Cytophaga johnsonae* have been shown to potently stimulate the release of TNF from a mouse macrophage cell line; concentrations as low as 50 pg/ml significantly stimulated TNF synthesis [159]. This highlights the potency of many of the bacterial proteins reported to induce cytokine synthesis.

7.4.2.2.3   Outer-membrane proteins

Approximately 50% of the dry mass of the outer membrane of Gram-negative bacteria consists of proteins and more than 20 immunochemically distinct proteins (termed outer-membrane proteins; OMPs) have been identified in *E. coli*. Apart from their structural role, OMPs have also been shown to have other functions, particularly with regard to transport, and have been classified as permeases and porins. Furthermore, several OMPs have been shown to be potent inducers of cytokine synthesis.

7.4.2.2.3.1   *Porins*

Porins are OMPs which form trimers that span the outer membrane and contain a central pore with a diameter of about 1 nm. These porins (e.g. OmpC and OmpF of *E. coli*) are permeable to molecules with molecular masses lower than approximately 600 Da. Apart from their transport functions, OMPs are major antigens and there are now several reports of their ability to stimulate cytokine synthesis.

Isolated porins from *Salmonella typhimurium* [160], *Yersinia enterocolitica* [161] and *Helicobacter pylori* [162] have been shown to stimulate monocytes and lymphocytes to release a range of pro-inflammatory and immunomodulatory cytokines including IL-1, IL-4, IL-6, IL-8, TNFα, GM-CSF and IFNγ. The porins from *Sal. typhimurium* showed similar dose responses to the LPS from this organism in their capacity to stimulate the release of TNF and IL-6, with a rectilinear dose response over the range 10 ng/ml to 1 μg/ml. In contrast, the porins were relatively inactive in stimulating IL-1α release from monocytes or IFNγ release from human lymphocytes [160]. The porins from *Y. enterocolitica* stimulated the release of IL-1α, IL-6 and TNFα from human monocytes in a dose-dependent fashion over the concentration range 100 ng/ml to 5 μg/ml with statistically significant release at 100 ng/ml, but they were less effective at stimulating the release of IL-8, IFNγ and GM-CSF [161]. In spite of inducing the above cytokines, the porins were unable to stimulate IL-3 and IL-4 release, even at concentrations as high as 20 μg/ml.

Injection of purified *Sal. typhimurium* porins into the paws of rats induced a dose-dependent oedema. Similar doses of LPS (0.3–30 μg) failed to produce an oedematous response. Oedema was unaffected by complement depletion but was slightly reduced by indomethacin and significantly reduced by dexamethasone. Rat peritoneal cells incubated with porins released histamine, but little prostacyclin, suggesting that porins have little ability to induce the prostanoid-producing enzyme cyclo-oxygenase [163]. Porins were also shown to kill D-galactosamine-sensitized LPS-responsive and LPS-unresponsive mice. A dose of 100 ng of porin was sufficient to kill 80–90% of animals, exhibiting a similar potency to that of LPS from *Sal. typhimurium*. The lethal effect of the porin preparation could be completely blocked by pre-administration of a neutralizing antiserum to TNFα but was not abolished by polymyxin B. Porins were also found to be pyrogenic in rabbits and to elicit a localized Schwartzman reaction when used as the sensitizing and eliciting agent. Both the fever and the Schwartzman reaction were unaffected by the administration of polymyxin B, indicating that LPS did not contribute to the biological responses [164].

The capacity of the porins to induce the synthesis of many pro-inflammatory cytokines, together with their stability to proteolysis, suggests that they could play a role in the virulence of bacteria. This capacity may also be the source of the ability of Neisserial porins to act as adjuvants [165]. Indeed, a recent report has revealed that porins from this bacterium can increase the surface expression of the co-stimulatory ligand B7-2 (CD86) on the surface of B-cells, which may explain the immunopotentiating ability of the neisserial proteins [166].

7.4.2.2.3.2   *Lipid A-associated proteins*

As has been described in Chapter 5, some methods of extracting LPS from bacteria, for example using trichloroacetic acid or butanol, result in an LPS–OMP complex to which the term 'endotoxin' should be applied. The proteins associated with the LPS are referred to as lipid A-associated protein (LAP) or endotoxin-associated protein (EAP) and are known to have biological activities distinct from those of LPS. For example, endotoxins, unlike protein-free lipopolysaccharides, were shown to be mitogenic for lymphocytes from LPS-unresponsive C3H/HeJ mice [167]. It was subsequently demonstrated that the B-cell mitogenicity of butanol-extracted endotoxin was attributable to the presence of LAP [168,169]. The properties of LAP have been extensively reviewed by Hitchcock and Morrison [170]. LAPs appear to be important virulence factors, as immunization of salmonella-hypersusceptible mice with LAP–LPS complexes, but not with LPS, protects against the lethality of *Sal. typhimurium* infection [171]. Furthermore, LAPs from several species have been shown to have potent adjuvant activity [172–174]. Thus, LAPs may be key mediators of leucocyte behaviour during infections and, indeed, Hogan and Vogel [175] have suggested that the LAPs represent a 'second signal' for the activation of macrophages. However, the role of this second signal in the responsiveness of the host to Gram-negative bacteria needs to be more fully investigated.

Johns and co-workers [176] reported that preparations of LAP from *Sal. typhimurium* have IL-1-like properties. This conclusion was based on the finding that injection of LAPs into mice induced the acute-phase reactant, serum amyloid A (SAA), a protein requiring a cytokine such as IL-1 for induction. However, unlike protein-free LPS, addition of LAP to macrophages did not induce 'SAA-inducing activity' in culture, an activity which was assumed to be due to IL-1. LAPs were also found to be active in the lymphocyte-activating factor (LAF) assay and to act as a co-stimulatory factor for the proliferation of resting human T-lymphocytes. LAPs were also shown by Johns et al. [177] to be capable of inducing the formation of granulopoietic colonies when added to human peripheral blood and bone marrow progenitor cells which had been depleted of accessory cells (monocytes, T- and B-lymphocytes). These workers also confirmed the great stability of the LAPs inasmuch as heating to 100°C for 30 min, or exposure to trypsin or pronase for 24 h at 37°C, did not decrease the biological activity. These findings have been largely confirmed by Porat et al. [178] who isolated LAPs from *E. coli* and found that the active protein had a molecular mass of 17 kDa. However, this group reported that, in contrast to the studies of Johns and co-workers [176,177], T-cells partially mediated the effect of the LAPs and that the formation of granulocyte/macrophage colonies induced by LAPs could be blocked by a neutralizing antiserum to IL-1β. It is not known whether these differences in biological activity are due to the fact that the LAPs are from different bacterial species.

Direct comparison of the cytokine-stimulating activity of LAPs and LPS has been reported for three bacterial species. Mangan and co-workers [179] have compared the capacities of LPS and LAP from *Sal. typhimurium* to stimulate IL-1 synthesis. Significant amounts of mRNA for IL-1β were induced in human monocytes exposed to 1 ng/ml LAP. Comparison of the activity of LAP and LPS revealed that the former was a more potent IL-1 stimulus. The authors of this book have compared the ability of LAP and protein-free LPS from *Actinobacillus actinomycetemcomitans*, an organism associated with the common inflammatory diseases of the gums — the periodontal diseases — to stimulate the release of IL-1β, IL-6 or TNFα from human monocytes or human gingival fibroblasts. LAP induced the release of IL-6, but not IL-1β or TNFα, from human gingival fibroblasts over the concentration range 10 ng/ml to 10 μg/ml with significant release at 10 ng/ml [180]. In contrast, LPS from this organism was unable to stimulate IL-6 release even at 10 μg/ml. At higher concentrations (100 ng/ml) the LAP also induced the release of IL-6 and IL-1β, but not TNFα, from human monocytes.

We have also demonstrated that LAPs from *Porphyromonas gingivalis*, one of the key organisms involved in the periodontal diseases, were potent stimulators of IL-6 release from human gingival fibroblasts [181]. Significant release of IL-6 was found at a concentration of 10 ng/ml, whereas LPS from this organism was one log order less active in this respect. Both the LAP and the LPS from this organism were equally active at inducing IL-6 release from myelomonocytic cells. The LAP fraction from this bacterium contained nine proteins. These have been fractionated and only the 17 kDa protein was shown to induce cytokine synthesis. This protein was sequenced and turns out to be an autolytic fragment of the key protease of this bacterium, known as R-1 protease. Of interest is the finding that this protease can proteolytically inactivate cytokines such as IL-1. Thus, this one protease seems to be capable both of inactivating cytokines and promoting their synthesis (L. Sharp, unpublished work).

In summarizing the very limited studies of this fascinating group of proteins, the claim by Johns and co-workers [176] that preparations of LAPs have IL-1-like activity deserves consideration. This was also the conclusion of our studies of a surface-associated fraction from the oral bacterium *A. actinomycetemcomitans*, which are discussed below [182]. This concept that bacterial proteins can mimic the actions of eukaryotic cytokines will be discussed in more detail in Chapter 9.

7.4.2.2.3.3   *Other outer-membrane proteins*

A preparation containing OMPs from the pathogenic bacterium *Shigella flexneri* was shown to induce murine macrophages to release TNFα and IL-6 [183]. The relationship between LPS and OMPs is known in some detail and has been described in various chapters (e.g. Chapter 6). However, the possible biological interactions between solubilized OMPs, LPS and other virulence factors remain to be clarified. For example, a 39 kDa OMP from *Proteus mirabilis*, which is mitogenic for B-cells [184], has been shown to inhibit the LPS-induced production of macrophage oxygen-derived free radicals as well as the LPS-induced synthesis of IL-1 by murine macrophages [185]. Thus, in addition to inducing cytokine synthesis, at least one OMP appears to be able to down-regulate bacterially induced cytokine synthesis. This capacity will be discussed in more detail in Section 4.3.

7.4.2.2.4   Fimbrial proteins

Fimbriae (also known as pili) are rod-shaped structures originating in the cytoplasmic membrane and are composed of a hydrophobic protein termed pilin. Their main function, which has been described in more detail in Chapters 2 and 3, is to enable the bacterium to adhere to host cells, or to other bacteria, by means of specific receptors. A bacterium may be able to elaborate a number of fimbriae, each with a specific adhesin at its tip, to enable it to adhere to particular host-cell receptors. Fimbriae are found mainly on Gram-negative bacteria although they have also been detected on some streptococci and actinomycetes.

Community-acquired urinary tract infections are normally caused by so-called uropathogenic strains of *E. coli* [186], which attach to bladder epithelial (uroepithelial) cells via their pili and induce these cells to secrete cytokines [187]. Injection of isolated *E. coli* pili into mice elicited local stimulation of IL-6 release [188] and, in *in vitro* studies, pili from *E. coli* were shown to stimulate epithelial cell lines to release IL-6 [189–191]. *E. coli* produces a number of fimbriae: type I fimbriae bind terminal mannose residues; S fimbriae recognize sialic acid; and P fimbriae (which are the most important in, for example, kidney infections) recognize α-D-Gal-(1→4)-β-D-Gal (globobiose), which is linked to ceramide on the outer face of the plasma cell membrane. P-fimbriated *E. coli* have been reported to be more potent inducers of IL-6 synthesis than isogenic non-fimbriated strains. Moreover, the lectin-binding domain of isolated P-fimbriae was shown to be required for IL-6 induction [189]. IL-6 synthesis by epithelial cells was reduced by treatment that inhibited fimbrial glycolipid receptor expression [192]. These findings suggest that fimbriae interact with cell-surface glycolipids to trigger cytokine synthesis. As we have reviewed in Chapter 6, ceramide is now thought to be involved in the intracellular signalling of LPS [193] and cytokines such as IL-1 [194]. Thus, it is interesting that Hedlund and co-workers [195] have reported that P-fimbriated, but not type I-fimbriated, *E. coli,* when cultured with human kidney epithelial cells, induced the release of ceramide and increased the phosphorylation of ceramide to ceramide 1-phosphate. This suggests that P-fimbriae can activate kidney cell cytokine production by a ceramide-signalling pathway which is distinct to that of LPS. The role of ceramide in eukaryotic cell signalling has been discussed in more detail in Chapter 6.

The oral bacterium *Porphyromonas gingivalis* has fimbriae which are able to stimulate IL-1β synthesis by various cell types including human gingival fibroblasts [196] and murine macrophages [197]. The fimbrial protein also induced accumulation of IL-1β mRNA in the macrophages of 'LPS-unresponsive' C3H/HeJ mice [197]. At a concentration of 4 μg/ml, fimbriae from this bacterium were able to induce IL-1β and GM-CSF gene expression in mouse embryonic calvarial bone cells and also to induce bone resorption [198]. Thus, it is possible that fimbriae from this, and other bacteria implicated in the periodontal diseases, could play a role in the alveolar bone destruction which is the hallmark of this disease.

Matsushita et al. [199] reported that a 55 kDa cell-surface protein, possibly a fimbrial protein, from the periodontopathogenic organism *Prevotella intermedia* was able to stimulate the release of IL-1α, IL-1β, TNFα, IL-6 and IL-8 from HPBMCs. Cytokine release was induced by as little as 0.1 μg/ml of this protein in all cases except for TNF, which was released only by concentrations greater than 1 μg/ml. The potency of the fimbriae was, generally, similar to that of LPS from this organism. Human gingival fibroblasts were less responsive to the fimbrial protein; even concentrations as high as 100 μg/ml stimulated the release of only low levels of IL-1β and IL-6. The protein was also able to induce secretion of IL-6 and TNFα from macrophages obtained from LPS-unresponsive C3H/HeJ mice, suggesting activation by an LPS-independent pathway.

In addition to the above, a 14 kDa fimbrial protein from *Sal. enteritidis* was shown to induce strong T-cell responses. This was illustrated by the stimulation of IL-2 release from T-cells of mice immunized with the organism [200].

Thus, fimbriae appear to play a role in induction of cytokine synthesis and this is obviously related, at least in the case of *E. coli*, with the binding to cell-surface receptors.

7.4.2.2.5   Unidentified surface-associated proteins

Part of the dogma of microbiology is that most, if not all, bacteria have an extracellular component, normally known as the capsule, which is composed of carbohydrates. Work from the laboratories of two of the authors (B. Henderson and M. Wilson) has shown that gently stirring bacteria in normal saline releases from the bacterial surface a surprising amount of protein with smaller amounts of carbohydrate. On the basis of electron microscopic examination of the cells before and after extraction, the eluted components are part of the loosely associated material present on the bacterial surface. For example, short-term saline extraction of the oral Gram-negative bacterium *Actinobacillus actinomycetemcomitans* releases up to 10% of the dry weight of the cell and the material released is 70% protein. If this material is separated on two-dimensional SDS/PAGE, then up to 60 different proteins or protein subunits can be identified [201] (Figure 7.5). While it is possible that a proportion of these proteins are the result of cell lysis, bacteria can easily tolerate exposure to normal saline and the bacteria appear fully viable after extraction. Thus, we believe that the bacterial surface is a distinct milieu that contains a collection of, possibly unique, proteins which play an important role in bacterial–bacterial and bacterial–eukaryotic communication.

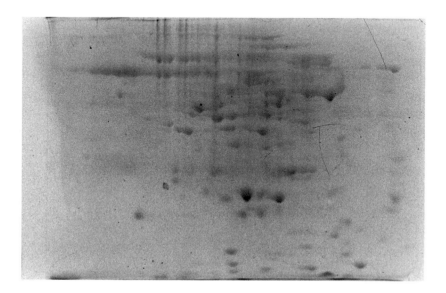

Figure 7.5 Two-dimensional SDS/PAGE of the surface-associated proteins of *Actinobacillus actinomycetemcomitans*

The material eluted from the surface of bacteria contains protein, carbohydrate and some lipid and, on the basis of protein and amino acid analysis, each bacterium has different patterns of proteins [202]. In order to be able to describe this material we have suggested an operational term — surface-associated material (SAM). It must be emphasized that this is an operational term and does not give any information about composition, and that each sample of SAM (even from the same bacterium) may have a different composition.

An interesting, although controversial, finding was that the SAM from *Actinobacillus actinomycetemcomitans* exhibited IL-1-like activity in bioassays for IL-1 [182]. SAM from a number of oral bacteria has subsequently been found to be capable of stimulating the release of cytokines (IL-1, IL-6, IL-8 and TNF) from various human cell populations, including monocytes, gingival fibroblasts and neutrophils ([180,202] and Reddi, K., Collins, P.D., Nair, S.P., Anderson, R., Meghji, S., Wilson, M. and Henderson, B., unpublished work; for reviews see [204,205]). Of particular interest was the finding that, with some oral bacteria, the SAM was considerably more potent than the corresponding LPS and could be as potent as the incredibly active hexa-acyl lipid A-containing LPS from *E. coli* (see Chapter 5 for a description of lipid A structure). For example, the SAM from *Actinobacillus actinomycetemcomitans* was able to stimulate IL-6 production by HPBMCs at a concentration of 10 pg/ml, the threshold concentration at which the *E. coli* LPS began to show activity [180].

Much of our attention has focused on the ability of the SAM (or components of this heterogeneous mixture) to stimulate IL-6 synthesis — a cytokine which has important biological and pathological actions [206]. The mechanism of induction of IL-6 transcription is also interesting. The available literature suggests that induction of IL-6 transcription requires the prior production and release of other pro-inflammatory cytokines such as IL-1 or TNFα [206]. Highly purified *E. coli* LPS, as expected, stimulated IL-6 secretion from human gingival fibroblasts, and such synthesis was completely blocked by adding neutralizing antibodies to IL-1 (α and β), IL-1ra or TNFα. The glucocorticoid dexamethasone (at 1 nM) also completely blocked LPS-induced IL-6 synthesis. In contrast, the SAM from *A. actinomycetemcomitans* stimulated human gingival fibroblast IL-6 synthesis, but such synthesis was not affected by neutralizing IL-1 or TNFα, and dexamethasone, even at high concentrations, could only inhibit IL-6 production by 50%. Using reverse transcriptase PCR to amplify the cytoplasmic mRNA for cytokines, it was found that LPS induced the synthesis of mRNA for IL-1 and IL-6, but that the SAM only induced mRNA for IL-6. The component responsible for IL-6 synthesis has been purified and found to be a 2 kDa peptide. These data suggest that this peptide is inducing IL-6 gene transcription in mesenchymal cells by a novel mechanism which does not involve the known controlling elements on the IL-6 gene [207]. It is of interest that when these bacteria were grown in liquid medium the IL-6-inducing component had a molecular mass of 60 kDa. Our interpretation of this is that the bacterium produces a binding factor so that this peptide is not lost into the medium (B. Henderson and M. Wilson, unpublished work). The possibility that this binding factor is the 60 kDa protein, chaperonin 60, needs to be considered.

In contrast to the potency of Gram-negative SAM to induce cytokine synthesis, our experience is that Gram-positive bacteria are much less active in this respect. Most of our experience is with *Staphylococcus aureus*, and very high concentrations of the SAM from this organism are required to induce cellular cytokine synthesis [208]. Indeed, we have recently shown that the SAM from this organism contains cytokine-inhibiting components and these will be discussed in section 7.4.3.

Other workers have also reported that soluble surface-associated proteins have the capacity to activate myelomonocytic cells and to induce cytokine synthesis. For example, Mai et al. [209] reported that *H. pylori*, an organism associated with gastric lesions, could stimulate human monocytes to express human leucocyte antigen D-region and the IL-2 receptor, and could also induce the synthesis of both IL-1 and TNF. As few as 1000 intact bacteria were able to induce significant cytokine synthesis. Extraction of LPS-free surface-associated proteins and LPS revealed that the former fraction was both significantly more potent and efficacious in stimulating cytokine synthesis. There has been increasing interest in this organism in recent years and a number of recent reports have suggested that *H. pylori* releases factors that can stimulate epithelial cells to produce leukotrienes and IL-8 [210] and also a protein which activates neutrophils [211]. As described for oral bacteria, the surface-associated protein fraction of *H. pylori* contained a number of distinct proteins and thus the potency of the active moiety is probably high. In an attempt to identify the component of *H. pylori* responsible for stimulating IL-8 release from epithelial cells, Huang et al. [212] concluded that a surface protein (possibly the CagA protein) was the most likely candidate.

A surface protein — the SR protein of *Streptococcus mutans* — has also been reported to stimulate human monocytes to release IL-1, IL-6 and TNF in the presence of polymyxin B [213].

7.4.2.2.6   Protein A
Most strains of *Staph. aureus* have a 42 kDa surface protein that is able to interact non-specifically with the Fc region of mammalian immunoglobulins. This protein, known as protein A, has also been shown to stimulate human mononuclear cells to release IL-1, IL-4, IL-6, TNFα and IFNγ and, in this study, was more active than other cell-wall constituents, including muramyl dipeptide, muramic acid and teichoic acid [214].

7.4.2.2.7   Molecular chaperones
All cells contain families of unrelated proteins of molecular masses ranging from 10 kDa to >100 kDa, whose function is to aid the correct folding and

transmembrane transport of proteins. These 'folding' proteins have been variously named heat-shock proteins, stress proteins or molecular chaperones. Using the correct terminology, constitutive proteins are classed as molecular chaperones and inducible proteins are termed heat-shock proteins/stress proteins. The main function of these proteins is to prevent the build-up of non-functional denatured proteins in cells. Most work has been done on three families of molecular chaperones of molecular masses 60, 70 and 90 kDa. To complicate matters, the 60 kDa protein is known as a chaperonin (abbreviated cpn) and requires an additional protein, known as cpn10, to produce a fully functional protein-folding unit.

Bacterial molecular chaperones are major immunogens and much of the immune response to infectious organisms is to these proteins [215]. It is only within the past few years that it has been realized that molecular chaperones have functions in addition to their intracellular protein-folding actions. This literature has been recently reviewed [216].

There are now a number of reports of the capacity of molecular chaperones to induce cytokine synthesis. Friedland et al. [217] reported that the 65 kDa heat-shock protein (cpn60) from *Mycobacterium leprae* induced the synthesis of mRNA for TNF in a human monocytic cell line at a concentration of 10 ng/ml. The protein was also capable of stimulating the release of IL-6 and IL-8 from monocytes at a concentration of 10 µg/ml. Another heat-shock protein from this organism, with a molecular mass of 71 kDa (hsp70), has also been shown to stimulate murine intraepithelial lymphocytes to release IL-3, GM-CSF, IFNγ and IL-6, but not IL-2, IL-4, IL-5 or TGFβ [218]. Subsequently, heat-shock proteins from a number of bacterial species (*E. coli*, *Legionella pneumophila*, *M. leprae*, *M. bovis*) have been shown to stimulate the accumulation of mRNA for IL-1α, IL-1β, IL-6, GM-CSF and TNFα in murine macrophages [219]. The activation of cytokine synthesis by molecular chaperones may be different from that induced by other stimuli such as LPS. For example, Peetermans et al. [220] have reported that the cpn60 from *Mycobacteria* stimulates human monocytes to produce pro-inflammatory cytokines, but fails to activate these cells. Mycobacterial cpn60 also causes the induction of various adhesion molecules (E-selectin, ICAM-1 and VCAM-1) on cultured human vascular endothelial cells [221]. However, the induction of adhesion molecules (which is normally due to cytokines such as IL-1 or TNF) could not be blocked by neutralizing antibodies to these cytokines. This suggests that this mycobacterial chaperonin could directly induce these vascular adhesion molecules and therefore that this molecule has inherent cytokine-like properties.

The authors' studies of molecular chaperones have largely concentrated on the *E. coli* cpn60 (a molecule also known as GroEL) and the cpn10 of *M. tuberculosis*. While cpn60 proved incapable of inducing a range of mesenchymal cells (fibroblasts, osteoblasts, etc.) to produce pro-inflammatory cytokines, it was found to be a potent inducer of human monocytes IL-6 synthesis. Stimulation of IL-6 was not blocked by neutralizing antibodies to CD14 (e.g. My4), demonstrating that activation of cytokine synthesis is not via an LPS-dependent pathway and suggesting that monocytes have a receptor for this intracellular protein-folding complex. *E. coli* cpn60 is composed of two stacked heptameric rings giving a complex of approximately 800 kDa. However, it appears that the complex is not required for cell activation, as heating or trypsinization of cpn60 has little effect on the cytokine-inducing activity of this molecule [222]. Mycobacterial cpn10 showed a similar pattern of cell activation, only acting on human monocytes. Surprisingly, the numbers of cells used was important, with 2×10^6 human monocytes having no activity, but 5×10^6 monocytes being responsive to cpn10. Native cpn10 is a dome-shaped heptamer with each subunit comprising approximately 100 amino acids. Using a series of truncated peptides we have shown that the activation of human monocyte pro-inflammatory cytokine synthesis is due to a linear set of residues in the roof and backbone of cpn10 (Reddi, K., Nair, S.P., Crean, S.J., Meghji, S., Poole, S., Wilson, M., Coates, A.R.M., Mascagni, P. and Henderson, B., unpublished work). This finding that small peptides can mimic the activity of intact cpn10 suggests that peptide antagonists could be developed.

The study of the extracellular roles of molecular chaperones is still in its infancy, but we believe that these molecules may play important roles in tissue pathology through their ability to induce cytokine synthesis. The molecular chaperones are ancient molecules and it is possible that the interaction between bacterial molecular chaperones and host cells is an ancient one which has only now been re-discovered.

7.4.2.3 Extracellular proteins as cytokine inducers
Bacteria export a range of proteins including toxins and proteases. The cytokine-modulating activity of such proteins will be discussed in this section.

7.4.2.3.1 Superantigens
The ability of certain bacterial exotoxins to activate large proportions of the peripheral T-lymphocyte populations of man and mouse, as opposed to the small numbers of lymphocytes activated by processed peptide antigens, led Kappler and Marrack to name them 'superantigens' [224]. While superantigens are classified as exotoxins their nature and mechanism of action is such that they will be dealt with separately from the other exotoxins in this section.

Superantigens are proteins which bind directly to MHC class II molecules without processing. Instead of binding in the groove of the MHC receptor (as peptide antigens do), superantigens bind to the lateral surfaces of the MHC protein and the Vβ region of the T-cell receptor. Each superantigen can bind to one or a few of the 20–50 Vβ proteins which exist in mice or humans, and thus a superantigen can stimulate between 2% and 20% of the T-cells present in an organism.

There has been much speculation about the role of superantigens, but no firm conclusions have been reached (see [225]). It is obvious that superantigens interfere, in a general sense, with the regulation of acquired immunity. Thus, in addition to activating large numbers of T-cells and associated B-cells they can cause the deletion of developing T-cells in the thymus. Whatever the ultimate role for superantigens turns out to be it is obviously an important bacterial defence mechanism, as many bacteria have now been shown to produce such proteins.

Most superantigens isolated to date are produced by *Staph. aureus* and by streptococcal species. Among the superantigens released by *Staph. aureus* are the group of enterotoxins consisting of six different serotypes (staphylococcal enterotoxins A–E), toxic shock syndrome toxin 1 (TSST-1) and an exfoliative toxin [226]. *Strep. pneumoniae* produces pyrogenic exotoxins, SPEA, B and C [226]. Two Gram-negative bacteria, *Yersinia enterocolitica* [227] and *Pseudomonas aeruginosa* [228], in addition to mycoplasmas [229], are also reported to produce superantigens. To date, most studies have concentrated on the superantigens from *Staph. aureus.* The intensive study of this organism owes much to the life-threatening condition, toxic shock syndrome, which is caused by a superantigen called TSST [230].

Injection of mice with a superantigen such as staphylococcal enterotoxin B (SEB) results in the rapid appearance of a variety of cytokines — TNF, IL-1, IL-6 and IFNγ — in the serum and, in D-galactosamine-sensitized mice, the injection of as little as 2 µg of SEB per mouse resulted in 50% mortality which could be blocked by a neutralizing antibody to TNFα [231]. The mechanism of cytokine generation *in vivo* is obviously complex, involving antigen-presenting cells, B-lymphocytes and T-lymphocytes. However, in cell culture it appears that binding of the superantigens to class II molecules is the trigger for cytokine production [232] and, in the case of staphylococcal enterotoxin A (SEA), that cross-linking of two MHC class II molecules by one superantigen is required for cytokine gene expression [233]. Cytokine synthesis, such as IL-1, is dependent on protein tyrosine phosphorylation [234]. The involvement of the MHC is clearly seen when class II-deficient mice are used, these animals being resistant to the effects of staphylococcal enterotoxins [235]. In studies where the dose-dependent activation of cytokine synthesis by SEA has been determined in cultures of murine monocytes, 200 ng/ml was sufficient to induce maximal production of TNFα with an IC_{50} of <50 ng/ml [236]. Another study, which examined staphylococcal enterotoxins B and E and TSST-1, reported that 1 ng/ml of these superantigens was sufficient to cause maximal production of pro-inflammatory cytokines [237]. TSST-1 and SEB, at a concentration of 100 ng/ml, have also been reported to stimulate the release of IL-12 from HPBMCs [238]. This is a key cytokine in inducing protective Th_1 immune responses to infecting organisms. In addition to inducing pro-inflammatory cytokines, Cavaillon and co-workers have demonstrated that the superantigenic exotoxin A and C from *Strep. pyogenes* can stimulate HPBMCs to produce the anti-inflammatory cytokines IL-10 and IL-1ra [239].

A number of investigations have also demonstrated that there are synergistic interactions between superantigens and endotoxins which could be important in the induction of tissue pathology [235,240,241].

7.4.2.3.2   Bacterial exotoxins

The previous section has dealt with one type of bacterial toxin. Of course, bacteria produce a wide variety of toxins and many of the most feared diseases of humanity — tetanus, botulinism, cholera, diphtheria, anthrax, plague — are caused by the actions of bacterial exotoxins. The prevalence and severity of these infectious diseases has resulted in bacterial exotoxins receiving a great deal of attention from the scientific community, and the molecular nature and cellular activity of a number of these toxins are now known in some detail (reviewed in [242]). Even Mims' finds it difficult to define the term toxin and it is helpful at this point to briefly review the mechanisms of bacterial toxins (see

Table 7.5 Some examples of the various classes of bacterial toxins

Type	Example	Bacterial source	Mechanism of action
Enzymic	α-Toxin (phospholipase C)	*C. perfringens*	Hydrolyses phosphorylcholine in cell membranes
	β-Toxin (sphingomyelinase)	*S. aureus*	Hydrolyses sphingomyelin in cell membranes
	Tetanus toxin	*C. tetanus*	Metalloproteinase cleaving synaptobrevins
Pore-forming	Listeriolysin	*L. monocytogenes*	Pore-forming
	α-Toxin	*S. aureus*	Pore forming
	δ-Toxin	*S. aureus*	Pore forming
	Pneumolysis	*S. pneumoniae*	Pore forming
A-B toxins (can both activate and inhibit host cell function)			
	Diphtheria toxin	*C. diphtheria*	ADP-ribosylates host elongation factor-2/stops protein synthesis
	Cholera toxin	*V. cholerae*	ADP-ribosylates host regulatory proteins disrupts cyclic AMP control
	Shiga toxin	*Sh. dysenteriae*	Cleaves host cell RNA stops protein synthesis
Superantigenic toxins	Enterotoxins	*S. aureus*	Activates T-cells bearing particular V_β subsets of receptors
Cell-cycle toxins	*Pasteurella multocida* toxin	*Pasteurella multocida*	Potent stimulator of cell proliferation
See Chapter 3 for a more detailed description of bacterial exotoxins.			

Table 7.5 and Chapter 3). Various excreted bacterial enzymes — proteases, hyaluronidase and lipases — which have a general damaging mode of action are classified as toxins. Other toxins are enzymes with specific substrates. For example, it has come as a surprise to find that the extremely potent Clostridial neurotoxins, which are the causative agents of tetanus and botulism, are zinc-dependent metalloproteinases whose targets are proteins of the muscle–nerve synaptic junction (e.g. synaptobrevin and syntaxin) responsible for neuro-transmitter release [243]. Metalloproteinases are increasingly being shown to be involved in the control of homeostatic and pathological states. Cell-surface metalloproteinases have recently been shown to catalyse the synthesis of mature TNFα from its precursor and this literature has been reviewed in Chapter 8. One very important eukaryotic cell-surface metalloendopeptidase is neutral endopeptidase (EC 3.4.24.11) which cleaves and inactivates a range of pharmacologically active oligopeptides, such as the enkephalins, bradykinin, calcitonin-related gene peptide (CGRP), endothelins, etc. It has been reported that homozygous transgenic knockouts lacking neutral endopeptidase are much more susceptible to endotoxin shock than non-transgenic animals, suggesting that the action of this enzyme is protective in sepsis [244].

Other toxins also have selective enzymic activities including ADP ribosyl transferase activity, adenylate cyclase activity and phospholipase activity. Some toxins act at the plasma membrane (e.g. cholera toxin), while others need to get into cells to promote their effects (e.g. diphtheria toxin and Shiga toxin). Anthrax toxin is an example of a enzymic toxin which enters cells by a fasci-nating mechanism. The toxin consists of three components, none of which are themselves toxic: oedema factor, protective antigen and lethal factor. The first two components are believed to contribute to different forms of the disease. Protective antigen binds to the cell membrane, and after proteolytic cleavage forms a binding site which can be competed for between lethal factor and oedema factor. If oedema factor binds to protective antigen it is internalized, and inside the cell it interacts with calmodulin to form an active adenylate cyclase which raises intracellular cyclic AMP levels. Cyclic AMP is a secreta-gogue, which probably explains the oedematous response to this toxin. Lethal factor, in contrast, when internalized (as it can be by a variety of cells) is selec-tively cytotoxic for macrophages, suggesting that these cells are essential for the lethal actions of this protein.

The plasma membrane of host cells is a major target of toxins. Substantial damage to the membranes of cells can compromise cell function and be lethal. Some toxins can act enzymically to destroy membranes. For example, *Clostridium perfringens* α-toxin is a phospholipase C which removes the phosphate head group from phosphatidylcholine and sphingomyelin of plasma membranes, resulting in cell damage. The other major mechanism of membrane damage utilized by bacteria is the formation of pore-forming toxins. A number are thiol-dependent, for example, the pneumolysin of *Strep. pneumoniae* and listeriolysin of *Listeria monocytogenes*. Others, such as the

Table 7.6 Bacterial exotoxins stimulating cytokine synthesis

Toxin	Cytokines induced	Minimal concentration required for induction	Remarks
Anthrax lethal toxin	IL-1, TNF	Atomolar	Most potent cytokine inducer
Anthrax oedema toxin	IL-6	Femtomolar	Inhibits LPS
Pneumolysin	IL-1, TNF	Femtomolar	High efficacy
C. difficile toxin B	IL-1, IL-6, TNF	Picomolar	Synergizes with LPS
S. aureus exotoxins	IL-1, TNF	Pico/nanomolar	Synergizes with LPS
Shiga-like toxins	IL-1, IL-6, TNF	Low nanomolar	Synergizes with cytokines
Streptococcal enterotoxins	IL-1, IFN, lympho-toxin		
Streptococcal mitogenic factor	IL-1, IFN, lympho-toxin		
Listeriolysin	IL-1		
Cholera toxin	IL-6, IL-10		Inhibits LPS
Pertussis toxin	IL-1, IL-4		Inhibits LPS
E. coli haemolysin	IL-1, TNF	Nanomolar	Releases preformed cytokines
S. aureus haemolysin	IL-1, TNF		Releases preformed cytokines
S. aureus β-toxin	IL-1, IL-6, sCD14		Releases preformed cytokines
Pseudomonas aeruginosa toxin A	IL-1		

haemolysin of *E. coli* and *Proteus vulgaris*, and the leukotoxin of *Pasteurella multocida*, contain nine amino acid repeats and are known as RTX (repeats in toxin) toxins. In addition to producing a wide variety of superantigens *Staph. aureus* produces a range of pore-forming toxins known as α-, δ- and γ-toxins, which have variations on the themes described for the other pore-forming toxins. The various classes of bacterial toxins are defined in Table 7.5.

A new facet of the study of bacterial toxins has been the discovery, and one which is still largely unrecognized, that they have the capacity to influence the control of cytokine networks. Whether this ability has any role in the control of bacterial virulence is open to speculation. The variety of bacterial exotoxins which have the capacity to stimulate, or inhibit, cytokine synthesis is shown in Table 7.6. It should be noted that a growing number of bacterial exotoxins appear to be more potent inducers of cytokine synthesis than they are toxins. Indeed, the most potent cytokine-inducing molecules turn out to be toxins, showing activity at incredibly low (femtomolar to atomolar) concentrations. The bacterial cytokine-inducing molecules which have been described in this chapter all appear to stimulate the synthesis of cytokines, presumably by increasing the rate of transcription or translation of the cytokine under study. Exotoxins also have this capacity but, in addition, have the ability to release IL-1β and TNFα from cultured monocytes/macrophages in a process which does not require gene transcription or translation.

7.4.2.3.2.1   *Membrane disrupting/pore-forming toxins*

E. coli haemolysin (ECH), an RTX haemolysin, is one of the most prevalent bacterial toxins, being produced by half of the enteropathogenic isolates of this bacterium. This protein produces pores in the cell membranes of granulocytes and monocytes. Purified ECH at nanomolar concentrations caused rapid depletion of the cellular ATP levels in human monocytes, leading to cell death. However, at subcytocidal doses (10-fold lower) there was a rapid release of large amounts of IL-1β from these monocytes. Such release was not due to the stimulation of protein synthesis, as it could not be blocked by transcription or translation inhibitors, suggesting that release was caused by the processing and release of preformed intracellular IL-1β precursor. Whether this is due to activation of ICE [245] or to some other, as yet undiscovered, pathway of IL-1β processing, has not been defined [246]. The α-toxin from *Staphylococcus aureus*, which has similar mechanistic properties to ECH, also caused the release of preformed IL-1β from cultured human monocytes at concentrations which are cytolytic. However, subcytolytic concentrations of this toxin resulted in the release of TNFα by an, as yet, unexplained mechanism [247]. ECH has also been reported to induce the release of a range of low-molecular-mass mediators (including free radicals, leukotrienes, histamine and serotonin) from mixed human blood cells (containing lymphocytes, monocytes and basophils) but, in contrast to the above reports, to inhibit the release of pro-inflammatory cytokines [248,249]. This difference in response to this toxin may relate to the nature of the cell populations used or to differences in the culture conditions used. Listeriolysin O (LLO), produced by *L. monocytogenes,* is a 60 kDa sulphydryl-activated pore-forming cytotoxin and a major virulence factor for this organism, which allows the bacterium to escape from phagocytic vesicles within cells, where it would be killed. This protein, in contrast to the toxins discussed above, has also been shown to potently induce murine macrophages to produce IL-1 mRNA and protein [250,251]. In a more recent study, exposure of murine peritoneal exudate and spleen cells to LLO resulted in the transcription of the genes for IL-1α, IL-6, IL-10, IL-12 and TNFα. The pore-forming and subsequent haemolytic action of LLO is known to be blocked by cholesterol. However, cholesterol had no effect on the ability of LLO to induce cytokine production. Thus, the cytokine-inducing activity of this protein does not seem to be directly related to pore-forming properties of LLO [252].

One of the most potent cytokine-inducing bacterial molecules described to date is the thiol-pore-forming toxin of *Strep. pneumoniae* known as pneumolysin. This is a 53 kDa protein with amino acid homology to LLO, described above. However, unlike these cytotoxins, pneumolysin is a cytoplasmic protein only released when bacteria lyse. Pneumolysin is an important virulence factor as shown by the loss of virulence of a pneumolysin-deficient mutant [253]. Purified pneumolysin has been shown to stimulate human monocytes to synthesize both IL-1β and TNFα and to induce synthesis of the former cytokine at low femtomolar concentrations. In addition to this incredible potency, pneumolysin was significantly more efficacious than LPS. Thus, at a concentration of 1 ng/ml this toxin induced the synthesis/release of three to four times the amount of IL-1 that 50 ng/ml LPS induced [254]. Pneumococcal pneumonia is characterized by one of the highest fevers, with core temperature rising to 40°C within 12 h. As *Strep. pneumoniae* has no endotoxin or other known pyrogens the explanation for the potent pyrogenic activity of this organism may be due to the potent ability of pneumolysin to induce the synthesis of pyrogenic cytokines.

As is patently obvious by now, *Staph. aureus* is a heavyweight producer of toxins. The β-toxin produced by this organism, which is a membrane-disrupting cytotoxin, was shown some three decades ago to be a sphingomyelinase [255]. This toxin appears to be involved in *Staph. aureus* infections in skin and in bovine mastitis. In a recent study, it was shown to have a remarkable selective cytotoxicity for human monocytes. Monocytes exposed to the toxin were rapidly killed and in the process released large amounts of IL-1β, IL-6 receptor and the LPS receptor CD14 [256]. These actions are reminiscent of the activities of the pore-forming toxins from *E. coli* and *Staph. aureus*.

It is surprising that these membrane-disrupting cytotoxic toxins also have the ability to release, or induce the synthesis of, pro-inflammatory cytokines. The incredible potency of pneumolysin as a cytokine-inducing agent has come as a surprise and it is legitimate to suggest that all pore-forming/membrane-disrupting bacterial toxins will have some degree of cytokine-inducing activity.

7.4.2.3.2.2 A-B toxins

A-B toxins include such well known examples as cholera toxin, diphtheria toxin, anthrax toxin, Shiga toxin and tetanus toxin. A number of these toxins ADP-ribosylate intracellular proteins; i.e. they remove the ADP-ribosyl group from NAD and attach it covalently to selected proteins (e.g. cyclic AMP regulatory protein or elongation factor-2), thus inhibiting or stimulating the actions of the modified proteins.

Various bacterial species produce a toxin able to be neutralized by antibodies to Shiga toxin (from *Shigella dysenteriae*) and, in consequence, have been termed Shiga-like toxins (SLTs). The SLTs from *E. coli* are also called verocytotoxins because of their actions on a cell line called *vero*. The SLTs are an example of an A-B toxin. The simplest type of A-B toxin (such as Shiga toxin) is produced as a single polypeptide chain which has one cell-binding and one enzymic site. Cleavage of the polypeptide following cell binding allows the enzymic portion of the toxin to enter the cell. In the case of Shiga toxin the enzymic portion acts to inhibit cellular protein synthesis. Infection with Shiga toxin-producing bacteria produces dysentery or haemorrhagic colitis, and patients are at risk of developing life-threatening renal and central nervous system complications.

SLTs bind to cell-surface glycolipids, globotriaosylceramide (Gb_3) or globotetraosylceramide (Gb_4) [257]. These are the same glycolipids that bind to bacterial fimbriae and which have been described in section 7.4.1.2. SLTs were shown to be cytotoxic for human vascular endothelial cells [258] and this suggests a mechanism to explain tissue pathology. However, the levels of toxin required to damage endothelial cells was higher than that expected to be found in patients. An answer to this conundrum came with the finding that the cytotoxic activity of SLTs was markedly potentiated (by up to six log orders) by pro-inflammatory cytokines [259,260]. SLTs are unable to stimulate endothelial cell cytokine synthesis, but they can stimulate the release of pro-inflammatory cytokines by murine and human macrophages [261–263]. Interestingly, the kinetics of TNF synthesis by murine macrophages exposed to LPS or SLTs was completely different. LPS rapidly induces LPS gene transcription and translation with mRNA levels peaking at 1 h. In contrast the levels of TNF mRNA in SLT-treated cells peaked after 6 h.

As mentioned above the SLTs bind to the neutral glycolipid Gb_3/Gb_4 receptor. It would appear that this is not the receptor on the monocyte which induces cytokine synthesis and that some other receptor mechanism is involved [262].

What is the mechanism of the cytokine-induced increase in the sensitivity of vascular endothelial cells to SLTs. One explanation is that the expression of Gb_3 on vascular endothelial cells can be up-regulated by pro-inflammatory cytokines [264]. A more detailed analysis of the mechanism of action of TNFα on vascular endothelial cells suggests that binding of this cytokine to the TNF-R55 increases the activity of a galactosyl transferase which incorporates galactose into ceramide-containing glycolipids [265]. Thus, it would appear that the SLTs have a dual mechanism of action as shown schematically in Figure 7.6. Release of the toxins induces the synthesis of pro-inflammatory cytokines which up-regulate the SLT receptor numbers on vascular endothelial cells, allowing these cells to then respond to the toxin. This is an excellent example of the bacterium getting the host to do the bacterium's dirty work.

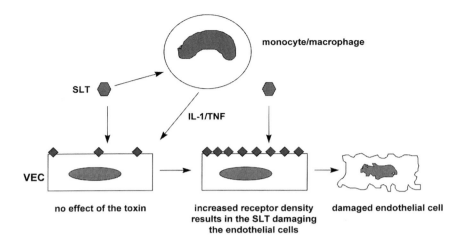

Figure 7.6 Interaction between the exotoxin, shiga-like toxin (SLT) and cyto-kines in the cytotoxic action of this molecule on vascular endothelial cells (VECs)
The toxin has little effect on cultured VECs. However, if the VECs are stimulated with pro-inflammatory cytokines they synthesize more of the globoseries receptors for SLT (◆) and become responsive to this toxin. *In vivo* the cytokines would be produced by leucocytes (probably monocytes or tissue macrophages) in response to SLT and the cytokines so-produced would then feedback on to the VECs and up-regulate the globoseries receptor for the toxin. Once sufficient numbers of receptors had been induced, the cells would be killed by the toxin.

Anthrax is caused by *Bacillus anthracis* and is transmitted largely by the spore form of this organism. The spores germinate at the site where they enter the body and produce the anthrax toxin whose subunits have been described earlier in this section. Oedema factor is presumably responsible for the local oedema. After local multiplication of bacteria they may spread to regional lymph nodes and then to the blood where they reach high concentrations generating the normally fatal systemic form of anthrax. This systemic form resembles a rapid form of septic shock and indeed the first overt sign of the disease in animals is death. It is thought that lethal factor is responsible for the shock-like fatal pathology. Lethal factor enters a variety of cells but is selectively cytotoxic for macrophages [266].

A role for macrophages and macrophage-derived cytokines in the lethal pathology of anthrax has recently been proposed. Mice died in response to the injection of protective antigen and lethal factor. However, depletion of macrophages by administration of silica rendered mice refractory to lethal factor. Sensitivity to this toxin could be restored by co-injection of lethal factor-sensitive (but not toxin-insensitive) macrophages. Administration of neutralizing antibodies to TNFα and IL-1 completely abrogated the lethal effects of this toxin. When murine macrophages were incubated with lethal factor it proved to be the most potent inducer of TNF synthesis yet reported, producing a marked bell-shaped dose response. Lethal factor was able to stimulate TNF synthesis at the incredibly low molar concentration of 12 atomolar (10^{-18} M) and maximal stimulation was seen with a concentration of 1.2 femtomolar (10^{-15} M). Thus, this toxin wins the cup for being the most active cytokine-inducing molecule yet described. The cytotoxic effects of lethal factor on macrophages, in contrast, only occurred when concentrations were raised to greater than 1 nM [267]. This is another example of the discrepancy that is increasingly being seen with toxins, between the concentration required to induce 'toxicity', and the lower concentrations required to induce cytokine synthesis.

It is not only lethal factor which induces cytokine synthesis. The oedema factor is also a potent inducer of the synthesis of IL-6 by human monocytes showing a remarkably steep dose–response relationship over the concentration range 1–4 ng/ml (11–44 picomoles). When oedema factor and LPS were incubated with monocytes no synergistic or inhibitory effects were noted. However, this toxin proved to be a potent inhibitor of TNFα synthesis [268], and the mechanism of action will be discussed in the next section on cytokine-inhibiting components from bacteria.

Cholera toxin, which is composed of an enzymic A subunit linked to five cell-surface-binding B subunits, like the oedema factor described above, is able to stimulate intracellular cyclic AMP levels in enterocytes, leading to massive loss of water and electrolytes. In addition, this toxin, unlike most other soluble proteins, is also a potent oral antigen which induces a Th_2-type cytokine profile in the Peyer's patches of the gastrointestinal tract [269–271]. The possibility exists that the adjuvant effect of cholera toxin is due to its ability to induce cytokine synthesis and there are now a number of reports of this toxin having this action. Thus, cholera toxin was found to enhance antigen presentation by macrophages and this was linked to increased production of both soluble and cell-associated IL-1 [272]. Direct injection of cholera toxin into murine ligated intestinal loops induced the transcription of IL-6 and IL-10, but not IFNγ, IL-1, IL-2, IL-4 or TNF [273]. The intestinal epithelial cell would be one of the first cells to encounter cholera toxin. Cultured intestinal epithelial cells stimulated with this toxin showed elevated IL-1 and IL-6 synthesis [274]. It was subsequently shown that cholera toxin rapidly stimulated rat intestinal epithelial cells to produce IL-6 mRNA and to secrete this cytokine. The secretion of IL-6 could be mimicked by dibutyryl cyclic AMP, suggesting that the IL-6-inducing action of cholera toxin is related to its ability to raise intracellular cyclic AMP levels [275]. The induction of IL-6 gene transcription and release by fibroblasts is linked to a cyclic AMP-dependent pathway in fibroblasts [276]. It has also been demonstrated that cholera toxin acts synergistically with the cytokines IL-1, TNFα and $TGF\beta_1$ in inducing IL-6 synthesis [275]. Thus, intestinal epithelial cells, on encountering cholera toxin, can produce large amounts of IL-6. Low concentrations of cholera toxin can also induce mast cells to release IL-6, but inhibit TNFα production [277]. The relevance of this finding to bacterial control of cytokine networks will be discussed in the next section on cytokine-network-inhibitory proteins. *E. coli* heat-labile enterotoxin, a closely related homologue of cholera toxin, has been shown to stimulate CD4 T-cells and induce IL-2 synthesis via its ganglioside-binding B subunits [278].

Pertussis toxin is produced by the bacterium *Bordatella pertussis*, the causative organism of whooping cough. This toxin also results in the activation of adenylate cyclase and has been reported to stimulate IL-4 production, which may explain its ability to up-regulate IgE responses [279]. *Pseudomonas aeruginosa* is an opportunistic pathogen which normally infects immunocompromised patients. It produces an A-B toxin (exotoxin A) which acts, like diphtheria toxin and the SLTs, to inhibit protein synthesis. Exotoxin A has been reported to induce murine peritoneal macrophages to secrete IL-1 [280].

7.4.2.3.2.3   *Other toxins*

The only other toxins which, to date, have been shown to be potent stimulators of cytokine production are toxins A and B from *Clostridium difficile*, which are associated with the symptoms of pseudomembranous colitis, an acute, often antibiotic-associated, inflammatory bowel disease. These are the largest toxins known, with molecular masses of 308 kDa (toxin A) and 270 kDa (toxin B) respectively. Toxin A is an enterotoxin which has mild cytotoxic activity and toxin B is a cytotoxin which does not elicit enterotoxicity [281]. They have been cloned and sequenced and are approximately 50% identical at the amino acid level and have similar primary structures. The C-termini of these toxins contain a region consisting of repeating oligopeptides and are involved in receptor binding. The N-termini of both toxins are thought to contain the toxin activity and be involved in cell entry [282]. Although these toxins bind to, and enter, cells they are not the classic A-B toxins. A receptor on rabbit ileal cell for toxin A has recently been identified as the disaccharidase sucrase-isomaltase [283].

These clostridial toxins have a unique mechanism of action in that they glucosylate a key residue (Thr-37) in the Rho family of low-molecular-mass GTP-binding proteins, thus affecting its activity. Rho proteins are involved in the receptor-mediated regulation of the actin cytoskeleton of eukaryotic cells and this explains the changes in the shape of cells exposed to these toxins [282].

In addition to the actions described, these toxins have been reported to act on leucocytes. For example the addition of clostridial toxins to human lymphocytes was shown to inhibit antigen- and mitogen-driven proliferation, but to leave IL-2 activation unaffected [284]. Toxin A at a concentration of 1 nM was reported to induce mouse peritoneal macrophages to release IL-1 [285]. This latter work has been verified using human monocytes and, in addition, it was shown that toxin B was a 1000-fold more active than toxin A (active at 10^{-12} M) in inducing the synthesis of the pro-inflammatory cytokines IL-1β, IL-6 and TNFα. The kinetics of cytokine production, and the positive effect of protein synthesis inhibitors, support the conclusion that this toxin is stimulating cytokine synthesis. Of interest was the finding that toxin B and LPS acted synergistically in the induction of cytokines [286]. The nature of the receptor on leucocytes transducing the activity of this toxin is still unresolved.

To conclude this section it is clear that many of the exotoxins produced by bacteria have the capacity to induce cytokine synthesis, and in a growing number of cases these toxins are more potent inducers of cytokine synthesis than they are cell toxins. Indeed, the most potent inducers of cytokine gene transcription are toxins, with anthrax toxin, lethal factor and pneumolysin showing activity at femto- to attomolar concentrations. Thus, it is possible that one key function of toxins, in addition to their accepted actions, is to influence local cytokine networks. In the case of the SLTs the effect of the toxin on vascular endothelial cells appears to require the induction of cytokines. It is likely that such synergy between toxins and cytokines is a common theme and one that needs to be explored in detail. Exotoxins can also synergize with other bacterial mediators such as LPS [241,286]. Again, the role of LPS and other CD14-related molecules in the actions of bacterial exotoxins is an area which would repay detailed exploration.

The proteins produced by bacteria which are capable of inducing pro-inflammatory cytokine production are listed in Table 7.7.

7.4.3.2.3 Bacterial proteinases and pro-inflammatory cytokines

Bacteria produce a very large number of proteinases with a wide range of substrate actions, and some of these enzymes function as exotoxins [287].

Table 7.7 Bacterial proteins stimulating cytokine synthesis

Protein	Cytokines induced	Lowest concentration inducing cytokines (ng/ml)
Bacterial exotoxins	IL-1, IL-6, IFN, TNF, lymphotoxin	Attomolar to nanomolar*
Protein A	IL-1, IL-6, TNF	1000
Fimbrial proteins	IL-1, TNF, GM-CSF	100–1000
Outer-membrane proteins	IL-1, TNF	100–1000
Molecular chaperones	IL-1, IL-3, IL-6, TNF	100
Porins	IL-1, IL-6, IL-8, IFN, TNF	10
Chaperonin 60	IL-1, IL-6	10
Superantigens (many)	IL-1, IL-6, TNF	1–10
Endotoxin-associated proteins	IL-1, IL-6, TNF	1
Lipoproteins (many)	IL-1, IL-6, TNF	0.5
Glycoproteins		0.05
Surface-associated proteins	IL-1, IL-6, IL-8, TNF	0.01

*See Table 8.6.

There is now growing evidence that bacterial proteinases can modulate cytokine networks and play important roles in up- or down-regulating inflammatory and immunological responses in infections. For example, exotoxin B (also known as streptopain), the conserved extracellular cysteine proteinase of *Strep. pyogenes*, can cleave pro-IL-1β between His-115 and Asp-116 to produce biologically active IL-1β [288]. This proteinase also cleaves a streptococcal C5a peptidase, producing a fragment which blocks C5a-induced neutrophil migration [289], and is able to produce pro-inflammatory kinins from the precursor H-kininogen [290]. This proteinase can therefore be equated to a pro-inflammatory signal.

There are also a number of examples of proteinases which have been shown to inactivate cytokines. *Pseudomonas aeruginosa* produces an alkaline phosphatase and an elastase, both of which can inactivate a range of human cytokines including IL-1, IL-2, IFNγ and TNFα [291]. A proteinase from *Legionella pneumophila* can inactivate IL-2 and cleave CD4 on human T-lymphocytes [292]. The authors of this book have recently shown that the oral Gram-negative bacterium *P. gingivalis* releases proteases that are capable of cleaving and inactivating IL-1, IL-1ra, and IL-6, even in the presence of serum proteins [293]. The active enzyme is believed to be R-1 protease which, as has been described earlier, undergoes autolytic breakdown producing a cytokine-inducing 17 kDa fragment.

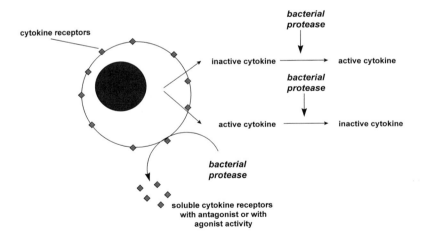

Figure 7.7 The possible roles for bacterial proteases in controlling cytokine networks

Such proteases could activate cytokines (such as IL-1β and TNFα) which are released from cells as inactive precursors requiring proteolytic processing. Equally, proteases could inactivate cytokines by proteolytic cleavage. Bacterial proteases have also been shown to release cytokine receptors from cell surfaces. This would have the effect of releasing cytokine-neutralizing (antagonist) proteins which would down-regulate the activity of the particular cytokines bound by the released receptors.

To complicate matters, a recent report has shown that certain bacterial metalloproteinases can liberate membrane IL-6 receptors (IL-6Rs). Soluble IL-6Rs can inactivate soluble IL-6 or, conversely, bind to IL-6R-negative cells and confer sensitivity to IL-6 by oligomerizing with gp130 (see Chaper 4) [294]. As IL-6 appears to have both pro- and anti-inflammatory properties the consequences of receptor release is unclear. The generality of this effect becomes a key question.

Thus, in considering the local cytokine networks produced by commensal or infectious bacteria, it is important to appreciate that if these organisms also release proteases this can have effects on cytokine networks. Such effects will obviously depend on the substrate specificity of the proteases. The possible ways in which proteases could modulate cytokine networks is represented diagrammatically in Figure 7.7.

7.4.3 Bacterial proteins that can inhibit pro-inflammatory cytokine networks

The major theme running through this book is that bacteria, and commensal bacteria in particular, can exert control over local cytokine networks. The previous sections have explained in some detail that bacteria contain many molecules, in addition to LPS, which can stimulate host cells to produce cytokines. Most of these studies have concentrated on cytokines which have pro-inflammatory actions. This bias represents a historical focus on such cytokines and the relative ease with which they can now be measured. However, with the finding that viruses produce cytokine-modulating proteins, there is the impetus to examine the ability of bacteria to (i) inhibit pro-inflammatory cytokine induction and (ii) stimulate the production of anti-inflammatory cytokines. Having established that viruses have this capacity it would be logical to conclude that bacteria, with their vastly larger genomes, would also be capable of producing cytokine-inhibitory proteins (or other molecules). However, it is only within the past few years that reports have appeared describing how bacteria produce proteins which can modulate pro-inflammatory cytokine networks. Of course, it must be realized that even LPS can induce anti-inflammatory cytokines such as IL-10 [295]. Indeed, it has recently been proposed that IL-10 and TGFβ may be responsible for the process of endotoxin desensitization [296], although the role of IL-10 is controversial [297].

The regulation of intracellular cyclic nucleotides is believed to be a major controlling factor in the synthesis of certain cytokines, particularly IL-6 and TNFα. Increased intracellular levels of cyclic AMP are thought to promote the synthesis of IL-6 but block the formation of TNF [276,298]. A number of bacterial toxins with ADP-ribosylating activity up-regulate the activity of adenylate cyclase and therefore raise intracellular cyclic AMP levels. Cholera toxin has been reported to be a potent stimulator of IL-6 production, to have some capacity to induce $TGF\beta_1$ transcription, but to inhibit the synthesis of

TNF [275,277]. Pertussis toxin inhibits the B-cell and macrophage response to LPS and is a potent inhibitor of IL-1 synthesis [277a]. Likewise, the oedema factor of anthrax toxin, which is an adenylate cyclase, stimulates monocyte IL-6 synthesis, but inhibits TNFα synthesis [268]. It is not clear what the overall response would be to a moiety which promoted IL-6 but inhibited TNFα synthesis. The role of IL-6 in inflammation and infection is still unclear. A recent study using IL-6 knockout mice has proposed that IL-6 has an overall protective effect in *E. coli* infections, but that it has no role in LPS-induced septic shock [299]. Thus, it is possible that exotoxins selectively inducing IL-6 synthesis could act to increase antibacterial actions of the host.

Exotoxin A, an ADP-ribosylating toxin of *Pseudomonas aeruginosa* which inhibits protein synthesis, inhibited lymphocyte proliferation and the synthesis of IL-1, TNFα, lymphotoxin and IFNγ at concentrations lower than that required to induce toxicity. The toxin was significantly more active in inhibiting IL-1α synthesis than in inhibiting the synthesis of IL-1β [300].

Pasteurella haemolytica produces an RTX leukotoxin which has immuno-suppressive properties [301] and down-regulates MHC class II expression on leucocytes [302]. These properties are likely to be related to some down-regulation of cytokine action.

Yersinia enterocolitica is a facultatively intracellular bacterium responsible for a range of food-borne diseases including gastroenteritis, invasive colitis and lymphadenitis. Treatment of mice with antibodies to IFNγ and TNFα exacerbates primary infections with this bacterium, demonstrating the role of these cytokines in the control of yersinia infections [303]. It has been shown that the survival and proliferation of *Y. enterocolitica* in the host depends upon a 70-kb plasmid encoding a number of released outer proteins, including a 41 kDa protein, YopB, which inhibits the synthesis of TNFα, while having no effect on the production of IL-1 or IL-6. YopB causes a significant decrease in the steady-state levels of TNFα in murine macrophages and may have a similar mechanism of action to that of oedema factor from anthrax toxin, in raising intracellular cyclic AMP levels. This has not yet been determined, however [304].

A protein from *Salmonella typhimurium* called *S. typhimurium*-derived inhibitor of T-cell proliferation (STI) has been reported to suppress T-cell proliferation, augmenting IFNγ and inhibiting IL-2 synthesis and also inhibiting IL-2 receptor and post-receptor events in CD3 lymphocytes [305]. *Brucella* species (which are intracellular organisms) have recently been reported to produce a high-molecular-mass protein which specifically inhibits TNFα synthesis by human, but not by murine, macrophages [306].

Cytokines are intimately involved in the process by which circulating leucocytes enter sites of inflammation/infection. It has been reported that the yeast-like organism *Cryptococcus neoformans* produces polysaccharides which induce the shedding of the adhesion protein, L-selectin and also TNF receptors from human neutrophils [307], and that surface-membrane proteins from *Helicobacter pylori*, *E. coli* and *Staphylococcus aureus* have similar actions [308].

Thus, bacteria involved in human infections can modulate cytokine synthesis and potentially down-regulate immune and inflammatory mechanisms. Is there any evidence that organisms of the normal commensal microflora can do the same? It must be borne in mind that, as such bacteria do not cause disease, they have been the subject of much less attention.

Klapproth and co-workers [309] have addressed the question of whether *E. coli* contains molecules able to regulate lymphocyte activation and cytokine production. Whole bacteria and lysates of bacteria from a range of *E. coli* strains were examined for their ability to inhibit cytokine production by peripheral blood mononuclear cells. Lysates from two pathogenic strains (out of 21 tested) inhibited mitogen-stimulated expression of mRNA for IL-2, IL-4, IL-5 and IFNγ and were able to inhibit lymphocyte proliferation. However, there was no inhibition of the intracellular levels of mRNA for the pro-inflammatory cytokines IL-1β, IL-6, IL-12 or RANTES or of the anti-inflammatory cytokine IL-10. Indomethacin did not block the cytokine inhibitory activity showing that the mechanism was not via prostaglandin-induced up-regulation of intracellular cyclic AMP levels. Heat treatment and dialysis suggested that activity was due to a protein with molecular mass >8 kDa, giving an approximate potency in the nanomolar range. Enteropathogenic strains of *E. coli* bind to intestinal epithelial cells and activate signal transduction pathways. This appears to be due to the secretion of proteins from the bacterium, and five distinct proteins have recently been identified in this process of epithelial cell activation. One of these has been shown to be the product of the *eae*B gene, known to be involved in signal transduction, and another was demonstrated to have significant similarity to the glycolytic enzyme glyceraldehyde-3-phosphate dehydrogenase (see the discussion below on the EGF receptor on *Mycobacterium tuberculosis*). The other three proteins of molecular mass 110, 40 and 25 kDa were unique [310]. The possibility exists that they are related to the immunosuppressive protein(s) described by Klapproth.

Of course these *E. coli* strains are not strictly commensal bacteria, being enteropathogenic, and their ability to produce proteins able to selectively inhibit lymphokine production, particularly IL-2, ties in with the colonic inflammation of the IL-2 knockout mouse [87]. However, this illustrates the point that bacteria can, by modulating cytokine networks, control local immune and inflammatory events.

The oral Gram-negative bacterium *A. actinomycetemcomitans*, which is a normal commensal organism but which is implicated in the pathology of the periodontal diseases [204], is a bacterium remarkable for the plethora of immunomodulatory factors that it produces. Our own studies have shown that this bacterium contains a number of potent, potentially immunomodulatory, proteins on its outer surface. This includes an 8 kDa protein which we have termed gapstatin, which is a potent inhibitor of cell (including lymphocyte) proliferation with a unique mechanism of action. This protein blocks cells in G_2 but acts in early S-phase by a mechanism which, we believe, involves inhibition of cyclin B synthesis ([311,312]; B. Henderson and M. Wilson, unpublished work). A 14 kDa cytoplasmic protein (which may be related to gapstatin) has recently been isolated from *A. actinomycetemcomitans* and shown to inhibit lymphocyte proliferation and suppress the production of IL-2, IL-4, IL-5 and IFNγ by ConA-stimulated murine CD4 lymphocytes [313,314]. This protein has been N-terminal sequenced and has no homology with any proteins in the sequence databases. It is possible that it may be related to the *E. coli* immunomodulatory protein. Another potential immunomodulatory factor that the authors have shown to be produced by this organism is the 2 kDa peptide found on the bacterium's surface and described earlier [207]. This protein directly induces the transcription for the IL-6 gene but does not induce the synthesis of IL-1 or TNF. As stated, IL-6 has recently been claimed to have a protective role in infection [300] and therefore such induction of IL-6 synthesis may act to down-regulate pathogenic cytokine networks.

In addition to these proteins, *A. actinomycetemcomitans* also produces a leukotoxin, which kills neutrophils and monocytes and is immunosuppressive [315], and a 60 kDa immunosuppressive protein which appears to act on B-lymphocytes [316].

In unpublished studies of the Gram-positive bacterium *Staphylococcus aureus*, the authors have found that it contains an unresolved activity on its surface which can block the binding of LPS to monocytes. We have also isolated a 39 kDa protein which has the ability to inhibit LPS- and IL-1-

Table 7.8 Bacterial proteins inhibiting cytokine synthesis

Protein	Cytokine whose synthesis is inhibited
Cholera toxin	TNF (not IL-6 or TGFβ)
Pertussis toxin	IL-1
Anthrax oedema toxin	TNF (stimulates IL-6)
Pseudomonas aeruginosa exotoxin A	IL-1, TNF, lymphotoxin, IFNγ
Botulinum toxin type D	TNF
Yersinia enterocolitica YopB	TNF (no effect on IL-1 or IL-6)
Proteus mirabilis 39 kDa protein	IL-1
S. typhimurium-derived inhibitor of T-cell proliferation (STI)	IL-2 (stimulates IFNγ)
Brucella-derived protein	TNF (not IL-1 or IL-6)
Enteropathogenic *E. coli* proteins	IL-2, IL-4, IL-5, IFNγ (not IL-1, IL-6, IL-12 or RANTES)
14 kDa *A. actinomycetemcomitans* protein	IL-2, IL-4, IL-5, IFNγ
S. aureus 39 kDa protein	LPS and IL-1-induced IL-6
Measles virus	IL-12

induced monocyte IL-6 production. The mechanism of action of this protein has not been defined.

The various proteins which act to inhibit cytokine synthesis are listed in Table 7.8.

7.5   Conclusions

As has been discussed extensively in other chapters, cytokines are key local hormones which act to maintain cellular and tissue homeostasis and which are also central regulators of host defences to infectious bacteria. In this chapter it has been clearly shown that bacteria contain an enormously wide range of components, of all classes of biomolecules, able to stimulate eukaryotic cells to synthesize and release cytokines. Such stimulation appears to occur by two distinct receptor-mediated events: CD14 and non-CD14. In addition, some of these non-CD14 cytokine-inducing components are extremely potent. One of the most surprising findings from the review of the literature is that the large group of bacterial exotoxins contain some of the most active cytokine inducers, demonstrating activity at concentrations in the femto- to atomolar levels. Indeed, it appears that all the different classes of exotoxins, which have widely different mechanisms of action, can promote cytokine release or synthesis. Moreover, many of these molecules are more active as cytokine inducers than they are as cell toxins.

Most of the published work has concentrated on the capacity of bacterial components to stimulate the synthesis of the pro-inflammatory cytokines IL-1, IL-8, TNFα and the less clearly defined IL-6, which has immunomodulatory actions including inhibition of inflammation. This would suggest that these components have a predisposition to induce inflammation. Having said this, it is worth stating the caveat that without having a global overview of the patterns of the cytokines that any particular bacterial component induces, it is difficult to clearly define it as pro- or anti-inflammatory. In the past few years a number of reports have begun to appear suggesting that certain bacteria produce proteins which can block the induction of pro-inflammatory cytokines. Thus, certain bacterial toxins can selectively inhibit TNF synthesis and other bacteria, including members of the normal commensal microflora, produce proteins that selectively inhibit lymphokine synthesis. It is probable that these proteins represent the tip of an iceberg of cytokine-inhibiting and cytokine-modulating molecules produced by bacteria. It is also likely that many of the molecules that bacteria produce will be recognized as cytokines. The interactions between bacterial components (modulins and bacteriokines) and host cells, in terms of controlling local cytokine networks, are shown schematically in Figure 7.8.

It is likely that many more surprises await us in terms of the interaction between cytokines and bacteria. Three recent example will be given. The first is the report that the presence of IL-10 on the surface of macrophages

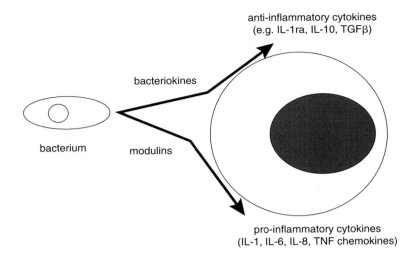

anti-inflammatory cytokines
(e.g. IL-1ra, IL-10, TGFβ)

bacteriokines

bacterium   modulins

pro-inflammatory cytokines
(IL-1, IL-6, IL-8, TNF chemokines)

Figure 7.8 The result of the interaction of bacterial modulins and bacteriokines with a host cell in terms of the patterns of cytokines induced

influences their ability to kill intracellular bacteria [317]. Direct interactions between cytokines and bacteria have been reported and in the second example it is now established that bacteria have receptors for a number of cytokines. For example, a receptor for EGF has been recognized on mycobacterial species including *Mycobacterium tuberculosis*. These bacteria respond to EGF by increasing their growth rate. This receptor has been isolated and the N-terminal sequence shows that it is the glycolytic enzyme glyceraldehyde-3-phosphate dehydrogenase [318]. The third report concerns the nature of the biological activity of defensins. These antibacterial peptides will be discussed in more detail in the final chapter. It has recently been proposed that neutrophil defensins HNP-1 and HNP-2 (human neutrophil peptides) are in fact chemokines which act to attract neutrophils to sites of infection [319].

This chapter has reviewed the literature of the non-LPS bacterial factors which induce cytokine synthesis and release, and has introduced the concept of cytokine-network-controlling molecules produced by bacteria. The role of such molecules in the control of host cytokine networks and the role of host molecules (mainly cytokines) in controlling bacterial function including the growth and activity of bacteria will be dealt with in the final chapter of this book, where we will attempt to synthesize the existing data of bacteria–host interactions vis-a-vis cytokines into a new and predictive scientific paradigm. In the next chapter the methods available, or in development, for the pharmacological regulation of cytokine synthesis or activity will be reviewed.

References

1. Janeway, C.A. (1992) The immune system evolved to discriminate infectious nonself from nonin-fectious self. Immunol. Today **13**, 11–16
2. Pugin, J., Heumann, D., Tomasz, A., Kravchenko, V.V., Akamatsu, Y., Nishijima, M., Glauser, M.P., Tobias, P.S. and Ulevitch, R.J. (1994) CD14 is a pattern recognition receptor. Immunity **1**, 509–516
3. Krieger, M. and Hertz, J. (1994) Structure and functions of multiligand lipoprotein receptors: macrophage scavenger receptors and LDL receptor-related protein (LRP). Annu. Rev. Biochem. **63**, 601–637
4. Elomaa, O., Kangas, M., Sahlberg, C., Tuukkanen, J., Sormunen, R., Liakka, A., Thesleff, I., Kraal, G. and Tryggvason, K. (1996) Cloning of a novel bacteria-binding receptor structurally-related to scavenger receptors and expressed in a subset of macrophages. Cell **80**, 603–609
5. Krieg, A.M., Yi, A.-Y., Matson, S., Waldschmidt, T.J., Bishop, G.A., Teasdale, R., Koretzky, G.A. and Klinman, D.M. (1995) CpG motifs in bacterial DNA trigger direct B-cell activation. Nature (London) **374**, 546–549
6. Smith, M.F., Eidlen, D., Arend, W.P. and Gutierrez-Hartmann, A. (1994) LPS-induced expression of the human IL-1 receptor antagonist gene is controlled by multiple interacting promoter elements. J. Immunol. **153**, 3584–3593
7. Weber-Nordt, R.M., Meraz, M.A. and Schreiber, R.D. (1994) Lipopolysaccharide-dependent induction of IL-10 receptor expression on murine fibroblasts. J. Immunol. **153**, 3734–3744
8. Brandtzaeg, P., Osnes, L., Øvstebø, R., Joø, G.B., Westvik, Å.-B. and Kierulf, P. (1996) Net inflam-matory capacity of human septic shock plasma evaluated by a monocyte-based target cell assay: identification of interleukin-10 as a major functional deactivator of human monocytes. J. Exp. Med. **184**, 51–60
9. Tannock, G.W. (1995) Normal Microflora. An Introduction to Microbes Inhabiting the Human Body. Chapman and Hall, London
10. Henderson, B. and Wilson, M. (1995) Modulins: a new class of cytokine-inducing, pro-inflam-matory bacterial virulence factor. Inflamm. Res. **44**, 187–197
11. Henderson, B. and Wilson, M. (1996) Cytokine induction by bacteria: beyond LPS. Cytokine **8**, 269–282
12. Henderson, B., Poole, S. and Wilson, M. (1996) Bacterial modulins: a novel class of virulence factors which cause host tissue pathology by inducing cytokine synthesis. Microbiol. Rev. **60**, 316–341
13. Henderson, B., Poole, S. and Wilson, M. (1996) Microbial/host interactions in health and disease: who controls the cytokine network. Immunopharmacology **35**, 1–21
14. Cossart, P., Boquet, P., Normark, S. and Pappuoli, R. (1996) Cellular microbiology emerging. Science **271**, 315–316
15. Orren, A. (1996) How do mammals distinguish between pathogens and non-self. Trends Microbiol. **4**, 254–257
16. Hostetter, M.K. (1996) An integrin-like protein in *Candida albicans*: implications for pathogenesis. Trends Microbiol. **4**, 242–246
17. Dimmock, N.J. and Primrose, S.B. (1994) Introduction to Modern Virology, 4th edn., Blackwell, Oxford
18. Gooding, L.R. (1992) Viral proteins that counter host immune defences. Cell **71**, 5–7
19. Gooding, L.R. (1994) Regulation of TNF-mediated cell death and inflammation by human adeno-viruses. Infect. Agents Dis. **3**, 106–115
20. Murphy, P.M. (1994) Molecular piracy of chemokine receptors by herpesviruses. Infect. Agents Dis. **3**, 137–154
21. Pickup, D.J. (1994) Poxviral modifiers of cytokine responses to infections. Infect. Agents Dis. **3**, 116–127
22. Smith, G.L. (1994) Virus strategies for the evasion of host response to infection. Trends Microbiol. **2**, 81–88

23. Spriggs, M.K. (1994) Cytokine and cytokine receptor genes 'captured' by viruses. Curr. Opin. Immunol. **6**, 526–529

24. Alcami, A. and Smith, G.L. (1995) Cytokine receptors encoded by poxviruses: a lesson in cytokine biology. Immunol. Today **16**, 474–478

25. DeChiara, T.M., Young, D., Semionow, R., Stern, A.S., Batula-Bernardo, C., Fiedler-Nagy, C., Kaffka, K.L., Kilian, P.L., Yamazaki, S., Mizel, S.B. and Lomedico, P.T. (1986) Structure-function analysis of murine interleukin-1: biologically active polypeptides are at least 127 amino acids long and are derived from the carboxyl terminus of a 270 amino acid precursor. Proc. Natl. Acad. Sci. U.S.A. **83**, 8303–8307

26. Thornberry, N.A., Bull, H.G., Calaycay, J.R. Chapman, K.T., Howard, A.D., Kostura, M.J., Miller, D.K., Molineaux, S.M., Weidner, J.R., Aunins, J., et al. (1992) A novel heterodimeric cysteine protease is required for interleukin-1β processing in monocytes. Nature (London) **356**, 768–774

27. Yuan, J., Shaham, S., Ledoux, S., Ellis, H.M. and Horvitz, H.R. (1993) The *C. elegans* death gene ced-3 encodes a protein similar to mammalian interleukin-1β converting enzyme. Cell **75**, 641–652

28. Vaux, D.L. and Strasser, A. (1996) The molecular biology of apoptosis. Proc. Natl. Acad. Sci. U.S.A. **93**, 2239–2244

29. Ray, C.A., Black, R.A., Kronheim, S.R., Greenstreet, T.A., Sleath, P.R., Salvesen, G.S. and Pickup, D.J. (1992) Viral inhibition of inflammation. Cowpox virus encodes an inhibitor of the interleukin-1β converting enzyme. Cell **69**, 597–604

30. Komiyama, T.K., Ray, C.A., Pickup, D.J., Howard, A. D., Thornberry, N.A., Petersen, E.P. and Salvesen, G. (1994) Inhibition of interleukin-1β converting enzyme by the cowpox virus serpin crmA. J. Biol. Chem. **269**, 19331–19337

31. Chua, T.P., Smith, C.E., Reith, R.W. and Williamson, J.D. (1990) Inflammatory response and the generation of chemoattractant activity in cowpox virus-infected tissues. Immunology **69**, 202–208

32. Frederickson, T.N., Sechler, J.M., Palumbo, G.J., Albert, J., Khairallah, L.H. and Buller, R.M. (1992) Acute inflammatory response to cowpox virus infection of the choroallantoic membrane of the chick embryo. Virology **187**, 693–704

33. Palumbo, G.J., Buller, R.M. and Glasgow, W.C. (1994) Multigenic evasion of inflammation by poxviruses. J. Virol. **68**, 1737–1749

34. Clouston, W.M. and Kerr, J.F. (1985) Apoptosis, lymphocytotoxicity and the containment of viral infections. Med Hypotheses **18**, 399–404

35. Clem, R.J. and Miller, L.K. (1993) Apoptosis reduces both the in vivo replication and the in vivo infectivity of a baculovirus. J. Virol. **67**, 3730–3738

36. Xue, D. and Horvitz, H.R. (1995) Inhibition of the *Caenorhabditis elegans* cell death protease CED-3 by a CED-3 cleavage site in baculovirus p35 protein. Nature (London) **377**, 248–251

37. Bump, N.J., Hackett, M., Hagunin, M., Seshagiri, S., Brady, K., Chen, P., Ferenz, C., Franklin, S., Ghayur, T., Li, P., Licari, P., Mankovich, J., Shi, L.F., Greenberg, A.H., Miller, L.K. and Wong, W.W. (1995) Inhibition of ICE family proteases by baculovirus antiapoptotic protein p35. Science **269**, 1885–1888

38. Rabizadeh, S., Lacount, D.J., Friesen, P.D. and Bredesen, D.E. (1993) Expression of the baculovirus p35 gene inhibits mammalian neural cell death. J. Neurochem. **61**, 2318–2321

39. Martinou, I., Fernandez, P.A., Missotten, M., White, E., Allet, B., Sadoul, R. and Martinou, J.C. (1995) Viral proteins E1B19K and p35 protect sympathetic neurones from cell death induced by NGF deprivation. J. Cell Biol. **128**, 201–208

40. Sugimoto, A., Friesen, P.D. and Rothman, J.H. (1994) Baculovirus p35 prevents developmentally-programmed cell death and rescues a CED-9 mutant in the nematode *Caenorhabditis elegans*. EMBO J. **13**, 2023–2028

41. Clem, R.J., Fechheimer, M. and Miller, L.K. (1991) Prevention of apoptosis by a baculovirus gene during infection of insect cells. Science **54**, 1388–1390

42. Crook, N.E., Clem, R.J. and Miller, L.K. (1993) An apoptosis-inhibiting baculovirus gene with a zinc finger-like motif. J. Virol. **67**, 2168–2174

43.   Cocchi, F., DeVico, A.L., Garzino-Demo, A., Arya, S.K., Gallo, R.C. and Lusso, P. (1995)
      Identification of RANTES, MIP-1α and MIP-1β as the major HIV-suppressive factors produced by
      CD8⁺ T cells. Science **270**, 1811–1815

44.   Bates, P. (1996) Chemokine receptors and HIV-1: an attractive pair. Cell **86**, 1–3

45.   Bleul, C.C., Farzan, M., Choe, H., Parolin, C., Clark-Lewis, I., Sodroski, J. and Springer, T.A. (1996)
      The lymphocyte chemoattractant SDF-1 is a ligand for LESTR/fusin and blocks HIV-1 entry.
      Nature (London) **382**, 829–833

46.   Neurath, A.R., Strick, N. and Sproul, P. (1992) Search for hepatitis B virus cell receptors reveals
      binding sites for interleukin-6 on the virus envelop proteins. J. Exp. Med. **175**, 461–469

47.   Smith, G.L. and Chan, Y.S. (1991) Two vaccinia virus proteins stucturally related to the inter-
      leukin-1 receptor and the immunoglobulin superfamily. J. Gen. Virol. **72**, 511–518

48.   Spriggs, M.K., Hruby, D.E., Maliszewski, C.R., Pickup, D.J., Sims, J.E., Buller, R.M.L. and Van Slyke, J.
      (1992) Vaccinia and cowpox viruses encode a novel secreted interleukin-1 binding protein. Cell
      71, 145–152

49.   Alcami, A. and Smith, G.L. (1992) A soluble receptor for interleukin-1β encoded by vaccinia virus.
      A novel mechanism for virus modulation of the host response to infection. Cell **71**, 153–167

50.   Colotta, F., Re, F., Muzio, M., Bertini, R., Polentarutti, N., Sironi, M., Giri, G., Dower, S.K., Sims,
      J.E. and Mantovani, A. (1993) Interleukin-1 type II receptor: a decoy target for IL-1 that is
      regulated by IL-4. Science **261**, 472–475

51.   Ben-Baruch, A., Michiel, D.F. and Oppenheim, J.J. (1995) Signals and receptors involved in the
      recruitment of inflammatory cells. J. Biol. Chem. **270**, 11703–11706

52.   Gura, T. (1996) Chemokines take center stage in inflammatory ills. Science **272**, 954–956

53.   Cook, D.N., Beck, M.A., Coffman, T.M., Kirby, S.L., Sheridan, J.F., Pragnell, I.B. and Smithies, O.
      (1995) Requirement for MIP-1α for an inflammatory response to viral infection. Science **269**,
      1583–1585

54.   Banchereau, J., Briere, F., Caux, C., Fluckiger, A.-C., Merville, P. and Rousset, F. (1996)
      Interleukin-10. In Therapeutic Modulation of Cytokines (Henderson, B. and Bodmer, M.W., eds.),
      pp. 317–331, CRC Press, Boca Raton

55.   Moore, K.W., Vieira, P., Fiorentino, D.F., Trounstine, M.L., Khan, T.A. and Mossmann, T.R. (1990)
      Homology of the cytokine synthesis inhibitory factor (IL-10) to the Epstein-Barr virus gene
      BCRF1. Science **248**, 1230–1234

56.   Hsu, D.H., de Waa- Malefyt, R., Fiorentino, D.F., Dang, M.N., Viera, P., de Vries, J., Spits, H.,
      Mossmann, T.R. and Moore, K.W. (1990) Expression of interleukin-10 activity by Epstein-Barr
      virus protein BCRF1. Science **250**, 830–832

57.   Vieira, P., de Waal-Malefyt, R., Dang, M.N., Johnson, K.E., Kastelein, R., Fiorentino, D.F., de Vries,
      J.E., Roncarolo, M.G., Mosmann, T.R. and Moore, K.W. (1991) Isolation and expression of human
      cytokine synthesis inhibitory factor (CSIF/IL-10) cDNA clones: homology to Epstein Barr virus
      open reading frame BCRF1. Proc. Natl. Acad. Sci. U.S.A. **88**, 1172–1176

58.   Rhode, H.J., Janssen, W., Rosen-Wolff, A., Bugert, J.J., Thein, P., Becker, Y. and Darai, G. (1993)
      The genome of equine herpesvirus type 2 harbors an interleukin-10 (IL-10)-like gene. Virus Genes
      7, 111–116

59.   Miyazaki, I., Cheung, R.K. and Dosch, H.-M. (1993) Viral interleukin-10 is critical for the induction
      of B cell growth transformation by Epstein-Barr virus. J. Exp. Med. **178**, 439–447

60.   Burdin, N., Peronne, C., Banchereau, J. and Rousset, F. (1993) Epstein-Barr virus-tranformation
      induces B lymphocytes to produce human interleukin-10. J. Exp. Med. **177**, 295–304

61.   Moore, K.W., O'Garra, A., de Waal-Malefyt, R., Vieira, P. and Mosmann, T.R. (1993) Interleukin-
      10. Annu. Rev. Immunol. **11**, 165–190

61a. Moore, P.S., Boshoff, C., Weiss, R.A. and Chang, Y. (1996) Molecular mimicry of human cytokine
      and cytokine response pathway genes by KSHV. Science **274**, 1739–1744

62.   Bell, S., Cranage, M., Borysiewicz, L. and Minson, T. (1990) Induction of immunoglobulin G Fc
      receptors by recombinant vaccinia virus expressing glycoproteins E and I of Herpes Simplex virus
      type 1. J. Virol. **64**, 2181–2186

63. Frey, J. and Einsfelder, B. (1984) Induction of surface IgG receptors in cytomegalovirus-infected human fibroblasts. Eur. J. Biochem. **138**, 213–216

64. McTaggart, S.P., Burns, W.H., White, D.O. and Jackson, D.C. (1978) Fc receptors induced by Herpes simplex virus. 1. Biologic and biochemical properties, Eur. J. Biochem. **121**, 726–730

65. Rawle, F.C., Tollefson, A.E., Wold, W.S. and Gooding, L.R. (1989) Mouse anti-adenovirus cytotoxic T lymphocytes: inhibition of lysis by E3gp19K but not E3 14.7K. J. Immunol. **143**, 2031–2037

66. Anderson, M., McMichael, A. and Peterson, P.A. (1987) Reduced allorecognition of adenovirus-2 infected cells J. Immunol. **138**, 3960–3966

67. McFadden, G. and Kane, K. (1994) How DNA viruses perturb functional MHC expression to alter immune recognition. Adv. Cancer Res. **63**, 117–209

68. Janeway, C.A. and Travers, P. (1997) Immunobiology: the Immune System in Health and Disease, 3rd edn., Blackwell, Oxford

69. Bloom, B.R., Modlin, R.L. and Salgame, P. (1992) Stigma variations: observations on suppressor T cells and leprosy. Annu. Rev. Immunol. **10**, 453–488

70. Mann, G.B., Fowler, K.J., Gabriel, A., Nice, E.C., Williams, R.L. and Dunn, A.R. (1993) Mice with a null mutation of the TGF-α gene have abnormal skin architecture and curly whiskers and often develop corneal inflammation. Cell **73**, 249–262

71. Luetteke, N.C., Qiu, T.H., Peiffer, R.L., Oliver, P., Smithies, O. and Lee, D.C. (1993) TGFα deficiency results in hair follicle and eye abnormailities in targeted and waved-1 mice. Cell **73**, 263–278

72. Shull, M.M., Ormsby, I., Kier, A.B., Pawlowski, S., Diebold, R.T., Yin, M., Allen, R., Sidman, C., Proetzel, G., Calvin, D., Annunziata, N. and Doetschman, T. (1992) Targeted disruption of the mouse transforming growth factor-β$_1$ gene results in multifocal inflammatory disease. Nature (London) **359**, 693–699

73. Kulkarni, A.B., Huh, C.-G., Becker, D., Geiser A., Lyght, M., Flanders, K.C., Roberts, A.B., Sporn, M.B., Ward, J.M. and Karlsson, S. (1993) Transforming growth factor-β$_1$ null mutation in mice causes excessive inflammatory response and early death. Proc. Natl. Acad. Sci. U.S.A. **90**, 770–774

74. Dalton, D.K., Pitts-Meek, S., Keshav, S., Figari, I.S., Bradley, A. and Stewart, T.A. (1993) Multiple defects in immune cell function in mice with disrupted interferon-γ genes. Science **259**, 1739–1742

75. Tanaka, T., Akira, S., Yoshida, K., Umemoto, M., Yoneda, Y., Shirafuji, N., Fujiwara, H., Suematsu, S., Yoshida, N. and Kishimoto, T. (1995) Targeted disruption of the NF-IL-6 gene discloses its essential role in bacterial killing and tumor cytotoxicity by macrophages. Cell **80**, 353–361

76. Rothe, J., Lesslauer, W., Lotscher, H., Lang, Y., Koebel, P., Kontgen, F., Althage, A., Zinkernagel, R., Steinmetz, M. and Bluethmann, H. (1993) Mice lacking the tumor necrosis factor receptor 1 are resistant to TNF-mediated toxicity but highly susceptible to infection by Listeria monocyto-genes. Nature (London) **364**, 798–801

77. Li, P., Allen, H., Banerjee, S., Franklin, S., Herzog, L., Johnston, C., Dowell, J., Paskind, M., Rodman, L., Salfeld, J., et al. (1995) Mice deficient in IL-1β converting enzyme are defective in production of mature IL-1β and resistant to endotoxic shock. Cell **80**, 401–411

78. Fantuzzi, G., Zheng, H., Faggioni, R., Benigni, F., Ghezzi, P., Sipe, J.D., Shaw, A.R. and Dinarello, C.A. (1996) Effect of endotoxin in IL-1β-deficient mice. J. Immunol. **157**, 291–296

79. Shornick, L.P., De Togni, P., Mariathasam, S., Goellner, J., Strauss-Schoenberger, J., Ferguson, T.A. and Chaplin, D.D. (1996) Mice deficient in IL-1β manifest impaired contact hypersensitivity to trinitrochlorobenzene. J. Exp. Med. **183**, 1427–1436

80. Kuida, K., Lippke, J.A., Ku, G., Harding, M.W., Livingstone, D.J., Su, M.-S.S. and Flavell, R.A. (1995) Altered cytokine export and apoptosis in mice deficient in interleukin-1β converting enzyme. Science **267**, 2000–2003

81. Zheng, H., Fletcher, D., Kozak, W., Jiang, M., Hofmann, K., Conn, C.A., Soszynski, D., Grabiec, C., Trumbauer, M.E., Shaw, A., et al. (1995) Resistance to fever induction and impaired acute phase response in interleukin-1β-deficient mice. Immunity **3**, 9–19

82. Fattori, E., Cappelletti, M., Cota, P., Sellitto, C., Cantoni, L., Carelli, M., Faggioni, R., Fantuzzi, G., Ghezzi, P. and Poli, V. (1994) Defective inflammatory response in IL-6-deficient mice. J. Exp. Med. **180**, 1243–1250

83. Libert, C., Takahashi, N., Canwels, A., Brouckaert, P., Bluethman, H. and Fiers, W. (1994) Response of interleukin-6-deficient mice to tumor necrosis factor-induced metabolic changes and lethality. Eur. J. Immunol. **24**, 2237–2242

84. Chai, Z., Gatti, S., Toniatti, C., Poli, V. and Bartfai, T. (1996) Interleukin (IL)-6 gene expression in the central nervous system is necessary for fever responses to lipopolysaccharide or IL-1β: a study on IL-6-deficient mice. J. Exp. Med. **183**, 311–316

85. Salgame, P.R., Abrams, J.S., Clayberger, C., Goldstein, H., Convit, J., Modlin, R.L. and Bloom, B.R. (1991) Differing lymphokine profiles of functional subsets of human CD4 and CD8 T cell clones. Science **254**, 279–289

86. Schorle, H., Holtsche, T., Hunig, T., Schimpl, A. and Horak, I. (1991) Development and function of T cells in mice rendered interleukin-2 deficient by gene targeting. Nature (London) **352**, 621–624

87. Sadlack, B., Merz, H., Schorle, H., Schimpl, A., Feller, A.C. and Horak, I. (1993) Ulcerative colitis-like disease in mice with a disrupted interleukin-2 gene. Cell **75**, 253–261

88. Weinberg, K. and Parkman, R. (1990) Severe combined deficiency due to a specific defect in the production of interleukin-2. N. Engl. J. Med. **322**, 1741–1743

89. Mizoguchi, A., Mizoguchi, E., Chiba, C., Spiekerman, G.M., Tonegawa, S., Nagler-Anderson, C. and Bhan, A.K. (1996) Cytokine imbalance and autoantibody production in T cell receptor-α mutant mice with inflammatory bowel disease. J. Exp. Med. **183**, 847–856

90. Kuhn, R., Lohler, J., Rennick, D., Rajewsky, K. and Muller, W. (1993) Interleukin-10-deficient mice develop chronic enterocolitis. Cell **75**, 253–261

91. Eckmann, L., Kagnoff, M.F. and Fierer, J. (1995) Intestinal epithelial cells as watchdogs for the natural immune system. Trends Microbiol. **3**, 118–120

92. Hurley, J.C. (1995) Endotoxemia: methods of detection and clinical correlates. Clin. Microbiol. **8**, 268–292

93. Cavaillon, J.-M. and Haeffner-Cavaillon, N. (1986) Polymyxin B inhibition of LPS-induced inter-leukin-1 secretion by human monocytes is dependent on the LPS origin. Mol. Immunol. **23**, 965–969

94. Barber, S.A., Perera, P.-Y. and Vogel, S.N. (1995) Defective ceramide response in C3H/Hej (LPSᵈ) macrophages. J. Immunol. **155**, 2303–2305

95. Soell, M., Diab, M., Haan-Archipoff, G., Beretz, A., Herbelin, C., Poutrel, B. and Klein, J. (1995) Capsular polysaccharide types 5 and 8 of Staphylococcus aureus bind specifically to human epithelial (KB) cells, endothelial cells and monocytes to induce release of cytokines. Infect. Immun. **63**, 1380–1386

96. Soell, M., Lett, E., Holveck, F., Scholler, M., Wachsmann, D. and Klein, J.-P. (1995) Activation of human monocytes by streptococcal rhamnose glucose polymers is mediated by CD14 antigen and mannan binding protein inhibits TNFα release. J. Immunol. **154**, 851–860

97. Takahashi, T., Nishihara, T., Ishihara, Y., Amano, K., Shibuyu, N., Moro, I. and Koga, T. (1991) Murine macrophage interleukin-1 release by capsular-like serotype-specific polysaccharide antigens of Actinobacillus actinomycetemcomitans. Infect. Immun. **59**, 18–23

98. Naito, Y., Ohno, J., Takazoe, I. and Okuda, K. (1992) The relationship between polysaccharide antigen and interleukin-1β producing activity in Porphyromonas gingivalis. Bull. Tokyo Dent. Coll. **33**, 187–195

99. Daley, L., Pier, G.B., Liporace, J.D. and Eardley, D.D. (1985) Polyclonal B cell stimulation and interleukin-1 induction by the mucoid exopolysaccharide of Pseudomonas aeruginosa associated with cystic fibrosis. J. Immunol. **134**, 3089–3093

100. Otterlei, M., Sundan, A., Skjak-Braek, G., Ryan, L., Smidsrod, O. and Espevik, T. (1993) Similar mechanisms of action of defined polysaccharides and lipopolysaccharides: characterization of binding and tumor necrosis factor alpha induction. Infect. Immun. **61**, 1917–1935

101. Mancuso, G., Tomasello, F., von Hunolstein, C., Orefici, G. and Teti, G. (1994) Induction of tumor necrosis factor alpha by the group- and type-specific polysaccharides from type III group B streptococci. Infect. Immun. **62**, 2748–2753

102. Schwab, J. (1993) Phlogistic properties of peptidoglycan-polysaccharide polymers from cell walls of pathogenic and normal flora bacteria which colonize humans. Infect. Immun. **61**, 4535–4539

103. Schwab, J. (1995) Bacterial cell wall-induced arthritis: models of chronic recurrent polyarthritis and reactivation of monoarticular arthritis. In Models and Mechanisms in Rheumatoid Arthritis (Henderson, B., Edwards, J.C.W. and Pettipher, E.R., eds.), pp. 439–454, Academic Press, London

104. Chedid, L. (1983) Muramyl dipeptides as possible endogenous immunopharmacological mediators. Microbiol. Immunol. **27**, 723–732

105. O'Reilly, T. and Zac, O. (1992) Enhancement of the effectiveness of anti-microbial therapy by muramyl peptide immunomodulators. Clin. Infect. Dis. **14**, 1100–1109

106. Heumann, D., Barras, C., Severin, A., Glauser, M.P. and Tomasz, A. (1994) Gram-positive cell walls stimulate synthesis of tumor necrosis factor alpha and interleukin-6 by human monocytes. Infect. Immun. **62**, 2715–2721

107. Weidemann, B., Brade, H., Reitschel, E.T., Dziarski, R., Bazil, V., Kusumoto, S., Flad, H.-D. and Ulmer, A.J. (1994) Soluble peptidoglycan-induced monokine production can be blocked by anti-CD14 monoclonal antibodies and by lipid A partial structures. Infect. Immun. **62**, 4709–4715

108. Timeerman, C.P., Matsson, E., Martinez-Martinez, L., DeGraf, L., Van Strijp, J.A.G., Verbrugh, H.A., Verhoef, J. and Fleer, A. (1993) Induction of release of tumor necrosis factor from human monocyte by staphylococci and staphylococcal peptidoglycan. Infect. Immun. **61**, 4167–4172

109. Lichtman, S.N., Wang, J., Schwab, J.H. and Lemasters, J.L. (1994) Comparison of peptidoglycan-polysaccharide and lipopolysaccharide stimulation of Kupffer cells to produce tumor necrosis factor and interleukin-1. Hepatology **19**, 1013–1022

110. Manthey, C.L., Qureshi, N., Stutz, P.L. and Vogel, S.N. (1993) Lipopolysaccharide antagonists block taxol-induced signalling in murine macrophages. J. Exp. Med. **178**, 695–702

111. Dokter, W.H.A, Dijkstra, A. J., Koopmans, S.B., Stulp, B.K., Keck, W., Halie, M.R. and Vellenga, E. (1994) G(Anh)MTetra, a natural bacterial cell wall breakdown product, induces interleukin-1β and interleukin-6 expression in human monocytes. J. Biol. Chem. **269**, 4201–4206

112. Dokter, W.H.A, Dijkstra, A. J., Koopmans, S.B., Mulder, A.B., Stulp, B.K., Halie, M.R., Keck, W. and Vellenga, E. (1994) G(Anh)MTetra, a naturally-occurring 1,6-anhydro muramyl dipeptide, induces granulocyte colony-stimulating factor expression in human monocytes: a molecular analysis. Infect. Immun. **62**, 2953–2957

113. Heiss, L.N., Moser, S.A., Unanue, E.R. and Goldman, W.R. (1993) Interleukin-1 is linked to the respiratory-epithelial cytopathology of pertussis. Infect. Immun. **61**, 3123–3128

114. Onta, T., Sashida, M., Fujii, N., Sugawara, S., Rikiishi, H. and Kumagai, K. (1993) Induction of acute arthritis in mice by peptidoglycan derived from gram-positive bacteria and its possible role in cytokine production. Microbiol. Immunol. **37**, 573–582

115. Riesenfeld-Orn, I., Wolpe, S., Garcia-Bustos, J.F., Hoffmann, M.K. and Tuomanen, E. (1989) Production of interleukin-1 but not tumor necrosis factor by human monocytes stimulated with pneumococcal cell surface components. Infect. Immun. **57**, 1890–1893

116. Tsutsui, O., Kikeguchi, S., Matsumara, K. and Kato, K. (1991) Relationship of the chemical structure and immunobiological activities of lipoteichoic acid from Streptococcus faecalis (Enterococcus hirae) ATCC 9790. FEMS Microbiol. Immunol. **76**, 211–218

117. Bhakdi, S., Klonisch, T., Nuber, P. and Fischer, W. (1991) Stimulation of monokine production by lipoteichoic acids. Infect. Immun. **59**, 4614–4620

118. Manthey, C.L. and Vogel, S.N. (1994) Interactions of lipopolysaccharide with macrophages. Immunol. Ser. **60**, 63–81

119. Standiford, T.J., Arenberg, D.A., Danforth, J.M., Kunkel, S.L., Van Otteren, G.M. and Strieter, R.M. (1994) Lipoteichoic acid induces secretion of interleukin-8 from human blood monocytes: a cellular and molecular analysis. Infect. Immun. **62**, 119–125

120. Mattssom, E., Verhage, L., Rollof, J., Fleer, A., Verhoef, J. and van Dijk, H. (1993) Peptidoglycan and teichoic acid from *Staphylococcus epidermidis* stimulates human monocytes to release tumour necrosis factor-α, interleukin-1β and interleukin-6. FEMS Immunol. Med. Microbiol. **7**, 281–288

121. Danforth, J.M., Strieter, R.M., Kunkel, S.L., Arenberg, D.A., Van Otteren, G.M. and Standiford, T.J. (1995) Macrophage inflammatory protein-1 alpha expression in vivo and in vitro: the role of lipoteichoic acid. Clin. Immunol. Immunopathol. **74**, 77–83

122. De Kimpe, S.J., Kengatharan, J.M., Thiemermann, C. and Vane, J.R. (1995) The cell wall components peptidoglycan and lipoteichoic acid from *Staphylococcus aureus* act in synergy to cause shock and multiple organ failure. Proc. Natl. Acad. Sci. U.S.A. **92**, 10359–10363

123. Kusunoki, T., Hailman, E., Juan, T.S.-C., Lichtenstein, H.S. and Wright, S.D. (1995) Molecules from *Staphylococcus aureus* that bind CD14 and stimulate innate immune responses. J. Exp. Med. **182**, 1673–1682

124. Sugiyama, A., Arakaki, R., Ohnishi, T., Arakaki, N., Daikuhara, Y. and Takada, H. (1996) Lipoteichoic acid and interleukin 1 stimulate synergistically production of hepatocyte growth factor (scatter factor) in human gingival fibroblasts in culture. Infect. Immun. **64**, 1426–1431

125. Zhang, Y., Broser, M., Cohen, H., Bodkin, M., Law, K., Reibman, J. and Rom, W. (1995) Enhanced interleukin-8 release and gene expression in macrophages after exposure to *Mycobacterium tuberculosis* and its components. J. Clin. Invest. **95**, 586–592

126. Moreno, C., Mehlert, A. and Lamb, J. (1988) The inhibitory effects of mycobacterial lipoarabinomannan and polysaccharides upon polyclonal and monoclonal human T cell proliferation. Clin. Exp. Immunol. **74**, 206–210

127. Sibley, L.D., Adams, L.B. and Krahenbuhl, J.L. (1990) Inhibition of interferon-gamma-mediated activation in mouse macrophages treated with lipoarabinomannan. Clin. Exp. Immunol. **80**, 141–148

128. Sibley, L.D., Hunter, S.W., Brennan, P.J. and Krahenbuhl, J.L. (1988) Mycobacterial lipoarabinomannan inhibits gamma interferon-mediated activation of macrophages. Infect. Immun. **56**, 1232–1236

129. Chan, J., Fan, X., Hunter, S.W., Brennan, P.J. and Bloom, B.R. (1991) Lipoarabinomannan, a possible virulence factor involved in persistence of *Mycobacterium tuberculosis* within macrophages. Infect. Immun. **59**, 1755–1761

130. Moreno, C., Taverne, J., Mehlert, A., Bate, C.A.W., Brealey, R.J., Meager, A., Rook, G.A.W. and Playfair, J.H.L. (1988) Lipoarabinomannan from *Mycobacterium tuberculosis* induces the production of tumour necrosis factor from human and murine macrophages. Clin. Exp. Immunol. **76**, 240–245

131. Barnes, P.F., Chatterjee, D., Abrams, J.S., Lu, S., Wang, E., Yamamura, M., Brennan, P.J. and Modlin, R.L. (1992) Cytokine production induced by *Mycobacterium tuberculosis* lipoarabinomannan: relationship to chemical structure. J. Immunol. **149**, 541–547

132. Dahl, K.E., Shiratsuchi, H., Hamilton, B.D., Ellner, J.J. and Toosi, Z. (1996) Selective induction of transforming growth factor β in human monocytes by lipoarabinomannan of *Mycobacterium tuberculosis*. Infect. Immun. **64**, 399–405

133. Hirsch, C.S., Hussain, R., Toossi, Z., Dawood, G., Shahid, F. and Ellner, J. (1996) Cross-modulation by transforming growth factor β in human tuberculosis: suppression of antigen-driven blastogenesis and interferon γ production. Proc. Natl. Acad. Sci. U.S.A. **93**, 3193–3198

134. Zhang, Y., Broser, M. and Rom, W.N. (1994) Activation of the interleukin-6 gene by *Mycobacterium tuberculosis* and lipopolysaccharide is mediated by nuclear factors NF-IL-6 and NF-κB. Proc. Natl. Acad. Sci. U.S.A. **91**, 2225–2229

135. Schlesinger, L.S., Hull, S.R. and Kaufman, T.M. (1994) Binding of the terminal mannosyl units of lipoarabinomannan from a virulent strain of *Mycobacterium tuberculosis* to human macrophages. J. Immunol. **152**, 4070–4079

136. Fenton, M.J. and Vermeulen, M.W. (1996) Minireview: immunopathology of tuberculosis: roles of macrophages and monocytes. Infect. Immun. **64**, 683–690
137. Berman, J.S., Blumenthal, R.L., Kornfeld, H., Cook, J.A., Cruikshank, W.W., Vermeulen, M.W., Chatterjee, D., Belisle, J.T. and Fenton, M.J. (1996) Chemotactic activity of mycobacteral lipoarabinomannans for human blood T lymphocytes in vitro. J. Immunol. **156**, 3828–3835
138. Adams, L. B., Fukutomi, Y. and Krahenbuhl, J.L. (1993) Regulation of murine macrophage effector function by lipoarabinomannan from mycobacterial strains with different degrees of virulence. Infect. Immun. **61**, 4173–4181
139. Roach, T.I.A., Barton, C.H., Chaterjee, D. and Blackwell, J.M. (1993) Macrophage activation: lipoarabinomannan from avirulent and virulent strains of *Mycobacterium tuberculosis* differentially induces the early genes c-fos, KC, JE and tumour necrosis factor-α. J. Immunol. **150**, 1886–1896
140. Roach, T.I.A., Chaterjee, D. and Blackwell, J.M. (1994) Induction of early response genes KC and JE by mycobacterial lipoarabinomannans: regulation of KC expression in murine macrophages by *Lsh/Ity/Bcg* (Candidate *Nramp*). Infect. Immun. **62**, 1176–1184
141. Zhang, Y., Doerfler, M., Lee, T., Guilleman, B. and Rom, W.M. (1993) Mechanisms of stimulation of interleukin-1β and tumour necrosis factor alpha by *Mycobacterium tuberculosis* components. J. Clin. Invest. **91**, 2076–2083
142. Peterson, P.K., Gekker, G., Hu, S., Sheng, W.S., Anderson, W.R., Ulevitch, R.J., Tobias, PS, Gustafson, K.V., Molitor, T.W. and Chao, C.C. (1995) CD14 receptor uptake of non-opsonized *Mycobacterium tuberculosis* by human microglia. Infect. Immun. **63**, 1598–1602
143. Espevik, T., Otterlei, M., Skjak-Braek, G., Ryan, L. and Wright, S.D. (1993) The involvement of CD14 in stimulation of cytokine production by uronic acid polymers. Eur. J. Immunol. **23**, 255–261
144. Otterlei, M., Varum, K.M., Ryan, L. and Espevik, T. (1994) Characterization of binding and TNFα-inducing ability of chitosans on monocytes: the involvement of CD14. Vaccine **12**, 825–832
145. Muhlradt, P.F. and Frisch, M. (1994) Purification and partial biochemical characterization of a *Mycoplasma fermentans*-derived substance that activates macrophages to release nitric oxide, tumor necrosis factor and interleukin-6. Infect. Immun. **62**, 3801–3807
146. Salman, M., Deutsch, I., Tarshis, Y., Naot, Y. and Rottem, S. (1994) Membrane lipids of *Mycoplasma fermentans*. FEMS Microbiol. Letts. **123**, 255–260
147. Kostyal, D.A., Butler, G.H. and Beezhold, D.H. (1994) A 48-kilodalton *Mycoplasma fermentans* membrane protein induces cytokine secretion by human monocytes. Infect. Immun. **62**, 3793–3800
148. Herbelin, A., Ruuth, E., Delorme, D., Michel-Herbelin, C. and Praz, F. (1994) *Mycoplasma arginii* TUH-14 membrane lipoproteins induce production of interleukin-1, interleukin-6 and tumor necrosis factor alpha by human monocytes. Infect. Immun. **62**, 4690–4694
149. Ma, Y., Seiler, K.P., Tai, K.-F., Yang, L., Woods, M. and Weis, J.J. (1994) Outer surface lipoproteins of *Borrelia burgdorferi* stimulate nitric oxide production by the cytokine-inducible pathway. Infect. Immun. **62**, 3663–3671
150. Radolf, J.D., Arndt, L.L., Akins, D.R., Curetty, L.L., Levi, M.E., Shen, Y., Davis, L.S. and Norgard, M.V. (1995) *Treponema pallidum* and *Borrelia burgdorferi* lipoproteins and synthetic lipopeptides activate monocytes/macrophages. J. Immunol. **154**, 2866–2877
151. Radolf, J.D., Norgard, M.V., Brandt, M.E., Isaacs, R.D., Thompson, P.A. and Beutler, B. (1991) Lipoproteins of *Borrelia burgdorferi* and *Treponema pallidum* activate cachectin/tumor necrosis factor synthesis: analysis using a CAT reporter construct. J. Immunol. **147**, 1968–1974
152. Weis, J.J., Ma, Y. and Erdile, L.F. (1994) Biological activities of native and recombinant *Borrelia burgdorferi* outer surface protein A dependence on lipid modification. Infect. Immun. **62**, 4632–4636
153. Norgard, M.V., Arndt, L.L., Akins, D.R., Curetty, L.L., Harrich, D.A. and Radolf, J.D. (1996) Activation of human monocytic cells by *Treponema pallidum* and *Borrelia burgdorferi* lipoproteins and synthetic lipopeptides proceeds via a pathway distinct from that of lipopolysaccharide but involves the transcriptional activator NF-κB. Infect. Immun. **64**, 3845–3852

154. Fikrig, E., Barthold, S.W., Kantor, F.S. and Flavell, R.A. (1990) Protection of mice against the Lyme disease agent by immunizing with recombinant OspA. Science **250**, 553–556

155. Anguita, J., Persing, D.H., Rincon, M., Barthold, S.W. and Fikrig, E. (1996) Effect of anti-interleukin 12 treatment on murine Lyme Borreliosis. J. Clin. Invest. **97**, 1028–1034

156. Hauschildt, S., Hoffman, P., Beuscher, H.-U., Dufheus, G., Heinreich, P., Wiesmuller, K.-H., Jumg, G. and Bessler, W.G. (1990) Activation of bone marrow-derived mouse macrophages by bacterial lipopeptide: cytokine production, phagocytosis and Ia expression. Eur. J. Immunol. **20**, 63–68

157. Shimizu, T., Iwamoto, Y., Yanagihara, Y., Kurimura, M. and Achiwa, K. (1994) Mitogenic activity and the induction of tumor necrosis factor by lipopeptide analogs of the N-terminal part of lipoprotein in the outer membrane of Escherichia coli. Biol. Pharm. Bull. **17**, 980–982

158. Dong, Z., Qi, X. and Fidler, I.J. (1993) Tyrosine phosphorylation of mitogen-activated protein kinases is necessary for activation of murine macrophages by natural and synthetic bacterial products. J. Exp. Med. **177**, 1071–1077

159. de Jong, D.M., Pate, J.L., Kirkland, T.N., Taylor, C.E., Baker, P.J. and Takayama, K. (1991) Lipopolysaccharide-like immunological properties of cell wall glycoproteins isolated from Cytophaga johnsonae. Infect. Immun. **59**, 2631–2637

160. Galdiero, F., Cipollaro de L'Ero, G., Bendetto, N., Galdiero, M. and Tufano, M. (1993) Release of cytokines induced by Salmonella typhimurium porins. Infect. Immun. **61**, 111–119

161. Tufano, M.A., Rossano, F., Catalanotti, P., Liguori, G., Marinelli, A., Baroni, A. and Marinelli, P. (1994) Properties of Yersinia enterocolitica porins: interference with biological functions of phagocytes, nitric oxide production and selective cytokine release. Institute Pasteur Res. Microbiol. **145**, 297–307

162. Tufano, M.A., Rossano, F., Catalanotti, P., Liguori, G., Capasso, C., Ceccarelli, T. and Marinelli, P. (1994) Immunobiological activities of Helicobacter pylori porins. Infect. Immun. **62**, 1392–1399

163. Galdiero, F., Tufano, M.A., Galdiero, M., Masiello, S. and DiRosa, M. (1990) Inflammatory effects of Salmonella typhimurium porins. Infect. Immun. **58**, 3183–3186

164. Galdiero, F., Sommese, L., Scarfogliero, P. and Galdiero, M. (1994) Biological activities — lethality, Schwartzman reaction and pyrogenicity — of Salmonella typhimurium porins. Microbiol. Pathogen. **16**, 111–119

165. Donnelly, J.J., Deck, R.R. and Liu, M.A. (1990) Immunogenicity of a Haemophilus influenzae poly-saccharide-Neisseria meningitidis outer membrane protein vaccine. J. Immunol. **145**, 3071–3079

166. Wetzler, L.W., Ho, Y. and Reiser, H. (1996) Neisserial porins induce B lymphocytes to express costimulatory B7-2 molecules and to proliferate. J. Exp. Med. **183**, 1151–1159

167. Skidmore, B.J., Morrison, D.C., Chiller, J.M. and Weigle, W.O. (1975) Immunologic properties of bacterial LPS II. The unresponsiveness of C3H/HeJ spleen cells to LPS-induced mitogenesis is dependent on the methods used to extract LPS. J. Exp. Med. **142**, 1488–1508

168. Goodman, G.W. and Sultzer, B.M. (1979) Endotoxin protein is a mitogen and polyclonal activator of human B lymphocytes. J. Exp. Med. **149**, 713–723

169. Sultzer, B.M. and Goodman, G.W. (1976) Endotoxin protein: a B cell mitogen and polyclonal activator of C3H/HeJ lymphocytes. J. Exp. Med. **144**, 821–827

170. Hitchcock, P.J. and Morrison, D.C. (1984) The protein component of bacterial endotoxins. In Handbook of Endotoxin Vol I Chemistry of Endotoxin (Rietschel, E.T., ed.), pp. 339–374, Elsevier, North Holland

171. Killion, J.W. and Morrison, D.C. (1986) Protection of C3H/HeJ mice from lethal Salmonella typhimurium LT2 infection by immunization with lipopolysaccharide-lipid-A-associated protein complexes. Infect. Immun. **54**, 1–8

172. Sultzer, B.M., Craig, J.P. and Castagna, R. (1985) The adjuvant effect of pertussis endotoxin protein in modulating the immune response to cholera toxoid in mice. Dev. Biol. Stand. **61**, 225–232

173. Sultzer, B.M., Craig, J.P. and Castagna, R. (1986) Endotoxin-associated proteins and their polyclonal and adjuvant activities. In Immunobiology and Immunopharmacology of Bacterial Endotoxins (Szentivanyi, A. and Friedman, H., eds.), pp. 435–447, Plenum, New York

174. Sultzer, B.M., Craig, J.P. and Castagna, R. (1987) Immunomodulation by outer membrane proteins associated with the endotoxin of Gram-negative bacteria. Prog. Leukocyte Biol. **6**, 113–123

175. Hogan, M.H. and Vogel, S.N. (1987) Lipid A-associated proteins provide an alternative 'second signal' in the activation of recombinant interferon-γ-primed, C3H/HeJ macrophages to a fully tumoricidal state. J. Immunol. **139**, 3697–3702

176. Johns, M.A., Sipe, J.D., Melton, L.B., Strom, T.B. and McCabe, W.R. (1988) Endotoxin-associated protein: interleukin-1-like activity on serum amyloid A synthesis and T lymphocyte activation. Infect. Immun. **56**, 1593–1601

177. Bjornson, B.H., Agura, E., Harvey, J.H., Johns, M., Andrews, R.G. and McCabe, W.R. (1988) Endotoxin-associated protein: a potent stimulus for human granulocytopoietic activity which may be accesory cell independent. Infect. Immun. **56**, 1602–1607

178. Porat, R., Yanoov, M., Johns, M.A., Shibolet, S. and Michalevicz, R. (1992) Effects of endotoxin-associated protein on hematopoiesis. Infect. Immun. **60**, 1756–1760

179. Mangan, D.F., Wahl, S.M., Sultzer, B.M. and Mergenhagen, S.E. (1992) Stimulation of human monocytes by endotoxin-associated protein: inhibition of programmed cell death (apoptosis) and potential significance in adjuvanticity. Infect. Immun. **60**, 1684–1686

180. Reddi, K., Wilson, M., Poole, S., Meghji, S. and Henderson, B. (1995) Relative cytokine-stimulating activities of the surface components of the periodontopathogenic bacterium *Actinobacillus actinomycetemcomitans*. Cytokine **7**, 534–541

181. Reddi, K., Poole, S., Nair, S., Meghji, S., Henderson, B. and Wilson, M. (1995) Lipid A-associated proteins from periodontopathogenic bacteria induce interleukin-6 production by human gingival fibroblasts and monocytes. FEMS Microbiol. Immunol. **11**, 137–144

182. Harvey, W., Kamin, S., Meghji, S. and Wilson, M. (1987) Interleukin-1-like activity in capsular material from *Haemophilus actinomycetemcomitans*. Immunology **60**, 415–418

183. Czarny, A., Witkowska, D. and Mulczyk, M. (1993) Induction of tumor necrosis factor and interleukin-6 by outer membrane proteins of Shigella in spleen cells and macrophages of mice. Arch. Immunol. Ther. Exp. **41**, 153–157

184. Korn, A., Kroll, H.-P., Berger, H.-P., Kahler, A., Hebler, R., Brauburger, J., Muller, K.-P. and Nixdorf, K. (1993) The 39-kilodalton outer membrane protein of *Proteus mirabilis* is an ompA protein and mitogen for murine B lymphocytes. Infect. Immun. **61**, 4915–4918

185. Weber, G., Link, F., Ferber, E., Munder, P.G., Zeitter, D., Bartlett, R.R. and Nixdorf, K. (1993) Differential modulation of the effects of lipopolysaccharide on macrophages by a major outer membrane protein of *Proteus mirabilis*. J. Immunol. **151**, 415–424

186. Hedges, S.R. and Svanborg, C. (1995) Urinary tract infection: microbiology, pathogenesis and host response. Curr. Opin. Infect. Dis. **8**, 39–42

187. Hedges, S., Agace, W. and Svanborg, C. (1995) Epithelial cytokine responses and mucosal cytokine networks. Trends Microbiol. **3**, 266–270

188. Linder, H., Engberg, I., Hoschutsky, H., Mattsby-Baltzer, I. and Svanberg, C. (1991) Adhesion-dependent activation of mucosal interleukin-6 production. Infect. Immun. **59**, 4357–4362

189. Hedges, S., Svensson, M. and Svanborg, C. (1992) Interleukin-6 response of epithelial cell lines to bacterial stimulation in vitro. Infect. Immun. **60**, 1295–1301

190. Kreft, B., Bohnet, S., Carstensen, O., Hacker, J. and Marre, R. (1993) Differential expression of interleukin-6, intercellular adhesion molecule 1 and major histocompatibility complex class II molecules in renal carcinoma cells stimulated with S fimbriae of uropathogenic *Escherichia coli*. Infect. Immun. **61**, 3060–3063

191. Svanborg, C., Agace, W., Hedges, S., Linder, H. and Svensson, M. (1993) Bacterial adherence and epithelial cell cytokine production. Zentralbl. Bakteriol. **278**, 359–364

192. Svensson, M., Lindstedt, R., Radin, N.S. and Svanborg, C. (1994) Epithelial glucosphingolipid expression as a determinant of bacterial adherence and cytokine production. Infect. Immun. **62**, 4404–4410

193. Wright, S.D. and Kolesnick, R.N. (1995) Does endotoxin stimulate cells by mimicking ceramide? Immunol. Today **16**, 297–302

194. Mathias, S.A., Younes, S.A., Kan, C.C., Orlow, I., Joseph, C. and Kolesnick, R.N. (1993) Activation of the sphingomyelin signalling pathway in intact EL4 cells and a cell-free system by IL-1β. Science **259**, 519–522

195. Hedlund, M., Svensson, M., Nilsson, Å., Duan, R-D. and Svanborg, C. (1996) Role of ceramide-signalling pathway in cytokine responses to P-fimbriated *Escherichia coli*. J. Exp. Med. **183**, 1037–1044

196. Ogawa, T., Kusumoto, Y., Uchida, H., Nagashima, S., Ogo, H. and Hamada, S. (1991) Immunobiological activities of synthetic peptide segments of fimbrial proteins from *Porphyromonas gingivalis*. Biochem. Biophys. Res. Commun. **180**, 1335–1341

197. Hanazawa, S., Murakami, Y., Hirose, K., Amano, S., Ohmori, Y., Higuchi, H. and Kitano, S. (1991) *Bacteroides (Porphyromonas) gingivalis* fimbriae activate mouse peritoneal macrophages and induce gene expression and production of interleukin-1. Infect. Immun. **59**, 1972–1977

198. Kawata, Y., Hanazawa, S., Amano, S., Murakami, Y., Matsumoto, T., Nishida, K. and Kitano, S. (1994) *Porphyromonas gingivalis* fimbriae stimulate bone resorption. Infect. Immun. **62**, 3012–3016

199. Matsushita, K., Nagaoka, S., Arakaki, R., Kawabata, Y., Iki, K., Kawagoe, M. and Takada, H. (1994) Immunobiological activities of a 55-kilodalton cell surface protein of *Prevotella intermedia*. Infect. Immun. **62**, 2459–2469

200. Ogunniyi, A.D., Manning, P.A. and Kotlarski, I. (1994) A *Salmonella enerididis* 11RX pilin induces strong T-lymphocyte responses. Infect. Immun. **62**, 5376–5383

201. Kirby, A.C., Meghji, S., Nair, S.P., White, P.A., Reddi, K., Nishihara, T., Nakashima, K., Willis, A.C., Sim, R., Wilson, M. and Henderson, B. (1995) The potent bone resorbing mediator of *Actinobacillus actinomycetemcomitans* is homologous to the molecular chaperone GroEL. J. Clin. Invest. **96**, 1185–1194

202. Reddi, K., Wilson, M., Nair, S.P., Poole, S. and Henderson B. (1996) Comparison of the pro-inflammatory cytokine-stimulating activity of the surface-associated proteins of periodontopathic bacteria. J. Periodont. Res. **31**, 120–130

203. Reference deleted.

204. Wilson, M. and Henderson, B. (1995) Virulence factors of *Actinobacillus actinomycetemcomitans* relevant to the pathogenesis of inflammatory periodontal diseases. FEMS Microbiol. Rev. **17**, 365–379

205. Wilson, M., Reddi, K. and Henderson, B. (1996) Cytokine-inducing components of periodon-topathogenic bacteria. J. Periodont. Res. **31**, 393–407

206. Akira, S., Taga, T. and Kishimoto, T. (1993) Interleukin-6 in biology and medicine. Adv. Immunol. **54**, 1–78

207. Reddi, K., Nair, S.P., White, P.A., Hodges, S., Tabona, P., Meghji, S., Poole, S., Wilson, M. and Henderson, B. (1996) Surface-associated material from the bacterium *Actinobacillus actinomycetem-comitans* contains a peptide which, in contrast to lipopolysaccharide, directly stimulates fibroblast interleukin-6 gene transcription. Eur. J. Biochem. **236**, 871–876

208. Nair, S.P., Song, Y., Meghji, S., Reddi, K., Harris, M., Ross, M., Poole, S., Wilson, M. and Henderson, B. (1995) Surface-associated proteins from *Staphylococcus aureus* demonstrate potent bone resorbing activity. J. Bone Miner. Res. **10**, 726–734

209. Mai, U.E.H., Perez-Perez, G.I., Wahl, L.M., Wahl, S.M., Blaser, M.J. and Smith, P.D. (1991) Soluble surface proteins from *Helicobacter pylori* activate monocytes/macrophages by a lipopolysaccharide-independent mechanism. J. Clin. Invest. **87**, 894–900

210. Lee, A.J., Fox, J. and Hazell, S. (1993) Pathogenicity of *Helicobacter pylori*: a perspective. Infect. Immun. **61**, 1601–1610

211. Evans, D.J., Evans, D.E., Takemura, T., Nakano, H., Lampert, H.C., Graham, D.V., Neil-Granger, D. and Kvietys, P.R. (1995) Characterization of a *Helicobacter pylori* neutrophil-activating protein. Infect. Immun. **63**, 2213–2220

212. Huang, J., O'Toole, P.W., Doig, P. and Trust, T.J. (1995) Stimulation of interleukin-8 production in epithelial cells by *Helicobacter pylori*. Infect. Immun. **63**, 1732–1738

213. Soell, M., Holveck, F., Scholler, M., Wachsman, D. and Klein, J.-P. (1994) Binding of *Streptococcus mutans* SR protein to human monocytes: production of tumor necrosis factor, interleukin-1, and interleukin-6. Infect. Immun. **62**, 1805–1812

214. Tufano, M.A., Cipollaro de Léro, G., Ianiello, R. and Galdiero, F. (1991) Protein A and other surface components of *Staphylococcus aureus* stimulate production of IL-1α, IL-4, IL-6, TNF and IFN-γ. Eur. Cytokine Netw. **2**, 361–366

215. Young, D.B. (1992) Heat shock proteins: immunity and autoimmunity. Curr. Opin. Immunol. **4**, 396–400

216. Henderson, B., Nair, S.P. and Coates, A.R.M. (1996) Review: molecular chaperones and disease. Inflamm. Res. **45**, 155–158

217. Friedland, J.S., Shattock, R., Remick, D.G. and Griffin, G.E. (1993) Mycobacterial 65-kDa heat shock protein induces release of pro-inflammatory cytokines from human monocytic cells. Clin. Exp. Immunol. **91**, 58–62

218. Beagley, K.W., Fujihasha, K., Black, C.A., Lagoo, A.S., Yamamoto, M., McGhee, J.R. and Kiyono, H. (1993) The *Mycobacterium tuberculosis* 71kDa heat shock protein induces proliferation and cytokine secretion by murine gut intraepithelial lymphocytes. Eur. J. Immunol. **23**, 2049–2052

219. Retzlaff, C., Yamomoto, Y., Hoffman, P.S., Friedman, H. and Klein, T.W. (1994) Bacterial heat shock proteins directly induce cytokine mRNA and interleukin-1 secretion in macrophage cultures. Infect. Immun. **62**, 5689–5693

220. Peetermans, W.E., Raats, C.J.I., Langermans, J.A.M. and van Furth, R. (1994) Mycobacterial heat shock protein 65 induces proinflammatory cytokines but does not activate human mononuclear phagocytes. Scand. J. Immunol. **39**, 613–617

221. Verdegaal, E.M.E., Zegveld, S.T. and van Furth, R. (1996) Heat shock protein 65 induces CD62E, CD106, and CD54 on cultured human endothelial cells and increases their adhesiveness for monocytes and granulocytes. J. Immunol. **157**, 369–376

222. Tabona, P., Rewddi, K., Khan, S., Nair, S.P., Crean, St. J., Meghji, S., Wilson, M., Preuss, M., Miller, M., Poole, S., Carne, S. and Henderson, B. (1998) Homogeneous *Escherichia coli* chaperonin 60 induces IL-1beta and IL-6 gene expression in human monocytes by a mechanism independent of protein conformation. J. Immunol., in the press

223. Reference deleted.

224. White, J., Herman, A., Pullen, A.M., Kubo, R., Kappler, J.W. and Marrack, P. (1989) The Vβ-specific superantigen staphylococcal enterotoxin B: stimulation of mature T cells and clonal deletion in neonatal mice. Cell **56**, 27–35

225. Schafer, R. and Sheil, J.M. (1995) Superantigens and their role in infectious disease. Adv. Pediatr. Infect. Dis. **10**, 369–390

226. Betley, M.J., Borst, D.W. and Regassa, L.B. (1992) Staphylococcal enterotoxins. Toxic shock syndrome toxin and streptococcal pyrogenic exotoxins: a comparative study of their molecular biology. Chem. Immunol. **55**, 1–35

227. Stuart, P.M. and Woodward, J.G. (1992) *Yersinia enterocolitica* produces superantigenic activity. J. Immunol. **148**, 225–232

228. Legaard, P.K., Legrand, R.D. and Misfeldt, M.L. (1991) The superantigen Pseudomonas exotoxin A requires additional functions from accessory cells for T lymphocyte proliferation. Cell. Immunol. **135**, 372–377

229. Cole, B.C., Kartchner, D.R. and Wells, D.J. (1989) Stimulation of mouse lymphocytes by a product derived from *Mycoplasma arthritidis*. VII. Responsiveness is associated with expression of a product(s) of the Vβ8 family present on the T cell receptor α/β for antigen. J. Immunol. **142**, 4131–4137

230. Kotb, M. (1995) Bacterial pyrogenic exotoxins as superantigens. Clin. Microbiol. Rev. **8**, 411–426

231. Miethke, T., Wahl, C., Echtenacher, B., Krammer, P., Heeg, K. and Wagner, H. (1992) T cell-mediated lethal shock triggered in mice by the superantigen staphylococcal enterotoxin B: critical role of tumor necrosis factor. J. Exp. Med. **175**, 91–98

232. Fuleihan, R., Mourad, W., Geha, R.S. and Chatila, T. (1991) Engagement of MHC-class II molecules by staphylococcal exotoxins delivers a comitogenic signal to human B cells. J. Immunol. **146**, 1661–1666

233. Mehindate, K., Thibodeau, J., Dohlsten, M., Kalland, T., Sekaly, R.-P. and Mourad, W. (1995) Cross-linking of major histocompatability complex class II molecules by staphylococcal entero-toxin A superantigen is a requirement for inflammatory cytokine gene expression. J. Exp. Med. **182**, 1573–1577

234. Scholl, P.R., Trede, N., Chatila, T.A. and Geha, R.S. (1992) Role of protein tyrosine phosphoryla-tion in monokine induction by the staphylococcal superantigen toxic shock syndrome toxin-1. J. Immunol. **148**, 2237–2241

235. Stiles, B.G., Bavari, S., Krakauer, T. and Ulrich, R.G. (1993) Toxicity of staphylococcal enterotox-ins by lipopolysaccharide: major histocompatibility complex class II molecule dependency and cytokine release. Infect. Immun. **61**, 5333–5338

236. Grossman, D., Lamphear, J.G., Mollick, J.A., Betley, M.J. and Rich, R.R. (1992) Dual roles for class II major histocompatibility complex molecules in staphylococcal enterotoxin-induced cytokine production and in vivo toxicity. Infect. Immun. **60**, 5190–5196

237. Sperber, K., Silverstein, L., Brusco, C., Yoon, C., Mullin, G.E. and Mayer, L. (1995) Cytokine secretion induced by superantigens in peripheral blood mononuclear cells, lamina propria lym-phocytes and intraepithelial lymphocytes. Clin. Diagn. Lab. Immunol. **2**, 473–477

238. Leung, D.Y.M., Gately, M., Trumble, A., Ferguson-Darnell, B., Schlievert, P.M. and Picker, L.J. (1995) Bacterial superantigens induce T cell expression of the skin-selective homing receptor, the cutaneous lymphocyte-associated antigen, via stimulation of interleukin-12 production. J. Exp. Med. **181**, 747–753

239. Muller-Alouf, H., Alouf, J.E., Gerlach, D., Ozegowski, J.-H., Fitting, C. and Cavaillon, J.-M. (1996) Human pro- and anti-inflammatory cytokine patterns induced by *Streptococcus pyogenes* erythro-genic (pyrogenic) exotoxin A and C superantigens. Infect. Immun. **64**, 1450–1453

240. Beezhold, D.H., Best, G.K., Bonventre, P.F. and Thompson, M. (1987) Synergistic induction of IL-1 by endotoxin and toxic shock syndrome toxin-1 using rat macrophages. Infect. Immun. **55**, 2865–2869

241. Henne, E., Campbell, W.H. and Carlson, E. (1991) Toxic shock syndrome toxin 1 enhances synthesis of endotoxin-induced tumor necrosis factor in mice. Infect. Immun. **59**, 2929–2933

242. Mims, C. A., Dimmock, N.J., Nash, A. and Stephen, J. (1995) Mims' Pathogenesis of Infectious Diseases, 4th edn., Academic Press, London

243. Montecucco, C. and Schiavo, G. (1994) Mechanism of action of tetanus and botulinum neurotox-ins. Mol. Microbiol. **13**, 1–8

244. Lu, B., Gerard, N.P., Kolakowski, L.F., Bozza, M., Zurakowski, D., Finco, O., Carroll, M.C. and Gerard, C. (1995) Neutral endopeptidase modulation of septic shock. J. Exp. Med. **181**, 2271–2275

245. Miller, D.K. (1996) Cytokine convertase inhibitors. In Therapeutic Modulation of Cytokines (Henderson, B. and Bodmer, M.W., eds.), pp. 143–170, CRC Press, Boca Raton

246. Bhakdi, S., Muhly, M., Korom, S. and Schmidt, G. (1990) Effects of *Escherichia coli* hemolysin on human monocytes. Cytocidal action and stimulation of interleukin-1 release. J. Clin. Invest. **85**, 1746–1753

247. Bhakdi, S., Muhly, M., Korom, S. and Hugo, F. (1989) Release of interleukin-1β associated with potent cytocidal action of staphylococcal alpha-toxin on human monocytes. Infect. Immun. **57**, 3512–3519

248. Konig, B. and Konig, W. (1993) Induction and suppression of cytokine release (TNF-alpha, IL-6, IL-1β) by *E. coli* pathogenicity factors (adhesins, alpha haemolysin). Immunology **78**, 526–533

249. Konig, B., Ludwig, A., Goebel, W. and Konig, W. (1994) Pore formation by the *Escherichia coli* alpha hemolysin: Role for mediator release from human inflammatory cells. Infect. Immun. **62**, 4611–4617

250. Tsukuda, H., Kawamura, I., Fujimara, T., Igarishi, K.-I., Arakawa, M. and Mitsuyama, M. (1992) Induction of macrophage interleukin-1 by *Listeria monocytogenes* hemolysin. Cell. Immunol. **140**, 21–30

251. Yoshikawa, H., Kawamura, I., Fujita, M., Tsukada, H., Arakawa, M. and Mitsuyama, M. (1993) Membrane damage and interleukin-1 production in murine macrophages exposed to listeriolysin O. Infect. Immun. **61**, 1334–1339

252. Nishibori, T., Xiong, H., Kawamura, I., Arakawa, M. and Mitsuyama, M. (1996) Induction of cytokine gene expression by listeriolysin O and roles of macrophages and NK cells. Infect. Immun. **64**, 3188–3195

253. Berry, A.M., Yother, J., Briles, D.E., Hansman, D. and Paton, J.C. (1989) Reduced virulence of a defined pneumolysin-deficient mutant of *Streptococcus pneumoniae*. Infect. Immun. **57**, 2037–2042

254. Houldsworth, S., Andrew, P.W. and Mitchell, T.J. (1994) Pneumolysin stimulates production of tumor necrosis factor alpha and interleukin-1β by human mononuclear phagocytes. Infect. Immun. **62**, 1501–1503

255. Doery, H.M., Magnusson, B.J., Cheney, I.M. and Gulasekharam, J. (1963) A phospholipase in staphylococcal toxin which hydrolyses sphingomyelin. Nature (London) **198**, 1091–1092

256. Walev, I., Weller, U., Strauch, S., Foster, T. and Bhakdi, S. (1996) Selective killing of human monocytes and cytokine release provoked by sphingomyelinase (beta-toxin) of *Staphylococcus aureus*. Infect. Immun. **64**, 2974–2979

257. O.Brien, A.D., Tesh, V.L., Donohue-Rolfe, A., Jackson, M.P., Olsnes, S., Sandvig, K., Lindberg, A.A. and Keusch, G.T. (1992) Shiga toxin: biochemistry, genetics, mode of action and role in pathogenesis. Curr. Top. Microbiol. Immunol. **180**, 67–94

258 Obrig, T.G., DelVecchio, P.J., Brown, J.E., Moran, T.P., Rowland, B.M., Judge, T.K. and Rothman, S.W. (1988) Direct cytotoxic action of Shiga toxin on human vascular endothelial cells. Infect. Immun. **56**, 2373–2378

259. Louis, C.B. and Obrig, T.G. (1992) Shiga toxin-associated hemolytic-uremic syndrome: combined cytotoxic effects of Shiga toxin, interleukin-1β, and tumor necrosis factor alpha on human vascular endothelial cells in vitro. Infect. Immun. **59**, 4173–4179

260. Tesh, V.L., Samuel, J.E., Perera, L.P., Sharefkin, J.B. and O'Brien, A.D. (1991) Evaluation of the role of Shiga and Shiga-like toxins in mediating direct damage to human vascular endothelial cells. J. Infect. Dis. **164**, 344–352

261. Barrett, T.J., Potter, M.E. and Strockbine, N.A. (1990) Evidence for participation of the macrophage in Shiga-like toxin II-induced lethality in mice. Microb. Pathog. **9**, 95–103

262. Tesh, V.L., Ramegowda, B. and Samuel, J.E. (1994) Purified Shiga-like toxins induce expression of proinflammatory cytokines from murine peritoneal macrophages. Infect. Immun. **62**, 5085–5094

263. Ramegowda, B. and Tesh, V.L. (1996) Differentiation-associated toxin receptor modulation, cytokine production, and sensitivity to Shiga-like toxins in human monocytes and monocytic cell lines. Infect. Immun. **64**, 1173–1180

264. Van der Kar, N.C.A.J., Monnens, L.A.H., Karmali, M.A. and van Hinsbergh, V.W.M. (1992) Tumour necrosis factor and interleukin-1 induce expression of the verocytotoxin receptor globotriaosylceramide on human endothelial cells. Implications for the pathogenesis of the hemolytic uremic syndrome. Blood **80**, 2755–2764

265. Van der Kar, N.C.A.J., Kooistra, T., Vermeer, M., Lesslauer, W., Monnens, L.A.H. and van Hinsbergh V.W. (1995) Tumor necrosis factor alpha induces endothelial galactosyl transferase activity and verocytotoxin receptors. Role of specific tumour necrosis factor receptors and protein kinase C. Blood **85**, 734–743

266. Friedlander, A.M. (1986) Macrophages are sensitive to anthrax lethal toxin through an acid-dependent process. J. Biol. Chem. **261**, 7123–7126

267. Hanna, P.C., Acosta, D. and Collier, R.J. (1993) On the role of macrophages in anthrax. Proc. Natl. Acad. Sci. U.S.A. **90**, 10198–10201

268. Hoover, D..L, Friedlander, A.M., Rogers, L.C., Yoon, I.-K., Warren, R.L. and Cross, A.S. (1994) Anthrax edema toxin differentially regulates lipopolysaccharide-induced monocyte production of

tumor necrosis factor alpha and interleukin-6 by increasing intracellular cyclic AMP. Infect. Immun. **62**, 4432–4439

269. Vajdy, M. and Lycke, N. (1992) Cholera toxin adjuvant promotes long-term immunological memory in the gut mucosa to unrelated immunogens after oral immunization. Immunology **75**, 488–494

270. Xu-Amano, B.J., Hiroshi, K., Jackson, R.J., Staats, H.F., Fujihashi, K., Burrows, P.D., Elson, C.O., Pillai, S. and McGhee, J.R. (1993) Helper T cell subsets for immunoglobulins A response: oral immunization with tetanus toxoid and cholera toxin selectively induces Th$_2$ cells in mucosa-associated tissues. J. Exp. Med. **178**, 1309–1320

271. Okahashi, N., Yamamoto, M., Vancott, J.L., Chatfield, S.N., Roberts, M., Bluethmann, H., Hiroi, T., Kiyono, H. and McGhee, J.R. (1996) Oral immunization of interleukin-4 (IL-4) knockout mice with a recombinant *Salmonella* strain or cholera toxin reveals that CD4$^+$ Th2 cells producing IL-6 and IL-10 are associated with mucosal immunoglobulin A responses. Infect. Immun. **64**, 1516–1525

272. Bromander, A., Holmgren, J. and Lycke, N. (1991) Cholera toxin stimulates IL-1 production and enhances antigen presentation by macrophages in vitro. J. Immunol. **146**, 2908–2914

273. Klimpel, G.R., Asuncion, M., Haithcoat, J. and Niesel, D.W. (1995) Cholera toxin and *Salmonella typhimurium* induce different cytokine profiles in the gastrointestinal tract. Infect. Immun. **63**, 1134–1137

274. Bromander, A.K., Kjerrulf, M., Holmgren, J. and Lycke, N. (1993) Cholera toxin enhances allograft presentation by cultured intestinal epithelial cells. Scand. J. Immunol. **37**, 452–458

275. McGee, D.W., Elson, C.O. and McGhee, J.R. (1993) Enhancing effect of cholera toxin on inter-leukin-6 secretion by IEC-6 intestinal epithelial cells: mode of action and augmenting effect of inflammatory cytokines. Infect. Immun. **61**, 4637–4644

276. Zhang, T., Lin, J. and Vilcek, J. (1988) Synthesis of interleukin-6 (interferon-β2/B-cell stimulatory factor 2) in human fibroblasts is triggered by an increase in intracellular cyclic AMP. J. Biol. Chem. **263**, 6177–6182

277. Leal-Berumen, I., Snider, D.P., Barajas-Lopez, C. and Marshall, J.S. (1996) Cholera toxin increases IL-6 synthesis and decreases TNF-α production by rat peritoneal mast cells. J. Immunol. **156**, 316–321

277a. Jakway, J.P. and DeFranco, A.L. (1986) Pertussis toxin inhibition of B cell and macrophage responses to bacterial lipopolysaccharide. Science **234**, 743–746

278. Nashar, T.O., Webb, H.M., Eaglestone, S., Williams, N.A. and Hirst, T.R. (1996) Potent immuno-genicity of the B subunits of *Escherichia coli* heat-labile enterotoxin: receptor binding is essential and induces differential modulation of lymphocyte subsets. Proc. Natl. Acad. Sci. U.S.A. **93**, 226–230

279. Mu, H.-H. and Sewell, W.A. (1993) Enhancement of interleukin-4 production by pertussis toxin. Infect. Immun. **61**, 2834–2840

280. Misfeldt, M.L., Legaard, P.K., Howell, S.E., Fornella, M.H. and Legrand, R.D. (1990) Induction of interleukin-1 from murine peritoneal macrophages by *Pseudomonas aeruginosa* exotoxin A. Infect. Immun. **58**, 978–982

281. Triadafilopoulos, G., Pothoulakis, C., O'Brian, M.J. and LaMont, J.T. (1987) Differential effects of *Clostridial difficile* toxins A and B on rabbit ileum. Gastroenterology **93**, 273–279

282. Aktories, K. and Just, I. (1995) Monoglucosylation of low-molecular-mass GTP-binding Rho proteins by clostridial cytotoxins. Trends Cell Biol. **5**, 441–443

283. Pothoulakis, C., Gilbert, R.J., Claradas, C., Castagliuolo, I., Semenza, G., Hitti, Y., Montcrief, J.C., Linevsky, J., Kelly, C.P., Nikulasson, S., et al. (1996) Rabbit sucrase-isomaltase contains a functional intestinal receptor for *Clostridial difficile* toxin A. J. Clin. Invest. **98**, 641–649

284. Daubener, W., Leiser, E., von Eichel-Streiber, C. and Hadding, U. (1988) *Clostridial difficile* toxins A and B inhibit human immune response in vitro. Infect. Immun. **56**, 1107–1112

285. Miller, P.D., Pothoulakis, C., Baeker, T.R., LaMoni, J.T. and Rothestein, T.L. (1990) Macrophage-dependent stimulation of T cell-depleted spleen cells by *Clostridium difficile* toxin A and calcium ionophore. Cell. Immunol. **126**, 155–163

286. Flegel, W.A., Muller, F., Daubener, W., Fischer, H-G., Hadding, U. and Northoff, H. (1991) Cytokine response by human monocytes to *Clostridium difficile* toxin A and B. Infect. Immun. **59**, 3659–3666

287. Hase, C.C. and Finkelstein, R.A. (1993) Bacterial extracellular zinc-containing metalloproteases. Microbiol. Rev. **57**, 823–837

288. Kapur, V., Majewsky, M.W., Li, L.L., Black, R.A. and Musser, J.-M. (1993) Cleavage of interleukin-1 beta (IL-1 beta) precursor to produce active IL-1 beta by a conserved extracellular cysteine protease from *Streptococcus pyogenes*. Proc. Natl. Acad. Sci. U.S.A. **90**, 7676–7680

289. Wexler, D.E., Chenoweth, D.E. and Cleary, P.P. (1985) Mechanism of action of the group A streptococcal C5a inactivator. Proc. Natl. Acad. Sci. U.S.A. **82**, 8144–8148

290. Herwald, H., Collin, M., Muller-Esterl, W. and Bjork, L. (1996) Streptococcal cysteine proteinase releases kinins: a novel virulence mechanism. J. Exp. Med. **184**, 665–673

291. Parmely, M.P., Gale, A., Clabaugh, M., Horvat, R. and Zhou, W.W. (1990) Proteolytic inactivation of cytokines by *Pseudomonas aeruginosa*. Infect. Immun. **58**, 3009–3014

292. Mintz, C.S., Miller, R.D., Gutgsell, N.S. and Malek, T. (1993) *Legionella pneumophila* protease inactivates IL-2 and cleaves CD4 on human T cells. Infect. Immun. **61**, 3416–3421

293. Fletcher, J., Reddi, K., Poole, S., Nair, S., Henderson, B. and Wilson, M. (1997) Interactions between periodontopathogenic bacteria and cytokines. J. Periodont. Res. **32**, 200–205

294. Vollmer, P., Walev, I., Rose-John, S. and Bhakdi, S. (1996) Novel pathogenic mechanism of microbial metalloproteinases: liberation of membrane-anchored molecules in biologically active forms exemplified by studies with the human interleukin-6 receptor. Infect. Immun. **64**, 3646–3651

295. Hart, P.H., Hunt, E.K., Bonder, C.S., Watson, C.J. and Finlay-Jones, J.J. (1996) Regulation of surface and soluble TNF receptor expression on human monocytes and synovial fluid macrophages by IL-4 and IL-10. J. Immunol. **157**, 3672–3680

296. Randow, F., Syrbe, U., Meisel, C., Krausch, D., Zuckerman, H., Platzer, C. and Volk, H.-D. (1995) Mechanism of endotoxin desensitization: involvement of interleukin-10 and transforming growth factor β. J. Exp. Med. **181**, 1887–1892

297. Berg, D.J., Kuhn, R., Rajewsky, K., Muller, W., Menon, S., Davidson, N., Grunig, G. and Rennick, D. (1995) Interleukin-10 is a central regulator of the response to LPS in murine models of endotoxic shock and the Shwartzman reaction but not endotoxin tolerance. J. Clin. Invest. **96**, 2339–2347

298. Renz, H., Gong, J.-H., Schmidt, A., Nain, M. and Gemsa, D. (1988) Release of tumor necrosis factor-α from macrophages: enhancement and suppression are dose-dependently regulated by prostaglandin E_2 and cyclic nucleotides. J. Immunol. **141**, 2388–2393

299. Dalrymple, S.A., Slattery, R., Aud, D.M., Krishna, M., Lucian, L.A. and Murray, R. (1996) Interleukin-6 is required for a protective immune response to systemic *Escherichia coli* infection. Infect. Immun. **64**, 3231–3235

300. Staugas, R.E.M., Harvey, D.P., Ferrante, A., Nandoskar, M. and Allison, A.C. (1992) Induction of tumor necrosis factor (TNF) and interleukin-1 by *Pseudomonas aeruginosa* and endotoxin A-induced suppression of lymphoproliferation and TNF, lymphotoxin, gamma interferon, and IL-1 production in human leukocytes. Infect. Immun. **60**, 3162–3168

301. Majury, A.L. and Shewen, P.E. (1991) Preliminary investigation of the mechanism of inhibition of bovine lymphocyte proliferation by *Pasteurella haemolytica* A1 leukotoxin. Vet. Immunol. Immunopathol. **29**, 57–68

302. Hughes, H.P.A., Campos, M., McDougall, L., Beskorwayne, T.K., Potter, A.W. and Babiuk, L.A. (1994) Regulation of major histocompatibility complex class II expression by *Pasteurella haemolytica* leukotoxin. Infect. Immun. **62**, 1609–1615

303. Autenrieth, I.B. and Heesemann, J. (1992) In vivo neutralization of tumor necrosis factor-alpha and interferon-gamma abrogates resistance to *Yersinia enterocolitica* infection in mice. Med. Microbiol. Immunol. **181**, 333–338

304. Beuscher, H.U., Rodel, F., Forsberg, A. and Rollinghoff, M. (1995) Bacterial evasion of host immune defence: *Yersinia enterocolitica* encodes a suppressor factor for tumor necrosis factor alpha expression. Infect. Immun. **63**, 1270–1277

305. Matsui, K. (1996) A purified protein from *Salmonella typhimurium* inhibits proliferation of murine splenic anti-CD3 antibody-activated T-lymphocytes. FEMS Immunol. Med. Microbiol. **14**, 121–127

306. Caron, E., Gross, A., Liautard, J.-P. and Dornand, J. (1996) *Brucella* species release a specific, protease-sensitive, inhibitor of TNF-α expression, active on human macrophage-like cells. J. Immunol. **156**, 2885–2893

307. Dong, Z.M. and Murphy, J.W. (1996) Cryptococcal polysaccharides induce L-selectin shedding and tumor necrosis factor receptor loss from the surface of human neutrophils. J. Clin. Invest. **97**, 689–698

308. Enders, G., Brooks, W., Jan, N.V., Lehn, N., Bayerdorffer, E. and Hatz, R. (1995) Expression of adhesion molecules on human granulocytes after stimulation with *Helicobacter pylori* membrane proteins: comparison with membrane proteins from other bacteria. Infect. Immun. **63**, 2473–2477

309. Klapproth, J.-M., Donnenberg, M.S., Abraham, J.M., Mobley, H.L.T. and James, S.P. (1995) Products of enteropathogenic *Escherichia coli* inhibit lymphocyte activation and lymphokine production. Infect. Immun. **63**, 2248–2254

310. Kenny, B. and Finlay, B.B. (1995) Protein secretion by enteropathogenic *Escherichia coli* is essential for transducing signals to epithelial cells. Proc. Natl. Acad. Sci. U.S.A. **92**, 7991–7995

311. Patel, M.D., Henderson, B., Galgut, P. and Olsen, I. (1994) Mechanism of the anti-proliferative action of *Actinobacillus actinomycetemcomitans* surface-associated material. J. Dent. Res. **73**, 793

312. White, P.A., Wilson, M., Nair, S.P., Kirby, A.C., Reddi, K. and Henderson, B. (1995) Characterization of an anti-proliferative surface-associated protein from *Actinobacillus actino-mycetemcomitans*. Infect. Immun. **63**, 2612–2618

313. Kurita-Ochiai, T., Ochiai, K. and Ikeda, T. (1992) Immunosuppressive effect induced by *Actinobacillus actinomycetemcomitans*: effect on immunoglobulin production and lymphokine synthesis. Oral Microbiol. Immunol. **7**, 338–343

314. Kurita-Ochiai, T. and Ochiai, K. (1996) Immunosuppressive factor from *Actinobacillus actino-mycetemcomitans* down regulates cytokine production. Infect. Immun. **64**, 50–54

315. Rabie, G., Lally, E.T. and Shenker, B.J. (1988) Immunosuppressive properties of *Actinobacillus actin-omycetemcomitans* leukotoxin. Infect. Immun. **56**, 122–127

316. Shenker, B.J., Vitale, L.A. and Welham, D.A. (1990) Immune suppression induced by *Actinobacillus actinomycetemcomitans*: effects on immunoglobulin production by human B cells. Infect. Immun. **58**, 3856–3862

317. Fleming, S.D. and Campbell, P.A. (1996) Macrophages have cell surface IL-10 that regulates macrophage bactericidal activity. J. Immunol. **156**, 1143–1150

318. Bermudez, L.E., Petrofsky, M. and Shelton, K. (1996) Epidermal growth factor-binding protein in *Mycobacterium avium* and *Mycobacterium tuberculosis*: a possible role in the mechanism of infection. Infect. Immun. **64**, 2917–2922

319. Chertov, O., Michiel, D.F., Wang, J.M., Tani, L., Murphy, W.J., Longo, D.L. and Oppenheim, J.J. (1996) Identification of defensin-1, defensin-2 and CAP37/azurocidin as T cell chemoattractant proteins released from interleukin-8-stimulated neutrophils. J. Biol. Chem. **271**, 2935–2940

8

Pharmacological modulation of cytokines

"Drugs can only repress symptoms, they cannot eradicate disease… There is at bottom only one genuinely scientific treatment for all disease, and that is to stimulate the phagocyte. Stimulate the phagocytes. Drugs are a delusion."

Sir Ralph Bloomfield-Bonnington
in The Doctor's Dilemma
by George Bernard Shaw

8.1   Introduction

It will now be clear to the reader that the term cytokine describes an enormous range of soluble and cell-bound proteins which have a myriad of functions and which can be considered as: (i) vital homeostatic local hormones; (ii) key defence proteins against infectious organisms; and (iii) mediators of the pathology of virtually all known human diseases. These various actions of cytokines have been reviewed in Chapter 4. Given these disparate actions, it will come as no surprise to learn that the world's pharmaceutical and biotechnology industries have a fascination with cytokines and are currently spending large sums of money to develop therapeutic agents that will selectively control cytokines. The impetus for much of the effort to develop cytokine therapeutics has been the often fatal condition known as septic shock, which can be caused both by Gram-negative and Gram-positive bacteria and, in which, contrary to Sir Ralph Bloomfield-Bonnington's theory, there has been too much stimulation of the phagocytes [1–3]. However, the pathology of all acute and chronic infectious diseases is driven by the overproduction of certain pro-inflammatory cytokines (IL-1, IL-8, TNF, etc.; see Table 8.1) and the pharmacological manipulation of cytokines could be of benefit for these diseases. The use of drugs capable of modulating specific cytokines in infectious diseases will become more important as the numbers of effective antibiotics progressively

Table 8.1 Cytokines and their involvement in human disease

Disease	Cytokines implicated in pathology
Septic shock	IL-1, IL-6, IL-8, TNF, IFNγ, (IL-10)
Leprosy	TNF, IFNγ
Tuberculosis	TNF, IFNγ
Malaria	IL-3, IFNγ
Allergy/asthma	IL-4, IL-5
Rheumatoid arthritis	TNF, IL-1, IL-6, chemokines, growth factors
Systemic lupus erythematosus	IL-2, TNF, IFNβ
Autoimmune diseases	IL-1, IL-2, TNF, IFNγ
Diabetes	IL-1, IFNγ
Ulcerative colitis	IL-1, IL-6, IL-8, IL-10, lymphokines
Leukaemia	IL-1, IL-4, IL-11
Psoriasis	IL-6, IL-8
Periodontal disease	IL-1, IL-2, IL-6, IL-8, TNF
Osteoporosis	IL-1, TNF, IL-6, LIF

shrink. This probable terminal decline in antibiotics means that other antimicrobial therapeutic modalities have to be explored. An obvious approach would be to therapeutically manipulate cytokine networks with such precision that the antimicrobial actions of specific cytokines within the network were maintained, but that the deleterious host responses caused by inappropriate cytokine production were minimized (Figure 8.1). This goal has not been properly formulated as yet and the aim of the pharmaceutical industry is still to inhibit the synthesis or activity of individual cytokines. This has resulted in intensive research being focused on a number of pro-inflammatory cytokines. As has been described in Chapter 7, Nature, as in so many other areas, has beaten man to the draw in developing anti-cytokine reagents, and is offering sage advice in terms of the approaches to take. The authors have reviewed the pharmacology of cytokine modulation in a number of recent reviews and books [4–7]. While much of this book has concentrated on cytokine networks, rather than on individual cytokines, the pharmaceutical industry has concentrated on specific, largely pro-inflammatory, cytokines as therapeutic targets. Currently, it is difficult enough to inhibit the production or action of individual cytokines, and the manipulation of discrete networks lies in the future. However, this is not to say that the removal of one pivotal cytokine may not have profound effects on cytokine network interactions. This is the expectation in the use of anti-TNF therapy for the treatment of rheumatoid arthritis (see later sections in this chapter).

As has been stated, septic shock is one of the conditions that has focused the pharmaceutical industry's interest on cytokine manipulation. However, many companies have a greater interest in non-infectious inflammatory diseases such as rheumatoid arthritis, atherosclerosis and asthma.

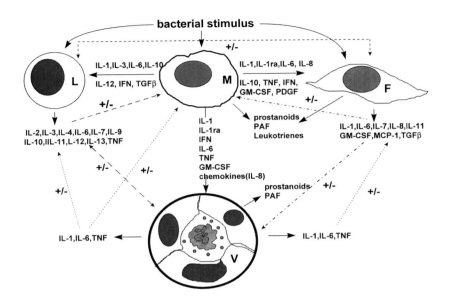

Figure 8.1 The interaction of bacterial factors (endotoxin, modulins, bacteriokines) with key cell populations including monocytes/macrophages (M), lymphocytes (L), fibroblasts (F) and vascular endothelial cells (V)
The lumen of the blood vessel also contains a polymorphonuclear leucocyte and platelets, which can also contribute to cytokine synthesis. Each cell population can make a wide array of cytokines (of which a selected number is shown). These cytokines can interact with the various cell populations depicted to up- or down-regulate cell activity. In addition, each of these cell populations can produce a range of other mediators which can also interact with the cells to up- or down-regulate $(+/-)$ specific cytokine synthesis. As an example of the latter, the production of products of the metabolism of cellular phospholipids [prostanoids, platelet-activating factor (PAF), leukotrienes] is denoted. This is the arena in which cytokine inhibitors must work in order to provide therapeutic benefit. For a full list of the abbreviations used, see page xiii.

8.2   Therapeutic modulation: where to intervene

The obvious starting point in the development of cytokine-modulating pharmacological agents is deciding possible points of intervention. Classic pharmacology was based upon the use of receptor agonists and antagonists. However, modern pharmacology now relies on a variety of approaches, and no area of pharmacology has been more catholic in its approach than cytokine pharmacology. Before describing the possible ways of interfering with cytokines it is important to establish one basic fact: cytokines are generally low-molecular-mass proteins which have no inherent functions (e.g. enzymic activity) and whose cellular actions are entirely dependent on the binding to high-affinity cell-surface receptors. It is this receptor binding and the associated post-receptor intracellular signalling and gene transcription which defines the biological activity of a cytokine and discriminates it from any other cytokine (see Chapter 4). There are at least seven distinct cytokine receptor

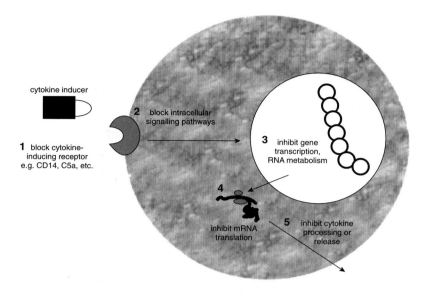

Figure 8.2 Schematic diagram showing the points at which it should be possible to intervene therapeutically to inhibit the synthesis of cytokines
The first point of intervention (1) is antagonism of the agonist inducing cytokine synthesis (e.g. endotoxin, other modulins, C5a, other cytokines). Anti-CD14 antibodies, CD14 inhibitors or excess CD14 can be used to block the biological actions of endotoxin. Following receptor binding it should be possible to selectively inhibit cell activation by use of inhibitors of intracellular signalling (kinase, phosphatase inhibitors, for example) (2). If the signal gets to the nucleus it should be possible to selectively block cytokine gene transcription or metabolism (trimming, splicing, cutting) of the mRNA (3). Inhibition of cytokine mRNA translation with antisense is a promising area of pharmaceutical development (4). With some cytokines such as IL-1, TNF and FGF, the protein must be proteolytically cleaved to release the active cytokine (5). This is another area of active pharmaceutical research. Finally, the mechanism of IL-1 release from cells is still not fully determined and may employ a novel pathway which could be a therapeutic target (5).

families and some of them, for example the receptors for IL-2 and IL-6, have multiple subunits and receptor occupancy, and signalling involves oligomerization (reviewed in [8,9] and Chapter 4). In global terms, cytokines, in order to exert biological activity, have to be induced (involving cell-membrane binding, intracellular signalling, gene transcription, mRNA translation, processing and release) and then bind to their specific cell receptor to activate the target cell. As illustrated in Figure 8.2 and defined in more detail in Table 8.2, each of these individual cellular actions is a possible point of therapeutic intervention. The discovery in recent years that certain cytokines (e.g. IL-1ra, IL-4, IL-10, IL-13, TGFβ) are able to inhibit the actions of pro-inflammatory cytokines, such as IL-1 and TNF, has also opened up the possibility of using such functional antagonist cytokines as drugs. Agents that can specifically induce the synthesis of these so-called anti-inflammatory cytokines would also be of potential benefit. To reiterate, there are two basic strategies in manipulat-

Table 8.2 Points of therapeutic intervention in controlling cytokines

1.	Antagonism of cytokine–receptor interactions
2.	Neutralization of soluble cytokines by soluble cytokine receptors, antibodies or other selective cytokine-binding/neutralizing mediators
3.	Selective inhibition of cytokine gene transcription (e.g. cyclosporin inhibits IL-2 transcription)
4.	Selective inhibition of cytokine gene mRNA translation (e.g. antisense drugs)
5.	Inhibition of the proteolytic processing of selected cytokines (IL-1β, TNFα, etc.)
6.	Selective inhibition of the export of certain cytokines (IL-1 being the major example)
7.	Inhibition of the specific or non-specific activation of cytokine-inducing cells
8.	Inhibition of intracellular transduction pathways of cytokine-responding cells
9.	Antagonist cytokines
10.	Bacterial inhibitors of cytokine synthesis or activity

ing cytokines. The first, and more classic, approach is to interfere with the action of the cytokine. This can be achieved by the antagonism of the cytokine receptor or by, in some other way, preventing the cytokine binding to its cellular receptor. At the present time this approach appears to have achieved most, and the use of neutralizing antibodies to cytokines is proving to have some clinical success. We will deal with this approach first. The second approach is to prevent the synthesis of specific cytokines by use of a range of strategies which have, to some extent, been tailored for the individual properties of certain cytokines. The three most interesting possibilities are IL-1 convertase inhibitors, antisense drugs and selective intracellular signalling inhibitors. However, all the methods in current use will be reviewed.

8.3   Pharmacological modulation of cytokines

In this section, the methods developed to inhibit the actions of selected cytokines will be reviewed. The starting point will be the attempts to develop antagonists of pro-inflammatory cytokines.

8.3.1   Development of cytokine receptor antagonists

In the late 1970s and early 1980s purified recombinant cytokines became available and the roles of IL-1 and IL-2, the first two cytokines to be discovered, began to be defined. Both cytokines were recognized as potential therapeutic targets in a variety of 'immunological' diseases (particularly rheumatoid arthritis and transplant rejection) and the first attempts were made to block their activity by producing receptor antagonists. Three types have been employed: (i) peptide-based antagonists; (ii) low-molecular-mass non-peptidic isosteres; and (iii) protein-based antagonists. Each of these strategies will be discussed in turn.

8.3.1.1 Peptide-based antagonists

In the mid-1980s, when this particular approach was developed, the only information available to the pharmacologist for modelling antagonists was the primary amino acid sequence of the target cytokines. Thus, it was a sensible strategy to determine if partial sequences would have any agonist or antagonist properties and this was applied most vigorously to IL-1. Human IL-1 is synthesized as a precursor of 271 residues (IL-1α; with the mature protein starting at position 119) or 269 residues (IL-1β; with the mature protein starting at residue 117) [10]. The three-dimensional crystal structure of the IL-1α and β molecules is known at a resolution as low as 0.2 nm, and both are approximately tetrahedral structures composed of a tightly packed core of β strands connected by less structured loops [11,12]. Mutagenesis studies suggest that residues involved in receptor binding and activation are scattered widely over the surface of the IL-1 molecules [13,14]. Thus, with hindsight, it would seem unlikely that partial structures would have any agonist function or be able to act in an antagonistic fashion. In spite of this, a number of reports of peptides with biological actions have appeared in the literature. For example, a 32-residue peptide containing the sequence 208–240 in IL-1β has been reported to have both pyrogenic and somnogenic activity [15]. A nonapeptide (residues 163–171) has been reported to have a range of biological actions including: stimulation of T-cell mitogenesis, adjuvant activity and stimulation of glycosaminoglycan synthesis. In contrast, this peptide lacks inflammogenic, pyrogenic or bone-resorbing activity [16–18]. One of the most surprising findings in recent years has been that many pro-inflammatory cytokines are hyperalgesic (i.e. they reduce the threshold for stimulation of pain-conducting nerves [19]). IL-1β is an incredibly potent hyperalgesic agent in the rat, and the surprising finding is that this activity can be blocked by tripeptide analogues of IL-1β — Lys-193-Pro-Thr-195 and Lys-D-Pro-Thr [20] — and by α-melanocyte-stimulating hormone (α-MSH) and stable analogues of this natural hormone, including Lys-11-Pro-Val-13 and Lys-D-Pro-Val [21,22]. Alpha-MSH and its analogues inhibit other responses of IL-1, including fever, production of the acute-phase response, thymocyte proliferation and neuroendocrine activity [23–26]. The C-terminal tripeptide of α-MSH (Lys-11-Pro-Val-13) can block the pyrogenicity of IL-1 [27] and appears to be anti-inflammatory [28,29]. The initial interpretation of these findings was that α-MSH and the IL-1β tripeptide were acting as receptor antagonists. However, it now transpires that this is not the case. For example these molecules do not inhibit all of the actions of IL-1 [20]. Further, Lys-D-Pro-Thr inhibited hyperalgesia evoked by cytokines other than IL-1 (e.g. TNF and IL-6) [30] and α-MSH inhibited fever induced by these cytokines [31]. These findings suggest that such peptides block the actions of IL-1 by a mechanism distinct from classic receptor antagonism. The actual mechanism of action is unclear, but is intriguing in relation to the increasing interest in the interactions between psychological stress and susceptibility to infections [32,33].

Could neuropeptides, induced as a result of stress, inhibit the host response to infectious micro-organisms, thereby decreasing the threshold for becoming infected? We will return to this topic in the final chapter.

The peptides described above do not express antagonist activity. An IL-1β C-terminal peptide (residues 237–269) has been reported to have antagonistic activity, inhibiting the binding of radiolabelled IL-1 to its receptor and blocking T-cell proliferation due to IL-1 [34]. No subsequent reports on this peptide have appeared. A decapeptide derived from the primary sequence (residues 34–43) of the growth factor TGFα has antagonist activity, blocking the mitogenic effects on fibroblasts in culture of TGFα and the related growth-promoting cytokine EGF [35]. Peptides based on a consensus sequence (WSXWS) found in the type I receptor superfamily (which includes IL-3, IL-4, IL-7, GM-CSF and erythropoietin; see Chapter 4) have been reported to inhibit the stimulatory activity of IL-3 and GM-CSF on basophils [36].

While these results are of interest there was a general consensus in the late 1980s that this peptide approach to cytokine therapeutics would not produce active drugs due to the metabolic instability and lack of bioavailability of most peptides. This judgement may still hold. However, it is interesting to examine current estimates of the sales of peptide drugs, such as luteinizing hormone-releasing hormone, which are thought to be around US$1 billion and to realize that many pharmaceutical companies have peptide-based drugs in clinical trial for diseases ranging from angina to wound healing [37].

Protein mutagenesis (deletional, insertional, site-directed) is a technique used extensively to investigate protein structure–function relationships and has been used to determine the residues in various cytokines involved in receptor binding. Some cytokine mutants have interesting receptor-binding properties. For example, IL-2 mutants with partial agonist and antagonist actions have been developed [38,39]. Substituting a tyrosine residue for an aspartate converts IL-4 into a potent antagonist [40]. It may be possible to use such mutant proteins in a therapeutic manner, although delivery of proteins is still a major problem.

8.3.1.2   IL-1 receptor antagonist

The nature and properties of the potent pro-inflammatory cytokine IL-1 have been discussed throughout this book. IL-1 is a molecule with catholic and potent actions, which plays a central role in host defence responses and in inducing tissue pathology, and is, as this chapter will describe, a carefully regulated cytokine. During the 1980s a number of groups reported the presence of IL-1 'inhibitors' in various biological fluids [41]. Arend in the U.S.A. [42] and Dayer in Switzerland [43] eventually succeeded in isolating a specific IL-1 inhibitor which is now known as interleukin-1 receptor antagonist (IL-1ra). Arend's group described a 22–25 kDa protein produced by human monocytes cultured on plastic wells coated with IgG. Dayer's group used urine from febrile patients with acute monocytic leukaemia and isolated an inhibitor of approximate molecular mass 25–35 kDa. Subsequently, both groups showed that their respective inhibitors blocked the binding of radiolabelled IL-1 to cells, thereby confirming that it functioned as an antagonist [44,45]. The human monocyte inhibitor was cloned and the purified mature recombinant protein, which comprises 152 amino acid residues (molecular mass 17.12 kDa unglycosylated and up to 26 kDa glycosylated), was shown to be a competitive antagonist of IL-1 [45,46]. The three-dimensional structure of IL-1ra resembles that of the other two forms of IL-1 [47]. Site-directed mutation of IL-1ra has shown that amino acid substitution of lysine to aspartic acid at position 145 produces a molecule with partial agonist activity [48]. The converse substitution of aspartic acid to lysine at the homologous region in IL-1β results in partial loss of agonist activity [49]. Of interest, in the context of the evolution of host defence systems and their control, is the estimation of the evolution of the IL-1 family. From sequence analysis and mutation rate calculations it has been estimated that the gene for IL-1ra diverged from IL-1 around 350 million years ago. It is estimated that the IL-1ra gene segregated about 50 million years before the divergence of IL-1 into the two forms we recognize today [50,51]. Thus, IL-1ra appeared on the scene during the carboniferous period as coal was being laid down and the first reptiles were appearing.

8.3.1.2.1 Mechanism of action of IL-1ra

IL-1ra inhibits the action of both IL-1α and IL-1β by competitive inhibition at the level of the cell-surface receptor. IL-1ra binds with similar affinity to IL-1 receptors but lacks intrinsic agonist activity, thus failing to trigger intracellular signalling and post-signalling events [42,45]. The biological control of IL-1 is further complicated by the finding that there are two distinct receptors for the IL-1 family. The human type I receptor is predicted from the cDNA sequence to be a protein of 569 amino acids consisting of a 20-residue signal peptide, an extracellular region of 317 residues, a single 22-amino-acid transmembrane residue and a cytoplasmic portion of 210 amino acids. On SDS/PAGE the molecular mass of this receptor is 80 kDa. The type I receptor is the one which transduces signals when it binds to IL-1. The second, or type II IL-1 receptor, is predicted from cDNA to contain 398 amino acids and only contains a 29-residue cytoplasmic tail. The type I and II receptors share 28% amino acid identity in their extracellular ligand-binding sequences. The type II receptor is not a signalling receptor and the current hypothesis is that it acts as a negative regulator of IL-1 at the cell surface. The type II IL-1 receptor can also be shed and, again, act as a negative regulator of IL-1 in a manner analogous to IL-1ra [52]. IL-1ra binds with equal avidity as IL-1α and IL-1β to the type I receptor [53], but binds with lower avidity to the type II receptor [54].

Table 8.3 IL-1ra inhibits IL-1 activity in *in vitro* models

Experimental model	Reference
Synovial fibroblast prostaglandin E_2 synthesis	[43]
Rodent bone resorption	[55]
Cartilage matrix degradation	[56]
Collagenase production	[57]
Endothelial cell adhesiveness for PMNs	[58]
Lymphocyte proliferation	[43]
Nitric oxide-dependent vasodilatation	[59]

Table 8.4 Disease-modifying actions of IL-1ra in animal models of disease

Model	Reference
LPS-induced intraperitoneal inflammation	[60]
LPS-induced pulmonary inflammation	[61]
Collagen-induced arthritis in mice	[62]
Antigen-induced arthritis in rabbits	[63,64]
Inflammatory bowel disease in rabbits	[65]
Graft-versus-host disease in mice	[66]
E. coli-induced septic shock in rabbits	[67,68]
E. coli septic shock in mice	[69]
E. coli septic shock in baboons	[70]
Staphylococcus epidermidis shock in rabbits	[71]
Klebsiella pneumoniae sepsis in rats	[72]

IL-1ra has been extensively studied *in vitro* and *in vivo* and the voluminous literature will only be sketched out here. By the time that IL-1ra had been cloned and expressed many *in vitro* assays of IL-1 activity had been devised. IL-1ra has been shown to block the actions of both forms of IL-1 in all such *in vitro* assays (Table 8.3). IL-1ra also blocks many forms of experimental inflammation in laboratory animals and some of these studies are highlighted in Table 8.4. Of particular interest is the ability of IL-1ra to prevent the lethal effects of exposure to LPS or to systemically administered bacteria in a range of species including rabbits, mice and primates. These various experimental models are believed to mimic the physiological effects which occur in patients suffering from septic shock caused by Gram-negative or -positive bacteria. The possibility that IL-1ra could be used to treat patients with septic shock was then tested in a range of clinical trials.

8.3.1.2.2   Clinical trials of IL-1ra in sepsis

One of the significant findings from the *in vitro* and experimental *in vivo* studies of IL-1ra was the large molar ratios of IL-1ra:IL-1 required to block the biological actions of IL-1. Synergen Ltd, who set up the clinical trials of IL-1ra, were therefore required to produce very large amounts of pharmaceutical grade IL-1ra for these studies and this was a major *tour de force* of bio-pharmaceutical manufacture. Administration of recombinant IL-1 to humans as a bone-marrow stimulant suggested that the maximum tolerated dose was 10 ng/kg. At this dose subjects experienced a range of symptoms including headache, muscle pain and malaise [73]. In a phase I trial 25 volunteers were administered recombinant IL-1ra by intravenous infusion in an escalating dose trial. Subjects received IL-1ra at doses from 1 to 10 mg/kg over a 3 h period without experiencing any haemodynamic problems. This study established the pharmacokinetics of IL-1ra with an alpha-phase serum half-life (i.e. the time taken for the IL-1ra to redistribute from the plasma to the tissues) of 21 min. The beta-elimination phase revealed a terminal serum half-life of 108 min in subjects with normal renal function [74]. Thus, volunteers can tolerate a million-fold more IL-1ra than IL-1. The major problem highlighted by this study was the short half-life of IL-1ra in plasma. A phase II open-label, randomized, placebo-controlled, multicentre trial of IL-1ra using 99 patients with sepsis syndrome was conducted in 1991 in the U.S.A. Patients were randomized into a placebo group or one of three treatment doses: 17 mg/h; 67 mg/h; or 133 mg/h. Patients were infused for 72 h and were then monitored over the next 28 days. The results of this trial were encouraging, with the 28-day all-cause mortality rate in the placebo group being 44%, and in the low-, medium- and high-dose groups equal to 32%, 25% and 16% respectively [75]. While the results of this phase II trial were very encouraging, two subsequent multicentre phase III trials, with around 1000 patients in each, have failed to show any efficacy of IL-1ra in patients with septic shock. Analysis of the clinical trial data suggests that only those patients with severe sepsis show any benefit from IL-1ra administration [75]. However, at the present time IL-1ra appears to have foundered on the rocks as far as sepsis therapy is concerned.

8.3.1.3 Conclusions

Cytokine receptors are generally present in low numbers on the cell surface, have a very high affinity for the agonist and make physical contact with the appropriate cytokine over a large proportion of the respective surfaces of the receptor and the cytokine. This is often talked of in terms of the cytokine making multi-point attachment with its receptor. This makes it very difficult to develop low-molecular-mass antagonists and, as has been described, for the moment there is a distinct lack of interest in the natural antagonist of IL-1. However, if it is not possible to antagonize the binding of particular cytokines to their receptor there are other ways of interfering with the cytokine–cell interactions.

8.3.2   Neutralization of soluble cytokines

Two major approaches have been taken to neutralizing cytokines which have been produced and released into the extracellular fluid. The first is the use of selective binding proteins, normally soluble forms of cytokine receptors, and the second the use of antibodies to neutralize the activity of selected cytokines.

8.3.2.1 Binding proteins and soluble receptors

As mentioned previously, there are numerous reports of the capacity of biological fluids (serum, plasma, tissue culture-enriched media, etc.) to inhibit the biological activity of selected cytokines [41]. Such inhibitory activity consists of molecules like IL-1ra, the blood protein α_2-macroglobulin [76], and cytokine receptors shed from the surfaces of cells. Under normal conditions, levels of cytokine receptors in biological fluids are generally low. However, it now appears that shedding of cytokine receptors is a normal phenomenon under conditions of stress. For example, administration of *E. coli* endotoxin to volunteers results, within 3 h, in a four- to fivefold increase in circulating levels of the 55 kDa (TNF-R55/TNFR1) and 75 kDa (TNFR75/TNFR2) receptors for TNFα [77]. This phenomenon is reflected in the increased levels of IL-2 receptors in the serum or urine of patients with cancer [78], arthritis [79] or AIDS [80], and the elevation of both forms of the TNF receptor in individuals with fever [81], leukaemia [82] or those undergoing haemodialysis [83].

Soluble cytokine receptors, because they bind with high affinity to cytokines, could act as inhibitors of cytokines found in the extracellular fluid and therefore be cytokine-modulating therapeutics. It is assumed that this is part of the normal action of shed receptors in controlling cytokine networks. There is now good experimental evidence that administration of soluble cytokine receptors can inhibit various disease-like states in animals, including rejection of heterotopic cardiac allografts in rats [84], experimental allergic encephalomyelitis in rats [85] and *E. coli* sepsis in baboons [86]. One of the disadvantages of soluble cytokine receptors is their short circulating half-life. One solution to this problem has been to construct chimaeric molecules composed of the receptor linked to the Fc region of human immunoglobulin. A construct of the T-lymphocyte CD4 receptor linked to an IgG heavy chain had a terminal half-life in rabbit blood of up to 48 h. This compares with approximately 15 min for the CD4 receptor alone [87]. Similarly, an IgG-TNF-R55 construct has improved pharmacokinetics [88] and such chimaeric receptors have been used for the treatment of patients with rheumatoid arthritis [89]. Synergen have used soluble TNF receptors in the treatment of septic shock, but the clinical trials failed to show any benefit [90]. The future potential of soluble receptors or receptor constructs as therapeutics for human diseases is likely to be limited as these constructs are expensive to produce and have to be given by injection. Such a therapeutic modality is most likely to be used in the treatment of acute conditions such as septic and toxic shock.

8.3.2.2 Cytokine-neutralizing monoclonal antibodies

Combinatorial chemistry is now being employed enthusiastically by the pharmaceutical industry to aid the production of selective binding agonists which

Table 8.5 Effects of neutralizing cytokines with neutralizing mAbs on experimental models

Cytokine neutralized	Disease model	Reference
IL-1	Antigen-induced arthritis	[92]
TNF	Collagen-induced arthritis	[93]
	Antigen-induced arthritis	[94]
	Septic shock	[95,96]
IL-3	Helminthic intestinal mastocytosis	[97]
IL-4	Parasite-induced IgE response	[98]
	Leishmaniasis	[99]
IL-5	Helminth-induced eosinophilia	[100]
	Experimental asthma	[101]
IL-6	Septic shock	[102]
IFNγ	Schwartzman reaction	[103]
	Cerebral malaria	[104]
PDGF	Atherosclerosis	[105]

can neutralize the biological activity of just about any biological molecule [7,91]. The immune system can be viewed as a stochastic combinatorial system designed to make the binding sites that we know as antibodies and T-cell receptors. Monoclonal antibodies (mAbs) have been raised to virtually all the known cytokines and the capacity of such antibodies to neutralize a wide range of experimental lesions in animals has been reported (Table 8.5). mAbs have certain advantages as therapeutics. Their natural milieu is the extracellular fluids and mucosal surfaces of the body and they exist in such environments for extended periods of time (the lifetime of human IgG in serum is 7–21 days). Antibodies can be produced with extremely high affinities for their antigens and the size of these Ig molecules can be modified by chemical or genetic manipulation. There is, however, one major disadvantage to the use of mAbs, namely the immune response they engender. mAbs are most easily raised in rodents (mice and rats) and administration of such antibodies to man results in a rapid anti-species (isotypic) response which, with murine antibodies, is known as the HAMA (human anti-mouse antibody) response. These anti-isotypic antibodies bind to the therapeutic mAb and prevent it working, and can, indeed, cause serious side-effects for patients including ana-phylaxis or serum sickness [106]. This problem of isotypic antibodies, which are directed against the Fc region of the antibody, has been circumvented, to a large extent, by the development of molecular biological techniques for making chimaeric or humanized forms of murine or rat antibodies. In chimaeric antibodies, the Fc portion of the murine antibody is substituted with a human Fc region. There are four classes of human IgG which differ in the biological activities of their Fc regions. For example, IgG_1 and IgG_3 are good complement fixers (i.e. activators) and may be useful if the chimaeric antibody is being used to kill the cells to which it binds. IgG_4 is a poor fixer of complement and is therefore a better choice for antibodies which are used to bind and inactivate cytokines. Thus, chimaeric antibodies limit the HAMA response, but there are still murine segments to which immune responses could be raised. To limit immunoreactivity it is possible to fully humanize murine antibodies. To do this the segments of the antibody molecule that confer specificity of binding and which are termed the complementarity-determining regions (CDRs) are 'grafted' into a human antibody, such that the antibody contains the mouse CDRs but the rest of the molecule is human [107]. Such fully humanized antibodies would not be expected to raise a significant immune response in humans and so would overcome the major obstacle to the therapeutic use of mAbs. Of course it is still possible that individuals administered such antibodies would raise either an anti-allotypic or anti-idiotypic response, particularly if the antibody in question was being administered chronically. A summary of the uses of, and responses to, mAbs in clinical trials is given in [108].

Anti-cytokine antibodies have been in clinical trial for the past 4 years. Most experience has come from the administration of anti-TNFα mAbs for the treatment of septic shock, inflammatory bowel disease and rheumatoid arthritis. Over the past decade, as has been described in earlier chapters, TNFα has been found to be a predominant cytokine in terms of its involvement in human pathology. Septic shock was the first disease for which a key role for TNFα was established and it is possible to treat experimental septic shock in many species by blocking the biological actions of this cytokine [109].

Murine mAbs to TNF have been used in three small clinical trials and showed some promise [110–112]. Since then, two large multicentre clinical trials, one in North America and one in Europe, have been completed. These studies compared a single administration of a murine mAb to TNF with a placebo in patients with septic shock. The data from the North American trial have been presented [113]. Patients who were in shock when administered the mAb showed a reduction in mortality. Those patients not in shock when administered antibody showed no clinical benefit. Thus, there appears to be a definite window of opportunity for the administration of antibodies to TNF in patients with septic shock.

Inflammatory bowel diseases (IBDs), which include Crohn's disease and ulcerative colitis, are serious conditions characterized by chronic inflammation of the intestinal mucosa and lamina propria and may be driven by bacterial factors. There is strong evidence for an involvement of TNFα in IBD. For example, there is significant elevation of the levels of TNF in the stools of patients with IBD [114]. In a preliminary study, a 12-year-old girl with unresponsive Crohn's disease showed complete endoscopic remission following treatment with an anti-TNF mAb [115]. A double-blind placebo-controlled clinical trial of an anti-TNF mAb has shown significant clinical benefit in patients with Crohn's disease [116].

Rheumatoid arthritis is a chronic inflammatory disease characterized by painful swollen joints and the progressive destruction of the articular cartilage and subchondral bone of the affected joints [117]. There is good evidence that TNFα is one of the major mediators of the pathology of this disease [118]. A number of clinical trials of chimaerized mAbs to TNF have been undertaken with patients suffering rheumatoid arthritis, and they all suggest that knocking out TNF can produce symptomatic relief [119–122]. However, it is not clear whether such antibody treatment will be able to control the tissue destruction which is the major problem in rheumatoid arthritis.

8.4   Inhibition of the synthesis, processing and release of cytokines

Thus far we have considered the manipulation of cytokines by antagonism of the receptors for specific cytokines or the inhibition of cytokines by soluble receptors or neutralizing mAbs. Another basic strategy for therapeutically modulating cytokines involves directly blocking their synthesis or, for certain cytokines, the additional steps (e.g. proteolytic processing) that are required before cytokines can be released from cells. Related to this is the strategy of manipulating intracellular signalling pathways activated by cytokines (Figure 8.2). These various points of intervention will be reviewed, starting with the cell activation that initiates the sequence of events that results in cytokine gene transcription (see Table 8.2).

8.4.1   Inhibition of cytokine induction

In most circumstances where there is an increase in the levels of particular cytokines there is an initiating event. In the case of septic shock, this event is the death of bacteria and the precipitous release of endotoxin [123] and its interaction, via CD14, with cytokine-producing cells. Of course, other stimulators of cytokine induction are known, including immune complexes, T-cell receptor binding, class II antigen receptor binding, integrin receptor binding, C5a, viruses, other bacterial constituents and cytokines themselves. In the context of this volume, we will concentrate on agents that may block the cytokine-inducing actions of endotoxin/LPS and their potential as therapeutics for treating Gram-negative septic shock.

8.4.1.1   Inhibition of LPS-induced cytokine synthesis

As has been discussed in earlier chapters, endotoxin and LPS are major cell signals for the induction of cytokine synthesis. There has therefore been an ongoing effort to develop agents that inhibit the capacity of LPS to stimulate cellular cytokine synthesis. Table 8.6 gives an overview of the possible means by which LPS-induced cell cytokine synthesis can be inhibited, and these various modalities will be discussed in turn.

Table 8.6 Modalities for inhibiting LPS-induced cytokine synthesis

Modality	Reference
Polymyxin B	[124,125]
LPS-neutralizing mAbs	[126–128]
Bactericidal/permeability-increasing protein	[129]
CAP 18	[130]
Endotoxin-neutralizing protein	[131]
CD14	[132]
Lipid A antagonists	[133,134]
Triacylglycerol-rich lipoproteins	[135]
IL-10	[136]
Cell signalling inhibitors	[137,138]

8.4.1.1.1   Polymyxin B

Polymyxin B is one of a series of polypeptide antibiotics produced by *Bacillus polymyxa,* a soil organism. This antibiotic has a potent bactericidal effect on some Gram-negative bacilli and also binds and inactivates certain, though not all, lipopolysaccharides. Unfortunately, polymyxin B administration can cause serious side-effects, including renal tubular damage, vertigo and paraesthesia. Although efforts have been made to produce less toxic versions of the polymyxins, they have not been successful [124].

8.4.1.1.2   Anti-LPS antibodies

The use of antibodies to neutralize the biological actions of cytokines has already been discussed and the problems and pitfalls highlighted. In the late 1980s and early 1990s there was considerable excitement over the use of neutralizing mAbs to LPS as therapeutics for septic shock. Biopharmaceutical companies such as Centocor, Xoma and Chiron invested heavily in the development of chimaerized mAbs to the lipid A portion of the LPS molecule. However, when such antibodies were tested in large-scale clinical trials they proved to be ineffective. Of course, this does not rule out the possibility that anti-LPS antibodies may be of use in the future. The potential use of anti-LPS antibodies is reviewed in [125].

8.4.1.1.3   Bactericidal/permeability-inducing protein

Studies of polymorphonuclear leucocytes have led to the isolation of a number of antibacterial proteins and peptides from the granules associated with these cells. The defensins are one family of small peptides (3–4 kDa) with antibacterial activity [139]. In recent years two other proteins with anti-endotoxic activity have been reported. The first is bactericidal/permeability-increasing protein (BPI), first discovered by Elsbach and Weiss [140]. This protein was shown to have potent antibiotic properties for Gram-negative organisms by binding to LPS [141,142]. BPI was cloned by Genentech [143] and was shown to have a full-length sequence of 456 amino acids with two glycosylation sites.

It was subsequently shown that the N-terminal region of this molecule carried all of the biological actions of BPI [144,145], and a cloned 23 kDa N-terminal fragment (rBPI$_{23}$) has been shown to retain the activity of the intact molecule [146,147]. The structural requirements of lipid A to bind to rBPI$_{23}$ have been examined by Gazzano-Santoro and co-workers [148] using a series of lipid A preparations from various bacteria as well as synthetic analogues. Lipid A molecules from *E. coli*, *Salmonella*, *Neisseria* and *Rhizobium* species all bound, and the nature of the negative charge on lipid A (which can be phosphate, carboxylate or sulphate) had no major effect on binding. However, as was expected, the lipid A acyl chains are important. Synthetic variants lacking the acyl chains failed to bind. Lipid IV$_A$, a precursor identical to *E. coli* lipid A, but lacking the 2' and 3' acyl chains, was the simplest structure found to bind to rBPI$_{23}$.

BPI and recombinant fragments have been tested for their capacity to inhibit experimental sepsis (Table 8.7) and in most models it appears to have therapeutic effects. One of the curious findings from these studies has been the elevation in the levels of serum TNFα in the blood of rats which have responded to BPI, and the interpretation of these studies was that BPI exerted its protective effect through a TNF-independent mechanism by inhibiting endotoxin-stimulated production of IL-6 [153]. Indeed, IL-6 has been reported to be required for a protective immune response to *E. coli* infection [154].

One of the major problems with the use of BPI as a therapeutic is its extremely short half-life in the circulation. One solution to this problem has been to produce chimaeric molecules consisting of BPI linked to lipopolysaccharide-binding protein (LBP). Such chimaeric molecules had biological activity and their half-life was considerably extended [155]. The report that LBP actually potentiates the biological activity of BPI [156] suggests that this process of chimaerization could be taken further to produce a long-lived, active therapeutic agent. The biology of BPI and related proteins has been discussed in more detail in Chapter 6. The biology of antibacterial peptides and proteins produced by multicellular organisms is discussed in Chapter 9.

Table 8.7 *In vivo* studies of BPI

Model	Result	Reference
Mouse endotoxin challenge	100% survival	[149,150]
Baboon live bacterial challenge	Decreased plasma LPS	
	No effect on mortality	[151]
Endotoxin challenge in rats	Significant blockade of haemodynamic	
	effects and inhibition of mortality	[152]
LPS-induced air pouch inflammation	Inhibited inflammatory effects of LPS	[153]

8.4.1.1.4   Cap 18

This is another protein from neutrophils with antimicrobial/LPS-binding properties. The cloned rabbit protein has a sequence containing 142 amino acids and a predicted molecular mass of 16.6 kDa [157]. The C-terminal domain (residues 105–142) contains an LPS-binding region, and synthetic peptides containing this segment of Cap 18 have the capacity to bind and inactivate the biological actions of LPS. Preliminary studies have shown that Cap 18 peptides are active against endotoxin-induced pathology in experimental animal models [158].

8.4.1.1.5   Endotoxin-neutralizing protein

The horseshoe crab (e.g. *Limulus polyphemus*) has been described in Chapter 5 as a source of haemolymph for the *Limulus* amoebocyte lysate (LAL) assay (see [159] for an up-to-date review of this assay). This haemolymph also contains endotoxin-neutralizing proteins of approximate molecular mass 12 kDa, which have been termed *Limulus* anti-LPS factors (LALF; the cloned protein being given the term ENP) and which are thought to be part of the crab's defence mechanism against endotoxin. LALF binds to the lipid A portion of LPS, blocking its biological actions on mammalian cells [160,161], and also inhibits the growth of Gram-negative bacteria [162]. LALF and ENP have been shown in addition to protect rabbits and rats from endotoxin shock and *E. coli* sepsis [163,164].

8.4.1.1.6   CD14

Chapter 6 provided a detailed discussion of the cellular and molecular biology of CD14 and its role in transducing the biological activity of LPS. CD14 is the major 'receptor' for LPS although it is not the signalling receptor. It has now been suggested, with some experimental backing, that the pathophysiological effects of LPS can be blocked by administering CD14 or by inhibiting the actions of CD14. For example, Haziot and co-workers have suggested that soluble CD14 could act as an antagonist of LPS and have shown that recombinant soluble CD14 (rsCD14) could inhibit TNFα synthesis by cultured human peripheral blood mononuclear cells [165] and could block LPS-induced lethality in mice [166,167].

8.4.1.1.7   Lipid A antagonists

The structure of lipid A and its structure–function relationships, including antagonist actions, have been detailed in Chapter 5. From a pragmatic pharmacological viewpoint the most sensible way of blocking LPS activity would be by using an antagonist of LPS (i.e. lipid A)-receptor binding. This has proved more difficult than expected and it is only within the past few years that stable antagonists have been developed and tested *in vivo*.

Three active disaccharide antagonists of LPS bioactivity have been described: (i) precursor lipid IV_A; (ii) partially deacylated LPS; and (iii) penta-acyl diphosphoryl lipid A derived from the LPS of *Rhodobacter sphaeroides* [168–170]. The first two structures only antagonize LPS action on human cells, whereas the *R. sphaeroides* lipid A (RSLA) acts as an antagonist for murine and human cells and has been shown to block LPS-induced cytokine induction by cultured macrophages [170,171] and LPS-induced lethality in mice [172,173]. However, it is unlikely that RSLA would be of pharmacological use, due to problems of extraction and the need for the final product to be of sufficient homogeneity and stability to meet with drug licensing requirements. Esai Pharmaceuticals have attempted to develop stable lipid A antagonists based on the proposed structure of RSLA [174,175]. This attempt revealed that the proposed structure of RSLA was incorrect [176] and that the synthetic product lacked agonist activity. This suggested that the agonist activity shown by RSLA may be due to small amounts of a potent agonist structure [177]. The final synthetic RSLA, while a good antagonist of lipid A, suffered stability problems upon storage with facile hydrolytic cleavage of the C-3 and C-3' acyl groups producing agonistic by-products [134,178]. To stabilize the molecule, the acyl linkages at C-3 and C-3' were replaced by ether linkages and the C-3,C-3' diether analogue retained the potent antagonist activity. The C-6' hydroxyl group was blocked by a methyl group to overcome chemical interaction at this site, leading to the final fully stable analogue, E5532. This analogue shows potent antagonist activity *in vitro*, and has been shown to protect mice from LPS-induced lethality and, when administered with an antibiotic, from lethal infection with *E. coli* [134,178].

8.4.1.1.8   Triacylglycerol-rich lipoproteins
It is well known that exposure to endotoxin results in the rise in plasma levels of triacylglycerols due to an increase in triacylglycerol-rich very-low-density lipoproteins (VLDLs) [179]. This change in plasma chemistry may be protective, as it has been demonstrated that VLDLs and chylomicrons bind and inactivate endotoxin and can protect animals against lethal endotoxin challenge [180,181]. In addition to neutralizing LPS, VLDL and chylomicrons act to increase its clearance from the plasma. Administration of triacylglycerol-rich lipoproteins to rats with caecal ligation and puncture, a septic model in which the gut contents produce a polymicrobial infection similar to that seen in human intra-abdominal sepsis, significantly improved survival compared with saline controls [135]. These data suggest a role for lipoproteins or lipid particles in the treatment of sepsis.

8.4.1.1.9   IL-10
The role of IL-10 in the control of endotoxin biology has been discussed in Chapter 6. While not directly interacting with LPS, exogenously administered IL-10 has been shown to be capable of inhibiting various forms of endotoxin shock and sepsis in experimental animals [136,182,183]. The role of so-called anti-inflammatory cytokines in the control of pro-inflammatory cytokine

network interactions and in the control of inflammation will be considered in a later section in this chapter.

8.4.1.1.10 Cell signalling inhibitors

One of the major growth areas in the study of LPS is the determination of the cellular signalling pathways responsible for the multiple action of this complex biological regulator. This has been reviewed in Chapter 6 and it is clear that LPS can stimulate most of the known cellular transduction pathways. The possibility of using inhibitors of intracellular signalling pathways to block the pathophysiological response to endotoxin is an attractive one and two reports have shown that this is possible. Tyrphostins are selective inhibitors of tyrosine kinases developed by Levitski and co-workers [184], who have demonstrated that the tyrphostin, AG 126, which blocks the LPS-induced tyrosine phosphorylation of a mitogen-activated protein kinase (specifically p42^{MAPK}), protects mice against LPS-induced lethality [137].

LPS stimulation of cells can also result in the generation of phosphatidic acid [185]. Lisofylline [(R)-1-5-hydroxyhexyl-3,7-dimethylxanthine], a metabolite of the rheological drug pentoxyfylline, is a potent inhibitor of phosphatidic acid formation, and administration to mice prevented the lethal effects of endotoxin exposure [185,186].

8.4.1.1.11 Inhibitors of lipid A synthesis

An exciting recent advance has been the report from Raetz's group at Merck of the development of hydroxamic acid inhibitors of a unique deacetylase involved in the synthesis of lipid A. Such inhibitors can kill *E. coli* and have been shown to be effective in treating mice with a lethal intraperitoneal dose of *E. coli* [186a].

8.4.2   Inhibition of cytokine synthesis

If it does not prove possible to block the activation of the cell, the next point at which the pharmacologist can intervene is via the complex set of events which leads from cell membrane activation to the synthesis of the cytokine, or cytokines. As can be seen in Figure 8.2, these events involve intracellular signalling pathways, gene activation and gene transcription, mRNA production and metabolism, protein translation and, finally, processing and release of the cytokine. In this section, the literature on non-specific inhibitors and on the use of oligonucleotide-based drugs will be dealt with. The possibility of selectively interfering with intracellular signalling will be left to the last section in the chapter (the methods used to inhibit cytokine activity are denoted in Figure 8.3).

8.4.2.1 Non-specific low-molecular-mass inhibitors of cytokine synthesis

An obvious goal for the pharmaceutical industry is to develop a low-molecular-mass cytokine inhibitor with pharmacokinetic properties requiring

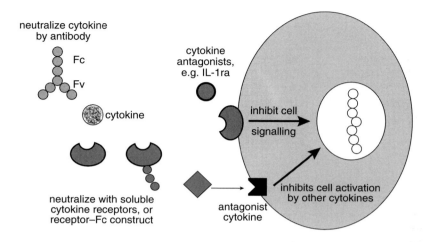

Figure 8.3 Methods of inhibiting cytokine activity
If it proves impossible to inhibit the synthesis of cytokines it is still possible to block their activity by targeting soluble molecules in the extracellular fluid. This can be done by raising monoclonal antibodies to the selected cytokine and screening for those antibodies which block its biological functions. Cloned, soluble forms of cytokine receptors can be used as high-affinity binding agents which act in a similar manner to antibodies. To extend the circulating lifetime of such receptors (which are normally rapidly cleared), constructs of the receptor with the Fc portion of human immunoglobulins have been prepared and shown to have significantly improved pharmacokinetics. Cytokine antagonists have proved difficult to produce but one natural antagonist (IL-1ra) has shown therapeutic potential, at least in animals. Another potential strategy is to use anti-inflammatory cytokines for therapy. This would include the use of IL-4, IL-10 or TGFβ. It would also be possible to inhibit the cellular response to cytokine by blocking the intracellular signalling processes.

the patient to take the drug no more than three or four times a day. Most of the large pharmaceutical companies have therefore screened their vast libraries of existing compounds and many other low-molecular-mass compounds in order to determine if any of these molecules have the requisite properties. This literature has been reviewed previously [5,187,188] and therefore will only be dealt with briefly in this chapter.

Steroids and compounds such as gold and penicillamine, which are used to treat rheumatoid arthritis, have been shown to inhibit the synthesis of pro-inflammatory cytokines, particularly IL-1 [187]. However, it is still not clear if this is the reason for the apparent efficacy of these drugs in treating chronic inflammation.

A range of other diverse compounds have been reported, over the past 10 years, to inhibit cytokine synthesis and these are listed in Table 8.8. Only those compounds which have shown activity *in vivo*, or are not described in the earlier reviews, will be discussed in this chapter.

John Lee and colleagues at SmithKline Beecham, King of Prussia, U.S.A. have pioneered the development of inhibitors of cytokine synthesis and their use in septic shock [189,190]. The initial approach was based on compounds

Table 8.8 Low-molecular-mass inhibitors of cytokine synthesis

Compound	Cytokines inhibited
Anti-arthritic drugs	
Gold	Various cytokines and lymphokines (at high concentrations)
Penicillamine	Various cytokines and lymphokines (at high concentrations)
Immunosuppressants	Selected lymphokines (IL-2, IL-4, etc.)
Cyclosporin	
FK506	
Rapamycin	
Vitamin D_3	IL-8
Vitamin K	IL-6
Glucocorticoids	IL-1, TNF, etc.
Prostanoids	IL-1, TNF
Taurolidine	IL-1
Pentamidine	IL-1
Tenidap (Pfizer)	IL-1
SK&F 105809	IL-1
E5090 (Tsukuba)	IL-1
IX 207-887 (Sandoz)	IL-1
RP54754 (Rhone–Poulenc Rorer)	IL-1
Pentoxyfylline (Hoechst)	TNF
E3330 (Tsukuba)	TNF
Thalidomide	TNF
SK&F 86002	TNF
SA 3443 (Sanen)	TNF

which inhibited cyclo-oxygenase and/or lipoxygenase and the first compound reported, SK&F 86002, inhibited IL-1 release from human monocytes at a concentration of 1–2 μM. This compound had protective effects in murine models of endotoxic shock and inhibited the synthesis of TNF in treated animals [191]. Another compound, SK&F 105809, also inhibited the symptoms of endotoxic shock in experimental animals [192], and compounds of this type are under ongoing study [189] and are discussed in more detail in section 8.5 of this chapter.

Pfizer in the U.S.A. have also been developing anti-cytokine therapeutics and have recently brought the compound Tenidap to the market for the treatment of rheumatoid arthritis [193]. Tenidap is an unusual compound with many potential inhibitory mechanisms (for example, it apparently inhibits cyclo-oxygenase and 5-lipoxygenase) and it has the following rank order of potency in inhibiting human monocyte cytokine synthesis: IL-6 >TNFα >IL-1, the IC_{50} for the inhibition of IL-6 being 2 μg/ml [194]. It is not clear, however, if Tenidap would have any beneficial actions in infectious diseases.

Sandoz have tested the compound IX 207-887, which is not an inhibitor of arachidonic acid metabolism, but does inhibit cytokine release from stimulated monocytes [195], in patients with rheumatoid arthritis. In a placebo-controlled, double-blind trial this compound showed impressive clinical benefits [196]. Surprisingly, Sandoz are not developing this agent.

A series of 2′-substituted chalcone derivatives have been reported to have inhibitory effects on cytokine biosynthesis; the mechanism of action not being understood [197]. High concentrations of chalcones can inhibit the lethal effects of endotoxin [198].

Three other low-molecular-mass inhibitors of cytokine synthesis have been reported to be active *in vivo*. The first is E5090 which inhibited LPS-induced pathology in a rat air pouch model [199]. The second is RP54745 which inhibited *ex vivo* production of peritoneal macrophage IL-1 synthesis induced by LPS [200]. The third is taurolidine, an analogue of the amino acid taurine, and used as an adjunct therapy in the treatment of infections, both as an antimicrobial agent and as an anti-toxin [201,202]. Taurolidine has been shown to inhibit LPS-induced IL-1 synthesis by peripheral blood mono-nuclear cells [203], which probably explains the earlier report of its anti-endotoxin activity in experimental animals [204].

A whole range of immunosuppressant drugs and agents also block the production of selected cytokines by leucocytes. This area has recently been reviewed [188] and will not be dealt with here.

8.4.2.2 Oligonucleotide-based cytokine inhibitors

Cytokines, like all other proteins, are translated from specific mRNA molecules which are, in turn, transcribed from specific genes. Over the past decade the capacity of oligonucleotides to hybridize specifically with mRNA or with DNA has been developed into a novel and exciting general mechanism

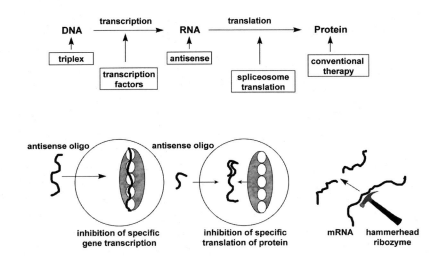

Figure 8.4 Mechanism of action of oligonucleotide-based drugs

Table 8.9 Antisense oligonucleotides inhibiting cytokine gene expression

Target	Chemistry	Target site	Cell type	Reference
IL-1α	P=O	AUG	Endothelial	[206]
IL-1β	P=S	AUG	PMN	[207]
	P=S	Coding		
IL-1 receptor	P=S/P=O	AUG/coding	Fibroblast	[208]
IL-2	P=S	AUG	T-lymphocyte	[209]
IL-4	P=O	AUG	B-lymphocyte	[210]
IL-6	P=O	Coding	Myeloma	[211]
IL-8	P=O	AUG	Melanoma	[212]
GM-CSF	P=O	AUG	Endothelial	[213]

Abbreviations used: AUG, oligonucleotide hybridizes to AUG translation initiation codon on mRNA; P=O, normal phosphate ester-containing oligonucleotide; P=S, phosphorothioate-type oligonucleotide.

for the specific inhibition of the synthesis of gene products. The exact mechanisms by which such oligonucleotides inhibit the synthesis of specific proteins is described schematically in Figure 8.4. Briefly, it is possible to use oligonucleotides to: (i) inhibit the transcription of specific genes; (ii) inhibit the translation of specific mRNAs; or (iii) destroy specific mRNA molecules and thus act in a similar manner to (ii). The use of oligonucleotides to control cytokine function has recently been reviewed [205] and the reader is referred to this source for further details. A growing number of cytokine genes have had their expression inhibited by oligonucleotide inhibitors and some representative examples are given in Table 8.9.

While, in theory, any gene product can be selectively blocked by oligonucleotide-based drugs (OBDs) there are major problems in delivery due to: (i) the labile nature of the phosphodiester bond in oligonucleotides; (ii) pharmacokinetic considerations; and (iii) the slow uptake into cells. Most studies have used oligonucleotides of between 12 and 25 residues and have overcome the first problem by modifying the nature of the bond linking nucleotides, with the phosphorothioate bond being used in place of the natural phosphodiester bond. Since the early 1990s there have been an increasing number of studies reporting on the *in vivo* pharmacology of OBDs (see [205]). In the context of cytokines and inflammation, mice injected with a phosphorothioate antisense oligonucleotide to the murine type I IL-1 receptor exhibited decreased neutrophil migration to skin sites injected with IL-1 [208]. An oligonucleotide to the cytokine- and LPS-inducible vascular adhesion receptor, ICAM-1, inhibited inflammatory changes in several model systems *in vivo,* most notably cardiac allograft rejection [214]. Perhaps the most exciting report is of the use of antisense to the transcription factor NF-κB to treat mice with experimental colitis caused either by instillation of 2,4,6,-trinitrobenzene sulphonic acid (TNBS) or inactivation of the gene for IL-10. NF-κB was found to be highly activated in the colons of these mice, and local administration of phosphorothioate antisense oligonucleotides, but not sense nucleotides, to NF-κB inhibited clinical and histological signs of disease. Antisense was more effective than glucocorticoids in treating the TNBS colitis. This report also highlights the role for NF-κB in the colitis caused by IL-10 knockout [215].

A recent report suggests that it is possible to use much shorter antisense oligonucleotides than normal: heptanucleotides, rather than those of 12–25 residues, still achieve selective and potent inhibition of protein synthesis [216]. This and other advances in targeting OBDs should see the OBD technology move from the laboratory and firmly into the clinic where it could have a major role to play in modulating inflammatory cytokine networks. Indeed, it is likely to be the only pharmacological modality which could possibly re-equilibrate tissue cytokine networks, producing the normal sets of networks found in healthy tissues.

8.4.2.3 Inhibition of cytokine processing
8.4.2.3.1 IL-1
A number of cytokines, most notably IL-1 and TNFα, are produced as pro-forms and require proteolytic processing in order for the active form of the molecule to be released (Figure 8.5). The nature of the enzymes involved in cytokine processing has recently been reviewed [217]. A number of proteinases, including bacterial proteinases (see Chapter 7), can convert the pro-form of IL-1β into a lower-molecular-mass active form. However, it was discovered that monocytes contain a specific enzyme which cleaves the pro-form of IL-1β at Asp-116-Ala-117. This enzyme was given the name pro-IL-1β-converting enzyme (ICE) [218,219]. In a superb piece of pharmaceutical scientific organization, scientists at the pharmaceutical company Merck Sharp and Dohme cloned, expressed, purified, assayed and developed potent inhibitors of ICE

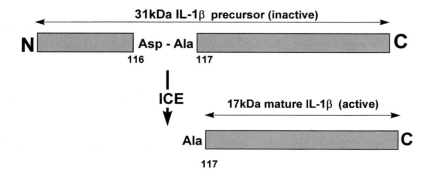

Figure 8.5 Mechanism by which the 31 kDa pro-form of IL-1β is converted into the active 17 kDa form
This occurs through the action of a specific cysteine proteinase, pro-IL-1β-converting enzyme (ICE), which cleaves the pro-form at between Asp-116 and Ala-117 to produce the biologically active IL-1β.

[220]. These studies showed that ICE was a novel cysteine protease composed of two subunits of 20 and 10 kDa with no sequence homology to any known cysteine proteinase. As this enzyme is required for the production of active IL-1β it is an obvious therapeutic target for conditions of excessive IL-1β production.

As described in Chapter 7, the cowpox virus contains a gene *crm*A that encodes a serpin-type proteinase inhibitor which selectively blocks the activity of ICE [221]. This gives added weight to the hypothesis that blockade of ICE could have therapeutic activity in inflammatory and infectious diseases [222]. A number of pharmaceutical companies have now produced selective inhibitors of ICE [223–225] and have tested them in various animal models including collagen-induced arthritis [226,227]. These compounds may well have efficacy in various infectious states where IL-1β overproduction is causing pathology.

8.4.2.3.2   TNFa

TNFα is synthesized as a 26 kDa precursor containing a 76-amino-acid N-terminal extension, and has to be processed to a 17 kDa mature form to express full biological activity. Pulse–chase experiments revealed that this processing occurred in a single step [228]. Mature TNF is cleaved at Ala-76-Val-77 and there is some evidence for serine proteases having a role in this processing [229]. Recently, a number of laboratories have identified a matrix metalloproteinase (MMP)-like enzyme responsible for the release of mature TNFα [230–232]. A number of companies have developed potent inhibitors of MMPs [233] and these have been used to determine if it is possible to block TNF processing *in vitro* and *in vivo*. A number of the potent hydroxamic acid inhibitors of metalloproteinases were found to inhibit the cleavage of precursor TNF and to selectively inhibit the cellular release of mature TNFα. These inhibitors had no effect on other cytokines such as IL-1β or IL-8 which also need proteolytic processing [231]. The hydroxamic acid compounds were also effective inhibitors of TNF production *in vivo* and were able to protect mice from LPS-induced lethality [230].

Thus, inhibitors of cytokine processing may have therapeutic potential in infectious and inflammatory diseases by targeting key pro-inflammatory cytokines.

8.5   Cytokine signal transduction as a therapeutic target

Signal transduction or intracellular signalling is the process by which cells integrate the multiple signals which they are constantly receiving from their environment. Such signals include both positive and negative regulatory information which must be internalized to produce specific patterns of gene transcription and regulation. This global regulation of cells is dependent on the patterns of activity of intracellular kinases and phosphatases. There are now a very large number of these enzymes (for example, the number of distinct kinase domain sequences is in the region of 400) [234], which fall into distinct families depending on the receptor linkage, structure and amino acid residues phosphorylated [235]. The signalling pathways believed to be involved in cells responding to LPS and endotoxin have been discussed in Chapter 6, as has the therapeutic effect of selective inhibitors of tyrosine kinases [236] and of the formation of phosphatidic acid [236] in mice with experimental endotoxic shock.

The pharmaceutical industry has viewed intracellular signalling as a target for the past decade [237], and with the increased understanding of the complexity of signal transduction processes has come the possibility of pharmacological selectivity. A large number of compounds are now known to act as inhibitors of both protein kinases and protein phosphatases and many of these are natural compounds and indeed some are the products of bacteria. Protein tyrosine kinases (Tks) appear to be major targets for bacterial secondary metabolites, with strains of *Streptomyces* producing aromatic compounds such as erbstatin, herbimycin and lavendustin, and *Pseudomonas* producing the inhibitor, genistein. Erbstatin has been used as the basis for the development of synthetic inhibitors of Tks termed tyrphostins [238]. Natural inhibitors of protein serine/threonine kinases include staurosporine, a fungal product, which is a potent inhibitor and has been used as the starting point for the synthesis of potent and highly selective inhibitors of protein kinase C (PKC) [239]. Another fungal product that is a promising novel immunosuppressant is rapamycin which inhibits a lipid kinase. A number of natural inhibitors of protein phosphatases have also been discovered. These include: the okadaic acid family, which are diarrhetic shellfish poisons; mycrocystins, potent liver toxins from blue-green algae; and cantharidin, the active principle of 'Spanish fly'. The immunosuppressants cyclosporine and FK506, the former being widely used in transplant rejection, are potent inhibitors of the calcium-activated protein phosphatase 2B (calcineurin). Inhibition of this enzyme results in selective inhibition of lymphokine (particularly IL-2) synthesis and suppression of T-lymphocyte responses (reviewed in [240]).

8.5.1   Phosphodiesterase inhibitors

Phosphodiesterases (PDEs) are a group of intracellular multi-isoform enzymes which hydrolyse cyclic AMP, thus inhibiting the actions of protein kinase A (PKA). Inhibition of these enzymes, therefore, maintains intracellular cyclic AMP concentrations and activated PKAs [241]. Different cell populations produce different classes of PDEs and the development of selective inhibitors of these isoforms is believed to hold the promise of selective pharmacological agents. The ability of bacterial exotoxins which raise intracellular levels of cyclic AMP to selectively inhibit LPS-induced TNFα synthesis (but not IL-6 synthesis) has been discussed in Chapter 7. Selective synthetic inhibitors of PDE IV, rolipram and RO-20-1724, have also been reported to inhibit LPS-induced TNFα synthesis by human monocytes, while having no effect on IL-1β or IL-6 synthesis. These PDE inhibitors acted to inhibit TNF mRNA synthesis and were also active if added after LPS stimulation, indicating that they acted both at the level of transcription and translation of this cytokine gene [242]. These PDE inhibitors could therefore be of use in situations where TNFα is the major pathological cytokine. Septic shock and cerebral malaria are two such conditions.

8.5.2   Tk inhibitors

Tks can be either receptor-linked and participate directly in transmembrane signalling, or intracellular and involved in signal transduction. These enzymes are now clearly seen to be pivotal in both processes, and enhanced activity of Tks is now implicated in cancer, atherosclerosis, psoriasis and many inflammatory conditions including septic shock (reviewed in [238]). The development of the tyrphostins by Levitzki and co-workers was the beginning of the search for selective inhibitors of Tks. A key paper reported on the development of inhibitors which discriminated between epidermal growth factor (EGF) receptor kinase and insulin receptor kinase [243]. As the EGF and related receptors may be involved in hyperproliferative diseases, including cancer, they have become a major target and a number of companies have programmes to develop EGF-Tk inhibitors. For example, CIBA–Geigy (now Novartis) have developed 4,5-dianilophthalimide which inhibits tumour growth promoted by EGF but not by the growth factor/oncogene platelet-derived growth factor [244]. The most potent compound reported to date is a specific inhibitor of the EGF receptor Tk synthesized by Parke–Davis (PD 153035), which has a K_i value of 5 pM [245]. As has been reviewed, Tk inhibitors of the tyrphostin class have been reported to inhibit experimental septic shock in mice. The most effective tyrphostins for the treatment of this murine form of septic shock are inhibitors of the tyrosine phosphorylation of the 42 kDa mitogen-activated protein kinase which, as described in Chapter 6, has recently been recognized as an important kinase in LPS-induced signalling in macrophages.

8.5.3   Serine/threonine kinases

Again, as described in Chapters 4 and 6, activation of receptor Tks can activate a kinase cascade which is generically termed the mitogen-activated protein kinase (MAPK) pathway. This pathway is composed of three protein kinases consisting of a serine/threonine protein kinase [MAP kinase kinase kinase (MAPKKK); e.g. Raf], which phosphorylates and activates another serine/threonine kinase (MAPKK), which in turn phosphorylates and activates a final serine/threonine kinase (MAPK). Other signals, including cytokines, can activate homologous systems, and the final MAPK or homologues activate, among other proteins, transcription factors such as c-Jun, Elk-1, NF-IL-6, etc. [246,247].

Workers from SmithKline Beecham Pharmaceuticals in the U.S.A. in the late 1980s, had developed cyclic imidazole compounds which were originally believed to be dual inhibitors of cyclo-oxygenase and 5-lipoxygenase. These compounds also inhibited the production of TNFα by LPS-activated monocytes and were shown to be effective inhibitors of experimental septic shock in rodents [191,192]. The capacity of these compounds to inhibit cytokine production led to them being termed CSAIDs (cytokine-suppressive anti-inflammatory drugs) [189,248]. These compounds appeared to inhibit the translation of pro-inflammatory cytokines and not their transcription. To identify the site of action of such compounds, they were labelled and incubated with cells and shown to bind to two closely related MAPK homologues, which have been termed CSBPs (CSAID-binding proteins), Binding to these proteins inhibited their kinase activity and this effect could be directly correlated with inhibition of cytokine synthesis, suggesting that these kinases were pivotal for cytokine production [249]. Parke–Davis Pharmaceuticals have also reported on the synthesis of a selective inhibitor of the MAPK-activating enzyme MAPK/ERK kinase without significant inhibition of MAPK itself [250]. The synthesis of inhibitors of the MAPK cascade therefore holds promise for the treatment of inflammatory and infectious conditions [251].

Probably the most exciting development in cytokine signalling is the discovery that occupation of cytokine receptors by the specific ligand results in the activation of a series of latent cytosolic transcription factors termed 'Signal Transducers and Activators of Transcription' (STATs), which are proposed to directly link cytokine cell-surface receptors to nuclear events. Upon occupancy of cytokine receptors the STATs become phosphorylated by a group of so-called Janus protein tyrosine kinases (JAKs), resulting in the former dimerizing and translocating to the nucleus where they recognize distinct *cis*-acting regulatory sequences in the DNA (see Chapter 4). Of importance is the discovery that STAT proteins only mediate the actions of particular cytokines [252]. Thus, it is possible that selective inhibitors of individual STATs could selectively block cytokine-mediated pathology, including that induced by infectious organisms.

As explained earlier, the end result of the activity of the cascades of kinases and phosphatases is the activation of transcription factors. These gene-modulating proteins are therefore therapeutic targets. E3330 is a synthetic compound produced by Tsukuba Pharmaceuticals, which was shown to inhibit the synthesis of TNFα [253]. The synthesis of this TNFα is regulated by the transcriptional activator NF-κB. It has now been demonstrated that this compound inhibits the transcription of the gene for TNFα by inhibiting the activation/nuclear translocation of NF-κB [254].

This short section gives an idea of the potential for the inhibition of intra-cellular signalling pathways in the control of cytokine action and cytokine synthesis. This promises to be a major area of pharmaceutical development for most human diseases including infectious and inflammatory conditions.

8.6   Antagonist cytokines

It will now be clear from the chapters on cytokines that the maintenance and regulation of cytokine networks depends on the mutual synergism or 'antagonism' of cytokines. The first cytokines to be discovered were the potent pro-inflammatory cytokines (IL-1, IL-6, TNF, etc.) which had apparently clear-cut actions on unstimulated cells. With increasing interest in these molecules it became clear during the late 1980s that some cytokines had the capacity to interfere with the synthesis or actions of these pro-inflammatory cytokines. The clearest example of this is the third member of the IL-1 family — IL-1ra — which acts as a true pharmacological antagonist of IL-1α and IL-1β. However, a number of other cytokines, including IL-4, IL-10, IL-13 and TGFβ$_1$, have been shown to be able to modulate the synthesis or actions of pro-inflammatory cytokines such that they have clear-cut anti-inflammatory and/or immunomodulatory activity. There is now a well established literature showing that a range of animal models of disease can be favourably modified by the administration of antagonist cytokines. This work has been reviewed by the authors [4–7] and will not be discussed in this section.

Although antagonist cytokines could have therapeutic benefits there are a number of reasons why their clinical use would be limited. The first is the fact that, being proteins, these cytokines would have to be administered systemically (e.g. by injection) and could not be given orally — the preferred route. Secondly, these cytokines would have a very short half-life in the circulation, which would obviously limit their therapeutic potential. The final problem is that, since they have a wide range of biological actions, administration of large amounts of cytokines is likely to be accompanied by unwanted side-effects. This has been seen in studies of the administration of cytokines such as IL-2 and TNF to patients with cancer.

While the administration of antagonistic cytokines may not be clinically useful, recent advances in pharmacological screening techniques have proposed a possible solution to these problems. One of the major problems in classic, synthetic, chemical-based pharmacology is generating enough novel compounds to produce new therapeutic agents. However, use of molecular biological techniques has resulted in the development of a technology called 'phage display', which has overcome this chemistry-based limitation. In this technique, small gene fragments coding for random peptides are fused to the coat proteins of filamentous bacteriophages, producing a library of peptide shapes. This library is then screened for its capacity to bind to a particular therapeutic target, which can be a cell receptor, enzyme, etc. Those phages displaying peptides with binding capacity can then be amplified and the DNA encoding the peptide sequenced, thereby allowing the peptide sequence to be elucidated. This technique enables enormous numbers ($>10^9$ peptides) to be screened — many, many log orders greater than could be produced by classic synthetic chemistry.

One of the major problems described earlier in this chapter was the nature of cytokine–receptor binding, which appeared to take place over a large surface area, making it unlikely that small molecules could ever be used as antagonists. However, recent reports using the technique of phage display have suggested that this problem may be solvable. The cytokine/hormone erythropoietin (EPO), a 34 kDa glycoprotein consisting of 165 amino acids which binds to a high-affinity surface receptor, is a primary regulator of erythropoiesis. Phage display has been used to generate a 20-residue cyclic peptide which binds to the EPO receptor with high affinity and which triggers the same intracellular signalling pathways as does the native protein. Furthermore, this peptide displays biological actions in whole animals [255,256]. This methodology has also been used to generate small peptide IL-1 receptor antagonists which have remarkably high affinity (2 nM) in blocking IL-1-driven responses in human cells [257].

This ability to use phage display to produce small peptides with potent agonist activity holds tremendous promise for the generation of physiological cytokine antagonists based on the structure of known antagonist cytokines. Such molecules by re-equilibrating, in a natural fashion, aberrant cytokine networks could have tremendous therapeutic potential in infectious and inflammatory diseases.

8.7   Bacteriokines

In Chapter 7, the recent finding that certain bacterial proteins can block the synthesis of pro-inflammatory cytokines was introduced. These molecules have been termed bacteriokines by the authors, and their potential as therapeutic agents has been suggested [258,259]. A very interesting recent finding has been a protein from *E. coli* which acts as an inhibitor of both tyrosine and serine/threonine kinases [260]. This potentially fascinating area of the use of one bacterial molecule to inhibit the activity of another will be discussed in more detail in Chapter 9. These preliminary findings suggest to the authors the potential that bacteriokines could be used to re-equilibrate aberrant cytokine networks and thus that they have therapeutic potential.

8.8   Conclusions

With the increasing problem of antibiotic resistance and the growing realization that bacteria may have the ability to become resistant to any novel antibacterial agent, the possibility of treating bacterial infections by targeting the, largely cytokine-driven, host response which produces tissue pathology is attractive. The major aim would be to limit the hosts' response to the infecting bacteria while preventing any deleterious pathology caused by an 'overspill' of pro-inflammatory cytokines. As can be seen from the literature reviewed in this chapter, this goal is rapidly becoming achievable.

References

1. Bone, R.C. (1991) Gram-negative sepsis: background, clinical features and intervention. Chest 100, 802–808
2. Bone, R.C. (1993) Gram negative sepsis: a dilemma of modern medicine. Clin. Microbiol. Rev. 6, 57–68
3. Bone, R.C. (1994) Gram positive organisms and sepsis. Arch. Intern. Med. 154, 26–34
4. Henderson, B. and Blake, S. (1992) Therapeutic potential of cytokine manipulation. Trends Pharmacol. Sci. 13, 145–152
5. Henderson, B. and Poole, S. (1994) Modulation of cytokine function: Therapeutic applications. Adv. Pharmacol. 25, 53–115
6. Henderson, B. (1995) Therapeutic modulation of cytokines. Ann. Rheum. Dis. 54, 519–523
7. Henderson, B. and Bodmer, M.W. (eds.) (1996) Therapeutic Modulation of Cytokines. CRC Press, Boca Raton
8. Nicola, N.A. (1994) Guidebook to Cytokines and their Receptors. Oxford University Press, Oxford
9. Ibelgaufts, H. (1995) Dictionary of Cytokines. VCH, Weinheim
10. Dinarello, C.A. (1996) Biological basis for interleukin-1 in disease. Blood 87, 2095–2147
11. Priestle, J.P., Schar, H.-P. and Grutter, M.G. (1989) Crystallographic refinement of interleukin-1 beta at 2.0Å resolution. Proc. Natl. Acad. Sci. U.S.A. 86, 9667–9971
12. Graves, B.J., Hatada, M.H., Hendrickson, W.A., Miller, J.K., Madison, V.S. and Satow, Y. (1990) Structure of interleukin-1 alpha at 2.7Å resolution. Biochemistry 29, 2679–2684
13. Auron, P.E., Quigley, G.J., Rossenwasser, L.J. and Gehrke, L. (1992) Multiple amino acid substitutions suggest a structural basis for the separation of biological activity and receptor binding in a mutant interleukin-1 beta protein. Biochemistry 31, 6632–6638
14. Simon, P.L., Kumar, V., Lillquist J.S., Bhatnagar, P., Einstein, R., Lee, J., Porter, T., Green, D, Sathe, G. and Young, P.R. (1993) Mapping of neutralizing epitopes and the receptor binding site of human interleukin-1 beta. J. Biol. Chem. 268, 9771–9779
15. Obal, F., Opp, M., Cady, A.B., Johannsen, L., Postlethwaite, A.E., Poppleton, H.M., Seyer, J.M. and Krueger, J.M. (1990) Interleukin-1α and an interleukin-1β fragment are somnogenic. Am. J. Physiol. 259, R439–R446
16. Antoni, G., Presentine, R., Perin, F., Mencioni, L., Villa, L., Censini, S., Ghiara, P., Valpini, G., Bassu, P. and Tagliabue, A. (1989) Interleukin-1 and its synthetic peptides as adjuvants for poorly immunogenic vaccines. Adv. Exp. Med. Biol. 251, 153–160
17. Boraschi, D., Antoni, G., Perni, F., Villa, L., Nencioni, L., Ghiara, P., Presentini, R. and Tagliabue, D. (1990) Defining the structural requirements of a biologically active domain of human IL-1β. Eur. Cytokine Network 1, 21–26
18. Lerner, U.H., Ljunggren, O., Dewhurst, F.E. and Boraschi, D. (1991) Comparison of human interleukin-1β and its 163-171 peptide in bone resorption and the immune response. Cytokine 3, 141–148
19. Dray, A. and Perkins, M. (1993) Bradykinin and inflammatory pain. Trends Neurosci. 16, 99–104
20. Ferreira, S.H., Lorenzetti, B.B., Bristow, A.F. and Poole, S. (1988) Interleukin-1β is a potent hyperalgesic agent antagonized by a tripeptide analogue. Nature (London) 334, 698–700
21. Follenfant, R.L., Nakamura-Craig, M., Henderson, B. and Higgs, G.A. (1989) Inhibition by neuropeptides of interleukin-1β-induced prostaglandin-independent hyperalgesia. Br. J. Pharmacol. 98, 41–43

22. Poole, S., Bristow, A.F., Lorenzetti, B.B., Gaines Das, R.E., Smith, T.W. and Ferreira, S.H. (1992) Peripheral analgesic activities of peptides related to α-melanocyte-stimulating hormone and inter-leukin-1β 193-195. Br. J. Pharmacol. **106**, 489–492

23. Cannon, J.G., Tatro, J.B., Reichlin, S. and Dinarello CA. (1986) α-Melanocyte-stimulating hormone inhibits immunostimulatory and inflammatory actions of interleukin-1. J. Immunol. **137**, 2232–2236

24. Daynes, R.A., Robertson, B.A., Cho, B.-H., Burnham, D.K. and Newton, R. (1987) α-Melanocyte-stimulating hormone exhibits target cell selectivity in its capacity to affect interleukin-1-inducible responses *in vivo* and *in vitro*. J. Immunol. **139**, 103–109

25. Robertson, B.A., Gahring, L.C. and Daynes, R.A. (1988) Neuropeptide regulation of interleukin-1 activities: capacity of α-melanocyte stimulating hormone to inhibit interleukin-1-inducible responses *in vivo* and *in vitro* exhibits target cell selectivity. Inflammation **10**, 371–385

26. Shalts, E., Feng, Y.-J., Ferin, M. and Wardlaw, S.-L. (1992) α-Melanocyte-stimulating hormone antagonizes the neuroendocrine effects of corticotrophin-releasing factor and interleukin-1α in the primate. Endocrinology **131**, 132–138

27. Richards, D.B. and Lipton, J.M. (1984) Effects of α-MSH11-13 (lysine-proline-valine) on fever in the rabbit. Peptides **5**, 815–817

28. Hiltz, M.E. and Lipton, J.M. (1989) Anti-inflammatory activity of a COOH-terminal fragment of the neuropeptide alpha-MSH. Res. Commun. **3**, 2282–2284

29. Hiltz, M.E. and Lipton, J.M. (1990) Alpha-MSH peptides inhibit acute inflammation and contact sensitivity. Peptides **11**, 979–982

30. Cunha, F.Q., Poole, S., Lorenzetti, B.B. and Ferreira, S.H. (1992) The pivotal role of tumour necrosis factor α in the development of inflammatory hyperalgesia. Br. J. Pharmacol. **107**, 660–664

31. Martin, L.W., Catania, A., Hiltz, M.E. and Lipton, J.M. (1991) Neuropeptide alpha-MSH antagonizes IL-6 and TNF-induced fever. Peptides **12**, 297–299

32. Alder, T. (1981) Psychoneuroimmunology. Academic Press, New York

33. Cohen, S. and Williamson, G.M. (1991) Stress and infectious diseases in humans. Psychol. Bull. **109**, 5–24

34. Palaszynski, E.W. (1987) Synthetic C-terminal peptide of IL-1 functions as a binding domain as well as an antagonist for the IL-1 receptor. Biochem. Biophys. Res. Commun. **147**, 204–211

35. Nestor, J.J., Newman, S.R., DeLeistro, B., Todaro, G.J. and Schreiber, A. B. (1985). A synthetic fragment of rat transforming growth factor α with receptor binding and antigenic properties. Biochem. Biophys. Res. Commun. **129**, 226–232

36. Bischoff, S.C., DeWeck, A. and Dahinden, C.A. (1992) Peptide analogues of consensus receptor sequences inhibit the action of cytokines on human basophils. Lymphokine Cytokine Res. **11**, 33–37

37. Kelley, W.S. (1996) Therapeutic peptides: The devil is in the detail. Biotechnology **14**, 28–31

38. Zurawski, S.M., Imler, S.M. and Zurawski, G. (1990) Partial agonist/antagonist mouse interleukin-2 proteins indicates a third component of the receptor complex functions in signal transduction. EMBO J. **9**, 3899–3905

39. Imler, J.-L. and Zurawski, G. (1992) Receptor binding and internalization of mouse interleukin-2 derivatives that are partial agonists. J. Biol. Chem. **267**, 13185–13190

40. Kruse, N., Tony, H.-P. and Sebald, W. (1992) Conversion of human interleukin-4 into a high affinity antagonist by a single amino acid replacement. EMBO J. **11**, 3237–3244

41. Larrick, J.W. (1989) Native interleukin-1 inhibitors. Immunol. Today **10**, 61–64

42. Arend, W.P., Joslin, F.G. and Massoni, R.J. (1985) Effects of immune complexes on production by human monocytes of interleukin-1 or an interleukin-1 inhibitor. J. Immunol. **134**, 3868–3875

43. Balavoine, J.-F., deRochemonteix, B., Williamson, K., Seckinger, P., Cruchaud, A. and Dayer, J.-M. (1986) Prostaglandin E$_2$ and collagenase production by fibroblasts and synovial cells is regulated by urine-derived human interleukin-1 and inhibitor(s). J. Clin. Invest. **78**, 1120–1124

44. Seckinger, P., Lowenthal, J.W., Williamson, K., Dayer, J.-M. and MacDonald, H.R. (1987) A urine inhibitor of interleukin-1 activity that blocks ligand binding. J. Immunol. **139**, 1546–1549

45. Hannum, C.H., Wilcox, C.J., Arend, W.P., Joslin, F.G., Dripps, D.J., Heimdahl, P.L., Armes, L.G., Sommer, A., Eisenberg, S.P. and Thompson, R.C. (1990) Interleukin-1 receptor antagonist activity of a human interleukin-1 inhibitor. Nature (London) **343**, 336–340

46. Eisenberg, S, P., Evans, R.J., Arend, W.P., Verderber, A., Brewer, M.T., Hannum, H.C. and Thompson, R.C. (1990) Primary structure and functional expression from complementary DNA of a human interleukin-1 receptor antagonist. Nature (London) **343**, 331–336

47. Eisenberg, S.P., Evans, R.J., Vigers, G.P.A., Bray, J, Caffes, T., Childs, J.D., Dripps, D.J, Thompson, R.C. and Brandhuber, B.J. (1993) Interleukin-1 receptor antagonist (IL-1ra) lacks one of the two receptor binding sites for IL-1β based on structure/function analysis. Lymphokine Cytokine Res. **12**, 376–382

48. Ju, G., Labriola-Thompkins, E., Campen, C.A., Benjamin, W.R., Karas, J., Polcinski, J., Biondi, D., Kaffka, K.L., Kilin, P.L., Eisenberg, S.P. and Evans, R.J. (1991) Conversion of the interleukin-1 receptor antagonist into an agonist by site-specific mutagenesis. Proc. Natl. Acad. Sci. U.S.A. **88**, 2658–2662

49. Gehrke, L., Jobling, S.O., Paik, S.L.K., McDonald, B., Rossenwasser, L.J. and Auron, P. (1990) A point mutation uncouples human interleukin-1β biological activity in receptor binding. J. Biol. Chem. **265**, 5922–5925

50. Eisenberg, S.P., Brewer, M.T., Verdeber, E., Heimdal, P., Brandhuber, B.J. and Thompson, R.C. (1991) Interleukin-1 receptor antagonist (IL-1ra) is a member of the IL-1 gene family. Evolution of a cytokine control mechanism. Proc. Natl. Acad. Sci. U.S.A. **88**, 5232–5236

51. Young, P.R. and Sylvester, D. (1989) Cloning of rabbit IL-1β. Differential evolution of IL-1α and IL-1β proteins. Protein Eng. **2**, 545–551

52. Symons, J.A. and Duff, G.W. (1990) A soluble form of the interleukin-1 receptor produced by a human B cell line. FEBS Lett. **272**, 133–136

53. Dripps, D.J., Brandhuber, B.J., Thompson, R.C. and Eisenberg, S.P. (1991) Interleukin-1 receptor antagonist binds to the 80kDa IL-1 receptor but does not initiate IL-1 signal transduction. J. Biol. Chem. **266**, 10331–10336

54. Granowitz, E.V., Clarke, B.D., Mancilla, A. and Dinarello, C.A. (1991) Interleukin-1 receptor antagonist competitively inhibits the binding of interleukin-1 to type II interleukin-1 receptor. J. Biol. Chem. **266**, 14147–14150

55. Seckinger, P., Klein-Nulend, J., Alander, C., Thompson, R.C., Dayer, J.-M. and Raisz, L.G. (1990) Natural and recombinant human interleukin-1 receptor antagonists block the effects of interleukin-1 on bone resorption and prostaglandin production. J. Immunol. **145**, 4181–4184

56. Smith, R.J., Chin, J.E., Sam, L.M. and Justen, J.M. (1991) Biologic effects of an interleukin-1 receptor antagonist on interleukin-1-stimulated cartilage erosion and chondrocyte responsiveness. Arthritis Rheum. **34**, 73–83

57. Arend, W.P., Welgus, H.G., Thompson, R.C. and Eisenberg, S.P. (1990) Biological properties of recombinant human monocyte-derived interleukin-1 receptor antagonist. J. Clin. Invest. **85**, 1694–1697

58. Carter, D.B., Deibel, M.R., Dunn, C.J., Tomich, C.-S., Laborde, J.L., Slightom, J.L., Berger, A.E., Bienkowski, M.J., Sun, F.F., McEwen, R.N., et al. (1990) Purification, cloning, expression and biological characterization of interleukin-1 receptor antagonist protein. Nature (London) **344**, 633–638

59. Petitclerc, E., Abel, S., DeBlois, D., Poubelle, P.E. and Marceau, F. (1992) Effect of interleukin-1 receptor antagonist on three types of responses to interleukin-1 in rabbit isolated blood vessels. J. Cardiovasc. Pharmacol. **19**, 821–826

60. McIntyre, K.W., Stepan, G.J., Kolinsky, K.D., Benjamin, W.R., Plocinski, J.M., Kaffka, K.L., Campen, C.A., Chizzonite, R.A. and Kilian, P.L. (1991) Inhibition of interleukin-1 (IL-1) binding and bioactivity in vitro and modulation of acute inflammation in vivo by IL-1 receptor antagonist and anti-IL-1 receptor monoclonal antibody. J. Exp. Med. **173**, 931–939

61. Ulich, T.R., Yin, S., Guo, K., del Castillo, J., Eisenberg, S.P. and Thompson, R.C. (1991) The intra-tracheal administration of endotoxin and cytokines. III. The interleukin-1 (IL-1) receptor antagonist inhibits endotoxin- and IL-1-induced acute inflammation. Am. J. Pathol. **138**, 521–524

62. Wooley, P.H., Whalen, J.D., Chapman, D.L., Berger, A.E., Aspar, D.G., Richard, K.A. and Staite, N.D. (1990) The effect of interleukin-1 receptor antagonist protein on type II collagen-induced arthritis in mice. Arthritis Rheum. **33**, S20

63. Lewthwaite, J., Blake, S.M., Hardingham, T.E. and Henderson, B. (1994) The effect of human inter-leukin-1 receptor antagonist on the induction phase of antigen-induced arthritis in the rabbit. J. Rheumatol. **21**, 467–472

64. Lewthwaite, J., Blake, S., Thompson, R.C., Hardingham, T.E. and Henderson, B. (1995) Anti-fibrotic action of interleukin-1 receptor antagonist in lapine monoarticular arthritis. Ann. Rheumatic Dis. **57**, 591–596

65. Cominelli, F., Nast, C.C., Clark, B.D., Schindler, R., Llerena, R., Eysselein, V.E., Thompson, R.C. and Dinarello, C.A. (1990) Interleukin-1 (IL-1) gene expression, synthesis and effect of specific IL-1 receptor blockade in rabbit immune complex colitis. J. Clin. Invest. **86**, 972–980

66. McCarthy, P.L., Abhyankar, S., Neben, S., Newman, G., Sieff, C., Thompson, R.C., Burakoff, S.J. and Ferrara, J.L.M. (1991) Inhibition of interleukin-1 by an interleukin-1 receptor antagonist prevents graft-versus-host diseases. Blood **78**, 1915–1918

67. Ohlsson, K., Bjork, P., Bergenfeldt, M., Hageman, R. and Thompson, R.C. (1990) Interleukin-1 receptor antagonist reduces mortality from endotoxin shock. Nature (London) **348**, 550–552

68. Wakabayashi, G., Gelfand, J.A., Burke, J.F., Thompson, R.C. and Dinarello, C.A. (1991) A specific receptor antagonist for interleukin-1 prevents *Escherichia coli*-induced shock in rabbits. FASEB J. **5**, 338–343

69. Alexander, H.R., Doherty, G.M., Buresh, C.M., Venyon, D.J. and Norton, J.A. (1991) A recombi-nant human receptor antagonist to interleukin-1 improves survival after lethal endotoxaemia in mice. J. Exp. Med. **173**, 1029–1032

70. Fischer, E., Marano, M.A., Van Zee, K.J., Rock, C.S., Hawes, A.S., Thompson, W.A., DeForge, L., Kenney, J.S., Remick, D.G., Bloedow, D.C., Thompson, R.C., Lowry, S.F. and Moldawer, L.L. (1992) Interleukin-1 receptor blockade improves survival and hemodynamic performance in *Escherichia coli* septic shock, but fails to alter host response to sublethal endotoxemia. J. Clin. Invest. **89**, 1551–1557

71. Aiura, K., Gelfand, J.A., Burke, J.F., Thompson, R.C. and Dinarello, C.A. (1993) Interleukin-1 (IL-1) receptor antagonist prevents *Staphylococcus epidermidis*-induced hypotension and reduces circulat-ing levels of tumor necrosis factor-α and IL-1β in rabbits. Infect. Immun. **61**, 3342–3350

72. Mancilla, J., Garcia, P. and Dinarello, C.A. (1993) The interleukin-1 receptor antagonist can either reduce or enhance the lethality of *Klebsiella pneumoniae* sepsis in newborn rats. Infect. Immun. **61**, 926–932

73. Walsh, C.E., Liu, J.M., Anderson, S.M., Russio, J.L., Nienhuis, A.S. and Young, M.S. (1992) A trial of recombinant interleukin-1 in patients with severe refractory aplastic anaemia. Br. J. Haematol. **80**, 106–110

74. Granowitz, E.V., Porat, R., Mier, J.W., Pribble, J.P., Stiles, D.M., Bloedow, D.C., Catalano, M.A., Wolff, S.M. and Dinarello, C.A. (1992) Pharmacokinetics, safety and immunomodulatory effects of human recombinant interleukin-1 receptor antagonist in healthy humans. Cytokine **4**, 353–360

75. Opal, S.M. (1996) Interleukin-1 receptor antagonist. In Therapeutic Modulation of Cytokines (Henderson, B. and Bodmer, M., eds,), pp. 198–219, CRC Press, Boca Raton

76. James, K. (1990) Interaction between cytokines and α-macroglobulin. Immunol. Today **11**, 163–166

77. Spinas, G.A., Keller, U. and Brockenhaus, M. (1992) Release of soluble receptors for tumor necrosis factor (TNF) in relation to circulating TNF during experimental endotoxaemia. J. Clin. Invest. **90**, 533–536

78. Semenzato, G., Foa, R., Agostini, C., Zambella, R., Trentin, L., Vinante, F., Benedetti, F., Chilosi, M. and Pizzolo, G. (1987) High serum levels of soluble interleukin-2 receptor in patients with chronic B lymphocytic leukaemia. Blood **70**, 396–400

79. Miossec, P., Elhamiani, M., Chichehian, B., DÁngeac, A.D., Sany, J. and Hirn, M. (1990) Interleukin-2 (IL-2) inhibitor in rheumatoid synovial fluid: Correlation with prognosis and soluble IL-2 receptor levels. J. Clin. Immunol. **10**, 115–120

80. Prince, H.E., Kleinman, S. and Williams, A.E. (1988) Soluble IL-2 receptor levels in serum from blood donors seropositive for HIV. J. Immunol. **140**, 1139–1141

81. Seckinger, P., Isaaz, S. and Dayer, J.-M. (1988) A human inhibitor to tumor necrosis factor α. J. Exp. Med. **107**, 1511–1516

82. Digel, W., Porzsolt, F., Schmid, M., Herrmann, F., Lessiauer, W. and Brockhaus, M. (1992) High levels of circulating soluble receptors for tumor necrosis factor in hairy cell leukaemia and type B chronic lymphocytic leukaemia. J. Clin. Invest. **89**, 1690–1693

83. Peetre, C., Thysell, H., Gribb, A. and Olsson, I. (1988) A tumor necrosis factor binding protein is present in human biological fluids. Eur. J. Haematol. **41**, 414–419

84. Fanslow, W.C., Sims, J.E., Sassenfeld, H., Morrisey, P.J., Gillis, S., Dower, S.K. and Widmer, M.B. (1990) Regulation of alloreactivity in vivo by a soluble form of the interleukin-1 receptor. Science **248**, 739–742

85. Jacobs, C.A., Baker, P.E., Roux, E.R., Picha, K.S., Toivola, B., Waugh, S. and Kennedy, M.R. (1991) Experimental autoimmune encephalomyelitis is exacerbated by IL-1α and suppressed by soluble IL-1 receptor. J. Immunol. **146**, 2983–2989

86. Van Zee, K.J., Kohna, T., Fischer, E., Rock, C.S., Moldawer, L.L. and Lowry, S.F. (1992) Tumor necrosis factor soluble receptors circulate during experimental and clinical inflammation and can protect against excessive tumor necrosis factor α in vitro and in vivo. Proc. Natl. Acad. Sci. U.S.A. **89**, 4845–4849

87. Capon, D.J., Chamow, S.M., Mordenti, J., Marsters, S.A., Gregory, T., Mitsuya, H., Byrn, R.A., Lucas, C., Wurm, F.M., Groopman, J.E., Broder, S. and Smith, D.H. (1989) Designing CD4 immunoadhesins for AIDS therapy. Nature (London) **337**, 525–531

88. Peppel, K., Crawford, D. and Beutler, B. (1991) A tumor necrosis factor (TNF) receptor-IgG heavy chain chimeric protein as a bivalent antagonist of TNF activity. J. Exp. Med. **174**, 1483–1489

89. Baumgartner, S., Moreland, L.W., Schiff, M.H., Tindall, E., Fleischmann, R.M., Weaver, A., Ettinger, R.E., Gruber, B.L., Katz, R.S., Skosey, J.L., et al. (1996) Double-blind placebo-controlled trial of tumor necrosis factor receptor (p80) fusion protein (TNFR:Fc) in active rheumatoid arthritis. Arthr. Rheum. **39**, S74

90. Stone, R. (1994) Search for sepsis drugs goes on despite past failures. Science **264**, 365–367

91. Harris, W.J. and Adair, J.R. (1997) Antibody Therapeutics. CRC Press, Boca Raton

92. van de Loo, F.A.J., Arntz, O.J., Otterness, I.G. and van den Berg, W.B. (1992) Protection against cartilage proteoglycan synthesis inhibition by anti-interleukin-1 antibodies in experimental arthritis. J. Rheumatol. **19**, 348–356

93. Thorbecke, G.J., Shah, R., Leu, C.H., Kuruvilla, A.P., Hardison, A.M. and Palladino, M.A. (1992) Involvement of endogenous tumor necrosis α and transforming growth factor β during induction of collagen type II arthritis in mice. Proc. Natl. Acad. Sci. U.S.A. **89**, 7375–7379

94. Lewthwaite, J., Blake, S., Hardingham, T., Foulkes, R., Stephens, S., Chaplin, L., Emtage, S., Catterall, C., Short, S., Nesbitt, A. et al. (1995) The role of TNFα in the induction of antigen-induced arthritis in the rabbit and the anti-arthritic effect of species-specific TNFα-neutralizing monoclonal antibodies. Ann. Rheum. Dis. **54**, 366–374

95. Beutler, B., Milsark, I.W. and Cerami, A.C. (1985) Passive immunization against cachectin/tumor necrosis factor protects mice from lethal effect of endotoxin. Science **229**, 869–871

96. Opal, S.M., Cross, A.S., Kelly, N.M., Sadoff, J.C., Bodmer, M.W., Palardy, J.E. and Victor, G.H. (1990) Efficacy of a monoclonal antibody directed against tumor necrosis factor in protecting neutropenic rats from lethal infection with Pseudomonas aeruginosa. J. Infect. Dis. **161**, 1148–1152

97. Madden, J.A.M., Urban, J.F., Ziltener, H.J., Schrader, J. W., Finkelman, F.D. and Katona, I.M. (1991) Antibodies to IL-3 and IL-4 suppress helminth-induced intestinal mastocytosis. J. Immunol. **147**, 1387–1391

98. Finkelman, F.D., Katona, I.M., Urban, J.F., Holmes, J., Ohara, J., Tung, A.S., Sample, J.V.G. and Paul, W.E. (1988) IL-4 is required to generate and sustain *in vivo* IgE responses. J. Immunol. **141**, 2335–2341

99. Chatelain, R., Varkila, K. and Coffman, R.L. (1992) IL-4 induces a Th2 response in *Leishmania major*-infected mice. J. Immunol. **148**, 1182–1187

100. Coffman, R.L., Seymour, B.W.P., Hudak, S., Jackson, J. and Rennick, D. (1989) Antibody to inter-leukin-5 inhibits helminth-induced eosinophilia in mice. Science **245**, 308–310

101. Elbon, C.L., Jacoby, D.B. and Fryer, A.D. (1995) Pretreatment with an antibody to interleukin-5 prevents loss of pulmonary M_2 muscarinic receptor function in antigen-challenged guinea pigs. Am. J. Respir. Cell Mol. Biol. **12**, 320–328

102. Fletcher-Starnes, H., Pearce, M.K., Teware, A., Yim, J.H., Zau, J.-C. and Abrams, J.S. (1990) Anti-IL-6 monoclonal antibodies protect against lethal *Escherichia coli* infection and lethal tumor necrosis factor and challenge in mice. J. Immunol. **145**, 4185–4191

103. Billiau, A., Heremans, H., Vanderckhove, F. and Dillen, C. (1987) Anti-interferon-γ-antibody protects mice against the generalized Schwartzman reaction. Clin. Exp. Immunol. **17**, 1851–1854

104. Grau, G.E., Heremans, H., Piguet, P.-F., Pointaire, P., Lambert, P.-H., Billiau, A. and Vassalli, P. (1989) Monoclonal antibody against interferon-γ can prevent experimental cerebral malaria and its associated overproduction of tumor necrosis factor. Proc. Natl. Acad. Sci. U.S.A. **86**, 5572–5574

105. Ferns, G.A.A., Raines, E.W., Sprugel, K.H., Motane, A.S., Reidy, M.A. and Ross, R. (1991) Inhibition of neointimal smooth muscle accumulation after angioplasty by an antibody to PDGF. Science **253**, 1129–1132

106. LoBuglio, A.F., Mansoor, N.S., Lee, J., Khazaeli, M.B., Carrano, R., Holden, H. and Wheeler, R.H. (1988) Phase I trial of multiple large doses of murine monoclonal antibody (C)17-1A. J. Natl. Cancer Inst. **80**, 932–936

107. Adair, J. (1992) Engineering antibodies for therapy. Immunol. Rev. **130**, 1–36

108. Egan, R.W., Cuss, F.M., Umland, S.P. and Chapman, R.W. (1996) Anti-interleukin-5 antibodies as therapeutic agents in asthma and other eosinophilic diseases. In Therapeutic Modulation of Cytokines (Henderson, B. and Bodmer, M.W., eds.), pp. 237–264, CRC Press, London

109. Bodmer, M.W. and Foulkes, R. (1996) TNFα neutralization by biological antagonists. In Therapeutic Modulation of Cytokines (Henderson, B. and Bodmer, M.W., eds.), pp. 221–236, CRC Press, Boca Raton

110. Exley, E.R., Cohen, J., Burman, W., Owen, R., Hansen, G., Lumley, J., Aulakh, J.M., Bodmer, M., Riddell, A., Stephens, S. and Perry, M. (1991) Monoclonal antibody to TNF in severe septic shock. Lancet **335**, 1275–1277

111. Vincent, J.L., Bakker, J., Marecauz, G., Schandene, L., Kahn, R.J. and Dupont, E. (1992) Administration of anti-TNF antibody improves left ventricular function in septic shock patients. Result of a pilot study. Chest **101**, 810–815

112. Fisher, C.J., Opal, S.M., Dhainaut, J.-F., Stephens, S., Zimmerman, J.L., Nightingale, P., Harris, S.J., Scheim, R.M.H., Panaek, E.A., Vincent, J.-L., et al. (1993) Influence of anti-TNF monoclonal antibody on cytokine levels in patients with sepsis. Crit. Care Med. **21**, 318–327

113. Cohen, J. and Carlet, J. (1996) INTERSEPT: an international, multicenter, placebo-controlled trial of monoclonal antibody to human tumor necrosis factor-alpha in patient with sepsis. Crit. Care Med. **24**, 1431–1440

114. Braegger, C.P., Nicholls, S., Murch, S.H., Stephens, S. and MacDonald, T.T. (1992) Tumour necrosis factor alpha in stool as a marker of intestinal inflammation. Lancet **339**, 89–91

115. Derkx, B., Taminiau, J., Radema, S., Stronkhorst, A., Wortel, C., Tygart, G. and Van Deventer, S. (1993) Tumour necrosis-factor alpha antibody treatment in Crohn's disease. Lancet **342**, 173–174

116. Targan, S.R. Hanauer, S.B., Ven Deventer, S.J.H., Mayer, L., Present, D.H., Braakman, T., DeWoody, K.L., Schaible, T.F. and Rutgeerts, P.J. (1997) A short-term study of chimeric monoclonal antibody cA2 to tumor necrosis factor alpha for Crohn's disease. N. Engl. J. Med. **337**, 1029–1035

117. McInnes, I.B. and Sturrock, R.D. (1995) Clinical aspects of rheumatoid arthritis. In Mechanisms and Models in Rheumatoid Arthritis (Henderson, B., Edwards, J.C.W. and Pettipher, E.R., eds.), pp. 3–24, Academic Press, London

118. Maini, R., Chu, C.Q. and Feldmann, M. (1995) Aetiopathogenesis of rheumatoid arthritis. In Models and Mechanisms in Rheumatoid Arthritis (Henderson, B., Edwards, J.C.W. and Pettipher, E.R., eds.), pp. 25–46, Academic Press, London

119. Brennan, F.M., Maini, R.N. and Feldman, M. (1992) TNFα — a pivotal role in rheumatoid arthritis. Br. J. Rheumatol. **31**, 293–298

120. Elliot, M.J., Maini, R.M., Feldman, M., Long-Fox, A., Charles, P., Katsikis, P., Brennan, F.M., Walker, J., Bijl, H. and Ghrayeb, B. (1993) Treatment of rheumatoid arthritis with chimaeric monoclonal antibodies to tumor necrosis factor-α. Arthritis Rheum. **36**, 1681–1690

121. Elliot, M.J., Maini, R.N., Feldman, M., Long-Fox, A., Charles, P., Bijl, H. and Woody, J.N. (1994) Randomized double-blind comparison of chimaeric monoclonal antibody to tumour necrosis factor (cA-2) versus placebo in rheumatoid arthritis. Lancet **344**, 1105–1110

122. Elliot, M.J. and Maini, R.N. (1995) Anti-cytokine therapy in rheumatoid arthritis. Baillieres Clin. Rheumatol. **9**, 633–652

123. Shenep, J.L., Barton, R.P. and Morgan, K.A. (1985) Role of antibiotic class in the rate of liberation of endotoxin during therapy for experimental Gram-negative bacterial sepsis. J. Infect. Dis. **151**, 1012–1018

124. Storm, D.R., Rosenthal, S.L. and Swanson, P.E. (1987) Polymyxin B and related peptide antibiotics. Annu. Rev. Biochem. **57**, 723–744

125. Rietschel, E.T., Brade, H., Holst, O., Brade, L., Muller-Loennies, S., Mamat, U., Zahringer, U., Beckmann, F., Seydel, U., Brandenburg, K., et al. (1996) Bacterial endotoxin: chemical constitution, biological recognition, host response, and immunological detoxification. Curr. Topics Microbiol. Immunol. **216**, 39–81

126. Ziegler, E.J., Fisher, C.J., Sprung, C.L., et al. (1991) Treatment of Gram-negative bacteremia and septic shock with HA-1A human monoclonal antibody against endotoxin. N. Engl. J. Med. **324**, 429–436

127. Greenman, R.L., Schei, R.M.H. and Martin, M.A. (1991) A controlled clinical trial of E5 murine monoclonal antibody to endotoxin in the treatment of Gram-negative sepsis. JAMA **266**, 1097–1102

128. Warren, H.S., Danner, L. and Munford, R.S. (1992) Anti-endotoxin monoclonal antibodies. N. Engl. J. Med. **326**, 1153–1157

129. Elsbach, P., Weiss, J., Franson, R.C., Beckerdite-Quagliata, A., Schneider, A. and Harris, L. (1979) Separation and purification of a potent bactericidal/permeability-increasing protein and a closely related phospholipase A2 from rabbit PMNs. Observation on their relationship. J. Biol. Chem. **254**, 11000–11009

130. Larrick, J,W., Hirata, M., Zheng, H., Straube, R.C., Sadoff, J.C., Foulke, G.E., Woitel, C.H., Fink, M.P., Dellinger, R.P., Teng, N.N.H., et al. (1993) A novel granulocyte-derived peptide with LPS-neutralizing activity. J. Immunol. **152**, 231–240

131. Alpert, G., Baldwin, G., Thompson, C., Wainwright, N., Novitsky, T.J., Gillis, Z., Parsonetet, J., Fleischer, G.R. and Siber, G.R. (1992) *Limulus* antilipopolysaccharide factor protects rabbits from meningococcal endotoxin shock. J. Infect. Dis. **165**, 494–500

132. Haziot, A., Rong, G.-W., Bazil, V., Silver, J. and Goyert, S.M. (1994) Recombinant soluble CD14 inhibits LPS-induced tumor necrosis factor-α production by cells in whole blood. J. Immunol. **152**, 5868–5876

133. Qureshi, N., Takayama, K. and Kurtz, R. (1991) Diphosphoryl lipid A from the nontoxic lipopolysaccharide of *Rhodopseudomonas sphaeroides* is an endotoxin antagonist in mice. Infect. Immun. **59**, 441–444

134. Christ, W.J., Asano, O., Robidoux, A.L.C., Perez, M., Wang, Y., Dubuc, G.R., Gavin, W.E., Hawkins, L.D., McGuiness, P.D., Mullarkey, M.A., et al. (1995) E5531, a pure endotoxin antagonist of high potency. Science **268**, 80–83

135. Read, T.E., Grunfeld, C., Kumwenda, Z.L., Calhoun, M.C., Kane, J.P., Feingold, K.R. and Rapp, J.H. (1995) Triglyceride-rich lipoproteins prevent septic death in rats. J. Exp. Med. **182**, 267–272

136. Howard, M., Muchamuel, T., Andrade, S. and Menon, S. (1993) Interleukin-10 protects mice from lethal endotoxemia. J. Exp. Med. **177**, 1205–1208

137. Novogrodsky, A., Vanichkin, A., Patya, M., Gazit, A., Osherov, N. and Levitzki, A. (1994) Prevention of lipopolysaccharide-induced lethal toxicity by tyrosine kinase inhibitors. Science **264**, 1319–1322

138. Rice, G.C., Brown, P.A., Nelson, R.J., Bianco, J.A., Singer, J.W. and Bursten, S.L. (1994) Protection from endotoxic shock in mice by pharmacological inhibition of phosphatidic acid. Proc. Natl. Acad. Sci. U.S.A. **91**, 3857–3861

139. Boman, H.G. (1995) Peptide antibiotics and their role in innate immunity. Annu. Rev. Immunol. **13**, 61–92

140. Weiss, J., Elsbach, P., Olsson, I. and Hodberg, H. (1978) Purification and characterization of a potent bactericidal and membrane active protein from the granules of human polymorphonuclear leukocytes. J. Biol. Chem. **253**, 2664–2672

141. Weiss, J., Muello, K., Victor, M. and Elsbach, P. (1984) The role of lipopolysaccharide in the action of the bactericidal/permeability-increasing neutrophil protein on the bacterial envelope. J. Immunol. **132**, 3109–3115

142. Farley, M.M., Shafer, W.M. and Spitznagel, J.K. (1988) Lipopolysaccharide structure determines ionic and hydrophobic binding of a cationic antimicrobial neutrophil granule protein. Infect. Immun. **56**, 1589–1592

143. Gray, P.W., Flaggs, G., Leong, S.R., Gumina, R.J., Weiss, J., Ooi, C.E. and Elsbach, P. (1989) Cloning of the cDNA of a human neutrophil bactericidal protein. J. Biol. Chem. **264**, 9505–9509

144. Ooi, C.E., Weiss, J., Elsbach, P., Frangione, B. and Mannion, B.A. (1987) A 25kDa NH$_2$-terminal fragment carries all the antibacterial activities of the human neutrophil 60kDa bactericidal/permeability-increasing protein. J. Biol. Chem. **262**, 14891–14894

145. Ooi, C.E., Weiss, J., Doerfler, M.E. and Elsbach, P. (1991) Endotoxin-neutralizing properties of the 25kD N-terminal fragment and a newly isolated 30kD C-terminal fragment of the 55-6-kD bactericidal/permeability-increasing protein of human neutrophils. J. Exp. Med. **174**, 649–655

146. Gazzano-Santoro, H., Parent, J.B., Grinna, L., Horwitz, T., Parsons, G., Theofan, P., Elsbach, J., Weiss, J. and Conlon, P.J. (1992) High affinity binding of the bactericidal/permeability-increasing protein and a recombinant amino-terminal fragment to the lipid A region of lipopolysaccharide. Infect. Immun. **60**, 4754–4761

147. Weiss, J., Elsbach, C., Shu, J., Castillo, L., Grinna, L., Horwitz, A. and Theofan, G. (1992) Human bactericidal/permeability-increasing protein and a recombinant NH$_2$-terminal fragment cause killing of serum-resistant gram-negative bacteria in whole blood and inhibit tumor necrosis factor release induced by bacteria. J. Clin. Invest. **90**, 1122–1130

148. Gazzano-Santoro, H., Parent, J.B., Conlon, P.J., Kasler, H.G., Tsai, C.-M., Lill-Elghanian, D.A. and Hollingsworth, R.I. (1995) Characterization of the structural elements in lipid A required for binding of a recombinant fragment of bactericidal/permeability-increasing protein rBPI[23]. Infect. Immun. **63**, 2201–2205

149. Marra, M.N., Thornton, M.B., Snable, J.L., Wilde, C.G. and Scott, R.W. (1994) Endotoxin-binding and -neutralizing properties of recombinant bactericidal/permeability-increasing protein and monoclonal antibodies. Crit. Care Med. **22**, 559–565

150. Fisher, C.J., Marra, M.N., Palardy, J.E., Marchbanks, C.R., Scott, R.W. and Opal, S.M. (1994) Human neutrophil bactericidal/permeability-increasing protein reduces mortality rate from endotoxin challenge: A placebo-controlled study. Crit. Care Med. **22**, 553–558

151. Rogy, M.A., Moldawer, L.L. and Oldenburg, H.S.A. (1994) Anti-endotoxin therapy in primate bacteremia with HA-1A and BPI. Ann. Surg. **220**, 77–85

152. Kohn, F.R. and Kung, A.H.C. (1995) Role of endotoxin in acute inflammation induced by Gram-negative bacteria: specific inhibition of lipopolysaccharide-mediated responses with an amino-terminal fragment of bactericidal/permeability-increasing protein. Infect. Immun. **63**, 333–339

153. Jin, H., Yang, R., Marsters, S., Ashjenazi, A., Bunting, S., Marra, M.N., Scott, R.W. and Baker, J.B. (1995) Protection against endotoxic shock by bactericidal/permeability-increasing protein in rats. J. Clin. Invest. **95**, 1947–1952

154. Dalrymple, S.A, Slattery, R., Aud, D.M., Krishna, M., Lucian, L.A. and Murray, R. (1996) Interleukin-6 is required for a protective immune response to systemic *Escherichia coli* infection. Infect. Immun. **64**, 3231–3235

155. Marra, M.N., Scott, R.W., Seilhammer, J.L. and Opal, S.M. (1996) The use of bactericidal/permeability-increasing protein and related protein as potential therapeutic agents for the treatment of endotoxin-related disorders. In Novel Therapeutic Strategies in the Treatment of Sepsis (Morrison, D.C. and Ryan, J.L., eds.), pp. 33–54, Marcel Decker, New York

156. Horwitz, A.H., Williams, R.E. and Nowakowski, G. (1995) Human lipopolysaccharide-binding protein potentiates bactericidal activity of human bactericidal/permeability-increasing protein. Infect. Immun. **63**, 522–527

157. Larrick, J.W., Hirata, M., Morgan, J.G. and Yen, M. (1991) Cloning of a cDNA for CAP18, a cationic lipopolysaccharide-binding protein. Biochem. Biophys. Res. Commun. **179**, 170–175

158. Larrick, J.W., Hirata, M., Balint, R.F., Huan, T.-H., Chen, C., Zhong, J. and Wright, S.C. (1996) Cap 18: a novel LPS-binding/antimicrobial protein. In Novel Therapeutic Strategies in the Treatment of Sepsis (Morrison, D.C. and Ryan, J.L., eds.), pp. 71–95, Marcel Dekker, New York

159. Hurley, J.C. (1995) Endotoxemia: methods of detection and clinical correlates. Clin. Microbiol. Rev. **8**, 268–292

160. Warren, H.S., Glennon, M.L., Wainwright, N., Amato, S.F., Black, K.M., Kirsch, S.J., Riveau, G.R., Whyte, R.I., Zapol, W.H. and Novitsky, T.J. (1992) Binding and neutralization of endotoxin by *Limulus* antilipopolysaccharide factor. Infect. Immun. **60**, 2506–2513

161. Desch, C.E., O'Hara, P. and Harlan, J.M. (1989) Antilipopolysaccharide factor from horseshoe crab, *Tachypleus tridentatus*, inhibits lipopolysaccharide activation of cultured human endothelial cells. Infect. Immun. **57**, 1612–1614

162. Morita, T., Ohtsubo, T., Nakamura, T., Tanaka, S., Iwanaga, S., Ohashi, K. and Niwa, M. (1985) Isolation and biological activities of *Limulus* anticoagulant (anti-LPS factor) which interacts with lipopolysaccharide (LPS). J. Biochem. (Tokyo) **97**, 1611–1620

163. Alpert, G., Baldwin, G., Thompson, C., Wainwright, N., Novitsky, J.J., Gillis, Z., Parsonnet, J., Fleischer, G.R. and Siber, G.R. (1992) *Limulus* antilipopolysaccharide factor protects rabbits from meningococcal endotoxin shock. J. Infect. Dis. **165**, 494–500

164. Saladino, R.A., Fleischer, G.R., Siber, G.R., Thompson, C. and Novitsky, T.J. (1996) Therapeutic potential of a recombinant endotoxin-neutralizing protein from *Limulus polyphemus*. In Novel Therapeutic Strategies in the Treatment of Sepsis (Morrison, D.C. and Ryan, J.L., eds.), pp. 97–110 Marcel Dekker, New York

165. Haziot, A., Rong, G.-W., Bazil, V., Silver, J. and Goyert, S.M. (1994) Recombinant soluble CD14 inhibits LPS-induced tumor necrosis factor-α production by cells in whole blood. J. Immunol. **152**, 5868–5876

166. Haziot, A., Rong, G.W., Lin, X.-Y., Silver, J. and Goyert, S.M. (1995) Recombinant soluble CD14 prevents mortality in mice treated with endotoxin (lipopolysaccharide). J. Immunol. **154**, 6529–6532

167. Leturcq, D.J., Moriarty, A.M., Talbott, G., Winn, R.K., Martin, T.R. and Ulevitch, R.J. (1996) Antibodies to CD14 protect primates from endotoxin-induced shock. J. Clin. Invest. **98**, 1533–1538

168. Lynn, W.A. and Golenbock, D.T. (1992) Lipopolysaccharide antagonists. Immunol. Today **13**, 361–371

169. Wang, M.-H., Flad, H.-D., Feist, W., Brade, H., Kusumoto, S., Rietschel, E.T. and Ulmer, A.J. (1991) Inhibition of endotoxin-induced interleukin-6 production by synthetic lipid A partial structures in human peripheral blood mononuclear cells. Infect. Immun. **59**, 4655–4664

170. Takayama, K., Qureshi, N., Beutler, B. and Kirkland, T.N. (1989) Diphosphoryl lipid A obtained from *Rhodopseudomonas sphaeroides* ATCC 17023 blocks induction of cachectin in macrophages by lipopolysaccharide. Infect. Immun. **57**, 1336–1338

171. Qureshi, N., Takayama, K. and Kurtz, R. (1991) Diphosphoryl lipid A obtained from the nontoxic lipopolysaccharide of *Rhodopseudomonas sphaeroides* is an endotoxin antagonist. Infect. Immun. **59**, 441–444

172. Zuckerman, S.H. and Qureshi, N. (1992) *In vivo* inhibition of LPS-induced lethality and TNF synthesis by *Rhodobacter sphaeroides* diphosphoryl lipid A is dependent on corticosterone production. Infect. Immun. **60**, 2581–2587

173. Qureshi, N., Hofman, J., Takayama, K., Vogel, S.N. and Morrison, D.C. (1996) Diphosphoryl lipid A from *Rhodobacter sphaeroides*: a novel lipopolysaccharide antagonist. In Novel Therapeutic Strategies in the Treatment of Sepsis (Morrison, D.C. and Ryan, J.L., eds.), pp. 111–131, Marcel Dekker, New York

174. Krauss, J.H., Seydel, U., Weckesser, J. and Mayer, H. (1989) Structural analysis of the nontoxic lipid of *Rhodobacter capsulatus* 37b4. Eur. J. Biochem. **180**, 519–526

175. Christ, W.J., Kawata, T., Hawkins, LD, Asano, O., Kobayashi, S. and Rossignol, D.P. (1992) Anti-endotoxin compounds and related molecules and methods. U.S. Patent application no. 935050

176. Christ, W.J., McGuinness, P.D., Asano, O., Wang, Y., Mullarkey, M.A., Perez, M., Hawkins, L.D., Blythe, T.A., Dubue, G.R. and Robidoux, A.L. (1994) Total synthesis of the proposed structure of *Rhodobacter sphaeroides* lipid A resulting in the synthesis of new potent lipopolysaccharide antagonists. J. Am. Chem. Soc. **116**, 3637–3638

177. Rose J.R., Christ, W.J., Bristol, J.R., Kawata, T. and Rossignol, D.P. (1995) Agonistic and antagonistic activities of bacterially-deived *Rhodobacter sphaeroides* lipid A: comparison with activities of synthetic material of the proposed structure and analogs. Infect. Immun. **63**, 833–839

178. Kawata, T., Bristol, J.R., Rose, J.R., Rossignol, D.P., Christ, W.J., Asano, O., Dubuc, G.R., Gavin, W.E., Hawkins, L.D., Lewis, M.D., et al. (1996) Specific lipid A analog which exhibits exclusive antagonism of endotoxin. In Novel Therapeutic Strategies in the Treatment of Sepsis (Morrison, D.C. and Ryan, J.L., eds.), pp. 171–186, Marcel Dekker, New York

179. Feingold, K.R., Staprans, I., Memon, R., Moser, A.H., Shigenaga, J.K., Doerrler, W., Dinarello, C.A. and Grunfeld, C. (1992) Endotoxin rapidly induces changes in lipid metabolism that produce hypertriglyceridemia: low doses stimulate hepatic triglyceride production while high doses inhibit clearance. J. Lipid Res. **33**, 1765–1776

180. Harris, H.W., Grunfeld, C., Feingold, K.R. and Rapp, J.H. (1990) Human very low density lipoproteins and chylomicrons can protect against endotoxin-induced death in mice. J. Clin. Invest. **86**, 696–702

181. Harris, H.W., Grunefeld, C., Feingold, K.R., Read, T.E., Kane, J.P., Jones, A.L., Eichbaum, E.B., Bland, G.F. and Rapp, J.H. (1993) Chylomicrons alter the fate of endotoxin, decreasing tumor necrosis factor release and preventing death. J. Clin. Invest. **91**, 1028–1034

182. Gerard, C., Bruyns, C., Marchant, A., Abramowicz, D., Vandenbeele, P., Delvaux, A., Fiers, W., Goldman, M. and Velu, T. (1993) Interleukin-10 reduces the release of tumor necrosis factor and prevents lethality in experimental endotoxemia. J. Exp. Med. **177**, 547–550

183. van der Poll, T., Marchant, A., Buurman, W.A., Berman, L., Keogh, C.V., Lazarus, D.D., Nguyen, L., Goldman, M., Moldawer, L.L. and Lowry, S.F. (1995) Endogenous IL-10 protects mice from death during septic peritonitis. J. Immunol. **155**, 5397–5401

184. Levitski, A. (1992) Tryphostins: tyrosine kinase blockers as novel antiproliferative agents and dissectors of signal transduction. FASEB J. **6**, 3257–3262
185. Bursten, S.L. and Harris, W.E. (1991) Rapid activation of phosphatidate phosphohydrolase in mesangial cells by lipid A. Biochemistry **30**, 6195–6203
186. Bursten, S.L., Harris, W.E. and Rice, G.C. (1996) Selective inhibition of phosphatidic acid synthesis: a novel approach to the treatment of sepsis and the systemic inflammatory response syndrome. In Novel Therapeutic Strategies in the Treatment of Sepsis (Morrison, D.C. and Ryan, J.L., eds.), pp. 199–226, Marcel Dekker, New York
186a. Onishi, H.R., Pelak, B.A, Gerckens, L.S., Silver, L.L., Kahan, F.M., Chen, M.-H., Patchett, A.A., Galloway, S.M., Hyland, S.A., Anderson, M.S. and Raetz, C.R.H. (1996) Antibacterial agents that inhibit lipid A biosynthesis. Science **274**, 980–982
187. MacKenzie, A.R. (1996) Low molecular weight inhibitors of cytokine synthesis. In Therapeutic Modulation of Cytokines (Henderson, B. and Bodmer, M.W., eds.), pp. 93–112, CRC Press, Boca Raton
188. Schmidt, J.A. and Bundick, R.V. (1996) Immunosuppressants as cytokine inhibitors. In Therapeutic Modulation of Cytokines (Henderson, B. and Bodmer, M.W., eds.), pp. 113–141, CRC Press, Boca Raton
189. Lee, J.C., Badger, A.M., Griswold, D.E., Dunnington, D., Truneh, A., Votta, B., White, J.R., Young, P.R. and Bender, P.E. (1993) Bicyclic imidazoles as a novel class of cytokine biosynthesis inhibitors. Ann. N.Y. Acad. Sci. **696**, 149–170
190. Lee, J.C. and Young, P.R. (1996) Role of CSB/p38/RK stress response kinase in LPS and cytokine signalling mechanisms. J. Leukocyte Biol. **59**, 152–157
191. Badger, A.M., Olivera, D., Talmage, J.E. and Hanna, N. (1989) Protective effect of SK&F 86002, a novel dual inhibitor of arachidonic acid metabolism in murine models of endotoxin shock. Circ. Shock **27**, 51–61
192. Olivera, D.L., Esser, K.M. and Lee, J.C. (1992) Beneficial effects of SK&F 105809, a novel cytokine suppressive agent, in murine models of endotoxic shock. Circ. Shock **37**, 301–306
193. Otterness, I.G., Bliven, M.L., Downs, J.T., Natoli, E.J. and Hanson, D.C. (1991) Inhibition of interleukin-1 synthesis by tenidap: a new drug for arthritis. Eur. Cytokine Netw. **3**, 277–283
194. Sipe, J.D., Bartie, L.M. and Loose, L.D. (1992) Modification of the proinflammatory cytokine production by the antirheumatic agents tenidap and naproxen. A possible correlate with clinical acute phase response. J. Immunol. **148**, 480–484
195. Schynder, J., Bollinger, P. and Payne, T. (1990) Inhibition of interleukin-1 release by IX-207-887. Agents Actions **30**, 350–361
196. Dougados, M., Combe, B., Beveridge, T., Bourdeux, I., Lallemand, A., Armor, B. and Sany, J. (1992) IX 207-887 in rheumatoid arthritis. A double blind clinical trial. Arthritis Rheum. **35**, 999–1006
197. Batt, D.G., Goodman, R., Jones, D.G., Kerr, J.S., Mantegna, L.R., McAllister, C., Newton, R.C., Nurnberg, S., Welch, P.K. and Covington, M.B. (1993) 2'-Substituted chalcone derivatives as inhibitors of interleukin-1 biosynthesis. J. Med. Chem. **36**, 1434–1442
198. Rao, M.N.A., Naidoo, L. and Ramaman, P.N. (1991) Antiinflammatory activity of phenyl styryl ketones. Pharmazie **46**, 542–543
199. Chiba, K., Goto, M. and Shirota, H. (1991) A novel inhibitor of IL-1 generation, E5090. In vitro inhibitory effects on the generation of IL-1 by human monocytes. Agents Actions (suppl.) **32**, 225–229
200. Folliard, F., Bousseau, A. and Terlain, B. (1992) RP54745, a potential antirheumatic compound. I. Inhibitor of macrophage stimulation and interleukin-1 production. Agents Actions **36**, 119–126
201. Brown, M.K., Leslie, G.B. and Pfirrmann, R.W. (1976) Taurolin, a new chemotherapeutic agent. J. Appl. Bacteriol. **41**, 363–366
202. McCartney, A.C. and Browne, M.K. (1988) Clinical studies on administration of taurolin in severe sepsis. A preliminary study. In Bacterial Endotoxins: Pathophysiological Effects, Clinical

Significance and Pharmacological Control (Levein, J., Bullen, H.R., ten Gate, J.W., van Deventer, S.J.H. and Sturk, A., eds.), pp. 361, Alan Liss, New York

203. Bedrosian, I., Sofia, R.D., Wolff, S.M. and Dinarello, C.A. (1991) Taurolidine, an analogue of the amino acid taurine, suppresses interleukin-1 and tumour necrosis factor synthesis in human peripheral blood mononuclear cells. Cytokine 3, 568–575

204. Pfirrmann, R.W. and Leslie, G.B. (1979) The anti-endotoxic activity of taurolin in experimental animals. J. Appl. Bacteriol. 46, 97–102

205. Bennett, C.F. and Crook, S.T. (1996) Oligonucleotide-based inhibition of cytokine expression and function. In Therapeutic Modulation of Cytokines (Henderson, B., and Bodmer, M.W., eds.), pp. 171–193, CRC Press, Boca Raton

206. Maier, J.A.M., Voulalas, P., Roeder, D. and Maciag, T. (1990) Extension of the life-span of human endothelial cells by an interleukin-1 antisense oligomer. Science 249, 1570–1574

207. Manson, J., Brown, T. and Duff, G. (1990) Modulation of interleukin-1β gene expression using antisense phosphorothioate oligonucleotides. Lymphokine Res. 9, 35–42

208. Burch, R.M. and Mahan, L.C. (1991) Oligonucleotides antisense to the interleukin-1 receptor mRNA block the effect of interleukin-1 in cultured murine and human fibroblasts and in mice. J. Clin. Invest. 88, 1190–1196

209. Stepkowski, S.M., Tian, L. and Kloc, M. (1993) Interleukin-2 antisense oligonucleotides inhibit T cell function. Trans. Proc. 25, 125

210. Bebernou, N., Matsiota-Bernard, P. and Guenounou, M. (1993) Effect of cytokine-specific antisense oligonucleotides on the immunoglobulin production by rat spleen cells in vitro. Biochimie 75, 55–60

211. Levy, Y., Tsapis, A. and Brouet, J.-C. (1991) Interleukin-6 antisense oligonucleotides inhibit the growth of human myeloma cell lines. J. Clin. Invest. 88, 696–699

212. Schadendorf, D., Moller, A., Algermissen, B., Worm, M., Sticherling, M. and Czarnetzki, B.M. (1993) IL-8 produced by human malignant melanoma cells in vitro is an essential autocrine growth factor. J. Immunol. 151, 2667–2675

213. Segal, G.M., Smith, T.D., Heinrich, M.C., Ey, F.S. and Bagby, G.C. (1992) Specific repression of granulocyte-macrophage and granulocyte colony-stimulating factor gene expression in interleukin-1-stimulated endothelial cells with antisense oligonucleotides. Blood 80, 609–616

214. Stepkowski, S.M., Tu, Y., Condon, T.P. and Bennett, C.F. (1994) Blocking of heart allograft rejection by intercellular adhesion molecule-1 antisense oligonucleotides alone or in combination with other immunosuppressive modalities. J. Immunol. 153, 5336–5346

215. Neurath, M.F., Pettersson, S., Meyer, K.-H., Zum Buschenfelde, M. and Strober, W. (1996) Local administration of antisense phosphorothioate oligonucleotides to the p65 subunit of NF-κB abrogates established experimental colitis in mice. Nature Medicine 2, 998–1004

216. Wagner, R.W., Matteucci, M.D., Grant, D., Huang, T. and Froehler, B.C. (1996) Potent and selective inhibition of gene expression by an antisense heptanucleotide. Nature Biotechnol. 14, 840–844

217. Miller, D.K. (1996) Cytokine convertase inhibitors. In Therapeutic Modulation of Cytokines (Henderson, B. and Bodmer, M.W., eds.), pp. 143–170, CRC Press, Boca Raton

218. Kostura, M.J., Tocci, M.J., Limjuco, G., Chin, J., Cameron, P., Hillman, A.G., Vhartrain, N.A. and Schmidt, J.A. (1987) Identification of a monocyte specific pro-interleukin-1β convertase activity. Proc. Natl. Acad. Sci. U.S.A. 86, 5227–5231

219. Black, R.A., Kronheim, S.R. and Sleath, P.R. (1989) Activation of interleukin-1β by a co-induced protease. FEBS Lett. 247, 386–390

220. Thornberry, N.A., Bull, H.G., Calaycay, J.R. Chapman, K.T., Howard, A.D., Kostura, M.J., Miller, D.K., Molineaux, S.M., Weidner, J.R., Aunins, J., et al. (1992) A novel heterodimeric cysteine protease is required for interleukin-1β processing in monocytes. Nature (London) 356, 768–774

221. Ray, C.A., Black, R.A., Kronheim, S.R., Greenstreet, T.A., Sleath, P.R., Salvesen, G.S. and Pickup, D.J. (1992) Viral inhibition of inflammation: cowpox virus encodes an inhibitor of the interleukin-1β converting enzyme. Cell 69, 597–604

222. Nicholson, D.W. (1996) ICE/CED3-like proteases as therapeutic targets for the control of inappropriate apoptosis. Nature Biotechnol. **14**, 297–301

223. Dolle, R.E., Hoyer, D., Prasad, C.V.C., Schmidt, S.J., Helaszek, C.T., Miller, R.E. and Ator, M.A. (1994) P_1 aspartate-based peptide α-(2,6-dichlorobenzoyl)oxymethyl ketones as potent time-dependent inhibitors of interleukin-1β-converting enzyme. J. Med. Chem. **37**, 563–569

224. Thornberry, N.A., Miller, D.K. and Nicholson, D.W. (1995) Interleukin-1β converting enzyme and related proteases as potential targets in inflammation and apoptosis. Perspectives in Drug Discovery **2**, 389–399

225. Milligan, C.E., Prevette, D., Yaginuma, H., Homma, S., Cardwell, C., Fritz, L.C., Tomaselli, K.J., Oppenheim, R.W. and Schwartz, L.M. (1995) Peptide inhibitors of the ICE protease family arrest programmed cell death of motorneurones *in vivo* and *in vitro*. Neuron **15**, 385–393

226. Miller, B.E., Krasney, P.A., Gauvin, D.M., Holbrook, K.B., Koonz, D.J., Abruzzese, R.V., Miller, R.E., Pagani, K.A., Dolle, R.E., Ator, M.A. and Gilman, S.C. (1995) Inhibition of mature IL-1β production in murine macrophages and a murine model of inflammation by WIN 67694, an inhibitor of IL-1β converting enzyme. J. Immunol. **154**, 1331–1338

227. Ku, G., Faust, T., Lauffer, L.L., Livingstone, D.J. and Harding, M.W. (1996) Interleukin-1β converting enzyme inhibition blocks progression of type II collagen-induced arthritis in mice. Cytokine **8**, 377–386

228. Kriegler, M., Perez, C., DeFay, K., Albert, I. and Lu, S.D. (1988) A novel form of TNF/cachectin is a cell surface cytotoxic transmembrane protein: ramifications for the complex physiology of TNF. Cell **53**, 45–53

229. Scuderi, P. (1989) Suppression of human leukocyte tumor necrosis factor secretion by the serine protease inhibitor p-toluenesulfonyl-L-arginine methyl ester (TAME). J. Immunol. **143**, 168–173

230. Mohler, K.M., Sleath, P.R., Fitzner, J.N., Cerreti, D.P., Alderson, M., Kerwar, S.S., Torrance, D.S., Otten-Evans, C., Greenstreet, T., Weerawarna, K., et al. (1994) Protection against a lethal dose of endotoxin by an inhibitor of tumour necrosis factor processing. Nature (London) **370**, 218–220

231. Gearing, A.J.H., Becket, P., Christodoulou, M., Churchill, M., Clements, J., Davidson, A.H., Drummond, A.H., Galloway, W.A., Gilbert, R., Gordon, J.L., Leber, T.M., Mangan, M., Miller, K., Nayee, P., Patel, S., Thomas, W., Wells, G., Wood, L.M. and Wooley, K. (1994) Processing of tumour necrosis factor-α precursor by metalloproteinases. Nature (London) **370**, 555–557

232. McGeehan, G.M., Becerer, J.D., Bast, R.C., Boyer, C.M., Champion, B., Connolly, K.M., Conway, J.G., Furdon, P., Karp, S., Kidao, S., et al. (1994) Regulation of tumour necrosis factor-α processing by a metalloproteinase inhibitor Nature (London) **370**, 558–561

233. Henderson, B. and Blake, S. (1994) Connective tissue destruction in rheumatoid arthritis: Therapeutic potential of metalloproteinase inhibitors. In Immunopharmacology of the Joints and Connective Tissues (Dingle, J.T. and Davies, M.E., eds.), pp. 199–223, Academic Press, London

234. Hardie, G. and Hanks, S. (eds.) (1995) The Protein Kinase Facts Book. Academic Press, London

235. Johnson, L., Noble, M.E.M. and Owen, D.J. (1996) Active and inactive protein kinases: structural basis for regulation. Cell **85**, 149–158

236. Bursten, S.L., Harris, W.E. and Rice, G.C. (1996) Selective inhibition of phosphatidic acid synthesis: a novel approach to the treatment of sepsis and the systemic inflammatory response syndrome. In Novel Therapeutic Strategies in the Treatment of Sepsis (Morrison, D.C. and Ryan, J.L., eds.), pp. 199–226, Marcel Dekker, New York

237. Brugge, J.S. (1993) New intracellular targets for therapeutic drug design. Science **260**, 918–919

238. Levitzki, A. and Gazit, A. (1995) Tyrosine kinase inhibition: an approach to drug development. Science **267**, 1782–1788

239. Davis, P.D., Hill, C.H., Lawton, G., Nixon, J.S., Wilkinson, S.E., Hurst, S.A., Keech, E. and Turner, S.E. (1992) Inhibitors of protein kinase C. 1. 2,3-bisarylmaleimides. J. Med. Chem. **35**, 177–184

240. MacKintosh, C. and MacKintosh, R.W. (1994) Inhibitors of protein kinases and phosphatases. Trends Biochem. Sci. **19**, 444–448

241. Krishna, T., Gristwood, R., Higgs, G.A. and Holgate, S.T. (1996) Phosphodiesterase inhibitors. In Therapeutic Immunology (Austen, F.K., Burakoff, S.J., Rosen, F.S. and Strom, T.B., eds.), pp. 170–178, Blackwell, Cambridge

242. Prabhakar, U., Lipshutz, D., Bartus, J.O., Slivjak, M.J., Smith, E.F., Lee, J.C. and Esser, M. (1994) Characterization of cAMP-dependent inhibitors of LPS-induced TNF alpha production by rolipam, a specific phosphodiesterase IV (PDE IV) inhibitor. Int. J. Immunopharmacol. **16**, 805–816

243. Yaish, P., Gazit, A., Gilon, C. and Levitzki, A. (1988) Blocking of EGF-dependent cell proliferation by EGF receptor kinase inhibitors. Science **242**, 933–935

244. Buchdunger, E., Trinks, U., Mett, H., Regenass, U., Muller, M., Meyer, T., McGlynn, E., Pinna, L.A., Traxler, P. and Lyndon, N.B. (1994) 4,5-Dianilophthalimide: a protein-tyrosine kinase inhibitor with selectivity for the epidermal growth factor receptor signal transduction pathway and potent in vivo antitumor activity. Proc. Natl. Acad. Sci. U.S.A. **91**, 2334–2338

245. Fry, D.W., Kraker, A.J., McMichael, A., Ambroso, L.A., Nelson, J.M., Leopold, W.R., Connors, R.W. and Bridges, A.J. (1994) A specific inhibitor of the epidermal growth factor specific tyrosine kinase. Science **265**, 1093–1095

246. Marshall, C.J. (1995) Specificity of receptor tyrosine kinase signalling: transient versus sustained extracellular signal-regulated kinase activation. Cell **80**, 179–185

247. Hill, C.S. and Treisman, R. (1995) Transcriptional regulation by extracellular signals: Mechanisms and specificity. Cell **80**, 199–211

248. Lee, J.C. and Adams, J.L. (1995) Inhibitors of serine/threonine kinases. Curr. Opin. Biotechnol. **6**, 657–661

249. Lee, J.C., Laydon, J.T., McDonnell, P.C., Gallagher, T.F., Kumar, S., Green, D., McNulty, D., Blumenthal, M.J., Heys, J.R., Landvatter, S.W., et al. (1994) A protein kinase involved in the regulation of inflammatory cytokine biosynthesis. Nature (London) **372**, 739–746

250. Dudley, D.T., Pang, L., Decker, S.J., Bridges, A.J. and Saltiel, A.R. (1995) A synthetic inhibitor of the mitogen-activated protein kinase cascade. Proc. Natl. Acad. Sci. U.S.A. **92**, 7686–7689

251. Lee, J.C. and Adams, J.L. (1995) Inhibitors of serine/threonine kinases. Curr. Opin. Biotechnol. **6**, 657–661

252. Schinder, C. and Darnell, J.E. (1995) Transcriptional response to polypeptide ligands: The Jak-STAT pathway. Annu. Rev. Biochem. **64**, 621–651

253. Miyamoto, K., Nagakawa, J., Hishinuma, I., Hirota, K., Yasuda, M., Katayama, K. and Yamatsu, I. (1992) Suppressive effects of E3330, a novel quinone derivative, on tumor necrosis factor-α generation from monocytes and macrophages. Agents Action **37**, 297–304

254. Goto, M., Yamada, K., Katayama, K.-I. and Tanaka, I. (1996) Inhibitory effect of E330, a novel quinone derivative able to suppress tumor necrosis factor-α generation, on activation of nuclear factor-κB. Mol. Pharmacol. **49**, 860–873

255. Wrighton, N.C., Farrell, F.X., Chang, R., Kashyap, A.K., Barbone, F.P., Mulcahy, L.S., Johnson, D.L., Barrett, R.W., Jollife, L.K. and Dower, W.J. (1996) Small peptides as potent mimetics of the protein hormone erythropoietin. Science **273**, 458–463

256. Livnah, O., Stura, E.A., Johnson, D.L., Middleton, S.A., Mulcahy, L.S., Wrighton, N.C., Dower, W.J., Jollife, L.K. and Wilson, I.A. (1996) Functional mimicry of a protein hormone by a peptide agonist: the EPO receptor complex at 2.8Å. Science **273**, 464–471

257. Yanofsky, S.D., Baldwin, D.N., Butler, J.H., Holden, F.R., Jacobs, J.W., Balasubramanian, P., Chinn, J.P., Cwirla, S.E., Peters-Bhatt, E., Whitehorn, E.A., et al. (1996) High affinity type I interleukin-1 receptor antagonists discovered by screening recombinant peptide libraries. Proc. Natl. Acad. Sci. U.S.A. **93**, 7381–7386

258. Henderson, B. and Wilson, M. (1996) Editorial. Homo bacteriens and a network of surprises. J. Med. Microbiol. **45**, 1–2

259. Henderson, B., Poole, S. and Wilson, M. (1996) Microbial/host interactions in health and disease: who controls the cytokine network? Immunopharmacology **35**, 1–21

260. Berger, SW.A., Rowan, K., Morrison, H.D. and Ziltener, H.J. (1996) Identification of a bacterial inhibitor of protein kinases. J. Biol. Chem. **271**, 24431–24437

9

Bacteria–cytokine interactions in health and disease: a new synthesis

> *"If microbes were capable of emotion*
> *they would rejoice each time an infant was born."*
>
> Tanner, G.W. (1995)
> *Normal Microflora.*
> *Chapman and Hall, London*

9.1   Introduction

The reader will now be aware of the complex cellular and molecular relation-ship which exists between bacteria, their molecular components (particularly LPS) and eukaryotic hosts. Cytokines play a central role in this relationship, acting both as the regulators of the host defence mechanisms against micro-organisms and as the major mediators of the pathology that is associated with infections. Readers will also be aware that cytokines and their receptors represent an interactive amplificatory system designed such that the induction of the synthesis of one cytokine inevitably results in the generation of a number of distinct cytokines, thus forming what is termed a cytokine network. This concept of the cytokine network is still in its infancy, and many questions, including the nature, function, stability and control of such networks, remain to be explored. Most information on cytokine networks has come from the study of cytokines in states of natural (e.g. autoimmune diseases) or experimental inflammation. Much less is known about the functions of cytokine networks in non-pathological tissues. Studies of the genetics of inflammatory (mainly autoimmune) diseases are showing the presence of disease-related polymorphisms (particularly in the regulatory elements) in the genes encoding the pro-inflammatory cytokines IL-1 and

TNF, as well as those encoding anti-inflammatory cytokines IL-1ra and IL-4 [1]. Such polymorphisms may be important in the susceptibility of individuals to particular diseases. Much less is known about the genetics of cytokines in the infectious diseases. However, this is obviously important and may lie at the heart of the known differences in the susceptibility of individuals to infections.

In addition to cytokine receptors, a range of other cell-surface receptors are involved in monitoring and controlling cytokine networks. The most obvious is the LPS receptor CD14. However, other non-cytokine receptors appear to play a role in controlling bacterially induced cytokine induction. A recent example is the finding that the T-cell co-stimulatory surface receptor CD28, which binds to B7 on antigen-presenting cells and is an obligatory binding event in lymphocyte activation, is an important factor in the pathology of the septic shock syndrome caused by *Staphylococcus aureus* toxin TSST-1. Genetic knockout of the CD28 receptor renders mice resistant to TSST-1-induced shock. This appears to be a result of the failure to secrete the potent pro-inflammatory cytokine TNFα [2]. The interaction of cytokines with such receptors is an obvious area of interest and growth. Other receptors, such as immunoglobulin Fc receptors, probably also play roles in controlling cellular cytokine synthesis in response to infecting micro-organisms.

With the rise in antibiotic resistance among pathogenic bacteria such as *Mycobacterium tuberculosis* and opportunistic pathogens such as *Staph. aureus,* there has been a resurgence of interest in the cellular biology of bacteria and, in contrast to the view that had prevailed in the period between the 1960s and the late 1980s, bacteria are increasingly being regarded as complex organisms. For example, they are now known to have multiple intra-cellular signalling pathways [3,4] with similarities to those found in eukaryotic cells. It is now also emerging that bacteria can exert significant control over eukaryotic cell function. Recent examples include the ability of bacteria to inhibit [5] or activate [6] selected protein kinases or induce mammalian cells to undergo apoptosis [7]. Furthermore, it is becoming evident that bacteria can directly interact with host cells and that these cell-to-cell interactions can modify the activity of the bacteria. For example, it has recently been demonstrated that the direct contact of bacteria with mammalian cells results in the up-regulation of selected bacterial virulence genes. Thus, upon contact with eukaryotic cells in culture, *Yersinia pseudotuberculosis* increased the rate of transcription of a group of highly regulated virulence proteins called Yops. In this experiment a luciferase gene was put under the control of the yop E promoter, so that cells transcribing this virulence system would also emit light. This study elegantly demonstrated that only those bacteria in contact with HeLa cells emitted light [8]. *Escherichia coli* has been shown to switch on its transcription of siderophore-producing proteins when it binds to the host cell receptor on uroepithelial cells [9]. In addition to the evidence for direct bacterial-to-eukaryotic cell communication with certain bacteria, it has recently been demonstrated that bacteria produce lysozyme-like enzymes.

These are proteins which, traditionally, have been viewed as part of the host's defence mechanisms against bacteria. Yet bacteria also produce such proteins, again highlighting the continuum that exists between prokaryotic and eukaryotic organisms [10]. It is this continuum which this chapter seeks to highlight.

Thus, we are living in an age where our view of micro-organisms and their relationships with their hosts is undergoing a Kuhnian paradigm shift. In this chapter we will bring together the strands of the argument for this new viewpoint on prokaryotic–eukaryotic interactions and the essential role of cytokines in this process. The recently described fever-inhibiting actions of vaccinia virus provides an excellent example of this new vision of host–microbial interactions [11].

9.2   The classic picture of bacteria–cytokine interactions

The classic view of the interaction between bacteria and the host organism has arisen from the study of the relatively small number of organisms that are capable of producing pathology in man or in experimental or domesticated animals. The bacteria causing human disease, and the pathology they produce, have been described in detail in Chapters 2 and 3. Bacterial infections, without exception, produce some degree of inflammation (the evolved response to infection), and the fevers associated with many infections point to the large amounts of pro-inflammatory cytokines generated. As has been reviewed in Chapter 3, the study of the pathological mechanisms of infections has focused mainly on bacterial exotoxins and on endotoxin, with the bacterium being regarded merely as a source of such potent disease-causing molecules. However, potential subtleties in the synthesis and actions of such pathogenic molecules has been largely ignored until recent years. In the past two decades increasing attention has been paid to the role of the local hormones (cytokines) which control host tissue homeostasis and which are involved in the mechanisms of innate and acquired immunity. As has been reviewed in the preceding chapters, much of the pathology of bacterial infections is now laid at the door of the bacterium's ability to overstimulate the synthesis of pro-inflammatory cytokines by the host. For decades almost the only cytokine-inducing component of bacteria recognized was endotoxin/LPS, a view reinforced by the focus of the pharmaceutical industry in the 1980s and 90s on Gram-negative septic shock. Thus, in terms of the interactions between bacteria and cytokines, a simple linear sequence of interactions (as shown in Figure 9.1) became the paradigm. Bacteria release factors (endotoxin) which then stimulate nearby cells to produce cytokines. It is the production of these cytokines which then awakens the host defence mechanisms of innate and acquired immunity. The bacterium is then largely relegated to being an inert vessel — the target of the host's defence systems. Of course, the cellular and molecular complexity of the reactions that have just been described are

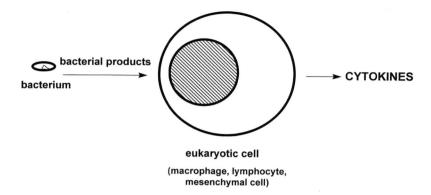

bacterial products

bacterium

CYTOKINES

eukaryotic cell
(macrophage, lymphocyte,
mesenchymal cell)

Figure 9.1 The conventional paradigm for the interaction between bacteria and host cells leading to the induction of cytokine synthesis
The bacterium releases products (e.g. endotoxin) which activate the cell to produce and secrete cytokines.

enormous and many distinct genes within cells have to be activated to produce these defence mechanisms. As described in detail in textbooks (for example [12,13]), the macrophage and the polymorphonuclear neutrophil (PMN) play the main role in the front-line innate defences against invading bacteria. The more complex and, in evolutionary terms, modern acquired system with its various lymphoid cells is only involved in combatting infections if they are chronic (i.e. intracellular) or involve reinfection and the involvement of a memory response. Acquired immunity requires a wide range of cytokines to 'drive the system' and, for example, the two distinct classes of CD4 T-lymphocytes defined by Marrack and Kappler — the Th_1 and Th_2 cells — are defined on the basis of the distinct patterns of cytokines they produce (Table 9.1; see Chapter 4). CD4 T-lymphocytes which do not produce either pattern of cytokines have been termed T_0 lymphocytes. The Th_1 and Th_2 lymphocytes have distinct and opposing functions. The Th_1 cell is designed to activate macrophages bearing intracellular parasites (the best examples being *Mycobacterium tuberculosis*, *Mycobacterium leprae* and *Leishmania* spp.), thereby enabling the macrophage to kill its unwelcome parasites. This CD4

Table 9.1 Pattern of cytokines produced by T-lymphocytes

Lymphocyte	Cytokines produced
Th_1	**IFNγ**, IL-2, IL-3, IL-10, TNFα, lymphotoxin, GM-CSF
Th_2	IL-3, **IL-4**, **IL-5**, IL-6, IL-10, IL-13
Th_0	Precursor cell from which Th_1 and Th_2 cells derive
	Th_0 cells produce all the above cytokines

Cytokines highlighted are the principle effector cytokines of each subset. The role of Th_1 and Th_2 cells in immunity is reviewed in [13a].

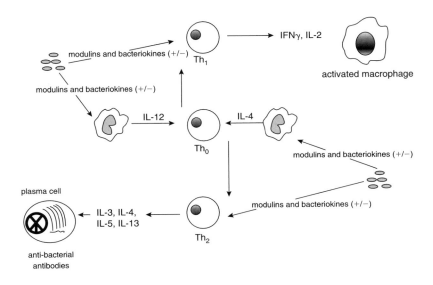

Figure 9.2 The interactions between bacteria and CD4 lymphocyte subsets
Bacteria directly acting or acting via the release of modulins and bacteriokines on non-lymphoid
cells (monocytes/macrophages, fibroblasts, etc.) or on lymphoid cells can generate the appropriate (or inappropriate) cytokines to induce the differentiation of the CD4 precursor cell (the Th_0
cell) into either the Th_1 macrophage-activating cell or the Th_2 B-lymphocyte helper cell. The
various cytokine-modulating components of the bacteria could also act on the mature CD4
T-cell subsets.

subset is therefore involved in delayed type hypersensitivity (DTH) responses
and is, in consequence, termed an inflammatory T-lymphocyte. In contrast,
the Th_2 cell has the function of stimulating cells involved in anti-parasite
defences. In particular, the Th_2 lymphocyte produces IL-4 and IL-5 and is
important in the production of IgE antibodies and the recruitment of
eosinophils. These processes are important in the defence against parasites such
as helminths. Thus, it is clear that these two arms of the CD4 system have
evolved to cope with distinct parasitic organisms. It is also noteworthy that
among the cytokines produced by both types of CD4 lymphocytes are those
with anti-inflammatory actions — IL-4, IL-10 IL-13 and TGFβ (Figure 9.2
and Table 9.1).

Infectious organisms entering the human body would trigger the appropriate population of CD4 cells, thus producing the appropriate panoply of
cytokines. This, in turn, would result in the appropriate form of immune
response which would eventually remove the organism. Evidence is now
accruing to suggest that infectious agents can modify the development of these
two distinct CD4 T-cell populations to enhance their survival. The earliest
evidence for this concept was the recognition of the clinical patterns of
leprosy. The healing (so-called tuberculoid) form of leprosy was found to be
associated with a strong DTH response, but patients had low levels of circulat-

ing antibodies. In contrast, in patients with uncontrolled (lepromatous) leprosy, the immunological pattern was the inverse of what has been described, with minimal DTH reactions and high antibody titres — exactly the immunological profile required to cope, not with an intracellular parasite, but with an extracellular one. The correspondence between susceptibility to infectious disease, and the profile of Th_1/Th_2 subsets produced by individuals, was first clearly defined with the causative agent of Leishmaniasis, *Leishmania major*, a parasite resembling trypanosomes, which lives within macrophages. This organism can either grow within the liver and spleen, producing visceral leishmaniasis, or within the skin, producing cutaneous leishmaniasis.

Th_1 and Th_2 lymphocytes do not derive from distinct lineages but develop from the same T-cell precursor, such development being determined by the signals produced in the cell's microenvironment. Cytokines play a major role in the development of these two distinct CD4 lineages. IL-12 is the major Th_1-inducing cytokine and IL-4 the key Th_2-inducing cytokine. The release of these cytokines from macrophages and other cells can be induced by a variety of undefined substances of microbial/parasitic origin, the best known being endotoxin. Understanding how soluble or cell-associated factors from bacteria and parasites control the differentiation and action of CD4 T-cells should be a major area of study. It is now accepted that resistance to many intracellular micro-organisms is linked to the induction of Th_1 responses, in particular the induction of the macrophage-activating cytokines IFNγ and TNFα. The growing number of bacterial proteins which can inhibit the synthesis of TNFα (Chapter 7, Table 7.8) is testimony to the importance of TNF in combatting bacterial infections.

With the recently derived knowledge that bacteria produce a very large number of cytokine-inducing components, it is certain that these molecules must play a major role in controlling the development of the Th_1/Th_2 lymphocyte subsets. Indeed, these molecules could be employed in confusing the immune system in terms of its committed response to particular infectious agents. Very little is known about the capacity of modulins or bacteriokines to stimulate macrophages or lymphocytes to produce specific cytokines. The ability of these molecules to induce the synthesis of IL-4 or IL-12 or indeed to mimic the biological activity of these cytokines, could markedly alter the differentiation of Th_1/Th_2 lymphocytes, thus modulating the immunological responses to bacteria. Over the past few years IL-12 has been shown to play a key role in the generation of protective Th_1 responses, and the literature on this particular role for this cytokine has been reviewed [14]. Such interaction may lie at the heart of the capacity of the normal microflora to live in harmony with the host. There is, therefore, a pressing need to determine how cytokine-inducing bacterial components can modulate the development of CD4 lineages, and also to discover the role that such interactions have in controlling the normal microflora and in the recognition of and response to infecting bacteria. If bacterial components can selectively modify Th_1/Th_2 responses then they may have useful roles in vaccines for infectious diseases.

The other major T-cell population — the CD8 cell — is a cytotoxic cell designed to kill virally infected host cells. The lymphoid- and myeloid-based host defence mechanisms are driven by a vast range of cytokines which act as signals for growth, differentiation and communication. These cytokines are vital for the homeostatic control of the immuno-inflammatory response and, as has been described in earlier chapters, they interact in complicated manners with themselves and with other proteins and non-proteinaceous mediators to produce complex controlling networks. Such networks also involve molecules such as growth hormone [15] or substance P [16], which have previously not been considered to play any role in bacterial infections. The interaction of such host molecules with bacteria is an area of microbiology which is only just being defined [17,18], but one which will have tremendous potential for understanding the cellular and molecular mechanisms of infections and, as a corollary, the mechanisms which allow multicellular organisms to support the vast numbers of bacteria that constitute the normal microflora.

The correct temporal pattern of myeloid and lymphoid cell activation and associated mediator production is required to combat the different forms of infection to which, for example, *Homo sapiens* is prey. Interference with any step in these molecularly complex systems can seriously disrupt the host defence mechanisms designed to protect us from micro-organisms. It has been appreciated for many years that, for example, the capsules of bacteria have the capacity to interfere with host defences. Thus, the capsule can prevent the activation of the complement cascade or, by resembling host tissues, can inhibit the production of bacteria-neutralizing antibodies [19]. The importance of the complement pathways in host defences is now being directly demonstrated by knockout studies. A good example is the C5a receptor-deficient mouse which is unable to clear *Pseudomonas aeruginosa* instilled into its lungs and therefore develops a lethal pneumonia [20].

In more recent years it has become apparent that, in addition to blockade of complement activation and antibody synthesis, micro-organisms have developed methods of neutralizing the biological actions of cytokines. This ability to block the biological actions of cytokines was first discovered by virologists and this literature is described in Chapter 7. However, there are now a growing number of examples of bacterial proteins and non-proteinaceous components able to selectively inhibit cytokine networks. Almost as a corollary to these studies, it has been found that, in certain of the cytokine knockouts produced in the past few years, the animals develop chronic inflammatory pathology which appears to be due to an inappropriate response to the normal microflora. It is these and related findings which, along with advances in our understanding of cytokine networks, have begun to define a new paradigm for the interactions between bacteria and host cells, and the cytokine-like molecules from both Kingdoms which control such interactions.

Innate immunity with its tears, skin acids and the acute-phase response (involving cytokines, complement proteins, acute-phase opsonins and myeloid cells) is believed to be the first line of defence against bacteria. However, while first discovered at the end of the 1970s, the large number of host antibacterial peptides which must play a significant role in dealing with both the normal microflora and with infectious organisms, is still largely ignored ([21–23] provide some recent reviews). There is also the suggestion that some of these peptides have cytokine-like actions [22] and that their synthesis can be induced by LPS [24]. These host antibacterial peptides will be described in more detail later in this chapter.

Since the beginning of the 1990s a growing number of reports have revealed the complex interactions which occur between bacteria and the multi-cellular host, and the role of cytokines in this process. This has led to the realization that bacteria produce a wide range of proteins and low-molecular-mass components which have the ability to modulate eukaryotic cell cytokine synthesis, and that such interactions form part of a super-network of interactions between prokaryotes and eukaryotes [17].

9.3   Modulins, virokines, bacteriokines and microkines

As detailed in Chapters 4–6, the discovery of mammalian cytokines arose as a consequence of the study of the pyrogenic actions of endotoxin and LPS. The discovery of endogenous pyrogen, later termed IL-1, was the forerunner of the enormous number of cytokines that have been discovered since the late 1970s. To give some idea of the enormity of these discoveries, Ibelgaufts [25], in his recently published *Dictionary of Cytokines*, estimated that he was dealing with some 200–300 cytokines and associated receptors and binding proteins. Many of these cytokines play key roles in the hosts' defences against bacteria. Studies by the authors [17] on cytokine-inducing components of bacteria suggested that bacteria could produce an inordinately large number of components able to induce cytokine synthesis. Having reviewed the literature it was concluded that there were a number of potentially distinct classes of cytokine-modulating molecules emanating from bacteria, and these have been variously termed modulins, virokines, bacteriokines and microkines (Table 9.2). The role of such molecules in the interactions between bacteria and eukaryotic cells is detailed in this section.

9.3.1   Modulins

LPS and the LPS fraction with associated outer-membrane proteins (known as endotoxin) has been a major reagent in the discovery of the many cytokines that are now known to exist. The complexity of the mechanisms used by LPS to stimulate mammalian cells, via circulating and cell-membrane proteins such as LBP and CD14, has been described in Chapter 6, and the consequences of this requirement to act via CD14 are still being examined. The role that CD14

Table 9.2 Definition of the terms 'modulins', 'bacteriokines', 'virokines' and 'microkines'

Term	Definition
Modulin	A bacterial molecule that can induce host cells to synthesize and secrete cytokines, thereby modulating the behaviour of these cells in such a way as to induce tissue pathology. As such bacterial molecules induce tissue pathology, they are part of the bacterial virulence mechanism and thus form the fifth type of virulence factor, the other four being adhesins, impedins, aggressins and invasins.
Virokine	A cytokine-like molecule produced by viruses and able to inhibit some aspect of the host defence response. Virokines include many soluble forms of cytokine receptors, inhibitors of IL-1β-converting enzyme and some anti-inflammatory cytokines. It is not clear if these molecules have been evolved by the virus independently or whether, as seems more likely, they have been 'captured' from eukaryotic cells.
Bacteriokine	Bacteriokines, like virokines, are cytokine-like molecules produced by bacteria. The key attribute of these molecules, in our definition, is that they have cytokine-like properties and have the capacity to down-regulate pro-inflammatory cytokine networks. Non-proteinaceous inhibitors of pro-inflammatory cytokine networks may also exist and the nomenclature would have to incorporate this in due course.
Microkine	Any cytokine-modulating molecule produced by micro-organisms.

plays as a pattern-recognition system will be described later in this chapter. In the 1980s, reports began to appear suggesting that bacterial components other than LPS could stimulate cytokine synthesis. At the present time a very large number of bacterial constituents and secreted products, many of them proteins, have been shown to have the capacity to stimulate mammalian cells to synthesize and/or secrete mainly pro-inflammatory cytokines. The authors of this book were the first to collate this diverse literature [17,26–29]. Pro-inflammatory cytokines are mediators of tissue pathology. We therefore argued that bacterial molecules inducing the synthesis of such cytokines had to be virulence factors. In Chapter 3 the four classes of bacterial virulence factors — adhesins, impedins, aggressins and invasins — have been defined and described. Bacterial factors which induced host cytokine synthesis did not seem to fit into these categories and so we defined a fifth group of bacterial virulence factor — the modulins. The definition of a modulin is as follows: a bacterial molecule which by virtue of its ability to induce cytokine synthesis is able to 'modulate' the behaviour of cells [17,26–29] (Table 9.2). LPS would therefore be defined as a modulin, as would many other molecules including a number of bacterial exotoxins. These bacterial cytokine-inducing molecules have been described in detail in Chapter 7.

LPS is a relatively inactive molecule in the absence of the LBP and CD14 provided by the host. As described in detail in Chapter 6 the role of soluble or membrane-bound CD14 appears to be either to increase the biological response to LPS or to lower the threshold for response. CD14 is a very unusual form of receptor. In the first place it does not have an intracellular signalling domain and, as discussed in Chapter 6, the true signalling component of the LPS receptor has still to be discovered. It is possible that CD14 acts like a number of cytokine receptors (e.g. IL-6) in that receptor occupancy triggers the oligomerization of adjacent signalling proteins (with IL-6 this would be gp130) and it is this oligomerization which activates cells. The second peculiarity of CD14 is that it binds to a very large number of bacterial components, mainly polysaccharide-containing molecules. This has suggested to Pugin and co-workers that CD14 is an example of a pattern-recognition receptor [30]. This concept of pattern-recognition receptors was introduced by Charles Janeway [31] to explain how the innate host defence systems could recognize the large number of different organisms which can invade multicellular eukaryotic organisms to cause disease. As explained in Chapter 6, other pattern recognition systems, such as those recognizing bacterial DNA and the scavenger receptors, may also act to recognize invasion of bacteria into the host. The modulins binding to CD14 stimulate cellular cytokine synthesis and this action would appear to be part of this recognition phenomenon. This raises the question of the role of the other various bacterial molecules, both macromolecules and small-molecular-mass compounds — such as the recently described macrophage-activating compound S-(2,3-dihydropropyl)-cysteine of *Mycoplasma fermentans* [32,33] — which stimulate cellular cytokine synthesis. It is possible that these additional bacterial components, which would be released from bacteria along with the CD14-stimulating molecules, could either act in concert with the CD14-dependent system (additively or synergistically) or act to oppose the activity of this system. These molecules may therefore be part of a larger interactive system between bacteria and host cells (Figure 9.3). This raises the question of how many eukaryotic cell-surface receptors there are for bacterial modulins? Are there superfamilies of receptors, by analogy with CD14, which recognize particular 'classes' of modulins, or has every modulin a distinct receptor? Given that the number of known modulins is increasing rapidly, there could be a large number of eukaryotic cell-surface receptors devoted to recognizing these bacterial cytokine inducers.

One surprising finding from reviewing the literature on cytokine-inducing bacterial components was the discovery that most of the known exotoxins of bacteria were capable of inducing the synthesis of pro-inflammatory cytokines. Indeed, the most potent modulins turn out to be bacterial exotoxins (see Chapter 3 for details of exotoxins and their biological actions). As described in detail in Chapter 7 the most potent exotoxin is anthrax lethal factor which is active at attomolar concentrations. Pneumolysin from *Streptococcus pneumoniae* is active at femtomolar concentrations. This is significantly more active than the lowest reported concentrations of LPS. Indeed, some of these exotoxins appear to be more potent cytokine inducers than they

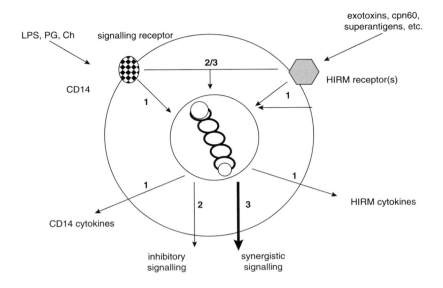

Figure 9.3 Potential interactions between the CD14 receptor system and the host immune response mediator (HIRM) receptors
Three possible sets of interactions exist: (1) the receptor systems do not interact at the post-receptor level; (2) they interact and have a negative interaction; or (3) they interact additively or synergistically. The number of possibilities will depend on the number of HIRM receptors on eukaryotic cells. Abbreviations used: Ch, carbohydrate; LPS, lipopolysaccharide; PG, prostaglandin.

are toxins. These findings add a new dimension to our understanding of bacterial exotoxins and strongly suggest that their activity is dependent on the ability to induce cytokine synthesis. The individual toxin may then synergize with the pattern of cytokines it induces. The best evidence in support of this hypothesis is the finding that Verotoxin (involved in the recent outbreaks of *E. coli* 0157 infection worldwide) causes the up-regulation of cytokine production and that it is these cytokines that induce the synthesis of receptors for the toxin, thereby allowing the toxin to bind to the target cells in sufficient amount to have biological effects [34]. It is not known whether this interaction between toxins and cytokines is a common 'motif' of the action of bacterial exotoxins (Figure 9.4).

9.3.2 Virokines, bacteriokines and microkines
One of the seminal discoveries which has led to the shift in our thinking about the interactions between micro-organisms and host cells has been the finding of cytokine-like proteins in the genomes of viruses. This literature has been reviewed in Chapter 7 and the term 'virokines' has been coined to describe these viral cytokines. One recent report has described four different cytokine-like genes in the genome of Kaposi's sarcoma-associated herpes virus (KSHV). This virus is related to the Epstein–Barr virus and herpes-virus saimari which,

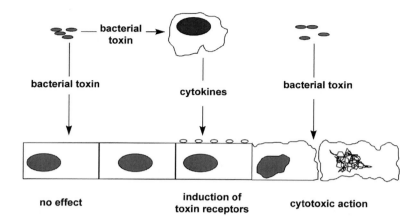

Figure 9.4 Scheme of the possible interactions which may occur between bacterial exotoxins and cytokines in order to produce the activity of the toxin

as has been described in Chapter 7, also contain virokines. In the case of KSHV, the cytokines encoded are: macrophage inflammatory protein (MIP)-I and MIP-II, IL-6 and interferon regulatory factor (IRF) [35]. In all examples of viruses containing virokines, the presence of these cytokine-like genes can be explained in terms of a defence mechanism against host systems designed to stop the virus replicating (see Chapter 7). For example, the presence of a functional IL-6 gene in KSHV is explained on the basis that IL-6 inhibits apoptosis and would therefore favour viral replication.

The molecular evolutionary relationship between these growing numbers of virokines and the cytokines they are mimicking has still not been deciphered and it is possible that many of these virokine-containing viruses have 'captured' the cytokine genes from the cells in which they replicate. This mechanism has been given the charming title of 'molecular piracy'. While it is currently perceived that the capture of cytokine genes can benefit the virus at the expense of the host, there are a number of examples which suggest that the presence of virokines can act in the hosts' favour as well. For example, as described in Chapter 7, vaccinia virus contains the gene product B15R which has 30% amino acid identity with the type I IL-1 receptor and binds IL-1β but not IL-1α. Disruption of this virokine gene inhibited the growth of the virus but enhanced the inflammatory response to virally infected cells and resulted in increased morbidity. Vaccinia virus also contains a gene, *crm*A, which encodes a serpin inhibitor of ICE. Thus, cells infected with vaccinia virus cannot produce the active IL-1β which would be utilized to induce local inflammation and activate the acute-phase response. Mutation of the *crm*A gene, to produce an inactive protein, resulted in a vaccinia virus which had a lower rate of replication, but at sites of infection there was a local increased inflammatory response. If the modified virus was injected intracranially, it killed more of the recipients than did the wild-type virus. One interpretation of these data is that the virus has evolved (or kidnapped) these anti-cytokine genes to limit the response of the host to its own immunoinflammatory defence systems [36,37]. This is not altruistic behaviour on the part of the virus, but makes evolutionary sense, in that killing the host is inimical to evolutionary survival.

If viruses are able to encode molecules which can disrupt cytokine networks this raises the question of whether the bacterial genome also contains similar genes, either as a result of evolutionary development or, possibly like viruses, by molecular piracy. Such molecules, if they exist, would have actions similar to those of virokines, i.e. they would act to inhibit the host immunoinflammatory responses to the bacterium. We have therefore given the term 'bacteriokine' [29] to such molecules, and to provide a generic term for molecules produced by micro-organisms which have the capacity to moderate immune and inflammatory mechanisms we have suggested the term 'microkine' [29].

Surprisingly little is known about the answer to this question of the existence of bacteriokines, as it is only within the past 3–4 years that bacterial proteins with cytokine-modulating actions have begun to be described. However, during this period around 15 proteins from a variety of bacteria have been demonstrated to inhibit the synthesis or actions of host cytokines. Many of these proteins are able to block the synthesis of pro-inflammatory cytokines such as TNF (as shown in Table 7.8). A few of these bacteriokines have been examined to determine something about their global ability to modulate cytokine networks. An undefined protein fraction from enteropathogenic strains of *E. coli* has been shown, using RT-PCR to identify cellular cytokine mRNAs, to inhibit the synthesis of lymphokines (including IL-2) while leaving the synthesis of pro-inflammatory cytokines intact [38]. A similar finding was made with a purified 14 kDa protein from *Actinobacillus actinomycetemcomitans* whose N-terminal sequence was defined [39]. These proteins may be exerting a specific effect on the acquired immune system while leaving the innate systems intact. Other bacteriokines may be targeted at the innate systems of immunity.

Another mechanism which has recently been found to be utilized by bacteria to control host responses is the use of bacterial proteinases. Proteinases are well-recognized factors in the pathology of a number of important idiopathic diseases of man, including emphysema, asthma, arthritis and cancer. A role for bacterial proteinases in nutrition and consequent bacterial growth has been recognized for some time. There is now evidence that bacterial proteinases can modulate inflammatory mechanisms by their released proteases. For example, bacterial proteinases can activate the kinin-generating cascade leading to the production of vasoactive peptides. On the other hand, bacterial proteinases can inhibit inflammation by inactivating the anaphylotoxin C5a [40]. As we have reviewed in Chapter 7, bacterial pro-teinases can proteolytically inactivate cytokines or can activate pro-forms of cytokines. Such proteinases can also cause the release of cytokine receptors from cells. The consequences of inactivating pro- or anti-inflammatory cytokines, activating pro-inflammatory cytokines and releasing soluble cytokine receptors (which would then act as cytokine antagonists), via bacterial proteinases, is almost impossible to define. One fascinating recent finding made by the authors is the cytokine-modulating actions of the R-1 proteinase of the oral Gram-negative bacterium *Porphyromonas gingivalis.* This proteinase is capable of proteolytically cleaving and inactivating cytokines such as IL-1 and IL-6. The proteinase is capable of self-cleavage and one of the cleavage products has itself the ability to stimulate cytokine synthesis. Thus, this one proteinase seems to carry with it the capacity to modulate cytokine networks (Sharp, L., Reddi, K., Wilson, M., Poole, S., Henderson, B., Curtis, M. and Tabona, P., unpublished work). These findings raise the possibility that bacterial proteinases could be therapeutic targets for the treatment of bacterial infections [42]. From the available evidence it would appear that bacterial pro-teinases can either be classified as modulins or bacteriokines or, indeed, both.

9.4   Cytokines and the normal microflora

Like History, which is the study of Kings and Queens, defeats and victories, leaving the ordinary man undocumented, so microbiology only documents those bacteria which cause disease to man or his domesticated animals, or which are used in industry. The authors contend that humanity has been guilty of hubris in its dealing with the world of micro-organisms. This began in the 1950s, with the belief that, with the discovery of antibiotics, infectious diseases were a thing of the past. The view also developed that bacteria were simple organisms incapable of cellular or molecular subtlety and by the 1970s and 1980s were only thought worth examining as workhorses for molecular biology. In the 1990s we are rapidly disabusing ourselves of these (almost Victorian) notions. We now realize that we have not beaten infections with antibiotics and that there is a recrudescence of many of the age-old infectious diseases. In addition, as described in Chapter 1 (Table 1.1) there is a constant stream of new diseases appearing. AIDS was the scourge of the 1970s and 80s and is still a major healthcare problem worldwide. In the 1990s the general population has uncovered an additional infectious 'organism' — the prion. In the U.K., as this book is being written, the population is under threat from an epidemic of spongiform encephalopathy due to the prion causing BSE. The prion appears to be the simplest form of infectious agent, being an infectious protein which causes pathology by misfolding identical host proteins. This shows that we still have many lessons to learn about infectious diseases and their mechanisms.

The rekindling of interest in microbiology has been sparked by discoveries in other spheres. For example, there has been an explosive growth of information on bacteria capable of living under extremes of conditions. These bacteria are known as extremophiles and are generally members of the third cellular kingdom Archaea. Such organisms are found in hot springs, in the mid-oceanic ridges and in Antarctica. Interest in such organisms is being fuelled by the pharmaceutical and biotechnology industries, particularly the need of the latter to produce patent-beating high-temperature polymerases to replace the *Taq* polymerase used in PCR. The studies of such organisms are revealing the role bacteria have played in the creation of the Earth as we know it. Fossil cyanobacteria have been dated back to 3.5 billion years and their photosynthetic capacity is believed to be responsible for the change of the Earth's atmosphere from a reducing environment containing volcanic gases to an atmosphere enriched in oxygen, thus allowing the evolution of life as we know it.

Homo sapiens is prey to infection by an estimated few hundred different micro-organisms [13,43]. However, the human body contains huge numbers of bacteria which live in apparent harmony with us. These bacteria, of which there may be up to 1000 different species, are variously called the 'normal flora or microflora', 'commensal micro-organisms' or 'indigenous microbiota'. The latter term is the most precise but the term, normal microflora, is in common usage and will be employed throughout this chapter (Figure 9.5). The normal microflora inhabits those parts of the body which are open to the environ-

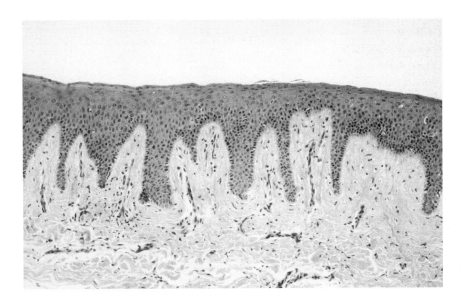

Figure 9.5 An epithelial surface on which the normal microflora lives and interacts

ment, including the skin, the various epithelial surfaces of the oral cavity, respiratory tract, gastrointestinal tract and the genital tracts. The ratio of the numbers of normal microflora to host cells at a single sample time is estimated to be 10:1 (i.e. 10^{14} bacteria to 10^{13} eukaryotic cells) [44]. The neonate is rapidly colonized by micro-organisms from the mother, and others assisting at the birth. Thereafter, there is an evolving succession of microbial colonizations of the various epithelial surfaces which presumably reaches a dynamic equilibrium at these various sites. It is surprising to think that our bodies, in terms of numbers of constituent cells, are more 'microbial' than 'mammalian'. It is possible to breed animals which do not contain this normal microflora. These animals are termed gnotobiotic (known life) and require to be kept under barrier conditions and fed germ-free diets throughout their lives. Such animals demonstrate a wide range of physiological, metabolic and immunological differences to their germ-laden counterparts. These differences are delineated in Table 9.3 which is taken from Tannock's recent monograph on the normal microflora.

The normal microflora is in dynamic equilibrium with the host and many members of this community are capable of causing disease if the conditions permit. This normally happens when an individual is in a debilitated state and nosocomial (hospital acquired) infections are a common sequelae of a run-in

Table 9.3 Some differences between germ-free and conventional rodents

Characteristic	Germ-free compared with conventional
Basal metabolic rate	Lower
Cardiac output	Less
Blood volume	Less
Small bowel mucosal surface area	Less
Small bowel mucosal turnover rate	Slower
Epithelial cell intracellular enzyme concentration	Higher
Lamina propria	Thinner
Small bowel wall mononuclear phagocytes	Less
Small bowel motility	Slower
Wet weight of caecum with contents	Greater
Urea in caecal contents	Present (none in conventional animals)
Ammonia in caecal contents	Less
Ammonia content in portal blood	Less
Iliocaecal lymph nodes	Smaller
Serum immunoglobulins	Less (some fractions absent)
Trypsin activity in large bowel contents	Present (inactivated in conventional animals)
Urobilinogens	Absent in germ-free animals
Coprastonol	Absent in germ-free animals
Short-chain fatty acids	Absent in germ-free animals
β-Glucuronidase	Absent in germ-free animals

Data from Tannock, 1995 [44].

with institutionalized medicine. However, such conditions cannot account for the set of diseases which, in aggregate, are known as the chronic inflammatory periodontal diseases (CIPDs) and which are second only to tuberculosis for worldwide prevalence (estimated at 15%) of a chronic bacterial disease. The periodontal diseases are caused by a number of the many Gram-negative bacteria which populate the oral cavity. The most common of the periodontal diseases is gingivitis which affects the whole population at one time or another. However, what leads to the development of the CIPDs, which are characterized by progressive destruction of the alveolar bone supporting the teeth, is not known.

In spite of this ability to respond to our own normal microflora, it is surprising that we, and multicellular eukaryotic species in general, have so little problem with these vast numbers of variegated micro-organisms. As has been described in detail in Chapter 7, all bacteria have a wide range of constituents which can induce mammalian cells to synthesize and secrete cytokines. The organisms which make up the normal microflora are no different in this respect. This raises the important question: how do multicellular organisms prevent themselves from responding to the normal microflora?

9.4.1   Interactions between the normal microflora and host epithelia

The normal microflora exists on a range of epithelial surfaces including the keratinized epithelium of the skin, the parakeratinized epithelia in the oral cavity (Figure 9.5) and the non-keratinized epithelia such as is found in the gastrointestinal tract. These epithelia are the first line of 'defence' against the commensal microflora. The term 'defence' is probably incorrect and the function of these cells is, we postulate, more likely to act to prevent the host responding to the bacteria constituting the normal microflora. This raises the obvious question: how do the epithelia do this? A clear problem, in this respect, is that it is now recognized that epithelial cells have the capacity to produce a wide range of cytokines, including the pro-inflammatory cytokines IL-1 and TNF, in response to bacterial constituents ([45] and Chapter 7). If cytokine synthesis were to occur in response to the large numbers of bacteria in their vicinity, it would result in the generation of inflammation at epithelial surfaces (Figure 9.6). The fact that this does not normally occur suggests that a more complex interaction is taking place between commensal bacteria and epithelial cells than has hitherto been recognized. We hypothesize that the answer to the question of how the normal microflora and host epithelia interact to prevent inflammation, is the presence, at epithelial surfaces, of a complex interactive network of cells (epithelial cells, other interacting eukaryotic cells and the bacteria) and mediators (bacteriokines, cytokines, other host factors including hormones, growth factors and the large number of antibiotic peptides which have been described in recent years) (Figure 9.7). What are the Popperian falsifiable criteria or predictions of such a hypothesis?

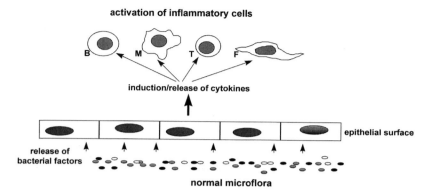

activation of inflammatory cells

induction/release of cytokines

epithelial surface

release of bacterial factors

normal microflora

Figure 9.6 The paradox of the interaction between the normal microflora and epithelia
Epithelial cells respond to bacterial constituents by producing cytokines. This ability acts as a warning of the presence of bacteria. However, if epithelia were to respond to their associated microflora they would produce inflammation. The question is how can epithelia interact with the normal microflora without becoming inflamed? Abbreviations used: B, B-lymphocyte; F, fibroblast; M, macrophage; T, T-lymphocyte.

An obvious prediction is that removal of one or other of the mediators described would lead to the generation of a pathological inflammatory state. Indeed, the first piece of evidence to suggest that host cytokine networks were involved in the control of the immune and inflammatory responses to the normal microflora was the reports that the knockout of the cytokines IL-2 or IL-10 in mice resulted in colitis or enterocolitis respectively [46–48]. These studies have been described in detail in Chapter 7. The key observation was that if IL-2-deficient mice were kept under germ-free conditions they did not develop the lethal colitis which affected congenic knockout mice with a normal microflora. This strongly suggests that the inflammation in the colons of these mice is due to an inflammatory response to the bacteria constituting the normal colonic microflora. Of interest, in this context, is a recent report which demonstrated that knockout of intestinal trefoil protein (a molecule with a characteristic three-leafed secondary structure), which appears to have protective and epithelial growth-promoting properties (and may therefore be a cytokine), also resulted in a lethal colitis [49]. It is of possible relevance that certain cytokines including IL-1, keratinocyte growth factor (KGF), fibroblast growth factors (FGFs) and the related oncogene Int-2, all have a β-trefoil structure [50]. These findings suggest that the absence of selected cytokines in the large and/or small intestine can result in a localized response to the normal microflora. If this is the case it would suggest that in non-transgenic animals the IL-2 (or IL-10) is involved: (i) in the down-regulation of inflammatory responses to the commensal microflora (and their released products/virulence factors); (ii) in the inhibition of the normal microflora from producing pro-

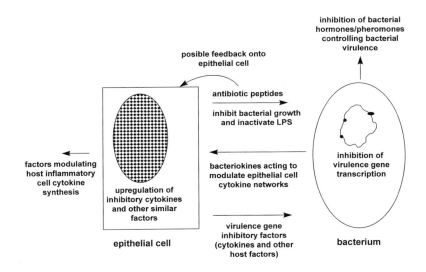

Figure 9.7 The role of bacterial anti-inflammatory proteins in the interaction between bacteria of the normal microflora and host
The synthesis and release of proteins which have such anti-inflammatory actions may mask the activity of pro-inflammatory bacterial constituents, thus allowing the organism to live in harmony with its host. In the example given, the bacterial pro-inflammatory factor is a CD14-binding moiety whose activity is blocked by a bacterial component, which either binds to CD14 and blocks this receptor or inhibits the synthesis of the pro-inflammatory cytokines induced by the CD14-binding event. The synthesis of such factors may be induced by host factors, and failure to produce the bacterial anti-inflammatory proteins would break this harmonious relationship and result in host responses. Bacterial meningitis may be the result of such a failure to produce the correct anti-inflammatory protein (see [114]).

inflammatory substances; or (iii) in a combination of both mechanisms (Figure 7.4). The cells producing the IL-2 and IL-10 are likely to be T-lymphocytes and macrophages, but not epithelial cells. The cells releasing the trefoil factor, for example, are goblet cells [49]. It is envisaged that the epithelial cell, acting as it does as a contact between the lumenal side of the gut and the mucosa, must play a role in controlling the synthesis of cytokines in the mucosa and must release (or allow the passage of) factors that can influence the disease-inducing potential of the normal microflora.

This hypothesis places the epithelial cells of the host at the centre of the control of the immunoinflammatory responses to the normal microflora. It also suggests a rethinking of our attitude towards inflammation. While John Hunter in the 18th Century stated that 'inflammation was a salutary process', most of us now regard it as a disease process. The processes of innate and acquired immunity are, it is assumed, constantly functioning to protect us from infecting micro-organisms without alerting us to this fact until they are overwhelmed and pathology ensues. However, we now have to consider the possibility that there is a constant level of inflammation going on at epithelial surfaces to allow the multicellular eukaryotic organism to cope with its own normal microflora. This inflammation is necessary for normal homeostasis and any interference with it could result in the host responding to its own normal microflora.

While the host may play a major role in controlling the immunoinflamma-tory responses to the normal microflora it is also likely that an enlargement of this hypothesis would take into account the findings over the past 5–6 years that micro-organisms can produce cytokine-like molecules which can control inflammatory responses. This has been most clearly defined with viruses which are now known to produce a wide range of soluble cytokine receptors, anti-inflammatory cytokines (IL-10) and inhibitors of cytokine convertases. Bacteria are now being shown to produce proteins and non-proteinaceous components which can block the synthesis or activity of cytokines. This literature has been reviewed in Chapter 7 and in [17,29]. For example, the authors have recently isolated and sequenced a novel 39 kDa protein from *Staphylococcus aureus* (termed SCIP; staphylococcal cytokine-inhibiting protein), which can prevent the activation of monocytes exposed to either LPS or IL-1. We have not ascertained the mechanism of action of this protein, but believe that it must inhibit LPS (i.e. CD14)-induced or IL-1-induced intracel-lular signalling pathways. Not all strains of *Staph. aureus* produce this protein and it is proposed that strains producing SCIP can block host responses to the pro-inflammatory components produced by the bacterium (Crean, St. J., Nair, S.P., Poole, S., Reddi, K., Taboni, P. and Henderson, B., unpublished work). However, failure to produce SCIP could render the organism likely to stimulate host defence responses and thus cause inflammation on epithelial surfaces leading to removal of the bacterium. We propose that the production of such anti-inflammatory proteins may be an important general mechanism for masking the pro-inflammatory effects of the normal microflora. Failure to produce such proteins could result in pathological consequences and this may be what occurs with the various bacteria known as opportunistic pathogens — organisms such as *Staph. aureus, Staph epidermidis, Strep. pneumoniae* and coliforms. It is also likely that the synthesis of such anti-inflammatory bacterial proteins may be influenced, or indeed directly controlled, by host factors. Then changes in the production of host factors (due to illness, stress or other causes) may directly or indirectly result in the down-regulation of bacterial anti-inflammatory factors and the generation of pathology (Figure 9.8).

At the time of writing 11 bacterial proteins have been reported to be capable of inhibiting the synthesis or action of pro-inflammatory cytokines (see Chapter 7, Table 7.8 and Figure 9.8). We believe that this will be the tip of an iceberg of bacterial proteins (bacteriokines) which are capable of controlling host cytokine networks. The induction of the synthesis of these proteins will, we predict, be dependent on host factors and will most likely be under some form of control by secreted products of other bacteria. It is important to

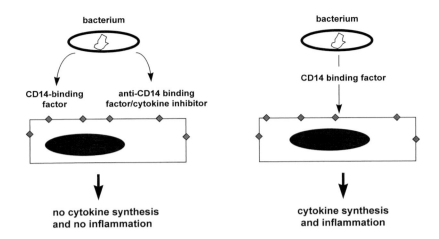

Figure 9.8 A proposal to account for the failure of epithelial surfaces in contact with the normal microflora to produce inflammation-inducing cytokines
A major proposed mechanism for inhibition of inflammation is the release of bacteriokines from bacteria. Such molecules act to modulate the cytokine-inducing activity of the epithelial cell, so that it does not produce pro-inflammatory cytokines, but produces instead anti-inflammatory cytokines, which inhibit the production of anti-inflammatory networks by neighbouring inflammatory cells. In turn, the epithelial cells produce antibiotic peptides which inhibit bacterial growth and inactivate LPS. These peptides may have cytokine-like properties and may feedback on to the producing cell. We also propose that the epithelial cells produce factors, probably cytokines (but also hormones and other cell metabolites), which inhibit the transcription of bacterial virulence genes.

remember that the normal microflora is a collection of mixtures of diverse Gram-positive and Gram-negative species which have 'evolved' to live in particular niches in the multicellular eukaryote (such as the oral cavity, gut, vagina, etc.; see Figure 2.1 in Chapter 2). An important prediction of our hypothesis is that the bacteriokines will be able to control not only the synthesis or activity of individual cytokines but will be capable of controlling tissue-specific cytokine networks. The authors envisage that bacteriokines, either individually, or perhaps in concert (i.e. bacteriokines from a number of bacteria in a particular habitat), can regulate the synthesis and action of cytokines in such a manner as to produce non-inflammatory cytokine networks. Any trend in the network towards pro-inflammatory activity, a situation which could be considered a state of disequilibrium, would be re-equilibrated by the bacteriokines. A further prediction which follows on from the previous statement is that bacteriokines, by controlling pro-inflammatory cytokine networks, will have potential as therapeutic agents and, once their mechanism of action is delineated, they could be mimicked by low-molecular-mass isosteres with better pharmacokinetic and pharmacodynamic properties. This opens up the possibility of developing truly effective inflammation-modulating drugs rather than the present goal of anti-inflammatory drugs.

Indeed, if our predictions are correct that there is a constant level of inflammation (possibly this should be termed 'productive inflammation') at epithelial surfaces, then agents which block inflammation may be deleterious and upset host–microflora interactions leading to pathology. Indeed, in a world in which ingestion of non-steroidal anti-inflammatory agents is commonplace, it is possible that much of the bacterial infection seen is an example of iatrogenic medicine. The secret of blocking inflammation associated with pathology may lie in re-equilibrating cytokine networks which are in disequilibrium.

9.4.2   Interaction between CD14 and bacteriokines/modulins

The role of CD14 as a pattern-recognition system warning the organism that it has been invaded by bacteria has been described in Chapter 7 and has briefly been alluded to earlier in this chapter. This CD14-dependent system also acts at mucosal surfaces as the LPS-unresponsive C3H/HeJ mouse remains chronically infected with Gram-negative bacteria while normal mice can clear such infections within 2 days [49]. The question of the nature of the receptors for the many other bacterial molecules (modulins and bacteriokines) which modulate cytokine induction arises at this point in relation to the CD14 system. The great unknown is the number of receptors that eukaryotic cells have for cytokine-inducing bacterial components. CD14 is the best studied eukaryotic receptor for bacterial components. Only a few other receptors for bacterial macromolecules have been defined. Bacterial superantigens, for example, bind to MHC class II receptors on T-cells. Certain bacterial exotoxins bind to defined receptors such as the globoseries (see Chapter 7). The type I fimbriae of *E. coli* have been shown to bind to two major glycoproteins (the uroplakins) on uroepithelial cells [52]. However, it is not known if these are signalling receptors. It is not possible to accurately predict if there are only a few pattern-recognition receptors for these non-CD14 modulins and bacteriokines, or whether there are a very large number of receptors evolved to cope with the mass of bacterial signalling molecules. Given that all eukaryotic cells contain constituents (mitochondria, molecular chaperones, etc.) from the two major microbial kingdoms (Bacteria and Archaea), and have evolved in competition, certainly with the Bacteria, for the past 500 million years, it would not be a surprise to find receptors for many bacterial constituents on these cells. The major question which arises is what is the interaction, if any, between the CD14 receptor and the various other receptors for bacterial constituents? These non-CD14 receptors, once occupied, result in the synthesis (modulins) or in the inhibition of synthesis (or action) of cytokines (bacteriokines). If these non-CD14 receptors are not also pattern-recognition receptors they certainly have a host-immune response modifying (HIRM) activity and we have named them as such in a recent article [17]. These HIRM receptors presumably have some form of interaction with CD14 which, depending on the receptor and agonist, may be additive, subtractive or positively or negatively synergistic (Figure 9.3). We propose that the HIRM

receptors are activated on epithelial surfaces to create the productive inflammatory state which allows the maintenance of the normal microflora without
induction of pathological inflammation. These receptors may also be activated
if bacteria enter into the tissues of the host, but the consequences of such
activation are more difficult to predict. It is possible that the inappropriate
activation of HIRM receptors in tissues may result in pathology. For example,
Gram-positive streptococci which reside in the mouth (e.g. *Streptococcus*
sanguis and *Streptococcus oralis*) have no well-described virulence factors.
However, if these bacteria gain access to the heart they can cause the life-
threatening condition of infective endocarditis. One possible reason for this is
that the proteins produced by these bacteria, and which modulate the oral
inflammatory response, may, in the tissues of the heart, give rise to a pathological response and so produce the set of pathological reactions known as
infective endocarditis.

As described in Chapter 6, the activity of LPS is dependent on a range of
serum factors including LBP, BPI, and soluble CD14. Do a similar set of interactions occur with the many other modulins and bacteriokines now known to
be produced and released from bacteria? Whether such interactions with
soluble proteins of the host are additive, synergistic, etc. will, we predict, be
the subject of much future study.

9.5   Interactions between cytokines, other host factors and bacteria

Thus far in the discussion we have focused on the interaction between bacterial
components and host cells and tissues and have only alluded to the possibility
that host molecules could feedback on to bacteria and modulate their function.
It is now being rapidly established that the interaction between bacteria and
relevant host cells can switch on virulence genes, and two pertinent recent
articles have been described at the beginning of this chapter [8,9]. Soluble
components of mammalian cells can also alter bacterial metabolism and a very
simple example of this is our finding that growing *Strep. sanguis*, in horse-
blood-containing agar, switches on the synthesis of a number of exported
proteins. The activity and nature of these proteins is currently being investigated. Our assumption is that in the presence of mammalian blood proteins the
bacterium recognizes that it is in a particular environment and switches on
relevant genes which may act as virulence factors.

Cytokines are major signalling proteins in eukaryotes. A key question is
whether these signals can be recognized by bacteria. However, before considering this question it is important to recognize that a fundamental measure in
microbiology — bacterial growth — is increasingly being re-evaluated and is
starting to appear as a much more complex control process than was
previously believed.

9.5.1   Control of bacterial growth

For the past century or more the basic technique in microbiology has been axenic (i.e. pure) culture and it has been assumed that each bacterium in culture can multiply independently of every other bacterium, given an appropriate supply of building blocks and the correct environmental conditions (pH, E_h, O_2, CO_2, etc.). Of course it has been recognized that pure cultures of bacteria rarely, if ever, occur in Nature. Mixed cultures, competing, red in tooth and claw, are the norm in the prokaryotic world and it is only in the past two decades or so that studies have begun to be made of mixed cultures.

While axenic culture, with its classic lag — exponential — stationary phase, is the foundation of microbiology, there has been a re-evaluation in recent years of bacterial growth and there is evidence that the assumptions of axenic culture are not wholly supportable and that bacterial growth depends on additional factors including growth factors produced by bacteria and by other cell populations with which bacteria associate. As an example, siderophores, which are iron-binding and transport molecules, act as growth factors in *Bacillus* cultures [53]. Many Gram-negative bacteria produce low-molecular-mass diffusible molecules known as acyl homoserine lactones which serve as signals for the process known as quorum sensing — a system for the density-dependent expression of specific sets of genes ([54] and Chapter 3). These acyl homoserine lactones are synthesized from S-adenosylmethionine and an acylated carrier protein by a specific enzyme, acyl homoserine lactone synthase [55]. N-(3-Oxohexanoyl)homoserine lactone has been shown to act as a growth factor for the regrowth of starved *Nitromonas europa* [56]. In a recent review on the subject of cell-to-cell communication in bacterial growth, Kaprelyants and Kell [18] coined the term 'microendocrinology' for the study of the factors produced by bacteria which regulate bacterial growth. These workers have reviewed the literature to show that a number of vertebrate-like hormones, including steroids and insulin, are associated with bacteria and that receptors for certain hormones including the pheromone-binding lipocalins are found in prokaryotes. Other hormone or transmitter-like molecules associated with higher organisms, including catecholamines, insulin, serotonin and a chorionic gonadotrophin-like ligand, have also been reported to stimulate the growth of particular bacteria [17].

9.5.2   Do cytokines have any control over bacterial growth?

Bacterial growth is clearly more than a simple linear response to the supply of nutrients and it appears increasingly likely that it can be controlled by specific bacterial and host factors. Do cytokines play any role in the growth of bacteria, and could these molecules act to control the numbers of commensal microflora at epithelial surfaces? Answers to these key questions have not yet been provided. In the past 5–6 years a trickle of studies have begun to establish the interactions between cytokines and bacteria in terms of the growth of the latter. These studies have focused largely on the effect of cytokines on intracellular bacteria, but there are a few studies of the actions of purified or cloned cytokines on the growth of free-living bacteria and on the expression of HIV.

HIV infection induces the synthesis of the pro-inflammatory cytokines IL-1 and TNF and this appears to be the beginning of a feedforward loop control system with both IL-1 [57] and TNF [58] having the capacity, albeit by different mechanisms, to promote the transcription of HIV. IL-1 and IL-6 (but not TNF) synergized in promoting HIV expression and the activity of IL-1 could be blocked by the anti-inflammatory cytokines TGFβ and IL-1ra [57].

While it was not unexpected that cytokines could promote the growth of viruses, the first indication that they could have an effect on the growth of bacteria was a report from Charles Dinarello's laboratory that IL-1 stimulated the growth of 'virulent' strains of *E. coli*. Dinarello is one of the pioneers of endogenous pyrogen/IL-1 research and his role in such research has been discussed in Chapters 5 and 6. Comparing six virulent and four avirulent strains of *E. coli*, the former showed enhanced log growth in terms of colony-forming units in the presence of IL-1 concentrations as low as 10 ng/ml. Avirulent strains did not respond to IL-1. IL-4 and TNF had no effect on the growth of these virulent strains. Of significance was the finding that the IL-1-induced bacterial growth could be almost totally abolished by IL-1ra. These findings suggested that the bacterial growth was not due to simply adding more protein to the cells and the inhibition of the growth-promoting effects of IL-1 by IL-1ra suggested that there was an IL-1 receptor involved. Using radioiodinated IL-1, it was found that the virulent strains of bacteria bound radioactivity in a cell density-dependent manner, while the avirulent stains did not bind the label. Binding of IL-1 to the bacteria was saturated at a concentration of 20 pg/ml. Labelled IL-1 could be competed by an excess of unlabelled IL-1 and the kinetics of desorption suggested that bacteria contained $(2-4) \times 10^4$ binding sites per bacterium, which is significantly greater than the numbers of IL-1 receptors present on human or murine cells [59]. These findings are extremely interesting and have been controversial as another group reported that they could not reproduce the results [60]. Dinarello then reported that freshly isolated virulent strains rapidly lost their ability to respond to IL-1, which may account for this inability to reproduce the results [61]. It also suggests that the induction of the IL-1 binding and transduction system is brought about by host factors and is rapidly lost in the absence of such signals. Another group, this time in England, reported that a virulent strain of *E. coli* responded by increasing its growth rate if incubated in the presence of IL-2 or GM-CSF. The effects of IL-2 could be blocked by heating the cytokine or by adding an IL-2-neutralizing antibody [62]. The possibility that this strain of *E. coli* contained receptors for IL-2 or GM-CSF was explored indirectly by the demonstration that the bacteria depleted the medium of IL-2. It was established that this depletion was not due to proteolytic cleavage of the IL-2 [63]. This bacterial growth-promoting effect of IL-2 could possibly explain the high levels of opportunistic infections found in patients receiving IL-2 for cancer or AIDS treatment [64,65]. IL-2 has also been reported to bind to the yeast *Candida albicans*, which is a major pathogen causing mucocutaneous infections. The bound IL-2 was still biologically active, suggesting that the fungal receptor was different from that used by mammalian cells [66].

Klimpel and co-workers [67,68] have reported that certain Gram-negative bacteria (*Salmonella typhimurium, Shigella flexneri* and *E. coli*) have receptors for TNFα. Radioiodinated TNFα bound to these bacteria and it could be competed by cold (i.e. unlabelled) TNFα but not TNFβ (lymphotoxin). Binding of TNF could also be inhibited by trypsin, treating the bacteria or heating them to 52°C for a few minutes, suggesting that the receptor was a protein; however, it was not affected by monoclonal antibodies to the 55 or 75 kDa TNF receptors. These antibodies also failed to bind to Western blots of the whole bacteria. Binding of TNF to bacteria appeared to enhance their uptake into HeLa cells or macrophages, but it was not ascertained whether the bound TNF was still biologically active. In addition, the potential growth-promoting properties of TNFα were not checked.

The most widespread chronic infectious disease is tuberculosis, in which the causative organism *Mycobacterium tuberculosis* resides in macrophages and produces granulomatous inflammation. As will be described, a number of studies have attempted to ascertain the effect of cytokines on the intracellular growth of mycobacteria. However, a recent report has established that *M. tuberculosis* and *M. avium,* when cultured extracellularly, can be stimulated to grow by recombinant human epidermal growth factor (EGF) with significant enhancement of growth occurring at concentrations of 50 ng/ml [69]. However, EGF had no influence on the growth of bacteria residing inside macrophages. Radioiodinated EGF bound to bacteria where it could be competed with unlabelled EGF, and Scatchard analysis revealed that the receptor on *M. avium* had a K_D of 2×10^{-10} M and that there were 450 ± 60 receptors per cell. The affinity of this bacterial EGF receptor is similar to that of the high-affinity 170 kDa EGF receptor found on mammalian cells [50]. Isolation and cloning of this mycobacterial EGF receptor revealed a 37 kDa protein with significant homology to the glyceraldehyde-3-phosphate dehydrogenase (GAPD) of group A streptococci. In streptococci, GAPD (which we all know as an enzyme of the glycolytic pathway) acts as a receptor for plasmin [70], fibronectin, lysozyme and cytoskeletal proteins [71]. Indeed, GAPD appears to have a wide variety of functions that do not require its enzymic actions.

The evidence is provisional, but suggests that bacteria can respond to cytokines in terms of their growth rates and also that some bacteria or strains of bacteria can synthesize specific receptors for cytokines. In the case of mycobacterial species the receptor is a high-affinity protein, and binding to this receptor induces increased growth rates. So far, no-one has shown that bacteria of the normal microflora contain cytokine receptors or respond to cytokines. Another possible interaction between cytokines and bacteria has been suggested by George [72]. Given that many cytokines bind to components of the extracellular matrix such as proteoglycans, George has hypothesized that bacterial surface carbohydrates may function as binding sites for cytokines.

In addition to the demonstration that they can bind to and modulate the growth of free-living bacteria, a number of studies have shown that cytokines can influence the growth of intracellular bacteria such as *Mycobacteria* spp., *Brucella* spp. and *Legionella* spp. Studies of the growth of *M. tuberculosis* or *M. avium* in human or murine macrophages has shown that cytokines such as IL-1, IL-3, IL-6 and TGFβ$_1$ can stimulate bacterial growth, while TNFα and IFNγ inhibit growth [63,73–76].

Intracellular growth of *Brucella abortus* in murine macrophages was not influenced by IL-1α, IL-4, IL-6, TNFα or GM-CSF. However, IL-2 or IFNγ caused a reduction in the numbers of intracellular bacteria [77]. TNFα also inhibited the growth of *Legionella pneumophila* in human peripheral blood monocytes. Inhibition of endogenous TNFα synthesis by pentoxyfylline caused an increased rate of bacterial proliferation [78].

It is clear that microbiology has barely scratched the surface of the study of the interactions of cytokines with bacteria. We predict that this will become a key area of modern cellular microbiology. Thus, at the epithelial cell surface we envisage that the following complex network of interactions exist: (i) signals (modulins and bacteriokines) passing from bacteria to epithelial and post-epithelial cells to modulate cytokine networks; (ii) cytokines and other mediators passing from the epithelial layer and cells behind this layer, acting to influence bacterial behaviour — largely acting as transcriptional control regulators; and (iii) a group of disparate peptides which are currently recognized for their antibiotic actions (Figure 9.7 and Table 9.4). The next section will briefly review the possible role of these peptides in controlling the normal microflora.

Table 9.4 Classification of antimicrobial peptides

Classification	Examples
Cysteine-rich, amphiphilic β-sheet peptides	β-Defensins, protegrins, tachyplesins
Cysteine-disulphide ring peptides with or without amphiphilic tails	Bactenecin, ranalexin, brevinins
Amphiphilic α-helical peptides without cysteine	Magainins, cecropins
Linear peptides with one or two predominant amino acids (e.g. proline or tryptophan)	Bac 5, Bac 7, PR39, indolicidin

9.5.3   Antibiotic peptides and the normal microflora

The majority of the multicellular animals inhabiting the Earth do not have an acquired immune response with its specialized gene-juggling T- and B-lymphocytes and the subpopulations of Th_1 and Th_2 cells. These animals have to rely on innate immune systems and, in evolutionary terms, this seems to have been sufficient for the past 500 million years. Innate immunity consists, at least in *Homo sapiens*, of various cell-based barriers, such as the skin and their associated antibacterial constituents including lactic acid, fatty acids, lysozyme, mucus and spermine. Once bacteria have breached the barrier, they come into contact with the cellular base of innate immunity — the phagocytosing/bactericidal neutrophils and monocytes/macrophages. These cells are activated by a plethora of cytokines and aided by a variety of soluble factors such as the complement pathway and the synthesis of the opsonizing/anti-proteolytic acute-phase proteins. In mammals, this innate system is deemed to be a stopgap measure before the acquired system kicks into action. However, for most of the Earths' inhabitants this is all they have.

How do insects defend themselves from bacterial infections and how do mammalian phagocytes kill bacteria? These questions have led to the discovery of a large number of invertebrate and vertebrate peptides which have antimicrobial activity and to the realization that innate defence systems have a direct antibiotic armament (see reviews [21–23] and [79–83]). Boman, the pioneer of this field, isolated the first insect antibiotic peptide in the early 1980s and named these molecules 'cecropins' [84]. The first mammalian cecropin was isolated from the small intestine of the pig in 1989 [85].

Antimicrobial peptides have now been found in mammals, birds, amphibia, insects and even plants. These peptides have diverse structures but most are cationic amphiphilic molecules which interact with microbial surfaces, often leading to the formation of pores or in some way causing membrane permeability and impairing the normal functioning of the cell. It is possible to group these molecules into four or five different structural groupings (Table 9.4). It is clear that these antibacterial peptides are active against Gram-negative and Gram-positive bacteria. Some of these peptides, such as the classic defensins, have a much wider range of activity being able to kill fungi, certain viruses and protozoans [20,23].

The role of the epithelial cell barrier in defending the organism against infection has been discussed earlier in this chapter and the possible interactions between epithelial cells and the normal microflora have been hypothesized. We now have to add to this picture the antibiotic peptides, as there is good evidence that epithelial cells from the gut, airway and tongue produce such antibiotic molecules [86–89]. The role of such peptides in controlling the normal microflora is an unexplored question and no gene knockouts have been reported. The only piece of evidence that such peptides may play a role in this control is provided by the experiment in which *Drosophila* were administered the protein synthesis inhibitor cycloheximide. This resulted in *E. coli*, which is part of the normal microflora of *Drosophila*, overgrowing and killing the insects [90]. Of course this experiment does not rule out other mechanisms for the bacterial overgrowth and death.

Recent studies of the antibiotic peptides of mammals have begun to show that there is a relationship between these peptides, LPS and cytokines. For example, there is growing evidence that many of the antibiotic peptides are inducible by LPS. Bovine tracheal epithelial cells demonstrated a marked up-regulation of the synthesis of tracheal antimicrobial peptides (TAP) when cells were exposed to LPS. These cells constitutively express CD14 and induction of TAP could be blocked by a monoclonal antibody to CD14 [24,88]. These antimicrobial peptides resemble cytokines in being inducible by LPS via CD14. Of interest in this context is the report that the silkworm *Bombyx mori* produces a bacterially inducible protein, Gram-negative bacteria-binding protein (GNBP), which has amino acid sequence similarity to CD14 and is recognized by at least one CD14 polyclonal antiserum [91]. Two recombinant insect antibiotic peptides, MBI-27 and MBI-28, have been shown to bind to LPS with similar affinity to that of polymyxin B and to be able to inhibit experimental septic shock in animals [92]. These peptides are considered to have pharmaceutical potential for the treatment of Gram-negative septic shock.

One of the repeatable findings in modern biology is the rapidity with which one group of defined proteins or peptides soon shows activity in other areas of biology. This multiplicity of function is now being observed with the antibiotic peptides. The finding that they can, like CD14, bind LPS, is one example of this multiplicity of action and role in controlling inflammation. Another activity which some of these peptides have overlaps with that of certain cytokines. Thus, the classic defensins, human neutrophil peptide (HNP)-1 and HNP-2 (but not HNP-3) and CAP37 [93,94], and the β-defensin, proBAC7 [95], have been reported to have chemotactic properties. Thus, certain of the antimicrobial peptides are chemokines. Interestingly, the defensins are chemotactic for T-lymphocytes and this finding ties together this arm of the innate immune response with the key cellular element of acquired immunity. This finding is interesting in light of recent discoveries that certain chemokines can block the uptake of HIV into CD4 lymphocytes (reviewed in detail in Chapter 7). Do antibiotic peptides have similar anti-retroviral activity? Other defensins have growth-factor-like properties [96,97]. Another fascinating action of antibiotic peptides is the potent inhibition of cortisol synthesis by cultured adrenal cells due to their ability to block ACTH (adrenocorticotrophic hormone) action at the receptor [98]. A recently recognized activity of the proline-arginine-rich antibiotic peptide, PR39, is that it inhibits the reactive oxygen-producing neutrophil NADPH oxidase, not directly, but by inhibiting assembly of the enzyme by interacting with the Src-homology 3 domains of a cytosolic component termed p47[phox] [99].

The defensins obviously have a range of actions on eukaryotic cell systems in addition to their antibiotic ones. It is obvious that further biological actions of these peptides will be discovered and we would predict that, at epithelial surfaces, they could form part of the interactive signalling which exists between the normal microflora and the host epithelial layer. It is interesting to note that in severe infections high levels of defensins are recorded in the blood [100].

In a world rapidly running out of antibiotics, the discovery of this large number of natural eukaryotic examples is a proverbial goldmine. Already a number of companies such as Magainin Pharmaceuticals, MicroLogix Biotech and Intrabiotics Pharmaceuticals are developing these peptides as pharmaceuticals [101].

9.6   Microbial autoimmunity?

Science is an amazingly recent creation of the human mind — and spirit — (this view is well argued by Cromer [102]) and one popular current view of the scientific method is largely based on the writings of the philosopher of science, Karl Popper, who has clearly tolled the death knell of scientific inductionism. Popper has provided an operational definition of the scientific method which involves the creation of hypotheses and their critical testing in order to try and falsify such hypotheses. Hypotheses which cannot be so tested are non-scientific. Furthermore, hypotheses carry with them various predictions. The more 'interesting' the hypothesis and the richer it is in predictions the better it is. Each prediction is a test of the hypothesis.

The authors of this book have proposed the hypothesis that the ability of multicellular organisms to support large numbers of bacteria on their epithelial surfaces is due to the existence of a network of chemical interactions between the normal microflora, epithelial cells and other cells (vascular endothelium, monocyte/macrophages and lymphocytes) associated with the epithelium. This implies that the body can recognize 'friend' from 'foe' in terms of normal microflora versus exogenous pathogens. It is obvious that this recognition can break down. The finding that a number of bacterial species, e.g. *Staph. aureus*, *Neisseria meningitidis*, *Haemophilus influenzae* and various streptococcal species, can be part of the normal microflora, but can also cause severe disease (e.g. meningitis; see Chapter 3 for a fuller discussion) is one example where this mechanism breaks down. It is not clear what bodily changes switch such bacteria from being harmless passengers to destructive hijackers, but it is important that we discover what they are and how they affect the bacterium.

One prediction from the hypothesis proposed is that there is a recognition system in existence to allow multicellular organisms to detect exogenous pathogens while recognizing the normal microflora as 'self'. If this was correct it would imply that there was a simple equivalent of the Janus system of mammalian acquired immunology which has to be able to identify self in order to recognize foreignness. The consequence of a breakdown in this recognition system is autoimmunity [103]. Self-recognition appears to be controlled by the dual processes of clonal deletion and inactivation of self-reactive lymphocytes. However, there is evidence that there is a constant background of autoimmunity with antibodies to many proteins being found in the serum. With bacterial infections, the response has to be limited to the offending organism without allowing the host to respond to the many other bacteria it carries. Thus, the innate system of immunity may have some degree of selectivity in terms of recognizing bacteria of the normal microflora as opposed to exogenous pathogens. Indeed, many of the most dangerous bacteria are those which produce exotoxins. The finding that most of these exotoxins are potent inducers of cytokine synthesis may provide a signature for recognition allowing that particular organism to become the target of attack. If such a self-recognition system existed could it produce 'microbial autoimmunity'? The opportunistic pathogens described earlier in this section (*Staph. aureus*, etc.) and the diseases they induce may be examples of this self-reactivity. Furthermore, if such a 'self-recognition' system existed, would it have any impact on the induction and pathogenesis of autoimmune disease proper?

A number of autoimmune diseases have been thought to be due to infectious organisms. The classic example is the induction of autoantibodies to heart, joints and kidney in individuals who are infected with *Streptococcus* species. Rheumatoid arthritis has, for many decades, been believed to be induced, or even driven [104], by some infectious agent. In recent years it has been suggested that this disease is due to cross-reactivity with the molecular chaperones of gut bacteria [105]. The belief in the infectious aetiology of the natural autoimmune diseases is bolstered by the various experimental forms of autoimmunity, which require that the autoantigen has to be administered as an emulsion with some bacterium, normally *M. tuberculosis*. In the absence of the bacteria no autoimmunity is induced [106]. While much attention has been paid to *M. tuberculosis* in experimental autoimmunity, it is unlikely to play a significant role in the autoimmune conditions found in Western populations in which, at least until relatively recently, tuberculosis was almost extinguished. Thus, if one were to look for bacterial culprits to account for autoimmune cross-reactivity, it would be more likely to be the bacteria which cause more common infections, for example the opportunistic pathogens that form part of the normal microflora. Induction of autoimmune T-cells is normally prevented because the self-antigen has to be presented along with the co-stimulatory molecule B7. In the absence of B7, antigen recognition leads to anergy or deletion of mature T-cells. However, as has been described in Chapter 7, porins from *Neisseria* species are able to induce B-lymphocytes to express B7 [107] and thus could provide this second signal for induction of self-reactive T-cells. Porins are also potent inducers of cytokine synthesis [108]. This idea has, in addition, been considered from the immunological viewpoint by Gianani and Sarvetnick [109], who have suggested that local induction of pro-inflammatory cytokines in infections plays a pivotal role in the loss of functional tolerance to self-antigens. The role of IL-12 in inducing Th_1 CD4 responses to infectious agents has been touched on earlier in this chapter [14]. It turns out that IL-12 is also a key factor in experimental autoimmunity. In diseases such as experimental allergic encephalomyelitis, TNBS (2,4,6-trinitrobenzene sulphonic acid)-induced colitis and diabetes in the non-obese diabetic mouse, neutralization of IL-12 alleviates the symptoms of disease, while administration of IL-12 exacerbates disease. Thus, in infections leading to the production of high levels of IL-12 and of B7, this combination (along with other bacterial factors) may be possible to trigger and perpetuate self-reactive T-cells. As we postulate that the normal microflora has a role to play in the control of cytokine networks then, indeed, these bacteria may have some controlling influence over immunological self-reactivity.

9.7   Conclusions

Research into the interactions between bacteria and cellular cytokines has been an area of relative activity over the past decade or so, and yet we still do not understand the complex orchestration which is taking place and which underpins much of our research into immune diseases generally [17]. We predict that there will be significant advances within this area. This will be helped by the outpouring of complete bacterial genome sequences and results from the Human Genome Project. The key observation and advance which has been highlighted in this book is that the micro-organism is not simply a passive stimulator of immune and inflammatory mechanisms, but is an active player producing a very wide range of molecules that can modulate the extremely complex cellular and molecular mechanisms which we group together as inflammation and immunity. Micro-organisms produce modulins, bacteriokines and virokines, collectively known as 'microkines'. These molecules, in aggregate, constitute a large number of genes and we believe we are only seeing the tip of the iceberg of the actual number of microkines. The virokines, which are immunoinhibitory cytokine-like molecules from viruses, may have been kidnapped from the eukaryotic host by an act (or acts) of 'molecular piracy'. The modulins and bacteriokines are unlikely to have the same source. As bacteria may have evolved before eukaryotic cells it is possible that these cytokine-modulating bacterial proteins may be the evolutionary ancestors of eukaryotic cytokines. It is now increasingly being recognized that bacteria are complex cells which interact with each other through proteinaceous [110] and non-protein [111] signals. An interesting example is the virulence locus in the opportunistic pathogen *Staph. aureus* which is controlled by an octapeptide pheromone [112]. These bacterial hormones or pheromones could be the evolutionary ancestors of our mammalian cytokines. It is not only micro-organisms that have developed strategies for modulating cytokine-controlled cell behaviour. *Leishmania* spp., which are obligate intracellular parasites discussed briefly earlier in this chapter, have the capacity to inactivate macrophages and thus live within them. Much of the inhibitory capacity of this organism is thought to be due to a surface lipophosphoglycan. This molecule can inhibit IL-1β gene transcription via a unique promoter gene silencer [113] and entry of *Leishmania* into macrophages is associated with the synthesis of TGFβ, a cytokine known to deactivate macrophages [114]. Lipophosphoglycan is a complex molecule consisting of four distinct domains. It has recently been demonstrated that one of these domains exerts an inhibitory effect on vascular endothelial cells preventing the up-regulation of endothelial cell adhesion (due to E-selectin, ICAM and VCAM) in response to endotoxin [115]. This finding probably accounts for the lack of vascular inflammation at the initial site of leishmanial infection. Thus, it is clear that at both ends of a spectrum of parasitic organisms — the viruses and protozoan parasites — mechanisms have evolved to inactivate or defeat cytokine-driven innate defence mechanisms. One expects as much of the bacteria.

As we approach the millennium we predict that a major growth area in microbiology and medicine will be fuelled by the growing incidence of antibiotic-resistant strains of bacteria and the realization that in certain diseases (e.g. septic shock and bacterial meningitis) antibiotics can cause the release of pro-inflammatory biological components [116]. Thus, the future treatment of infectious diseases may involve a combination of antibiotics and cytokine-modulating adjunctive therapy. The use of a variety of cytokine-modulating modalities has been attempted in the treatment of septic shock and the general failure of such agents has been reported in Chapters 6 and 8. Other life-threatening bacterial diseases may benefit from the use of anti-cytokine agents, and a recent review [117] has highlighted the adjunctive anti-cytokine therapy used or proposed for use in the treatment of the very serious infection, bacterial meningitis. Various cytokine-blocking agents have been shown to have therapeutic benefit in the treatment of experimental meningitis. The only anti-cytokine agent used clinically in the treatment of bacterial meningitis is the glucocorticoids. These agents have promising effects on the neurological and audiological complications of bacterial meningitis [117]. It is believed that the use of adjunctive anti-cytokine therapeutics will become standard for the treatment of severe infectious conditions to limit the damage due to overstimulation of the innate and acquired immune systems. Indeed, if resistance to antibiotics increases to the ultimate extent, then renormalization of cytokine networks allowing clearance of the bacteria without damaging the patient will become the only treatment of choice. This is why it is so crucially important to understand the essence of bacteria–cytokine interactions in health and disease.

References

1. Daser, A., Mitchison, H., Mitchison, A. and Muller, B. (1996) Non-classical MHC genetics of immunological disease in man and mouse. The key role of pro-inflammatory cytokine genes. Cytokine **8**, 593–597
2. Saha, B., Harlan, D.M., Lee, K.P., June, C.H. and Abe, R. (1996) Protection against lethal toxic shock by targeted disruption of the CD28 gene. J. Exp. Med. **183**, 2675–2680

3.   Norris, V., Grant, S., Freestone, P., Canvin, J., Sheikh, F.N., Toth, I., Trinei, M., Modha, K. and Norman, R.I. (1996) Minireview — Calcium signalling in bacteria. J. Bacteriol. **178**, 3677–3682

4.   Kenelly, P.J. and Potts, M. (1996) Minireview — Fancy meeting you here! A fresh look at 'prokaryotic' protein phosphorylation. J. Bacteriol. **178**, 4759–4764

5.   Berger, S.A., Rowan, K., Morrison, H.D. and Ziltener, H.J. (1996) Identification of a bacterial inhibitor of protein kinases. J. Biol. Chem. **271**, 23431–23437

6.   Ireton, K., Payrastre, B., Chap, H., Ogawa, W., Sakaue, H., Kasuga, M. and Cossart, P. (1996) A role for phosphoinositide 3-kinase in bacterial invasion. Science **274**, 780–782

7.   Chen, Y., Smith, M.R., Thirumalai, K. and Zychlinsky, A. (1996) A bacterial invasin induces macrophage apoptosis by binding directly to ICE. EMBO J. **15**, 3853–3860

8.   Pettersson, J., Nordfelth, R., Dubinina, E., Bergman, T., Gustafsson, M., Magnusson, K.E. and Wolf-Watz, H. (1996) Modulation of virulence factor expression by pathogen target cell contact. Science **273**, 1231–1233

9.   Zhang, J.P. and Normark, S. (1996) Induction of gene expression in *Escherichia coli* after pilus-mediated adherence. Science **273**, 1234–1236

10. Mushegian, A.R., Fullner, K.J., Koonin, E.V. and Nester, E.W. (1996) A family of lysozyme-like virulence factor in bacterial pathogens of plants and animals. Proc. Natl. Acad. Sci. U.S.A. **93**, 7321–7326

11. Alcami, A. and Smith, G.K. (1996) A mechanism for the inhibition of fever by a virus. Proc. Natl. Acad. Sci. U.S.A. **93**, 11029–11034

12. Mims, C.A., Playfair, J.H.L., Roitt, I.M., Wakelin, D., Williams, R. and Anderson, R.M. (1993) Medical Microbiology. Mosby, Hong Kong

13. Mims, C., Dimmock, N., Nash, A. and Stephen, J. (1995) Mims' Pathogenesis of Infectious Diseases, 4th edn., Academic Press, London

13a. Abbas, A.K., Murphy, K.M. and Sher, A. (1996) Functional diversity of helper T lymphocytes. Nature (London) **383**, 787–793

14. Seder, R.A., Kelsall, B.L. and Jankovic, D. (1996) Differential roles for IL-12 in the maintenance of immune responses in infectious versus autoimmune diseases. J. Immunol. **157**, 2745–2748

15. Gonzalo, J.A., Mazuchelli, R., Mellado, M., Frade, J.M.R., Carrera, A.C., von Kobbe, C., Merida, I. and Martinez-A, C. (1996) Enterotoxin septic shock protection and deficient T helper 2 cytokine production in growth hormone transgenic mice. J. Immunol. **157**, 3298–3304

16. Kincy-Cain, T. and Bost, K.L. (1996) Increased susceptibility of mice to *Salmonella* infection following in vivo treatment with the substance P antagonist Spantide II. J. Immunol. **157**, 255–264

17. Henderson, B., Poole, S. and Wilson, M. (1996) Microbial/host interactions in health and disease: who controls the cytokine network. Immunopharmacology **35**, 1–21

18. Kaprelyants, A.S. and Kell, D.B. (1996) Do bacteria need to communicate with each other for growth? Trends Microbiol. **4**, 237–242

19. Orren, A. (1996) How do mammals distinguish between pathogens and non-self? Trends Microbiol. **4**, 254–258

20. Hopken, U.E., Lu, B., Gerard, N.P. and Gerard, C. (1996) The C5a chemoattractant receptor mediates mucosal defence to infection. Nature (London) **383**, 86–89

21. Zasloff, M. (1992) Antibiotic peptides as mediators of innate immunity. Curr. Opin. Immunol. **4**, 3–7

22. Martin, E., Ganz, T. and Lehrer, R.I. (1995) Defensins and other endogenous peptide antibiotics of vertebrates. J. Leukoc. Biol. **58**, 128–136

23. Boman, H.G. (1995) Peptide antibiotics and their role in innate immunity. Annu. Rev. Immunol. **13**, 61–92

24. Diamond, G., Russell, J.P. and Bevins, C.L. (1996) Inducible expression of an antibiotic peptide gene in lipopolysaccharide-challenged tracheal epithelial cells. Proc. Natl. Acad. Sci. U.S.A. **93**, 5156–5160

25. Ibelgaufts, H. (1995) Dictionary of Cytokines. VCH, Weinheim

26. Henderson, B. and Wilson, M. (1995) Modulins: a new class of cytokine-inducing, pro-inflamm-atory bacterial virulence factor. Inflamm. Res. **44**, 187–197

27. Henderson, B. and Wilson, M. (1996) Cytokine induction by bacteria: beyond LPS. Cytokine **8**, 269–282

28. Henderson, B., Poole, S. and Wilson, M. (1996) Bacterial modulins: a novel class of virulence factors which cause host tissue pathology by inducing cytokine synthesis. Microbiol. Rev. **60**, 316–341

29. Henderson, B. and Wilson, M. (1996) *Homo bacteriens* and a network of surprises. J. Med. Microbiol. **45**, 1–2

30. Pugin, J., Heumann, D., Tomasz, A., Kravchenko, W., Akamatsu, Y., Nishijima, M., Glauser, M.P., Tobias, P.S. and Ulevitch, R.J. (1994) CD14 is a pattern recognition receptor. Immunity **1**, 509–516

31. Janeway, C.A. (1992) The immune system evolved to discriminate infectious nonself from non-infectious self. Immunol. Today **13**, 11–16

32. Muhlradt, P.F., Meyer, H. and Jansen, R. (1996) Identification of S-(2,3-dihydroxypropyl)cystein in a macrophage-activating lipopeptide from *Mycoplasma fermentans*. Biochemistry **35**, 7781–7786

33. Frisch, M., Grndchandt, G. and Muhlradt, P.F. (1996) *Mycoplasma fermentans*-derived lipid inhibits class II major histocompatibility complex expression without mediation by interleukin-6, inter-leukin-10, tumor necrosis factor, transforming growth factor β, type I interferon, prostaglandins or nitric oxide. Eur. J. Immunol. **26**, 1050–1057

34. Louis, C.B. and Obrig, T.G. (1992) Shiga toxin-associated hemolytic-uremic syndrome: combined cytotoxic effects of Shiga toxin, interleukin-1β, and tumor necrosis factor alpha on human vascular endothelial cells in vitro. Infect. Immun. **59**, 4173–4179

35. Moore, P.S., Boshoff, C., Weiss, R.A. and Chang, Y. (1996) Molecular mimicry of human cytokine and cytokine response pathway genes by KSHV. Science **274**, 1739–1744

36. Spriggs, M.K., Hruby, D.E., Maliszewski, C.R., Pickup, D.J., Sims, J.E., Buller, R.M.L. and Van Slyke, J. (1992) Vaccinia and cowpox viruses encode a novel secreted interleukin-1 binding protein. Cell **71**, 145–152

37. Alcami, A. and Smith, G.L. (1992) A soluble receptor for interleukin-1β encoded by vaccinia virus. A novel mechanism for virus modulation of the host response to infection. Cell **71**, 153–167

38. Klapproth, J.-M., Donnenberg, M.S., Abraham, J.M., Mobley, H.L.T. and James, S.P. (1995) Products of enteropathogenic *Escherichia coli* inhibit lymphocyte activation and lymphokine production. Infect. Immun. **63**, 2248–2254

39. Kurita-Ochiai, T. and Ochiai, K. (1996) Immunosuppressive factor from *Actinobacillus actino-mycetemcomitans* down regulates cytokine production. Infect. Immun. **64**, 50–54

40. Travis, J., Potempa, J. and Maeda, H. (1995) Are bacterial proteinases pathogenic factors? Trends Microbiol. **3**, 405–407

41. Reference deleted.

42. Barrett, J.F. and Isaacson, R.E. (1995) Bacterial virulence as a potential target for therapeutic intervention. Annu. Rep. Med. Chem. **30**, 111–118

43. Salyers, A.A. and Whitt, D.D. (1994) Bacterial Pathogenesis: A Molecular Approach. ASM Press, Washington

44. Tannock, G.W. (1995) Normal microflora: an introduction to microbes inhabiting the human body. Chapman and Hall, London

45. Hedges, S.R., Agace, W.W. and Svanborg, C. (1995) Epithelial cytokine responses and mucosal cytokine networks. Trends Microbiol. **3**, 266–270

46. Sadlack, B., Merz, H., Schorle, H., Schimpl, A., Feller, A.C. and Horak, I. (1993) Ulcerative colitis-like disease in mice with a disrupted interleukin-2 gene. Cell **75**, 253–261

47. Mizoguchi, A., Mizoguchi, E., Chiba, C., Spiekerman, G.M., Tonegawa, S., Nagler-Anderson, C. and Bhan, A.K. (1996) Cytokine imbalance and autoantibody production in T cell receptor-α mutant mice with inflammatory bowel disease. J. Exp. Med. **183**, 847–856

48.   Kuhn, R., Lohler, J., Rennick, D., Rajewsky, K. and Muller, W. (1993) Interleukin-10-deficient mice develop chronic enterocolitis. Cell **75**, 253–261

49.   Mashimo, H., Wu, D.-C., Podolsky, D.K. and Fishman, M.C. (1996) Impaired defence of intestinal mucosa in mice lacking intestinal trefoil factor. Science **274**, 262–265

50.   Nicola, N.A. (1994) Guidebook to Cytokines and their Receptors. Oxford University Press, Oxford

51.   Reference deleted.

52.   Wu, X.-R., Sun, T.-T. and Medina, J.J. (1996) *In vitro* binding of type I-fimbriated *Escherichia coli* to uroplakins Ia and Ib: Relation to urinary tract infections. Proc. Natl. Acad. Sci. U.S.A. **93**, 9630–9635

53.   Aaronson, S. (ed.) (1981) Chemical Communication at the Microbial Level. CRC Press, Boca Raton

54.   Williams, P., Bainton, N.J., Swift, S., Chhabra, S.R., Winson, M.K., Stewart, G.S.A.B., Salmond, G.P.C. and Bycroft, B.W. (1992) Small molecule-mediated density-dependent control of gene expression in prokaryotes: bioluminescence and the biosynthesis of carbapenem antibiotics. FEMS Microbiol. Lett. **100**, 161–168

55.   Schaefer, A.L., Val, D.L., Hanzelka, B.L., Cronan, J.E. and Greenberg, E.P. (1996) Generation of cell-to-cell signals in quorum sensing: Acyl homoserine lectone synthase activity of a purified *Vibrio fischeri* LuxI protein. Proc Natl. Acad. Sci. U.S.A. **93**, 9505–9509

56.   Cooper, M., Batchelor, S.M. and Prosser, J.I. (1995) In The Life of a Biofilm (Winpenny. J.W.T., Handley, P., Gilbert, P. and Lappin-Scott, H., eds.), pp. 93–96, Bioline, Cardiff

57.   Poli, G., Kinter, A.L. and Fauci, A.S. (1994) Interleukin-1 induces expression of the human immunodeficiency virus alone and in synergy with interelukin-6 in chronically infected UI cells: Inhibition of inductive effects by the interleukin-1 receptor antagonist. Proc. Natl. Acad. Sci. U.S.A. **91**, 108–112

58.   Osborne, L., Kunkel, S. and Nabel, G.J. (1989) Tumor necrosis factor alpha and interleukin-1 stimulate human immunodeficiency virus enhancer by activation of the nuclear factor kappa B. Proc. Natl. Acad. Sci. U.S.A. **86**, 2336–2340

59.   Porat, R., Clark, B.D., Wolf, S.M. and Dinarello, C.A. (1991) Enhancement of growth of virulent strains of *Escherichia coli* by interleukin-1. Science **254**, 430–432

60.   Kim, K.S. and Le, J. (1992) IL-1β and *Escherichia coli*. Science **258**, 1562

61.   Porat, R., Clark, B.D., Wolff, S.M. and Dinarello, C.A. (1992) IL-1β and *Escherichia coli*. Science **258**, 1562–1563

62.   Denis, M., Campbell, D. and Gregg, E.O. (1991) Interleukin-2 and granulocyte-macrophage colony-stimulating factor stimulate growth of a virulent strain of *Escherichia coli*. Infect. Immun. **59**, 1853–1856

63.   Denis, M., Campbell, D. and Gregg, E.O. (1991) Cytokine stimulation of parasitic and microbial growth. Res. Microbiol. **142**, 979–983

64.   Murphy, P.M., Lane, H.C., Gallin, J.I. and Fauci, A.S. (1988) Marked disparity in incidence of bacterial infections in patients with the acquired immune deficiency syndrome receiving interleukin-2 or interferon gamma. Annu. Intern. Med. **108**, 36–41

65.   Maoleekoonpairoj, S., Mittelman, A., Savona, S., Ahmed, T., Puccio, C., Gafney, E., Skelos, A., Arnold, P., Coombe, N., Baskind, P. and Arlin, Z. (1989) Lack of protection against bacterial infections in patients with advanced cancer treated by biologic response modifiers. J. Clin. Microbiol. **27**, 2305–2311

66.   Treseler, C.B., Mariarz, R.T. and Levitz, S.M. (1992) Biological activity of interleukin-2 bound to *Candida albicans*. Infect. Immun. **60**, 183–188

67.   Luo, G., Niesel, D.W., Shaban, R.A., Grimm, E.A. and Klimpel, G.R. (1993) Tumor necrosis factor alpha binding to bacteria: Evidence for a high affinity receptor and alterations in bacterial virulence properties. Infect. Immun. **61**, 830–835

68.   Klimpel, G.R. and Niesel, D.W. (1993) Tumor necrosis factor receptors on microorganisms. US Patent 5270038

69. Bermudez, L.E., Petrofsky, M. and Shelton, K. (1996) Epidermal growth factor-binding protein in *Mycobacterium avium* and *Mycobacterium tuberculosis*: a possible role in the mechanism of infection. Infect. Immun. **64**, 2917–2922

70. Lottenberg, R., Broder, C.C., Boyle, M.D.P., Kain, S.J., Schroeder, B.L. and Curtiss, R. (1992) Clonig, sequence analysis and expression in *Escherichia coli* of a streptococcal plasmin receptor. J. Bacteriol. **174**, 5204–5210

71. Pancholi, V. and Fischetti, V.A. (1992) A major surface protein on group A streptococci is a glyceraldehyde 3-phosphate dehydrogenase with multiple binding activities. J. Exp. Med. **176**, 415–426

72. George, A.J.T. (1994) Letters: Surface-bound cytokines — a possible effector mechanism in bacterial immunity. Immunol. Today **15**, 88–89

73. Denis, M. and Gregg, E.O. (1990) Recombinant tumour necrosis factor decreases whereas recombinant interleukin-6 increases growth of a virulent strain of *Mycobacterium avium* in human macrophages. Immunology **71**, 139–141

74. Denis, M. (1991) Growth of *Mycobacterium avium* in human monocytes: identification of cytokines which reduce and enhance intracellular microbial growth. Eur. J. Immunol. **21**, 391–395

75. Denis, M. (1991) Modulation of *Mycobacterium lepraemurium* growth in murine macrophages: beneficial effects of tumor necrosis factor alpha and granulocyte macrophage colony-stimulating factor. Infect. Immun. **59**, 705–707

76. Shiratsuchi, H., Johnson, J.L. and Ellner, J.J. (1991) Bidirectional effects of cytokines on the growth of *Mycobacterium avium* within human monocytes. J. Immunol. **146**, 3165–3170

77. Jiang, X. and Baldwin, C.L. (1993) Effects of cytokines on intracellular growth of *Brucella abortus*. Infect. Immun. **61**, 124–134

78. Matsiota-Bernard, P., Lefebre, C., Sedqui, M., Cornillet, P. and Guenounou, M. (1993) Involvement of tumor necrosis factor alpha in intracellular multiplication of *Legionella pneumophila* in human monocytes. Infect. Immun. **61**, 4980–4983

79. Boman, H.G. and Hultmark, D. (1987) Cell-free immunity in insects. Annu. Rev. Microbiol. **41**, 103–126

80. Boman, H.G. (1991) Antibacterial peptides: key components needed in immunity. Cell **65**, 205–207

81. Hultmark, D. (1993) Immune reactions in Drosophila and other insects — a model for innate immunity. Trends Genetics **9**, 178–183

82. Good, J. (ed.) (1995) Antimicrobial Peptides. Ciba Foundation Symposium No. 186, Wiley, Chichester

83. Boman, H.G. (1994) Cecropins: antibacterial peptides from insects and pigs. In Phylogenetic Perspectives in Immunity. The Insect Host Defence System (Hoffman, J., Natori, S. and Janeway, C., eds.), pp. 24–37, Landes Biomedical Publishers, Austin, Texas

84. Steiner, H., Hultmark, D., Engstrom, A., Bennich, H. and Boman, H.G. (1981) Sequence and specificity of two antibacterial proteins involved in insect immunity. Nature (London) **292**, 246–248

85. Lee, J.-Y., Boman, A., Sun, C., Anderrson, M., Jornvall, H., Mutt, V. and Boman, H.G. (1989) Antibacterial peptides from pig intestine: isolation of a mammalian cecropin. Proc. Natl. Acad. Sci. U.S.A. **86**, 959–962

86. Selsted, M.E., Miller, S.I., Henschen, A.H. and Ouellette, A.J. (1992) Enteric defensins: antibiotic peptide components of intestinal host defence. J. Cell Biol. **118**, 929–936

87. Diamond, G., Jones, D.E. and Bevins, C.L. (1993) Airway epithelial cells are the site of expression of a mammalian antimicrobial peptide. Proc. Natl. Acad. Sci. U.S.A. **90**, 4596–4600

88. Russell, J.P., Diamond, G., Tarver, A.P., Scanlin, T.F. and Bevins, C.L. (1996) Coordinate expression of two antibiotic genes in tracheal epithelial cells exposed to the inflammatory mediators lipopolysaccharide and tumor necrosis factor alpha. Infect. Immun. **64**, 1565–1568

89. Schonwetter, B.S., Stolzenberg, E.D. and Zasloff, M.A. (1995) Epithelial antibiotics induced at sites of inflammation. Science **267**, 1645–1648

90. Flyg, C., Dalhammer, G., Rasmuson, B. and Boman, H.G. (1987) Insect immunity. Inducible antibacterial activity in *Drosophila*. Insect Biochem. **17**, 153–160

91.   Lee, W.-J., Lee, J.-D., Kravchenko, V.V., Ulevitch, R.J. and Brey, P.T. (1996) Purification and molecular cloning of an inducible Gram-negative bacteria-binding protein from the silkworm *Bombyx mori*. Proc. Natl. Acad. Sci. U.S.A. **93**, 7888–7893

92.   Gough, M., Hancock, R.E.W. and Kelly, N.H. (1996) Antiendotoxin activity of cationic peptide antimicrobial agents. Infect. Immun. **64**, 4922–4927

93.   Territo, M.C., Ganz, T., Selsted, M.E. and Lehrer, R. (1989) Monocyte chemotactic activity of defensins from human neutrophils. J. Clin. Invest. **84**, 2017–2020

94.   Verbanac, D., Zanetti, M. and Romeo, D. (1993) Chemotactic and protease-inhibiting activities of antibiotic peptides precursors. FEBS Lett. **317**, 255–258

95.   Chertov, O., Michiel, D.F., Xu, L., Wang, J.M., Murphy, W.J., Longo, D.L., Taub, D.D. and Oppenheim, J.J. (1996) Identification of defensin-1, defensin-2 and CAP37/azurocidin as T cell chemoattractant proteins released from interleukin-8-stimulated neutrophils. J. Biol. Chem. **271**, 2935–2940

96.   Murphy, C.J., Foster, B.A., Mannis, M.J., Selsted, M.E. and Reid, T.W (1993) Defensins are mitogenic for epithelial cells and fibroblasts. J. Cell Physiol. **155**, 408–413

97.   Gallo, R.L., Ono, M., Povsic, T., Page, C., Eriksson, E., Klagsbrun, M. and Bernfield, M. (1994) Syndecans, cell surface heparan sulfate proteoglycans, are induced by a proline-rich antimicrobial peptide from wounds. Proc. Natl. Acad. Sci. U.S.A. **91**, 11035–11039

98.   Zhu, Q. and Solomon, S. (1992) Isolation and mode of action of rabbit corticostatic (antiadrenocorticotropin) peptides. Endocrinology **130**, 1413–1423

99.   Shi, J., Ross, C.R., Leto, T.L. and Blecha, F. (1996) PR-39, a proline rich antibacterial peptide that inhibits phagocyte NADPH oxidase by binding to Src homology 3 domains of p47^{phox}. Proc. Natl. Acad. Sci. U.S.A. **93**, 6014–6018

100. Panyutich, A.V., Panyutich, E.A., Krapivin, V.A., Baturevich, E.A. and Ganz, T. (1993) Plasma defensin concentrations are elevated in patients with septicaemia or bacterial meningitis. J. Lab. Clin. Med. **122**, 202–207

101. Kelley, K.J. (1996) Using host defences to fight infectious diseases. Nature Biotechnol. **14**, 587–590

102. Cromer, A. (1993) Uncommon Sense: The Historical Nature of Science. Oxford University Press, Oxford

103. Rose, N.R. and MacKay, I.R. (eds.) (1992) The Autoimmune Diseases II, 2nd edn. Academic Press, Orlando

104. Rook, G.A.W., Lydyard, P.M. and Stanford, J.L. (1993) A reappraisal of the evidence that rheumatoid arthritis and several other idiopathic diseases are slow bacterial infections. Ann. Rheum. Dis. **52**, S30–S38

105. Kaufmann, S.H.E. (1990) Heat shock proteins and the immune response. Immunol. Today **11**, 129–136

106. Henderson, B., Edwards, J.C.W. and Pettipher, E.R. (eds.) (1995) Mechanism and Models in Rheumatoid Arthritis. Academic Press, London

107. Wetzler, L.W., Ho, Y. and Reiser, H. (1996) Neisserial porins induce B lymphocytes to express costimulatory B7-2 molecules and to proliferate. J. Exp. Med. **183**, 1151–1159

108. Galdiero, F., Cipollaro de L'Ero, G., Bendetto, N., Galdiero, M. and Tufano, M. (1993) Release of cytokines induced by *Salmonella typhimurium* porins. Infect. Immun. **61**, 111–119

109. Gianani, R. and Sarvetnick, N. (1996) Viruses, cytokines, antigens, and autoimmunity. Proc. Natl. Acad. Sci. U.S.A. **93**, 2257–2259

110   Kaiser, D. and Losick, R. (1993) How and why bacteria talk to each other. Cell **73**, 873–885

111. Swift, S., Stewart, G.S.A.B. and Williams, P. (1996) The inner workings of a quorum sensing signal generator. Trends Microbiol. **4**, 463–466

112. Ji, G., Beavis, R.C. and Novick, R.P. (1995) Cell density control of staphylococcal virulence mediated by an octapeptide pheromone. Proc. Natl. Acad. Sci. U.S.A. **92**, 12055–12059

113. Hatzigeorgiou, D.H., Geng, J., Zhu, B., Zhang, Y, Liu, K., Rom, W.M., Fenton, M.J., Turco, S. and Ho, J.L. (1996) Lipophosphoglycan from *Leishmania* suppresses agonist-induced IL-1β gene expression in human monocytes via a unique promoter sequence. Proc. Natl. Acad. Sci. U.S.A. **93**, 14708–14713

114. Barral-Netto, A., Barral, A., Brownell, C.E., Skeiky, Y.A.W., Ellingsworth, L.R., Twardzik, D.R. and Reed, S.G. (1992) Transforming growth factor-β in leishmanial infection: a parasite escape mechanism. Science **257**, 545–548

115. Ho, J.L., Kim, H.-K., Sass, P.M., He, S., Geng, J., Xu, H., Zhu, B., Turco, S.J. and Lo, S.K. (1996) Structure-function analysis of *Leishmania* lipophosphoglycan: distinct domains that mediate binding and inhibition of endothelial cell function. J. Immunol. **157**, 3013–3020

116. Friedland, I.R., Jafari, H., Ehrett, S., Rinderknecht, S., Paris, M., Coulthard, M., Saxen, H., Olsen, K. and McCracken, G.H. (1993) Comparison of endotoxin release by different antimicrobial agents and the effect on inflammation in experimental *Escherichia coli* meningitis. J. Infect. Dis. **168**, 657–662

117. Van Furth, A.M., Roord, J.J. and Van Furth, R. (1996) Role of pro-inflammatory and anti-inflammatory cytokines in pathophysiology of bacterial meningitis and effect of adjunctive therapy. Infect. Immun. **64**, 4883–4890

Subject index